DISCOVERY *of* AUSTRALIA'S FISHES

DISCOVERY *of* AUSTRALIA'S FISHES

A History of Australian Ichthyology to 1930

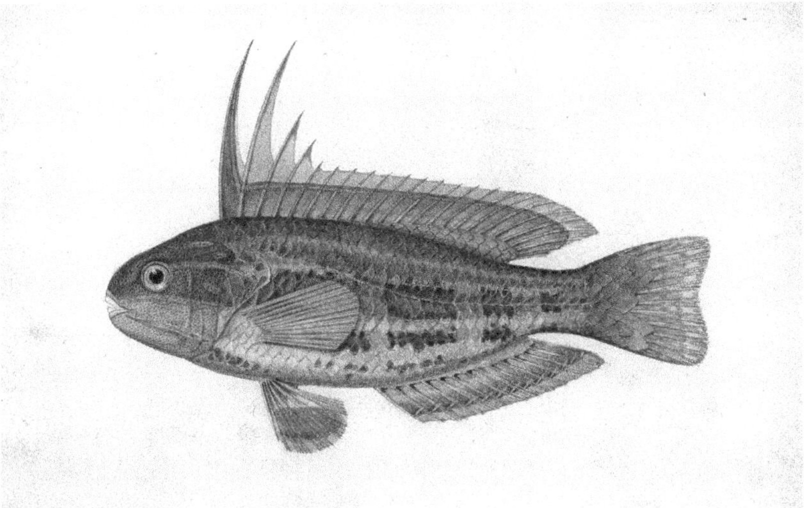

BRIAN SAUNDERS

CSIRO PUBLISHING

National Library of Australia Cataloguing-in-Publication entry

Saunders, Brian (Brian Greig)

Discovery of Australia's fishes : a history of Australian ichthyology to 1930 / by Brian Saunders.

9780643106703 (hbk.)
9780643106710 (epdf)
9780643106727 (epub)

Includes bibliographical references and index.
Ichthyology – Australia – History.
Ichthyologists – Biography.

597.0994

Produced with the support of the South Australian Museum.

Government of South Australia

South Australian Museum

Published by
CSIRO PUBLISHING
36 Gardiner Road, Clayton VIC 3168
Private Bag 10, Clayton South VIC 3169
Australia
Telephone: [+613] 9545 8555
Local call: 1300 788 000 (Australia only)
Fax: +61 3 9662 7555
Email: csiropublishing@csiro.au
Web site: www.publishing.csiro.au

Front cover: Rainbow Fish, *Heteroscarus acroptilus*, of southern Australia painted by Edgar Waite for the frontispiece of his fish catalogue. Courtesy of the South Australian Museum Archives.
Back cover: Juvenile White Ear, *Parma microlepis*, from Sydney Harbour, drawn by George French Angas in 1858. Courtesy Museum Board of South Australia. Spiny-tailed Leatherjacket, *Acanthaluteres brownii*, from King George Sound, drawn by Ferdinand Bauer, 1801. Courtesy Natural History Museum.

Set in Adobe Garamond Pro 9.5/13.6
Cover design by Alicia Freile, Tango Media
Typeset by Oryx Publishing Pty Ltd
Printed by Ingram Lightning Source

Feb26_RP_ILS

Contents

Preface

Recent histories of ichthyology are generally concerned with its development as one discipline among the biological sciences, or appear as an introductory component in books of general ichthyology and in regional faunal works. Scattered through the literature are circumscribed chronological accounts of the discoveries made within limited geographical areas.

In this book I have reviewed the early development of ichthyology in Australia, with emphasis on the discovery of the Australian fauna. I have taken ichthyology to indicate the scientific knowledge of fishes, thus largely excluding aboriginal lore and the 'popular' observations of early settlers. I have treated the scientific publications in greater detail although, admittedly, some of the early published work was not very scientific. I have also traced the history of modern ichthyology in order to show the background of the science into which observations on the Australian fauna were integrated.

Many of the fishes of northern Australia are distributed through greater or lesser parts of the Indian and Western Pacific oceans. These species were frequently taken initially by collectors who worked and travelled outside Australia so I have considered their work briefly, together with that of the zoologists who received their collections.

Those who worked on the fishes usually depended upon collectors, amateur or professional. These collectors were often acute observers but sometimes relatively undiscriminating. Their role in the provision and selection of specimens was significant and has been documented where known. I have included the work done by Australian fish workers on the fishes of other countries, particularly that which is concerned with the fishes of New Zealand, New Guinea and Antarctica. Edgar Waite, in particular, sandwiched a lengthy sojourn in New Zealand between his appointments in Sydney and Adelaide. I follow him to New Zealand and review his activities there.

This account ceases at 1930, a somewhat arbitrary limit, but useful in that 1930 was the year of completion of publication of the first 'modern' checklist of Australian fishes. The 1920s saw the deaths of the three men who were the first who could be reasonably termed, in a modern sense, Australian ichthyologists; the annotated checklist (McCulloch 1929–30) can be seen as the culmination of their taxonomic endeavours. By this time, the general composition of the fauna was documented.

Sources

Good general introductions to the history of ichthyology are found in the first volume of Baron Cuvier's *Histoire Naturelle des Poissons* (1828) and in Albert Günther's *Introduction to the Study of Fishes* (1880a), although much of the material in the latter is derived from the former. I have used both as sources of biographical information on early naturalists and collectors.

The most useful summary of ichthyology in Australia is Gilbert Whitley's *Survey of Australian Ichthyology* (1964). His two short works on the early history of Australian zoology (Whitley 1970; Whitley 1975) have also been useful. Bibliographies for Allan McCulloch (Whitley 1926a), J. D. Ogilby (Whitley 1926b) and Edgar Waite (Hale 1928) provided an effective guide to their contributions.

Collection Building in Ichthyology and Herpetology (Pietsch and Anderson 1997) contains accounts of the accomplishments and lives of some of the ichthyologists who worked in Australia, the United States and elsewhere.

Introductory material in regional faunal accounts commonly includes historical information; for instance, the summary of ichthyological discovery in *The Fishes of New Guinea* (Munro 1967).

The most important sources for this history are the publications of the zoologists who did the work and, fortunately, most of these papers are readily available. The earliest relevant works are those of European zoologists and they commonly made brief reference to their own sources of specimens and information. The two landmark works on fish classification written in the 19th century are the 22-volume *Histoire Naturelle des Poissons* (Cuvier and Valenciennes 1828–1849) and the eight-volume *Catalogue of the Fishes in the British Museum* (Günther 1859–1870). Scattered through these works is an enormous amount of information on the sources used in their compilation. Cuvier, in particular, had a most immediate sense of history in writing his account; he discussed the ideas of his predecessors and the sources of his specimens. The zoologists residing in Australia who studied fishes usually published in Australian journals once these were established, so this work is particularly accessible.

Many of Australia's fishes were first described in Europe. The history of their discovery and description is often complex, dependent upon and integrated with the history of exploration of Australia and the Pacific. Journals, logs and reports derived from those voyages which were orientated towards natural history discovery provide data on the discovery of Australian fish.

Australian archives, museum fish departments and state libraries contain unpublished manuscripts, letters, notebooks and diaries which offer a great deal of help in understanding the way in which early work in Australia was accomplished. The single most useful source for the latter years of the study, informative for the period 1893 to 1927, is the diary of Edgar Waite, kindly made available to me by his grandson, Duncan Waite.

I have also consulted biographies, bibliographies and obituaries of the naturalists involved, and historical works on the history of institutions and periodicals. Histories of the Australian natural history museums in Sydney (Whitley's unpublished history; Strahan 1979), Melbourne (Pescott 1954; Rasmussen 2001), Brisbane (Mather 1986) and Adelaide (Hale 1956) have been particularly helpful.

The *Zoological Catalogue of Australia, Volume 35, Fishes* (2006) by Douglass Hoese, Dianne Bray, John Paxton and Gerald Allen is a key to the taxonomic history of Australian fishes. More general information on the systematics and taxonomy of fishes is derived mostly from Nelson (2006), Eschmeyer (1998) and 'FishBase'.

Notes on method

The main account is divided into three periods: to 1800, 1800 to 1870, and 1870 to 1930. The first period is sketched; the second attended to in more detail; the last in most detail. There are several reasons for this unequal treatment, the more cogent being the increasing significance of the later work, improved access to source materials for more recent studies, and the fact that no detailed account of the last period has yet been published. Biographic themes have had precedence over temporal concerns – it has sometimes been important to follow a worker's achievements from one period into another.

The approach I have taken has been generally biographical, with most emphasis being placed upon the work of individuals. Alternative methods may have been, where feasible, to follow the work of institutions, to trace the resolution of problems within groups of fishes, or to use a strict chronology. I tried each of these but they made the account even more disjointed than the present rather discursive approach.

In the last period (1870 to 1930) there is a little repetition, introduced so that the 'life and work' of each of the three most significant characters (Ogilby, Waite and McCulloch) can be read independently.

Both common and scientific names of fishes are usually quoted. The name of the author of the original description and the date of publication of that description are sometimes included, and enclosed in parentheses when the species is presently placed in a different genus to that used by the original author. For example, Cuvier placed all whiting in the genus *Sillago*. The Spotted Whiting is now placed in *Sillaginodes* so it is cited as '*Sillaginodes punctata* (Cuvier, 1829)' whereas the Sand Whiting remains '*Sillago ciliata* Cuvier, 1829' as it is still placed in the genus used by Cuvier in his original description. A full citation for each species is appended in a classification of the fishes mentioned in the text.

I have used square brackets in association with scientific names to indicate changes in taxonomy. For example, in the section on Broussonet: 'Oxeye Herring, *Clupea* [now *Megalops*] *cyprinoides*' indicates that this fish was placed in the genus *Clupea* by Broussonet and that it is now placed in *Megalops*. In another context square brackets are used to insert the present author's comments among text quoted from other authors. Wherever practicable, captions for the figures take the place of footnotes.

A much larger number of figures would have been desirable, preferably illustrating all of the fish mentioned; this was not possible. Most of the species not here illustrated can be found in Allen *et al.* 2002 (freshwater fish), Allen 2009 (tropical Australia), Gomon *et al.* 2008 (southern Australia excepting Tasmania), Last *et al.* 1983 (Tasmania), Kuiter 1993 (south-eastern Australia), Paulin and Roberts 1992 (New Zealand), Gon and Heemstra 1990 (Antarctica) and Last and Stevens 2009 (sharks and rays). The most comprehensive single source is *Swainston's Fishes of Australia* (Swainston 2011).

There are numerous omissions from this account. Some are due to my judgment as to the relative importance of individuals and their work, some to the limitation of available space, and doubtless there are some due to oversight.

Abbreviations of sources

ADB	Australian Dictionary of Biography
AMA	Australian Museum Archives
DB	Deutsche Biographie
DAB	Dictionary of American Biography
DNB	Dictionary of National Biography
DNZB	Dictionary of New Zealand Biography
DPIQ	Department of Primary Industries, Queensland
DSB	Dictionary of Scientific Biography
FDAM	Fish Department of Australian Museum
IDSAM	Ichthyology Dept South Australian Museum
ML	Mitchell Library
MV	Museum Victoria
NDB	Neue Deutsche Biographie
NHMA	Natural History Museum Archives
OED	Oxford English Dictionary
QMA	Queensland Museum Archives
SAMA	South Australian Museum Archives
SRSA	State Records of South Australia
WD	Waite Diary, with or without date

Acknowledgements

Thanks to Bernadette, David and Richard for their advice, patience and proofreading.

My thanks go also to Nick Alexander of CSIRO Publishing; Kate Alport, curatorial officer of the then Anthropology Archive of the South Australian Museum; Kathy Buckley, Queensland Museum library and archives; Patricia Egan, Australian Museum archives; Jill Evans, South Australian Museum library; Barry Farrell, Amateur Fishermen's Association of Queensland; Vanessa Finney, Australian Museum archives; Lea Gardam, South Australian Museum archives; the late John Glover, previously curator of fishes at the South Australian Museum; Dr Martin Gomon, curator of fishes, Museum Victoria; Richard and Margaret Hogge for information on Haslar Hospital; Andrew Isles, Natural History Books, Melbourne; Jeff Johnson, collection manager at the Queensland Museum ichthyology department; Mark McGrouther, collection manager at the Australian Museum ichthyology department; Jane Meadows, John Oxley Library, Brisbane; Professor Suzanne Miller, Director of the South Australian Museum; Dr John Paxton, ichthyologist at the Australian Museum; Dr Barry Russell, formerly of the Northern Territory Museum and Art Gallery; Trevor Scott, previously curator of fishes at the South Australian Museum; Terry Sim, previously collection manager of fishes at the South Australian Museum; Rosanne Walker, Basser Library, Canberra; Julie Wood, Mitchell Library, Sydney.

Duncan Waite, grandson of the zoologist Edgar Waite, kindly offered an extended loan of his grandfather's diaries with permission to quote them in this publication.

The staff of the following institutions were unfailingly helpful: State Library of South Australia; Barr Smith Library at Adelaide University; State Records of South Australia; Macleay Museum at Sydney University; Queensland Department of Primary Industries; Queen Victoria Museum archive, Launceston.

Thanks to the institutions whose pre-1930 journal issues have provided some of the monochrome illustrations, including the Australian Museum (Records of the Australian Museum, Australian Museum Magazine), Linnean Society of New South Wales (Journal and Proceedings of the Linnean Society of New South Wales); Queensland Museum (Memoirs of the Queensland Museum); South Australian Museum (Records of the South Australian Museum).

Thanks to the following sources whom I attempted to contact to request permission for publication of an image but without success: Journal of the Canadian Medical Association for a portrait of John Richardson published in 1969; E J Brill, Leiden for a portrait of Pieter Bleeker published in 1964; J Cramer, Weinheim for a portrait of Karl Klunzinger published in 1964.

Brian Saunders
Coffin Bay
South Australia

PART ONE
(to 1800)

European ichthyology was in a rudimentary state when its practitioners began to encounter Australian fish. The ichthyology of classical times was based on the limited fauna of the Mediterranean and this was only slightly extended by contact with the shore fishes of the north-east Atlantic. Information about the great diversity of fishes in the Indo-Pacific and the tropical Atlantic came gradually to the centres of learning in Europe. By the closing years of the 18th century there were great numbers of specimens, particularly from the Indian and Pacific Oceans, and some of these were taken in Australian seas. The very early settlement of Australia was so beset with problems that natural history observations were sparse and subject to analysis only when specimens and notes were sent back to England. There are brief accounts of some of the many early naturalists who described fishes whose range includes northern Australia – their names are encountered as authors of the initial descriptions of many of our northern species.

Early ichthyology to the time of Ray and Willughby

The Greek philosopher and natural historian, Aristotle (384–322 BC), recorded many pertinent observations of the fishes of the Mediterranean and left among his works enough information to constitute a rudimentary system of ichthyology. However, this system naturally contained a great number of errors of fact and of interpretation. Until Renaissance times classical knowledge of this kind was challenged only intermittently and at the risk of incurring the disapproval of the institutions of the day. It was generally thought sufficient to study the works of the ancient authors without recourse to independent observation. Thus ichthyology, in common with other branches of the natural and physical sciences, was not well served by direct observation. There is little evidence of original work in ichthyology until the 16th century when Belon, Salviani and Rondelet studied and wrote on the fishes of the Mediterranean and Europe.

Frenchmen Pierre Belon (1517–1564) and Guillaume Rondelet (1507–1566), and the Italian Hippolyte Salviani (1514–1572) cultivated a special interest in fishes. All three published on fishes in the 1550s and their writings have been preserved. Their work, and that of their contemporaries, was still partly fettered by the classical writings but established the standard for ichthyology for another century.

In England John Ray (1627–1705), a theologian, and his friend and colleague Francis Willughby (1635–1672) classified the plants and animals known to them. Their work depended upon their own observations of the structure of organisms and was largely free from the prejudices and distortions of classical authors. They established a system of ichthyology that was a useful basis for further study. To Cuvier (1828), the master of ichthyology a century and a half later, Willughby was '*le grand ichtyologiste anglais*' (DSB).

Willughby and Ray published *De Historia Piscium* (1686) in which a rational system of classification was proposed, dependent upon the nature of the skeleton, the arrangement of the fins and the form of the supporting fin rays. The relationships defined in this work did not extend to a generic level but a step towards a recognisably modern system had been taken.

Artedi, Linnaeus and the *Systema Naturae*

The Swedish botanist, Carl Linné ('Linnaeus') (1707–1778), collected and collated data on plants and animals. Fortunately he derived most of his information on fishes, as presented in his *Systema Naturae*, from an enthusiastic and gifted fellow-countryman and friend, Peter Artedi (1705–1735). Artedi was attracted to the study of fishes largely to the exclusion of other animals; such a degree of specialisation, uncommon at this time, was a great advantage. He collected and organised all of the available literature from the time of Aristotle to his own day. He founded a system of classification based on the published work of Willughby and Ray, and on his own studies of the morphology and anatomy of the 500 or so species of fish available to him. He studied some major collections, notably those of Hans Sloane in London (which later provided the nucleus for the fish collection of the British Museum) and Albertus Seba in Amsterdam (see page 8). Artedi's work included representatives of most of the major fish groups and he stated clearly the criteria used to separate and classify each group, and each species within its group.

Artedi drowned at the age of 30 but had already established a logical system of ichthyology: he based his system on physical characteristics that were easy to observe and this encouraged the rational placement of new species of fish within his scheme by future students. Linnaeus published Artedi's own work as the *Ichthyologia* (1738) and used the classification, with some minor modifications, in his *Systema Naturae*. He had the good judgment to leave his friend's work largely intact. Linnaeus modified and augmented this work through 12 editions of the *Systema*, the tenth of which has been made the starting point for modern nomenclature and classification (Wheeler 1962).

Linnaeus' work represented a great simplification and rationalisation of the data that had been published on the variety of living organisms. The immediate results were to provide a structure to the knowledge that was accumulating and to provoke more scientific and popular interest in botany and zoology. The *Systema Naturae* was the main framework which naturalists of the English and French explorations of the late 18th century used for classifying the organisms they discovered.

Thirteen editions of the *Systema Naturae* appeared between 1735 and 1789. Although the tenth edition (1758) is taken as the beginning of standard binomial nomenclature, the twelfth edition (1766) included many more species and a somewhat different, though generally unimproved, system for the classification of fishes. The 13th edition (1789) appeared 11 years after Linnaeus' death; it was prepared and much expanded by Johann Friedrich Gmelin (1748–1804) but the work was very uneven in quality due to Gmelin's limited experience in natural history (he was Professor of Chemistry at Gottingen) and to the large number of contributors who worked on this edition (Cuvier 1828).

The tenth edition of the *Systema Naturae* included more than 50 species of fish that have since been recorded from Australian seas but none of these was collected in Australia. There was a significant representation of large, wide-ranging sharks and their attendant fishes. For instance, the White Shark (*Carcharodon carcharias*) from 'Europe', the Pilotfish (*Naucrates ductor*) and Suckerfish (*Remora remora*) from the 'Indian Ocean'. Relatively few species of fish have circumglobal distributions but pelagic species such as Sunfish (*Mola mola*), Flyingfish (*Exocoetus volitans*), Swordfish (*Xiphias gladius*), some Tunas (Scombridae) and a Dolphinfish (*Coryphaena hippurus*) were known to Linnaeus, as was the nearly ubiquitous Sea Mullet (*Mugil cephalus*). Among the more passive pelagics he included a few that have wide distributions such as the large Unicorn Leatherjacket (*Aluterus monoceros*) from 'Hong Kong' and the Sargassum Fish (*Histrio histrio*); Linnaeus did not specify a locality for the Sargassum Fish but noted 'open sea swimming among algae'. The *Systema Naturae* also included fishes from India, China, the Red Sea and the East Indies; a few common tropical Australian species were thus named: Lionfish (*Pterois volitans*) from 'Indonesia', Moorish Idol (*Zanclus cornutus*) from 'The Indies', Moon Wrasse (*Thalassoma lunare*) from 'India', Surgeonfish (*Acanthurus nigricans*) from the 'Red Sea'. A few more 'Australian' species were added to the 1766 edition, including the Tailor (*Pomatomus saltatrix*) from 'Carolina' and the Stonefish (*Synanceia horrida*) from 'India'. Thus Linnaeus, who at this time knew little of Australia, nevertheless recorded more than 1% of the 4500 fishes now listed as Australian.

Linnaeus initially used 61 genera for his fishes, based on Artedi's definitions, but these were soon found to be insufficient. New genera were proposed for fishes not known to Artedi; some were adopted by Linnaeus, others by his successors, such as Gronovius of Leyden (1730–1777) (Wheeler 1958). Keeping pace with new generic definitions, which were sometimes ill-conceived, soon became difficult although the works of Cuvier, and later of Günther, tended to stabilise this level of nomenclature.

Linnaeus had a most practical view of the value of early publication of new discoveries, being aware that unpublished work would lead to duplication of effort and great confusion. He urged publication on Solander, who was dealing with the specimens of Cook's first voyage, but the bulk of the work generated by the specimens and observations on the *Endeavour* voyage was too great to be handled rapidly. Solander was still struggling with it when he died ten years after his return.

Fishes of tropical seas were brought back to the few naturalists prepared to describe and classify them. The new discoveries were usually incorporated in the Linnaean system and magnificent illustrations were prepared for some of the more lavish publications. At first the artistic elements of illustration took precedence over accuracy but appreciation of the value of precise and uniform representation was soon to dictate the format of scientific illustration in the natural sciences, including ichthyology.

Some of the French philosopher-naturalists of the mid to late 18th century worked as though their 'natural' systems of classification were based on a form of descent of species, one from another. This was a rejection of the basis of Linnaean classification which was essentially a system of pigeon-holing: good for filing but not conducive to an appreciation of the relationships between different species. The situation among the fishes was rather better than that of other groups since Artedi had considered a large number of characters in constructing his classification.

Linnaeus' students, their voyages of discovery, and their French contemporaries

The influence of Linnaeus was manifold: not only did he provide a taxonomic framework and stimulate new work by proposing new schemes of classification, he also trained and encouraged young naturalists, some of whom joined voyages of exploration. He called them his apostles, and not without reason. Their expertise so obviously enhanced the results of expeditions that it became normal to expect naturalists to participate. This made it natural for the British and French naval voyages to include scientific observers in later years.

Several of Linnaeus' students made significant contributions to ichthyology and several worked in the Indo-Pacific region, here collecting many species whose range includes the seas of Australia. Daniel Solander (see page 17), among the best remembered in the present day, accompanied Cook and Banks on Cook's first voyage in the Pacific. Peter Forsskål undertook the greater part of the natural history work on a Danish expedition to the Middle East, reporting on the fish fauna of the Red Sea. Carl Thunberg and Pehr Osbeck collected in the East Indies, China and Japan.

Thunberg

Carl Peter Thunberg (1743–1828), a Swedish doctor and student of Linnaeus, undertook several expeditions in South Africa. He was employed as a surgeon by the Dutch East India Company and travelled in their employ in 1775, collecting in Japan, Java and Ceylon. He took part in trading for the Company from the station at Nagasaki but shared the fate of the rest of this post, being confined for most of his 15 months in Japan to the little island of Deshima in Nagasaki Harbour (Blunt 1971). This was prior to Perry's opening of Japan so Thunberg's cramped position was nevertheless privileged, the Dutch being Japan's sole trading partners at this time; he did attain some limited access to the natural history of the region. Like most of Linnaeus' pupils, Thunberg had studied the classification of animals as well as plants, and his publications included several papers on fishes from the East Indies and Japan. He returned to Uppsala, Sweden where he became Director of the museum and assumed Linnaeus' position as Professor of Botany. His work included the original descriptions of several species that were later found around northern Australia; for example, the Striped Catfish, *Silurus* [now *Plotosus*] *lineatus* described from China.

Osbeck

Pehr Osbeck (1723–1805) was a ship's chaplain who described a number of fishes from his travels in China and Japan; he sailed from Gothenburg in 1750, returning to Sweden in 1752. Among the fishes he described were several that extend into the northern seas of Australia; for example, the Scribbled Leatherjacket, *Balistes* [now *Aluterus*] *scriptus* from China.

Forsskål

King Frederick V of Denmark sponsored a scientific exploration of Arabia in 1761 and Peter Forsskål (1732–1763) was appointed naturalist to the expedition on the recommendation of Linnaeus. Forsskål, born in Finland, was a naturalist of genius with a wide knowledge and enormous enthusiasm. He died during this expedition but made a remarkable contribution, even in this short time, to several aspects of natural history. His work included collections of, and observations on, fishes of the Red Sea. This was the first time that the shore fishes of the Indo-Pacific region had been carefully sampled, albeit from its periphery. Forsskål's descriptions were published after his death by his friend Karsten Niebuhr (Forsskål 1775), the sole survivor of this Danish expedition. The proportion of endemic species among the shore fishes of the Red Sea is considerable, 10% to 15% (Briggs 1974; Randall 1983), but the majority have a wide distribution in the Indo-Pacific. His descriptions of fishes well known in northern Australia include the Mangrove Jack, *Sciaena argentimaculata* [now *Lutjanus argentimaculatus*]; Giant Trevally, *Scomber* [now *Caranx*] *ignobilis* and Tarwhine, *Sparus* [now *Rhabdosargus*] *sarba.*

Forsskål described nearly 200 species as new and nearly one third of these have been collected from the shores of northern Australia, where many are commonly encountered. He established many new genera in the course of his work although only a few of these are still in use (such as *Acanthurus, Siganus, Torpedo*). Forsskål attracted criticism from Günther and others by the use of 'barbarisms' such as *Abudefduf*, derived from an Arabic name, for a genus of Damselfishes. To the conservative Günther only Latin or latinised Greek was acceptable.

French contemporaries of Linnaeus

French travellers and naturalists collected and sent home descriptions and specimens from the Indian Ocean. These also came eventually to the notice of the industrious Bloch and Lacépède, and later to Cuvier. This was also a time in France for heroic encyclopaedic works on natural history, such as those of Bonnaterre and Buffon mentioned below.

Commerson

Philibert Commerson (1727–1773), born in Ain, France, undertook medical studies but was primarily interested in natural history. He sailed as naturalist on Louis de Bougainville's voyage in 1766, visiting the east coast of South America, Tierra del Fuego, Tahiti, the islands of Melanesia and Java. Bougainville circumnavigated the world on this voyage but Commerson stayed in Mauritius, working on the natural history, until his death. Commerson's observations, descriptions and drawings of the fishes of Mauritius were accurate; however, he used descriptive phrases instead of the Linnaean binomial format and his work was not published. More than 20 years after his death, Commerson's manuscripts were found by Duméril, then Lacépède's assistant, and made available to the ichthyologists Bloch and Lacépède, and later to Cuvier. Thus Commerson's work was incorporated, with acknowledgement, in the ambitious projects undertaken by these naturalists.

Lacépède published over 30 of Commerson's descriptions, but most of these are regarded as invalid since the names used were not constructed according to the binomial system. His name forms the specific epithet of some Australian fish, such as the Spanish Mackerel, *Scomberomorus commerson* (Lacépède, 1800).

Sonnerat

Pierre Sonnerat (1749–1814) travelled to Mauritius in 1767. With Commerson, he explored Mauritius, Réunion and Madagascar between 1768 and 1771, collecting and drawing various plants and animals, including fishes. He then sailed on a naval vessel as clerk and naturalist, and took part in expeditions to New Guinea, and later to India and China; he published accounts of his travels. Some of Sonnerat's fishes were described by Cuvier and Valenciennes and his name is encountered as the specific epithet in some Australian fishes; for instance the Tomato Rockcod, *Cephalopholis sonnerati* (Valenciennes, 1828).

Buffon

Georges Louis Buffon (1707–1788) was a wealthy and ambitious naturalist, superintendent of the Jardin du Roi in Paris in the reign of Louis XV. Buffon built the Jardin into a scientific centre of importance, increased the natural history collections and undertook an encyclopaedic work, *Histoire naturelle, générale et particulière, avec la description du cabinet du roi* (1749–1804). He attempted to solve problems in classification by making new observations taken direct from the organisms he studied. His concepts of the significance of fossils to geology and the history of life were realistic and contrasted sharply with the views held in his time. The theologians of the day considered banning his *Histoire* but Buffon's diplomacy placated them and his publications were a popular success. He befriended the young Comte de Lacépède who completed the *Histoire*. Lacépède was 32 at the time of Buffon's death in 1788 and proceeded with his own work on fishes (see page 10).

Bonnaterre

Pierre Bonnaterre (1751–1804), an ordained priest with an enthusiasm for natural history, contributed the volumes on ichthyology to an ambitious French encyclopaedia of natural history, published between 1782 and 1832. He collected illustrations of vertebrates and insects for the encyclopaedia and also described many of the fishes. His name appears in Australian ichthyology mainly as the author associated with descriptions of several tunas and sharks, generally collected in the Mediterranean. His work also included descriptions of two Australian sharks: Spotted Wobbegong, *Orectolobus maculatus* and Epaulette Shark, *Hemiscyllium ocellatum* from the 'south seas'. The latter was collected at the Endeavour River, Queensland by Cook's expedition, drawn by Spöring (see page 16), and eventually described by Bonnaterre in 1788.

Seba, Dutch naturalists and the stadholder's cabinet

Albertus Seba (1665–1736), an apothecary of Amsterdam, was one of the world's greatest private collectors of natural history specimens. The importance of his work was assured when he commissioned naturalists to study his collection and artists to illustrate a series of books on the objects he had gathered. These drawings were published, together with descriptive text, in a four volume work: *Loccupletissimi Rerum Thesauri Accurata Descriptio*, 1734–1765 (a modern reprint of the plates was published in 2005 as *Cabinet of Natural Curiosities*). Many of the fishes depicted were from the Indo-Pacific and a few of these are found around the shores of northern Australia.

The Dutch tradition of natural history collection and voyaging in the 17th and 18th centuries, particularly associated with their settlements in the East Indies, presented opportunities to bring back impressions of the fishes. Nieuhof, Vlamingh, Renard, Valentijn and others collected and published some of this new information. Most of their colourful illustrations were fanciful in the extreme, unlike the more accurate work of Isaac Lamotius (previously attributed to Vlamingh) (Pietsch 1995). Maarten Houttuyn (1720–1798), a Dutch naturalist, published descriptions of fishes from Japan in 1782; several of his species have been found in northern Australian waters.

The prince of the Netherlands, 'the stadholder', accumulated a large natural history collection, which included the skins of many fishes from the Dutch East Indies. The 'stadholder's cabinet', often referred to in Cuvier's work, was confiscated by the French revolutionary army during the invasion of the Netherlands in 1795. The collection was taken to the Muséum National d'Histoire Naturelle in Paris (Cuvier 1995). Only when the manuscripts and collections reached Bloch and Lacépède, in post-Linnean times, did this early work begin to take on a wider significance.

Bloch and Schneider, Pallas, Lacépède, Broussonet

Bloch and Schneider

In the late 18th and early 19th centuries the influx to Europe of new species, particularly from the Indo Pacific region, provided a great stimulus for the study of these unfamiliar forms and for the revision of fish classification. This is seen in the work of European zoologists at the time, particularly Bloch and Lacépède (and later Cuvier and Müller), who received and actively sought new species for study from overseas collectors and from the examination of older collections and literature.

In Berlin, a well-to-do physician, Marcus Elieser Bloch (1723–1799), studied first the fishes of his country's rivers and then fishes from foreign parts (Bloch 1785–1795) (see colour plate 1, upper). His early work included the description of some Indo-Pacific fishes whose distribution includes tropical Australia; for example, the Thicklip Wrasse, *Labrus* [now *Hemigymnus*] *melapterus*, described by Bloch (1791) from Japanese specimens. He constructed a new classification partly on the basis of his own observations but was also rather uncritical in accepting the work of other authors; he set out to produce a massive, illustrated book. Bloch did not begin writing on fishes until he was 56 and died while the major work was in its early stages; his associate and student, philologist and naturalist Johann Gottlob Theaenus Schneider (1750–1822), completed the task, publishing *Systema Ichthyologiae Iconibus* (1801) (see colour plate 1, centre and lower). Bloch and Schneider are the authors of some well-known Australian species, including the Clown Triggerfish, *Balistoides conspicillum* of coral reefs; Comb Wrasse, *Coris picta*, an eastern Australian endemic; and the Stargazer, *Kathetostoma laeve*, widespread in temperate Australia. Many of Bloch's original type specimens have survived (Paepke 1999), including the fish he used to describe the familiar Barramundi, *Lates calcarifer*. Approximately 100 species described by Bloch have since been recorded from the coast of northern Australia.

The *Systema Ichthyologiae Iconibus* is of special consequence to the history of Australian ichthyology since Schneider had access to the descriptions of New Zealand fishes in manuscript form by Johann Forster, the principal naturalist who accompanied Cook on his second Pacific voyage. Forster's authorship of species within this work is generally acknowledged and his specific names are accepted. Among the Australian fishes described in this work from New Zealand specimens are the Snapper, *Chrysophrys auratus* and the Hagfish, *Eptatretus cirrhatus*.

Pallas

Peter Simon Pallas (1741–1811), a German zoologist, modified the Linnean system of classification and published ambitious works on botany, zoology and other topics. He worked in Russia for over 40 years, travelled extensively, and described the natural history of many regions. Pallas also received specimens from Japan and Indonesia, including many fish. Among these were the remarkable Slingjaw Wrasse, *Sparus* [now *Epibulus*] *insidiator* and several other species whose range includes the coral reefs of northern Australia.

Lacépède

The system of Bloch and Schneider was succeeded by that of Bernard Germain Etienne de Lacépède (1756–1826). Lacépède's interests were very wide: he published on music, electricity and many aspects of natural history. Georges Buffon in his latter years sought Lacépède's assistance with his *Histoire Naturelle*, a major project that had occupied Buffon for many years. Lacépède was to prepare volumes on the reptiles and fishes and Buffon found him a position as keeper at the Cabinet du Roi, associated with the Jardin des Plantes. Lacépède completed a great work of compilation and original description that was of major importance to ichthyology: his *Histoire Naturelle des Poissons* was published in five volumes from 1798 to 1803.

Lacépède had become involved in the politics of the day, but also helped to plan the reorganisation of the Jardin des Plantes in Paris as the Muséum d'Histoire Naturelle, an institution which was intended to rival London's British Museum. Lacépède was obliged to leave Paris in 1793, just before legislation establishing the Museum was passed, and he remained in exile during the Terror, continuing his labours on the *Histoire Naturelle des Poissons*. This work included new descriptions of many fish from the Indo-Pacific that are

Blue Blanquillo, *Labrus* [now *Malacanthus*] *latovittatus* (top) was based on a figure by Commerson; Hump-headed Wrasse, *Labrus* [now *Coris*] *aygula* (centre); Ringed Slender Wrasse, *Labrus* [now *Hologymnosus*] *annulatus* (bottom). All are common northern Australian reef fishes with wide Indo-Pacific distributions, described by Lacépède. From Lacépède 1801. Courtesy Museum Board of South Australia.

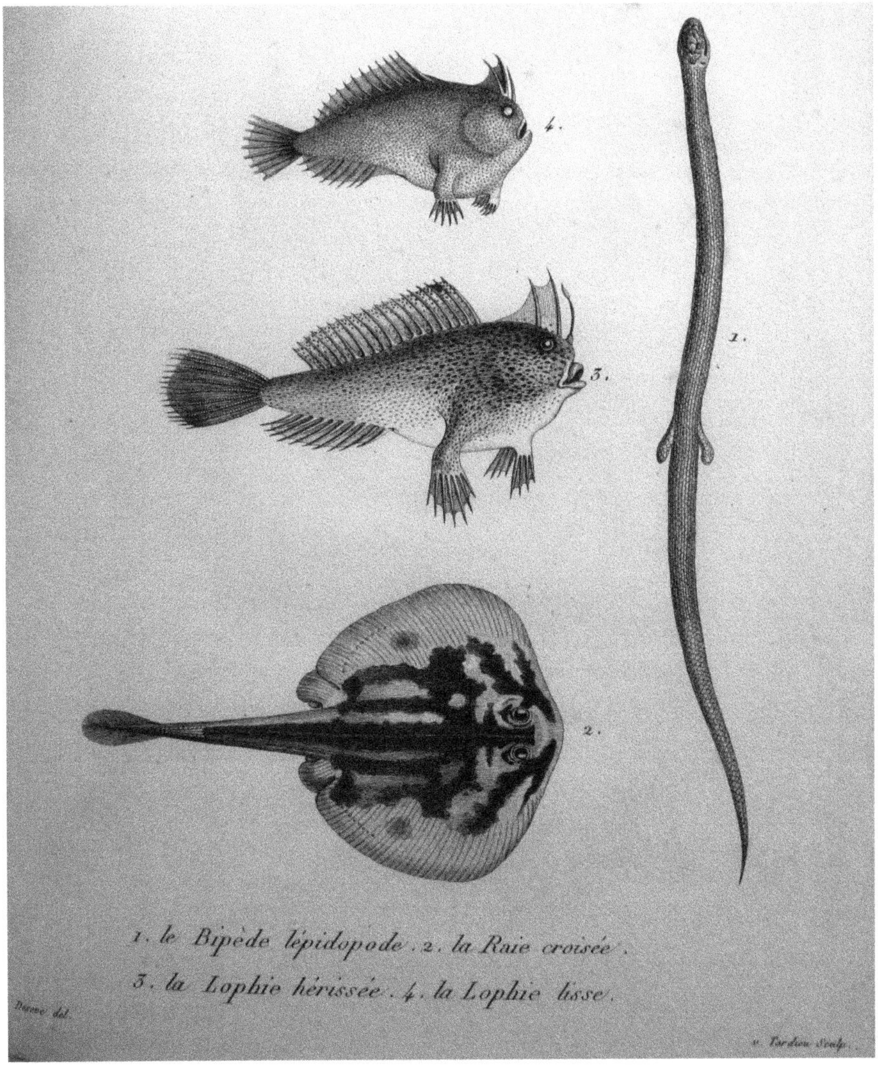

Lacépède described new animals from southern Australia: the Banded Stingaree, *Raja cruciata* [now *Urolophus cruciatus*] (lower) and the Spotted Handfish, *Lophius* [now *Brachionichthys*] *hirsutus* (centre). He named a second Handfish species *la Lophie lisse*, latinised as *Lophius laevis*, but Latreille had used this name for another Anglerfish a little earlier, Lacépède's fish now being known as *Sympterichthys unipennis* (Cuvier, 1817) (upper). The same plate illustrates the common Scalyfoot Lizard, *Pygopus lepidopus* (Lacépède, 1804). From Lacépède 1804. Courtesy Museum Board of South Australia.

well known in Australia. For instance, the Jungle Perch, *Centropomus* [now *Kuhlia*] *rupestris*; Lacépède based this species on Commerson's description of a specimen from Réunion. On returning to Paris in 1794, he was appointed to a professorial position at the Muséum d'Histoire Naturelle and resumed an active life in politics under the Napoleonic regime.

After completing his *Histoire Naturelle des Poissons* Lacépède continued to describe new species of fish, including a Stingaree, *Urolophus cruciatus* and the Sea Dragon, *Phyllopteryx taeniolatus* from Péron's collections in Australian seas (Lacépède 1804).

The accuracy, and hence value, of his *Histoire Naturelle des Poissons* was compromised due to Lacépède's isolation from much of the literature of the time, and even from some of his own notes and collections, during the revolution in France. He made no effort to find 'natural' relationships, believing that an artificial system based on externally observable features was desirable. Cuvier and others were later very critical of the work but it contained a significant quantity of original observation and marked a real advance over the earlier texts of ichthyology.

Broussonet

Pierre Marie Auguste Broussonet (1761–1807), professor at Montpellier, produced some brief papers on fishes and sharks, apparently planning a large work of which he published only a small part. The latter had the imposing title of *Ichthyologia sistens Piscium descriptiones et icones, decas I* (1782) and comprised descriptions of ten species from Cook's first voyage, made available to Broussonet by Joseph Banks. Broussonet made several visits to London, initially in 1780–82 when he had access to Banks' collections and library (Whitehead 1969). Broussonet may have begun studying Cook's fishes under Solander's supervision but, after Solander's death in 1782, he apparently abandoned the project.

Several fishes collected from the central Pacific, and common in northern Australia, were first described by Broussonet. Some other specimens and drawings which had been received from Banks at the same time eventually came to Cuvier from Montpellier and were incorporated in his major work.

Oxeye Herring, *Clupea* [now *Megalops*] *cyprinoides*, described by Broussonet from Vanuatu, is a common fish of northern Australian estuaries and widely distributed in the Indo-Pacific. From Broussonet 1782. Courtesy Museum Board of South Australia.

The British Museum before 1800 (Shaw)

In London, Hans Sloane (1660–1753) had been physician to the governor of Ireland and was elected president of the Royal Society. He had a great interest in natural history and his publications included a work on the fishes of Jamaica, following Willughby's system. He had amassed impressive collections of plants, animals and minerals and planned that, when he died, these collections would become the nucleus of a museum for the entertainment and enlightenment of the public. This was the foundation of the British Museum. Sloane's will provided for curation and administration of the new museum but, ill-advisedly, collections of art were combined with those of natural history. The politics of supervision of the museum were such that the board of directors was soon composed largely of landed gentlemen and clergy. This led to the deterioration of the natural history collections and a poor rate of acquisition. Early curators, 'keepers', of the zoological material received little by way of funds or encouragement, to the detriment of the museum (Gunther 1980).

Botany at this time was protected by the interest of Joseph Banks but his concern for natural history did not, unfortunately, extend to an active involvement in the curation of animal specimens. For many years the status of zoology at the new museum was very low. The arts of preservation were not well understood and fishes, many specimens of which were in the form of skins, were particularly vulnerable to damage from damp, mould and insects.

George Shaw (1751–1813) became an assistant to Edward Gray, keeper of the natural history collections, in 1791. Shaw had been trained in Oxford for the church and later in Edinburgh in medicine. He practised medicine and had written zoological works, but his talents lay more in popularising natural science than in the demanding organisational work of an effective museum curator. He wished to use the museum premises to hold public lectures but was forbidden to do so by the trustees of the museum. Shaw was keeper of the natural history collections from 1807 until his death in 1813.

Shaw published a number of natural history texts, the most ambitious being *General Zoology*, 18 tomes that appeared between 1801 and 1813, including two volumes on fishes. In his *Zoology of New Holland* (1794) Shaw described some fish but only a Shovelnose Ray, *Raja* [now *Aptychotrema*] *rostrata* from Botany Bay was a new species for Australia; the description seems to have been derived from Banks' manuscript. In the same work he illustrated and described the Old Wife as *Chaetodon constrictus* [now *Enoplosus armatus*] but it had already been described and named in John White's *Journal of a Voyage to New South Wales* (1790) (see page 21). Shaw's *Naturalists' Miscellany* (1790-1813), with illustrations by the artist Frederick Polydore Nodder, included descriptions of fishes from Cook's voyages, apparently based on the actual specimens rather than drawings (see colour plate 2, upper).

Although Shaw was active and enthusiastic, he had little interest in caring for the collections: during his term as keeper they suffered badly. His neglect showed in the lack of new acquisitions as well as in the deterioration of existing specimens. In *The Founders of Science at the British Museum 1753–1900*, A.E. Gunther states:

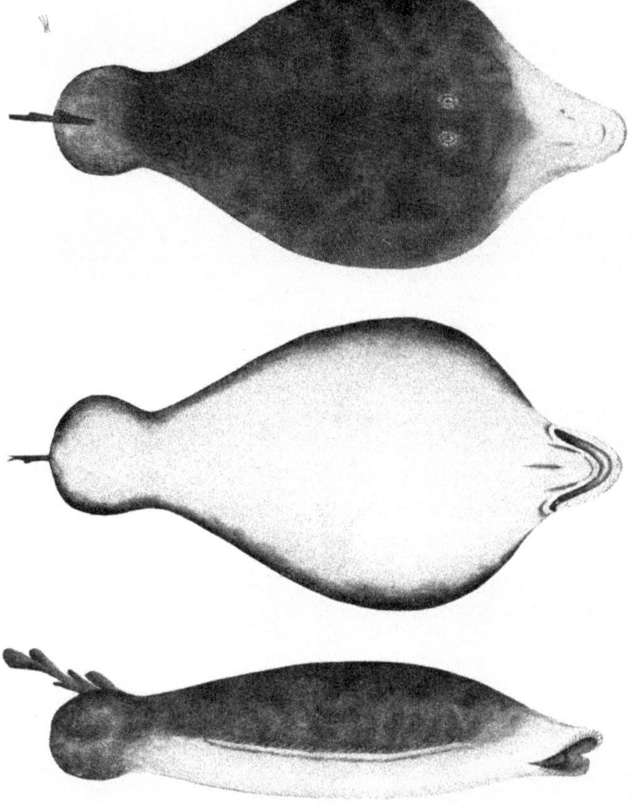

Shaw and Nodder illustrated a supposed Anglerfish, *Lophius monopterygius* in the *Naturalists' Miscellany* (1795). Nearly 150 years later Gilbert Whitley identified it as an Electric Ray, the so-called Coffin Ray [now *Hypnos monopterygium*]. These creatures become remarkably swollen and distorted after a day or so on the beach, hence Shaw's confusion. From Whitley 1940.

> The period of his [Shaw's] keepership ... was one of transition between the eighteenth century conception of what a keeper should do, and that of the nineteenth. In the eighteenth century a keeper's work was done if he kept the specimens labelled and arranged in their cases, and had made the inventory the trustees required. (Gunther 1980)

Recognition of the modern triad of obligations in museum work (curation, research and education) was still a long way off.

After Shaw, ichthyology attracted little interest from the workers at the British Museum until about 1830, when John Gray (see page 43) began a series of publications on fishes.

Voyages to Australia before 1800

Ichthyology, in a modern sense, began in Australia with early voyages of discovery. The Dutch and Portuguese mariners apparently found little of interest to them on Australian shores and their references to natural history were rudimentary.

Dampier

William Dampier (1651–1715) brought back and published a coherent account of some of the plants and animals of Australia (Dampier 1703). He visited the shores of Western Australia twice in the 17th century (in 1688 and again in 1699) and prepared crude illustrations of a variety of plants and animals, including a few fish. There has been some disagreement over which species he represented but some of the families can be fairly confidently identified: the Flying Fish (Exocoetidae) and Suckerfish (Echeneidae) are unequivocal; the drawing of a Dolphinfish (*Coryphaena hippurus*) is recognisable; his 'Monk Fish' appears to be a Flathead (Platycephalidae); and his 'Fish of the Tunny kind' resembles a Spanish Mackerel (*Scomberomorus* sp.). He refers to Willughby's *Ichthyographia* (1685) and *Historia Piscium* (1686), certainly the best ichthyology texts of the time, which had illustrations and descriptions of the first three of the above (Whitley 1970; George 1999).

Cook's first voyage

The English accomplished little more until naturalists and natural history artists were provided for voyages of exploration and settlement. With James Cook's surveys there accumulated, for the first time, a useful body of knowledge concerning the plants and animals of Australia. There was a remarkable delay in publishing this material which meant that little information was available to guide settlers or zoologists at the time of the First Fleet and for many years thereafter.

Fishes were not neglected on Cook's first voyage: Cook himself, Banks and the artist Parkinson all made reference to various species as they encountered them. When the crew caught fishes they were seeking food, not scientific specimens. This is reflected in journals kept by members of the expedition, in which the stingrays of Botany Bay (Stingray Harbour for a short time) were described. The edible qualities of the rays were certainly appreciated. Cook described the stingrays and also made the first reference to leatherjackets (Monacanthidae) in Australia. His sailors had seen similar fish in the Caribbean.

Banks

Joseph Banks (1743–1820) inherited a large fortune which left him free to pursue his interest in natural history, accumulating a great deal of knowledge and an incomparable library. He had a particular interest in botany but informed himself generally through the works of Linnaeus and Buffon. His first overseas collecting was carried out on an expedition to Labrador in 1766 and this provided experience which he used to good effect when preparing to travel with Cook. It was Banks who chose the staff and materials that resulted in the collection and preservation of by far the best natural history specimens that had ever been brought back from a long voyage.

Banks' observations were not often related to fishes but he noted the 'singular Phenomenon' of little Mud-skippers (gobies of the subfamily Oxudercinae) among the mangroves at Thirsty Sound in Queensland:

> Upon seeing us he immediately fled from us leaping as nimbly as a frog by the help of his breast fins ... if seen in the water he often leaped out and proceeded upon dry land & where the water was filld with small stones standing above its surface would leap from stone to stone rather than go into the water.
> (Brunton 1998)

The natural history priorities of the expedition, which obviously represent the priorities of Banks himself, are clear with regard to the drawings. The record of plants was to be as complete as possible and the animals whose colour might be lost or altered before they could be attended to had the next priority. For this reason fishes were prominent among the groups that were well represented by the drawings. Parkinson left a large number of zoological drawings in outline with colour notes so that they could be completed later. The drawings were in one sense supplementary to descriptions that were made by Solander (see page 17).

When the *Endeavour* returned to England there was a great deal of attention to preparation of the plant drawings with the clear intention of publication. The work was not completed and Banks became occupied with other projects. The animal drawings received little attention for many years.

Cook's artists

Two of the artists on this voyage drew Australian fishes: Parkinson and Spöring. Sydney Parkinson (c.1745–1771) was born in Edinburgh where he learnt drawing. At the age of about 20 he moved to London where he supported himself by giving tuition in drawing. His talents came to the attention of Joseph Banks who set him to work drawing, among other subjects, preserved specimens of animals from his 1766 expedition to Labrador and Newfoundland. He is thought to have been responsible for most of the zoological drawings from the *Endeavour* trip.

Herman Diedrich Spöring (c.1733–1771) was born in Abo, Sweden (now Turku, Finland) and came to London at the age of about 22. He worked as a watchmaker for 11 years and then found employment as a clerk for Daniel Solander, an Assistant Keeper in the Natural History Department of the British Museum. His clerical work continued on the voyage but his talents as an artist were soon exploited when Parkinson was overloaded with work. The land expeditions at Botany Bay collected large numbers of new plants so Spöring drew some of the animals, including the impressive stingrays.

The drawings of fish by Cook's artists, although often incomplete, were accurate: the species can generally be identified readily. Some of these drawings were used for the description of new species by Bloch and Schneider in their *Systema Ichthyologiae Iconibus*. The drawings made by Parkinson and Spöring are preserved at the British Museum and some have been published; in particular, in Peter Whitehead's *Forty Drawings of Fishes made by the Artists who accompanied Captain James Cook on his Three Voyages to the Pacific* (1968). A few of the fishes from Australia are reproduced but most of the drawings are of fishes taken in New Zealand and Tahiti. In 1986 Alwyne Wheeler of the British Museum published part three of a *Catalogue of the Natural History Drawings Commissioned by Joseph Banks on the Endeavour Voyage 1768–1771 held in the British Museum (Natural History)*. This publication, which unfortunately presents only a small selection of the actual drawings, deals with the zoology; ten of the fish drawings on this list appear to have been made in Australia.

Several sharks and rays were drawn by Spöring and the rest by Parkinson. Spöring's depiction of the Eastern Fiddler Ray, *Raja* [now *Trygonorrhina*] *fasciata* was used by Müller and Henle (1841) to illustrate their description of this common Australian ray. A number of southern Australian forms were drawn from New Zealand specimens, including the Eastern Australian Salmon, *Arripis trutta* (Forster, 1801); Bloch and Schneider (1801) published Forster's description. Parkinson had illustrated the species as *Mulloides sapidissimus* from a New Zealand specimen taken on Cook's first voyage.

Spöring suffered from malaria and dysentery following the sojourn in Batavia and died two days before his fellow artist, Sydney Parkinson, on the return journey.

Solander

Daniel Solander (1736–1782) was born in Piteå in the north of Sweden. He attended the university at Uppsala where he became a student of Linnaeus. From 1763 he worked at the British Museum, cataloguing the collections of animals, herbarium specimens and fossils. Solander was an experienced and able naturalist whose friendship with Joseph Banks arose from their common enthusiasm for natural science. In London, and on the voyage with Cook, Solander earned the enduring confidence of Banks.

Solander has suffered much criticism over the years because he published very little of the natural history of Cook's first voyage. However, he worked persistently at a card catalogue of specimens, making accurate and lengthy descriptions, and this has remained of value. Once again, Banks expected him to concentrate on the massive botanical collection and this he did, but not to the exclusion of work on the zoology. He died of a stroke at the age of 49 (Duyker 1998).

Solander's descriptions and drawings, although never published, were well organised and provided a significant resource for naturalists. Many of those who worked on fishes in the years following the expedition made use of this material: Müller and Henle in their work on the plagiostomes, Bloch and Schneider, John Richardson, and John Gray. Sarah Bowdich (1791–1856), an accomplished traveller, author and artist, copied Parkinson's drawings, and those of Georg Forster, for Cuvier and Valenciennes to use in preparing their *Histoire Naturelle des Poissons*.

Some authors who used Solander's manuscript descriptions recognised his contribution formally in the specific name: the Wahoo, *Acanthocybium solandri* (Cuvier, 1832), Gemfish, *Rexea solandri* (Cuvier, 1832) and a Toadfish, *Canthigaster solandri* (Richardson, 1844).

Cook's second voyage: Johann and Georg Forster

Banks and Solander did not accompany Cook on his second voyage but Johann Reinhold Forster (1727–1798), a German naturalist and linguist, was engaged in their place. Johann Georg Adam Forster (1754–1794), his son, then learning the art of scientific illustration, was taken as an assistant.

In 1772 the Forsters travelled with Cook in the *Resolution*; they did not come to Australia but collected fishes in New Zealand, many of which occur also in Australia. The descriptions and illustrations prepared by the Forsters were not published with the results of the voyage but, in 1801, some of their work was incorporated in Bloch and Schneider's *Systema Ichthyologiae Iconibus*. The Forsters had serious financial problems and this prompted Johann Reinhold to sell his son's paintings to Banks, who made them available for study. Georg Forster published an account of the Pacific voyage: *A Voyage round the World in His Britannic Majesty's Sloop, Resolution, commanded by Capt. James Cook, during the years 1772–5* (Forster 1777). Johann Forster (1788) published some results of the natural history of the voyage, including some fish descriptions. The elder Forster's manuscript descriptions of animals seen on the voyage were finally published in 1844, edited by M.H.C. Lichtenstein, Curator of the Berlin Museum.

Cook had a second vessel on this voyage, the *Adventure*, in the charge of Captain Tobias Furneaux. The *Adventure* was separated from the *Resolution* on the way to New Zealand and Furneaux visited Bruny Island, coasting the east shore of Tasmania before setting out for New Zealand. The naturalists were all on the *Resolution* so the Tasmanian fishes that were seen did not come to the attention of the scientific observers. Furneaux refers merely to 'Dog fish ... fish not unlike Spratts ... trout'.

Cook's third voyage

On Cook's third and final voyage he visited Tasmania and referred, in his journal, to taking large numbers of fish in Adventure Bay with the seine net, most of them Elephant Fish (*Callorhinchus milii*). William Anderson, who had sailed as surgeon's chief mate on the second voyage, was Cook's surgeon on the last expedition. He made natural history notes and was favourably impressed by the variety and number of fish taken, remarking particularly on the Flathead (*Platycephalus bassensis*). Anderson might have accomplished more but died of 'consumption' on the voyage home.

William Ellis was aboard the *Discovery* on Cook's third voyage as surgeon's second mate. Fifteen of his 115 paintings are of fishes but these have apparently not been used by ichthyologists (Whitehead 1968). Ellis died in 1785 in a fall from the mainmast of a ship at Ostend.

Vancouver

George Vancouver, who had participated in Cook's second and third voyages, commanded a Royal Navy expedition (1791–1795) in the *Discovery*, charting the west coast of North America, the Hawaiian Islands and south-western Australia. His naturalist was Archibald Menzies (1754–1842), an Assistant Surgeon in the Royal Navy with a knowledge of botany, who owed his position in the expedition to Joseph Banks (Anderson 1960). Vancouver spent two weeks at King George Sound in south-western Australia in October 1791. While Menzies investigated the botany, fishing parties were sent out and he became intrigued by the catch. He remarked:

> Fish we have already observed were not very plentiful and we in some measure ascribed their scarcity to the number of large Sharks which frequented the Sound. Among those taken in the Seine or with Hook and line were the Sur mullet [*Upeneichthys vlamingii*], the common Mackrel [*Scomber australasicus*][1]... several species of Bream ... (Menzies journal, ML)

Menzies described two of the 'Bream', both probably wrasse, although his descriptions were not published. Gilbert Whitley (1956) has given an account of Menzies' fish.

1 Both *Upeneichthys vlamingiii* and *Scomber australasicus* were described 30 years later by Cuvier from specimens collected at King George Sound by Quoy and Gaimard.

Collections in Australia before 1800

The establishment of the British settlement at Port Jackson might have marked the natural beginning of a steady accumulation of knowledge of Australian fishes but, in the absence of a professional naturalist in the new colony, this was not the case. The fishery was important from the beginning as the stores brought were adequate for only a short time, the quality of the land was unknown and the time of arrival of further supplies from England was uncertain. When the second fleet arrived, after an interval of almost two and a half years, it brought many more mouths to feed in the form of convicts, free settlers, administrators and soldiers. The settlement had to become self supporting in a very short period of time. Although the fishery was of great importance for the first few years it was not well organised and the necessary expertise was not forthcoming from England.

Popular names for the new fish were often adapted from those of British fish to which they had some more or less superficial similarity, or grew locally from seamen's or settlers' usage. Watkin Tench in his *Narrative of the Expedition to Botany Bay* (1789) refers to a fish '... to which, from the form of a bone in the head resembling a helmet, we have given the name of light horseman' (the familiar Snapper, *Chrysophrys auratus*). The Old Wife (*Enoplosus armatus*) and Sergeant Baker (*Latropiscis purpurissatus*) were names used by seamen for quite different fishes they had seen in other parts of the world. Salmon and whiting had British namesakes that were only very distantly related. Mullet were mullet wherever you were. Later, aboriginal names were sometimes used: Congolli, Barramundi, Barracouta and Callop, to name a few. Locality names were commonly attached to the more familiar designations such as Murray Cod, Macquarie Perch, Tasmanian Trumpeter. Locally derived descriptive names were common: Flathead, Yellowbelly, Tommy Rough.

The scientific naming of these new fishes proceeded slowly as the specimens were taken back to museums in Europe, often indifferently preserved, and might wait many years to receive attention from specialists able to describe them. Some of the early collections were lost at sea, or in museums, or disappeared into private collections. The British Museum was, at this time, by no means the obvious destination for zoological specimens from abroad.

Zoological exploration also proceeded slowly. However, the governor, Arthur Phillip, published *Voyage to Botany Bay* (1789) and included references to the natural history of Port Jackson. Among these were descriptions, with accompanying illustrations, of the Port Jackson Shark (*Heterodontus portusjacksoni*) and its dentition; and of the Wobbegong (*Orectolobus maculatus*), a large, benthic shark.

Phillip was the first of many governors to support the sciences in Australia by encouraging natural history collecting, the formation of learned societies and museums, and occasionally by their own publications. Several are mentioned below: Grey (South Australia), Franklin (Tasmania), Hunter, King and Denison (New South Wales). John Hunter (1737–1821) drew some Australian plants and animals (Calaby 1989). These were by no means rigorous scientific illustrations, but remain of interest.

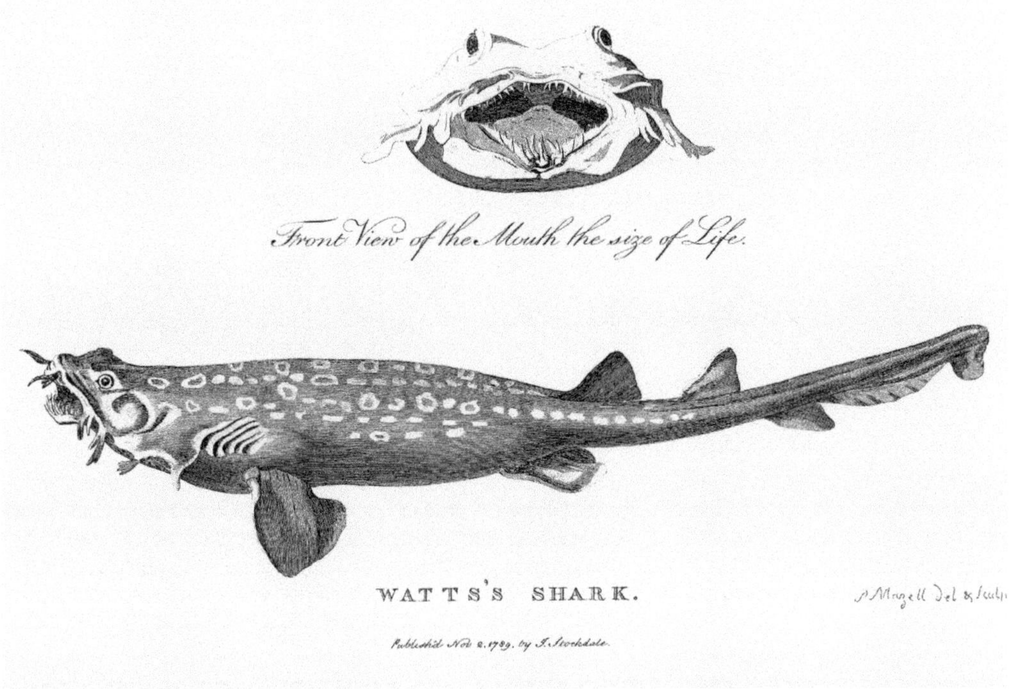

The Spotted Wobbegong, *Orectolobus maculatus* (Bonnaterre, 1788) was called 'Watts's Shark' by Phillip who wrote '... after having lain on the deck for two hours seemingly quiet, on Mr Watts's dog passing by, the shark sprang upon it with all the ferocity imaginable, and seized it by the leg, nor could the dog have disengaged himself had not the people near at hand come to his assistance.' From Phillip 1789.

Captain John White (1757–1832), surgeon-general with the First Fleet, was instructed to give some attention to natural productions. He collected and preserved specimens, including fish, and provided notes. The collections were sent to London where his journal was prepared for publication. The fish were drawn and George Shaw (see page 13) provided identifications and latin descriptions. White's *Journal of a Voyage to New South Wales*, published in London in 1790, includes these brief descriptions and rather crude illustrations. A seahorse was among these illustrations but no description was included. Sixty-five years later Pieter Bleeker (see page 83) based a very brief description on this illustration, simply contrasting a few of its features with those of the Bigbelly Seahorse, *Hippocampus abdominalis*, also known from Port Jackson.

Shaw at the British Museum received fish specimens from the new settlement. Most of his work on them was published in White's *Journal*, the *Naturalists' Miscellany* (1789–1813) (see page 13) and in volume 5 of his *General Zoology* (1804).

F.A.A. Meyer in Leipzig identified, described and named animals from the Australian collections of White and Phillip in 1793, including the Port Jackson Shark, *Squalus* [now *Heterodontus*] *portusjacksoni*.

Captain John White appears as the author of some fish taken around Port Jackson: Old Wife, *Chaetodon* [now *Enoplosus*] *armatus* (top left), Rough Leatherjacket, *Balistes granulata* [now *Scobinichthys granulatus*] (bottom left) and Eastern Frogfish, *Lophius* [now *Batrachomoeus*] *dubius* (bottom right). White's Seahorse, *Hippocampus whitei* Bleeker, 1855 (top right) was based on the illustration in White's publication. Extracts from plates 25 and 30, from White 1790.

PART TWO
(1800 to 1870)

From 1800 to 1870 ambitious exploring voyages were conducted by France and England; Australian seas were often visited. Instructions for the collection of natural history specimens were issued, and naturalists and artists were generally appointed to accompany the explorers. This period also saw some early work carried out in the Australian settlements, although most of the specimens thought to be of interest were still dispatched to museums and collectors in Europe and England.

Flinders (Bauer, Brown) and Baudin (Péron and Lesueur)

Voyage of the *Investigator* (commander: Matthew Flinders) 1801–1803

Matthew Flinders and George Bass had circumnavigated Tasmania in 1798 and Flinders was anxious to carry out a much more extensive survey. He communicated with Joseph Banks who supported his plans and gained cooperation from the British Admiralty. Banks also organised the scientific staff for the expedition. The naturalists who were aboard the *Investigator* with Flinders in 1802 were able to collect plants and animals on the voyage around Australia. Flinders was pleased and impressed by the dedication and energy of his naturalists, writing to Banks '... it is fortunate for science that two men of such assiduity & abilities as Mr Brown & Mr Bauer have been selected: their application is beyond what I have been accustomed to see' (quoted in Watts *et al.* 1998). Flinders himself participated in some of the collecting excursions and seemed generally well disposed towards the efforts of his naturalists. However, Robert Brown, an army surgeon and experienced botanist who was in charge of the collectors and illustrators, was a little less pleased with his young captain. He wrote to one of his patrons, C.T. Grenville:

> Captain Flinders, who does not rate the importance of such collections very high, thought, I suppose, he did enough in affording me opportunities of landing at our different anchorages. The trouble of ordering boxes to be made and the occasional employment of his carpenters in that business, he does not seem to have reckoned on. However, as I would rather attribute his conduct to his inexperience in such matters than to any other causes, I think a few words from Sir Joseph will set him right in his notions of collectors and collections. (quoted in Hamilton and Bruce 1998)

Brown's published results from these collections, and from his work in Australia after the circumnavigation had been completed, were remarkable: the first clear picture of the vegetation of Australia emerged. Brown later became Curator and Librarian for Joseph Banks and made the collections available to British and overseas researchers.

Ferdinand Lucas Bauer (1760–1826) was born in Austria and became an accomplished botanical artist. His work, and that of his brother Franz, was known to Joseph Banks. Franz Bauer painted botanical subjects at Kew Gardens for 50 years and Ferdinand, then 41, was recommended by Banks for the post of natural history artist on the *Investigator*. He drew hundreds of the plants and animals that were discovered on this voyage. Once again, the work was unpublished.

The zoological results from the *Investigator* were less impressive and the ichthyological results were sparse. However, a small collection of fishes was made and Bauer's magnificent drawings provided information that was to be used to good effect by John Richardson (see page 46).

A recent appreciation of Bauer's magnificent drawings was published in *An Exquisite*

Eye (Watts *et al.* 1998). Here are included several beautiful and accurate fish drawings, among them that of a male Spiny-tail Leatherjacket with full adult colours. This drawing was in Robert Brown's possession; he made it available to John Richardson who based his description of the species on this drawing alone, naming it *Aleuterius* [now *Acanthaluteres*] *brownii*, (see colour plate 2, lower).

The description of this leatherjacket appeared in 1846 in Richardson's *Ichthyology of the Voyage of H.M.S. Erebus & Terror*, a work in which the author made good use of drawings from several sources (see page 51). In the same work, Richardson described yet another leatherjacket species, which he named *Aleuterius baueri* [the Pygmy Leatherjacket, now known as *Brachaluteres jacksonianus* (Quoy and Gaimard, 1824)] in recognition of the artist, on the basis of another Bauer drawing:

> This species is named in honour of Mr. Ferdinand Bauer, being founded on one of his admirable drawings in Dr. Brown's possession. The figure presents the careful execution and felicity of touch which characterize Mr. Bauer's pencil, and looking to his known scrupulous accuracy in details, it may be considered as the representation of a generic form not yet described, in which the characters of several groups of Plectognathi are combined. (Richardson 1844–1848)

The latter comment refers to the ability of this little fish to inflate its body in the manner of the toadfish and to the lack of an external ventral fin rudiment.

Early French exploration

The French who had visited Australia prior to Baudin's expedition had not been very successful in their natural history investigations. La Pérouse, whose company included astronomers, geographers, artists and no less than five naturalists, had been lost after meeting the 'first fleet' at Botany Bay in 1788, although some of his collections had been dispatched to France from Java. The fate of this expedition remained a mystery until the wrecks of the ships were located at Vanikoro four decades later. In 1791 two ships, the *Recherche* and the *Espérance*, carried an expedition led by Admiral Bruni D'Entrecasteaux to search for La Pérouse. Jacques Julien de La Billardière, naturalist to the expedition, made extensive collections and observations while in Tasmania, and later in south-western Australia. Many of the members of this company died on the return journey, including d'Entrecasteaux himself, and the expedition virtually ceased to exist. With England and France then at war, La Billardière was separated from his collections; they were returned to him later through the influence of Joseph Banks, ever a good advocate for botanists (Dodge 1971).

Voyage of *Géographe* and *Naturaliste* (Commander: Nicolas Baudin) 1800–1804

Nicolas Baudin (1754–1803), sailing the south coast of Australia at the same time as Flinders, had naturalists aboard who collected an enormous amount of information. The expedition was well equipped through the planning of Baudin himself and through the ambition of the Emperor, who strove to excel in all of his projects. Baudin commanded the *Géographe* and Emmanuel Hamelin the *Naturaliste*. The expedition set out from Le

Havre in October 1800. The brothers Henri and Louis de Freycinet were ensigns at the beginning of the voyage, Henri on the *Géographe* and Louis on the *Naturaliste*. Louis played an important part in completing the publication of an account of the voyage and later in commanding another voyage to the Pacific and Australia.

Initially there was a large complement of scientists but trials and squabbles in the early part of the journey caused many of them to abandon the expedition at Mauritius. François Péron had been appointed zoological assistant initially and his importance was increased enormously by the default of the majority. Baudin had initially chosen two young men, Charles Lesueur and Nicolas-Martin Petit, who had signed on for the voyage as assistant gunners, to illustrate his log-book. Their work was praised by the naturalist, Bory de Saint-Vincent[2], one of those who left Baudin at Mauritius. However, he regretted that, although the work was accurate and attractive, the illustrations were not presented in such a way as to satisfy the demands of scientific illustration. This was corrected by some supervision from Péron as the voyage proceeded.

François Péron (1775–1810) was given a rudimentary education by the parish priest of Cérilly, his small home town. France was in turmoil through revolution and war and, when Péron was 17 and about to enter the Church, he was obliged to enlist in the army. He was captured by the Imperial forces and in prison the sight of his right eye, which had always been very poor, was lost altogether. He was set free late in 1794 and, by the age of 20, he was the secretary at the town hall in Cérilly.

Shortly after this a lawyer of Cérilly provided money for Péron to attend a medical school in Paris. Here he studied the natural sciences and literature and came to the attention of Lacépède and de Jussieu, influential exponents of natural history, who supported him in his application to join Baudin's voyage to the Pacific.

Péron sailed with Baudin in the *Géographe* and demonstrated a disturbing insight into his strengths and weaknesses in a private summary of his own character:

> Irresponsible, scatter-brained, argumentative, indiscreet, too absorbed in my own opinions, incapable of ever giving way for any reason of expediency, I can make enemies and alienate my best friends. These faults are the consequence of my education and the independent life I have led. I know that they tarnish the good qualities that I may possess, but such is the power of habit, that my efforts to overcome them have thus far been in vain. Nevertheless, while I reproach myself for them, I do not blush for them in the slightest. I feel that my heart is a stranger to the harm that I may have done, and my regret for it exonerates me before the tribunal of my conscience. These failings are made up for by my good qualities. Virtuous, sensitive, generous, I have never knowingly hurt anyone ... I still retain all the friends I have had, whether at school, in the army or in Paris. (Laissus 1988)

Péron's position in the scheme of the expedition was initially modest: he was to assist in anthropological and ethnological observations. The desertions at Mauritius, followed by the

2 Jean Baptiste Bory de Saint-Vincent, French naturalist and soldier, edited the *Dictionnaire Classique d'Histoire Naturelle*, 1822–1831 in which some Australian fishes were described, including the Elephant Fish, *Callorhinchus milii* from southern Australia.

deaths of the two remaining zoologists early in the voyage, threw Péron into prominence as the only person trained in any degree for zoological work. His relationship with Baudin deteriorated badly and the young zoologist's enthusiasm on more than one occasion led to delays that the captain found intolerable. Baudin responded by reprimanding Péron and imposing conditions on his further collecting forays, which Péron largely ignored. 'Indiscreet, too absorbed in my own opinions, incapable of ever giving way for any reason of expediency' indeed; and his subsequent behaviour suggests that the tribunal of his conscience remained untroubled.

Nonetheless, it was the activities of Péron, including his influence on his devoted friend Lesueur, that now stand out as of the first and lasting importance in this voyage. But it is still possible to sympathise with Nicolas Baudin who, with the wind rising off an unfamiliar and uninhabited shore, and the naturalist's party again neglecting the agreed rendezvous, alone bore the responsibility for command.

Charles-Alexandre Lesueur (1778–1846), born at Le Havre, France, was 22 when he began drawing for Baudin. Jacqueline Bonnemains (1988), in her biographical sketch of Lesueur, mentions searching for evidence of instruction in drawing in the records of Lesueur's education, but without a definitive result.

His collaboration with Péron was most beneficial to them both: Lesueur learnt natural history in the field and the rigours of scientific illustration while Péron had an able illustrator and field assistant. The zoologist in whom Baudin placed most reliance was René Maugé, an associate from his earlier travels, but he died at Oyster Bay in Tasmania in February 1802. This left Péron and Lesueur to undertake the bulk of the work of zoological collecting, preservation, identification, description and illustration. Petit had initially concentrated on landscape illustration, but later shared some of the work on animals.

In July 1801 in Shark Bay Péron (1807) contrasted the 'hideous sterility of the continent' with the prolific and diverse productions of the waters. He and Lesueur collected a specimen that they regarded as warranting a new genus of triggerfish (Balistidae) and Péron classified it according to the ichthyological system of 'mon illustre maître M. de Lacépède.' Lacépède's publications at this time supplied the latest classification of fishes. De Freycinet was astonished by the toadfish (Tetraodontidae) of a bay and islet at Dirk Hartog Island which he named 'Baie et Ilot des Tétrodons, from the great number of fishes of that genus that we found there ...' Later, the voyagers were impressed by the fishes of King George Sound but disappointed with Kangaroo Island waters where they reported abundant large sharks.

The expedition returned to France in March 1804 but publication was again a difficulty. Baudin had died of tuberculosis in Mauritius on the return voyage and could not defend the expedition from its detractors, prominent among whom were the scientists who had abandoned him in Mauritius. Péron and Lesueur worked on the collections that they brought back but initially without support. The Muséum d'Histoire Naturelle de Paris received the enormous zoological collections from Baudin's voyage and the naturalists of the Museum were delighted. Baron Cuvier declared that 'Messrs Péron and Lesueur alone will have made known more new creatures than all the travelling naturalists of recent times put together.' (Laissus 1988). Sadly, this was not the case. However, after a submission to Napoleon from the prominent naturalists of the Institute, Péron and Lesueur were commissioned and subsidised to prepare an official account of the voyage. By this time, they had been back in France for over two years and Péron's health was deteriorating. He published volume one of his account of the voyage in 1807 but, in 1810, after a long illness and to Lesueur's great grief, Péron died in Cérilly at the age of 32. Lack of prompt financial subsidy from the government, Péron's poor health and Lesueur's lack of experience ensured that only a few aspects of the collections were ever properly prepared for publication (DSB; Duyker 2006; Péron 1807).

After Péron's death, Louis de Freycinet acceded to a government request to complete the account of the voyage (Péron and de Freycinet 1816). Lesueur resented this, but the long delays make it understandable enough.

Péron's description of a Sevengill Shark, *Squalus* [now *Notorynchus*] *cepedianus* from Adventure Bay, Tasmania was included in the first volume of the account of the voyage, published in 1807, so he is the author of this species. After Péron's death, Lesueur (1822) published a description of the Tiger Shark, *Squalus* [now *Galeocerdo*] *cuvier*, apparently using Péron's notes on a specimen from the north-west coast of Australia; both Péron and Lesueur appear as authors of this species. Lesueur (1824) also described the Banded Toadfish, *Batrachus* [now *Halophryne*] *diemensis* from a Tasmanian specimen. This authorship barely hints at the contribution to Australian ichthyology made by Péron and Lesueur but generally realised through the publications of other authors.

In 1815, Lesueur was engaged by Willliam Maclure, a geologist, to participate in scientific travels in North America. When his time with Maclure was finished, Lesueur decided to settle in the United States and earned his living as a drawing master. His researches were wide ranging but ichthyology was central to his interests. In America, he had the satisfaction of being able to publish some of his discoveries as he made them. He described many new species of fish, mostly between 1817 and 1829, and he is still regarded as the author of nearly 30 of these species. He concentrated on freshwater fishes of the eastern United States but his descriptions included several stingrays and a sturgeon. His magnificent coloured drawings illustrated these original descriptions.

However, little of Lesueur's work from Baudin's voyage, which is still preserved in Le Havre, France, has even now been published. Gilbert Whitley (1975) gives an account of some of Lesueur's paintings. He was able to inspect them in the museum at Le Havre where a collection of Lesueur's 'Faune Australienne' is kept.

The two most successful voyages of natural history discovery in Australia up to this time were certainly those of Baudin and Flinders, but the participants in neither voyage had their due. Baudin had died of tuberculosis and had been harshly criticised by members of his expedition. Napoleon, preoccupied by his plans to invade Britain, condemned Baudin's conduct of the voyage and paid little heed to the requirements of Péron and Lesueur; the naturalists were without the government support essential in preparing the results of their endeavours for publication. The return of Brown and Bauer in 1805 aroused little interest in a Britain that was glorying in Nelson's victory, and dramatic demise, at the Battle of Trafalgar. Flinders himself, obliged to seek shelter at Ile-de-France (Mauritius) when his ship required repairs on the journey back to England, was detained by the governor, eventually being released after six and a half years.

French voyages (1817 to 1840)

The French were remarkably active in the Pacific throughout this period, conducting at least nine major expeditions. Several of these were significant from an ichthyological viewpoint, as informed naturalists were present to oversee the collection and preservation of specimens. These men, generally French naval surgeon-naturalists, also took a personal interest in seeing that the findings were published.

In 1817, the year following completion of his account (begun by Péron) of Baudin's expedition, Louis de Freycinet set out for the Pacific on a voyage that lasted over three years. He commanded the corvette *Uranie* and, having experienced the eccentricities and misconduct of professional naturalists on Baudin's voyage, appointed two naval surgeons to carry out the natural history observations. These were Jean-René Quoy and Joseph Paul Gaimard.

Quoy

Jean-René Quoy (1790–1869) was born in France, son of a surgeon. At 16 he entered the School of Naval Medicine at Rochefort and from 1807 served as surgeon on ships during the Napoleonic Wars. He was instructed to collect specimens of natural history during the French assault on the island of Réunion (then the Ile de Bourbon) and this may have stimulated his interest in natural science. After the demise of Napoleon, Quoy was appointed surgeon-general of the *Uranie* under the command of Louis de Freycinet.

After completing his contribution to the account of the Freycinet expedition's zoology Quoy was appointed professor of anatomy at the School of Naval Medicine but then applied for a place with Dumont d'Urville on the *Astrolabe*. His colleague, Gaimard, already had the position of surgeon-major on this expedition so Quoy was appointed zoologist. The loss of some of the collections and drawings on the previous expedition when the *Uranie* was wrecked made Quoy particularly cautious on this latter voyage; he and Gaimard sent collections, descriptions and drawings back to France from the major ports of call and duplicated much of the material. He appreciated Cuvier's work in progress (see page 37) and was particularly attentive to the collection and drawing of fishes. Quoy brought back over 4000 sketches, including many of his own, and deposited the collections at the Muséum d'Histoire Naturelle.

Georges Cuvier died in 1832 and Quoy had been led to expect a professorial position curating molluscs and zoophytes, groups to which he had paid special attention on the voyage of the *Astrolabe*. However, Achille Valenciennes was appointed and Quoy resigned his endeavours in natural history in favour of the Naval Health Service where he served for over 20 years, retiring in 1858 (DSB).

Gaimard

Joseph Paul Gaimard (1796–1858), born in France, entered the naval medical school at Toulon in 1816 and became a naval surgeon. He joined Freycinet's expedition on the

Uranie as surgeon and naturalist and collaborated with Quoy in producing the report on the expedition. He was in the South Pacific again between 1826 and 1829, participating in Dumont d'Urville's expedition, again with his colleague Jean René Quoy.

Gaimard's initiative and accuracy of observation were evident in his studies of cholera in western Russia, and later in Paris in 1832. He published a paper on this disease which is regarded as a classic work. In 1835 Gaimard led a group of scientists aboard the *Recherche* on a voyage to Iceland and Greenland; and later led explorations in Lapland and Spitsbergen. His work was painstaking and accurate (DSB).

Voyage of the *Uranie* and *Physicienne* (commander: Louis de Freycinet) 1817–1820
The expedition was well equipped and de Freycinet defied naval tradition by taking his young wife, Rose. Rose de Freycinet kept her own diary which was published in 1927; an English translation, entitled *A Woman of Courage. The Journal of Rose de Freycinet on her Voyage around the World 1817–1820* was published in 1996. This diary gives an oblique view of the expedition: Louis, her husband, is squarely in the centre of her world, but she made her own observations and judgments on the people and phenomena that she encountered.

The *Uranie* arrived in Shark Bay, Western Australia, in 1818 with the main purpose of continuing the survey work started by Baudin. Large natural history collections were made, including fish that were unknown to science. The expedition proceeded by way of Waigeo in Indonesia, Guam and Hawaii, from each of which significant fish collections were made. They arrived at Port Jackson at the end of 1819.

Governor Lachlan Macquarie entertained Louis and Rose in Sydney, where the *Uranie* stayed for a month. The naturalists were active and Quoy participated in an exploration through the Blue Mountains to the Bathurst Plains. The collections included fishes from Port Jackson and some were described as new species.

Freycinet returned via Cape Horn but the *Uranie* struck a submerged reef near shore in the Falklands and was, at great hazard, run onto the beach. The ship could not be repaired so de Freycinet's people began to build a small boat. While the work was in progress, two other vessels appeared and the French were taken off by Captain Galvin in the *Mercury*. He had put in to the Falklands for repairs and was carrying arms for rebels in Chile; however, Freycinet was able to hire his help for the expedition. En route for Montevideo, de Freycinet bargained further with Galvin who eventually agreed that the French could purchase the vessel. Freycinet renamed it the *Physicienne* and in it the expedition returned to France in 1820, complete with large natural history collections. The wreck at the Falklands resulted in the loss of many specimens, notes and drawings and perhaps resulted in some of the confusion over localities which surfaced in the publications. For instance, two Australian leatherjackets (Monacanthidae) were attributed to Mauritius and the eastern Australian Blackfish, *Boops* [now *Girella*] *tricuspidata* to Shark Bay.

Quoy and Gaimard published their account of the expedition in 1824 in the zoological part of Louis de Freycinet's report (de Freycinet 1824–25). This account included descriptions of a number of fish characteristic of Australian shores; Quoy and Gaimard also

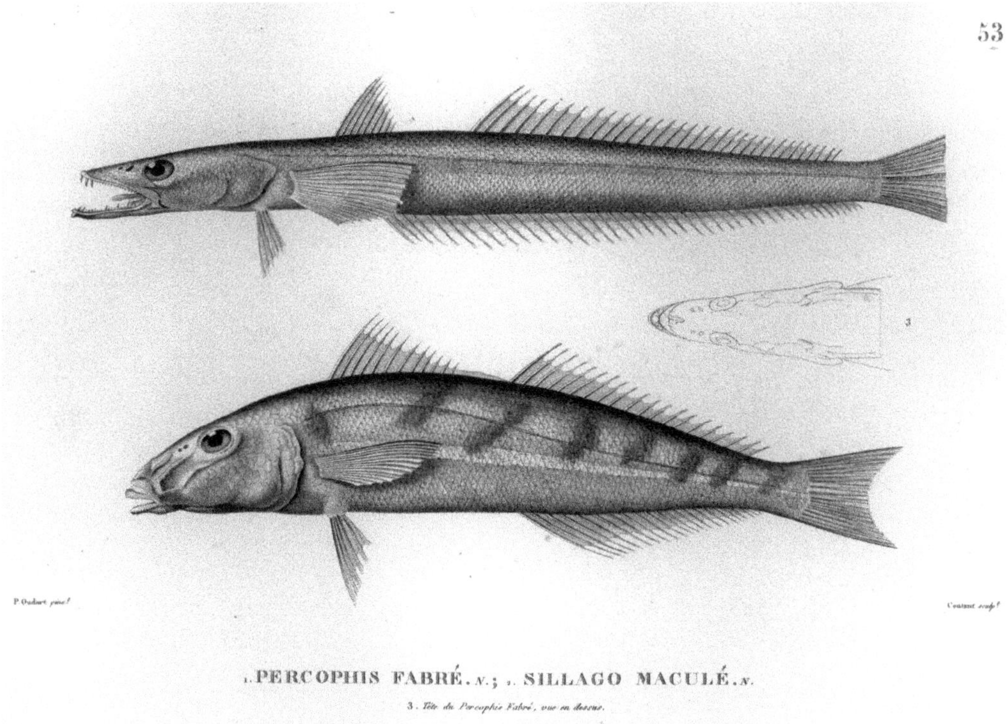

The Trumpeter Whiting, *Sillago maculata* (bottom) was described from a Sydney specimen. Above this, a new species of fish from Rio de Janeiro is illustrated (*Percophis brasiliensis*) (top). SLSA: Rare Book Room 508.3d++-Zoologie/par J.R.Quoy et J.P. Gaimard (folio atlas), 1824, Plate 53 Poissons, image courtesy of the State Library of South Australia.

provided Cuvier with a wealth of specimens, descriptions and drawings of the fishes they had encountered (see colour plate 3).

Voyage of the *Coquille* (commander: Louis-Isadore Duperrey) 1822–25

In 1822 Duperrey set out in the *Coquille* with two surgeons, René Primevère Lesson and Prosper Garnot, both of whom had studied medicine and natural history. The expedition had been planned with the assistance of Dumont d'Urville who was among the officers.

The senior surgeon-naturalist, Garnot, suffered from dysentery and left the expedition at Port Jackson. However, Lesson was also a talented naturalist whose account was published in 1830–31. Duperrey's own account was never completed due to the revolution of 1830.

Lesson

René Primevère Lesson (1794–1849) served in the French Navy during the Napoleonic Wars as a surgeon. He was enlisted as pharmacist and botanist with Duperrey in 1822 and was expected to assist Garnot in collecting natural history specimens.

Lesson's work on the vertebrates of the collections occupied several years, during which he also published several works on ornithology and some papers on fish, including the description of a New Zealand seahorse (Lesson 1818) and contributions to Bory de St Vincent's *Dictionnaire Classique d'Histoire Naturelle* (1829–30). His report on the *Coquille* zoology included the description of new species of fish, several of which were later found to comprise part of the tropical Australian fauna.

Cuvier and Valenciennes benefited from the work of Lesson and Garnot; the *Histoire Naturelle des Poissons* is sprinkled with references to their specimens. A few such references are paraphrased below:

- *Pelates quadrilineatus* sent by Péron to 'le Cabinet du Roi' and reported from Port Jackson by Lesson and Garnot. (Cuvier and Valenciennes 1829).
- *Trigla* [now *Chelidonichthys*] *kumu* brought back by Lesson and Garnot from New Zealand and described in *Zoologie du Voyage de la Coquille* - Parkinson drew this fish on Cook's first expedition and named it *Trigla papilionacea*, which appears under a handsome drawing in Banks' library - inhabitants of New Zealand call it 'koumou' (Cuvier and Valenciennes 1829).
- *Pagrus guttulatus* [now *Chrysophrys auratus*] recorded by the late Péron from the coasts of New Holland - found again by Quoy and Gaimard at the anchorage in King George Sound and at Jervis Bay - Lesson and Garnot have taken it from the river mouths in New Zealand (Cuvier and Valenciennes 1830).

Lesson pursued his career in the navy, becoming Chief Pharmacist at Rochefort in 1839.

Voyage of the *Astrolabe* (commander: Dumont d'Urville) 1826–29

Quoy and Gaimard were again chosen as naturalists on Dumont d'Urville's voyage of the *Astrolabe*, in 1826. One of the main objectives of the voyage was to attempt, once again, to determine the fate of La Pérouse and his party; the vessel was the *Coquille*, the same as commanded by Duperrey the previous year, but d'Urville renamed it *Astrolabe* after La Pérouse's ship.

The two surgeon-naturalists produced a well illustrated, accurate account of the ichthyology of the voyage (Quoy and Gaimard 1834). The descriptions and drawings from this second expedition incorporated greater anatomical detail than those of the *Uranie* report. Most of the new species described were collected in Mauritius and Guam, but they also described *Malacanthus* [now *Siphonognathus*] *radiatus* from South Australia (see colour plate 4).

Again the two naturalists shared their collections with the authors of the *Histoire Naturelle des Poissons* and were duly acknowledged; for example, Valenciennes (1839) described *Julis* [now *Coris*] *auricularis* from a specimen 'MM Quoy and Gaimard have dispatched from the Port of King George, New Holland, during the expedition of M. Dumont d'Urville ...' (Cuvier, 1829). Quoy and Gaimard collected the Spotted Whiting, *Sillago* [now *Sillaginodes*] *punctata* in King George Sound and the Silver Whiting, *Sillago bassensis* in Westernport, Victoria. Cuvier (1829) described both of these species and they were illustrated by Quoy and Gaimard (1834).

Voyage of the *Astrolabe* and *Zélée* (commander: Dumont d'Urville) 1837–40

D'Urville set out again in the *Astrolabe* in 1837, this time with the surgeon-naturalists Jacques Bernard Hombron (1798–1852) and Honoré Jacquinot (1815–1887). A second vessel was used, the *Zélée*, in the charge of Charles Jacquinot (Honoré's elder brother) who had travelled with Duperrey and with d'Urville on the earlier voyages in the Pacific. Collections included a few fish from Hobart and from New Zealand, but most of their specimens were taken north of Australia. In 1840 they returned to France. D'Urville began his report of the voyage but was killed in Paris in a railway accident in 1842. His account of the voyage was completed by Dumoulin and Jacquinot.

The reptiles and fishes were dealt with by Jacquinot and Alphone Guichenot (see page 82) in Dumont d'Urville's account (1853). Hombron and Jacquinot collected fish from a variety of localities, particularly Malaysia, Fiji, Samoa, Borneo and the Caroline Islands. An Eastern Australian Salmon from Tasmania was named *Centropristes tasmanicus* [synonym of *Arripis trutta*]. Their illustration shows a discoloured, shrunken specimen.

Voyage of the *Bonite* (commander: Auguste-Nicolas Vaillant) 1836–37

Joseph Fortuné Théodore Eydoux (1802–1841) and Louis François Auguste Souleyet (1811–1852) were surgeon-naturalists who had worked together on a circumnavigation in *La Favorite* in 1830–32 and again in 1836–37 in *La Bonite.* In the *Bonite* the young surgeons were accompanied by the experienced natural history collector and botanist Charles Gaudichaud-Beaupré, who had been with de Freycinet in the *Uranie*. Although Eydoux died in 1841, only a few years after their return, Souleyet completed the work in both their names (Eydoux and Souleyet 1850). Most of their more interesting fish collections were made in Hawaii and Macao; Valenciennes (1839) acknowledged their contribution: 'Once again we owe thanks to M. Gaudichaud and to MM. Eydoux and Souleyet, who have made large collections during the voyage of the Bonite.'

The ichthyological (and other) results of these French expeditions can be contrasted with those of Cook and of Baudin mainly in that the results were made accessible. The work of Quoy, Gaimard and Lesson was particularly significant in the history of the discovery of Australia's fish. The workers who followed them could not afford to ignore these early publications if they were to practise effective taxonomy.

Darwin (Jenyns)

Britain's Royal Navy was prominent in conducting voyages of scientific discovery in the 19th century and a series of successful endeavours constitute prominent milestones in oceanographic and natural history discovery. Of the greatest importance were the voyages of the *Beagle*, the *Erebus* and *Terror* and, in the latter part of the century, the *Challenger*.

The itinerary of the *Beagle* on the voyage from December 1831 to October 1836 included a visit to King George Sound in the south of Western Australia. The naturalist was, of course, Charles Robert Darwin (1809–1882) and he did not neglect fish. His collection of fishes was attended to on his return by the Reverend Leonard Jenyns (1800–1893), the vicar of Swaffham Bulbeck, Cambridgeshire. Jenyns was known as an experienced field naturalist with a special interest in fish (Barrett and Freemen 1986) and had been invited by Fitzroy, commander of the *Beagle*, to accompany him as naturalist on the voyage. His refusal led to Darwin being offered the position.

Darwin collected fish throughout the cruise, generally preserving them in spirits, but keeping a smaller number as skins only. The greatest number of specimens was taken on the east and west coasts of South America. He provided notes on the fishes collected, including colour descriptions, and Jenyns approved of his systematic approach:

> The particular locality assigned to each species respectively in the following work may be relied upon as correct; pains having been taken by Mr Darwin to affix a small ticket of tin, with a number stamped upon it, to each specimen, and to enter a note immediately in the manuscript catalogue, having the same number attached. In only three or four instances these tickets were found wanting, on arrival of the collection in this country. (Jenyns 1842)

Darwin's notes on fish '... caught by net in Princess Royal Harbour [Albany, Western Australia]' include a reference to a leatherjacket's parasite: 'colourless. inhabiting a most enormous cavity ... in flesh on sides & behind the gills & extending apparently to membrane of abdominal viscera. - whole animal, excepting tip of tail included in cavity.' (Darwin 1836) This parasite came to the attention of other observers at about the same time (see pages 53 and 60).

About half the specimens collected were too damaged in the course of the voyage to be useful, and this included many of those taken in King George Sound, Western Australia. Several fishes from the Sound were described as new and three of these are still recognised as valid species. Another fish, the Common Galaxias, *Galaxias maculatus* that is found in southern Australia and was described by Jenyns, was collected by Darwin from 'a fresh-water brook, in Hardy Peninsula, Tierra del Fuego...' and from 'streamlets and creeks high up the river of Santa Cruz in Patagonia ...'; Darwin also took this species '... in fresh-water in the Bay of Islands in New Zealand.' (Jenyns 1842). Jenyns gave the

PLATE XIV

Caranx declivis

The Common Jack Mackerel, *Caranx* [now *Trachurus*] *declivis* collected by Charles Darwin at King George Sound, Western Australia and described by Leonard Jenyns. From Jenyns 1842.

not inconsiderable synonymy of this species a flying start by describing the New Zealand and South American specimens under different names.

Jenyns worked with great care and was repeatedly exhorted by Darwin to complete the report on his fishes. He used the methods of Cuvier and paid his respects:

> The method of description, and the mode of computing the fin-ray formula, will be found conformable to the plan adopted in the *Histoire des Poissons* of Cuvier and Valenciennes; a work which, in so many respects, must always serve as a model to labourers in this department of zoology. (op. cit.)

In Jenyns' final report on the *Beagle* fishes the descriptions were systematic and detailed, and supported by 29 plates.

The greatest significance of this voyage to ichthyology, as to all other branches of biology, lay in the observations which Darwin was able to make on the variation in animals and plants in the parts of the world visited. His conclusions were eventually published in *On the Origin of Species* (Darwin 1859) and altered the course of the biological sciences. The objectives of the study of living organisms and their fossilised ancestors were abruptly clarified. In ichthyology, the vast collection of data, extended and summarised so ably in the contemporary work of Cuvier and Valenciennes, would eventually be ordered more rationally through the medium of evolutionary theory.

Cuvier and Valenciennes, and their contributors

Cuvier

Georges Cuvier (1769–1832) was born in Montbéliard and studied for the ministry. He was a private tutor on the Normandy Coast for several years from 1788 where he had opportunities to establish and indulge an enthusiasm for marine biology and palaeontology. In 1795 he was appointed assistant professor of comparative anatomy at the Muséum National d'Histoire Naturelle in Paris. He was highly regarded by Napoleon, but remained professor of comparative anatomy after the restoration of the monarchy until his death. Cuvier was made Chancellor of the University of Paris and, on the accession of King Louis-Philippe, was made a peer of France. As Baron Cuvier, he became minister of the interior. Apart from the enormous volume of his scientific work, he was active in planning national education policy and in the French Protestant Church (DSB).

Cuvier's greatest contributions to zoology resulted from his expertise in comparative anatomy and palaeontology. In Cuvier's day it was widely believed that species did not become extinct but he established the fact and importance of extinction, although he thought that new species were formed through episodes of creation rather than evolution. He undertook accurate reconstructions of fossil animals and was a pioneer of functional morphology.

In the *Régne Animal* (1816, and later editions) Cuvier described the structural relationships of animal groups and derived therefrom a scheme of classification. The work included a structure for the classification of fishes, defining orders, families and genera. The first edition included the original description of the Tasselled Leatherjacket, *Balistes penicilligerus* [now *Chaetodermis penicilligera*] 'from Australian Seas'. In later editions many Australian fishes were described (see colour plate 5, upper).

Prior to his major work on fishes, Cuvier revised some small groups in separate publications, describing a few new species and including Australian specimens. He treated Anglerfishes (Lophiiformes) (Cuvier 1817) and Porcupinefishes (Diodontidae) (Cuvier 1818) in this way.

Cuvier refined this system over many years and the results were firstly a more rational approach to classification, based largely on consideration of a wide variety of morphological and anatomical characters; and secondly a great collection of data. Both were presented in the form of the 22 volume *Histoire Naturelle des Poissons* (1828–1849). Cuvier had an able assistant in Achille Valenciennes and the latter continued the work after Cuvier's death in 1832.

Valenciennes

Achille Valenciennes (1794–1865), born in Paris, was attached to the Muséum d'Histoire Naturelle throughout his working life (Monod 1963). At the age of 18 he was assistant and preparator to Etienne Geoffroy Saint-Hilaire and later to Lamarck and Lacépède. He appears to have worked with Cuvier from about 1815 and was an able and active assistant throughout the production of the *Histoire*. After Cuvier's death in 1832, just 20 years after Valenciennes began work at the Museum, the latter was appointed a Professor of Zoology, a position he held for the rest of his life. He worked on the *Histoire* for a further 17 years after Cuvier's death.

Histoire Naturelle des Poissons

In the *Histoire* Cuvier revised the classification which he had published in his *Régne Animal* (1816), giving a precis of the new classification at the end of the first volume of the *Histoire*. The systematic part of the work was divided into 'books' and 'chapters'. The first volume contained books one (history of ichthyology) and two (nature and organisation of fishes). Thereafter 18 books, one for each 'family' of fishes, were distributed through the remaining 21 volumes. Each book was divided into chapters, and each of these represented a genus or group of related genera.

The materials available to Cuvier and Valenciennes were varied, consisting mostly of old collections and manuscripts, relatively fresh material from recent expeditions (often made more valuable by the notes or publications of the collectors), and older published work. It was all grist to Cuvier's mill. He prided himself on being more discriminating than previous compilers in culling their redundant descriptions but his own work inevitably contains many similar instances of error. He was very effective in public relations and was able to instil some of his own ardour into his collectors in a manner reminiscent of Linnaeus. The *Histoire* was an effective means of encouraging his disciples as he interpreted and published their discoveries; he was scrupulous and generous in acknowledgment, frequently naming species after his contributors.

Reading the *Histoire* gives a sense of immediacy to the discoveries being made at the time. For instance, in paraphrase: 'Lesson and Garnot are occupied at present on the fish part of the voyage of Duperrey ... a great number of beautiful species perfectly represented.' (Cuvier 1828); 'Also begun are the ichthyological part of the voyage of M. Ruppel and that of M. Ehrenberg into Nubia, Abyssinia and the Red Sea. These two works rich in all branches of natural history do great honour to the naturalists who have gone to collect new observations with such courage, and who report them with such accuracy and sagacity.' (op. cit.); '*Macquaria* ... a single species, *Macquaria australasica* [Macquarie Perch], recently found in the River Macquarie, which is itself a recent geographic discovery ... thus we are pleased to acknowledge the general to whom we owe this extension of knowledge of the globe ... first specimens of this fish from Lesson and Garnot, naturalists on Duperrey's expedition ... the fish was taken at Bathurst, a new town founded on the River Macquarie.' (Cuvier 1830).

Cuvier was usually mild in his criticism of the work of contemporaries but scathing with that of some of his predecessors. Lacépède in particular is criticised for inadequate description and the multiplication of species. Cuvier praised 'Admiral Corneille de Vlaming, who collected and represented a great number of fishes, and merits not to be forgotten' and refers to Ruysch, Valentyn and Renard as 'his three miserable copyists' (Cuvier 1829). With reference to the description of *Trachichthys australis*, collected for Surgeon White in Sydney

Harbour and described briefly by George Shaw, then curator of the British Museum, Cuvier said only that 'Unhappily the description of Shaw is not as complete as could be desired.' He also complained of 'naturalists who refuse to erect new genera when they find "beings" that require them, so placing the new "beings" arbitrarily: these species are bounced from genus to genus.' (Cuvier 1829)

The older collections were searched for useful material: the Cabinet du Roi (often as skins); the stadholder's (Dutch) collection (see page 8); collections of other workers, particularly Bloch and the more recent Paris collections of Péron and Lesueur. Cuvier visited London, and Valenciennes visited London and Berlin, to consult the museum collections and their curators. Among the older manuscripts were those of Commerson; and of Solander, available from Banks' library through the botanist, Robert Brown.

Some names recur frequently in Cuvier's credits for fishes from the Indo-Pacific, particularly Delalande, Leschenault, Péron, Quoy and Gaimard, Dussumier, Kuhl and van Hasselt, Ehrenburg, Rüppell.

Dussumier

Jean Jacques Dussumier (1792–1883), a ship-owner and merchant from Bordeaux, was referred to by Cuvier (1828) as 'a young man who has already made several voyages in his own ships to China and India' and credited him with sending collections of fishes from Malabar and the Seychelles to the Muséum National d'Histoire Naturelle in Paris. Dussumier also donated paintings that he had commissioned at Canton of 'several beautiful Chinese fishes' – a reference to a manuscript illustrating 25 fishes, 13 of which are identified with reference to the text of the *Histoire Naturelle des Poissons*.

Delalande

Pierre Antoine Delalande (1787–1823) had been an assistant naturalist to Etienne Geoffroy Saint-Hilaire and so was a most informed collector. He travelled to Brazil, Cape Verde and the Cape of Good Hope and brought back extensive natural history collections, including large numbers of fishes. His South African fishes included species well known in southern Australia, such as the Mulloway, *Argyrosomus japonicus* and Barracouta, *Thyrsites atun*. Delalande died of 'fever' while editing the description of his travels. Valenciennes described the Yellowtail Kingfish, *Seriola lalandi*, from Delalande's specimens from Brazil.

Leschenault

Jean Baptiste Leschenault (1773–1826) travelled as botanist on Baudin's expedition to the Pacific in the *Geographe*. He became Director of the botanical gardens at Pondicherry, India in 1816 (Cuvier 1995) and made large natural history collections, including many fishes which he sent to Cuvier.

Ehrenberg

Christian Gottfried Ehrenberg (1795–1876) was sent by the Royal Academy of Sciences of Berlin and, with the support of Alexander von Humboldt, on an expedition through Libya to

Egypt, Arabia and the Red Sea coast of Ethiopia. He and his companion, Friedrich Wilhelm Hemprich, were away from 1820 to 1825 and made observations on many branches of natural science. Ehrenberg's work on the fishes of the Red Sea was of particular value to Cuvier and Valenciennes and is quoted many times, and with well-placed confidence, in the *Histoire Naturelle des Poissons*.

Rüppell

Eduard Rüppell (1794–1884), a German naturalist and explorer was sent by a scientific institution, the Senckenbergischen Naturforschenden Gesellschaft of Frankfurt, to study natural history in North Africa. He collected fish from the Gulf of Suez and from Djedda from 1826 (Botros 1971), reported his work in *Fische des Rothen Meeres* (Rüppell 1828), and provided Cuvier with specimens. Cuvier (1830) wrote in the *Histoire*: 'M. Rüppell, who is finishing the ichthyological part of his voyage, has asked us to report several of his fishes from the Red Sea, to allow comparison with ours for the sake of the synonymy.' Rüppell returned to the Red Sea in 1831 and made further collections, sending back preserved fish, skins and skeletons. He worked on this material on his return and published the results, which included the establishment of 12 new genera, most of which are still recognised as valid. Over 40 of his new Red Sea species are also found in northern Australian waters. An example is the huge Humphead Maori Wrasse, *Cheilinus undulatus* Rüppell, 1835, a prominent inhabitant of Australian coral reefs.

Kuhl and van Hasselt

Heinrich Kuhl (1797–1821), another German naturalist, was appointed to the Natuurkundige Commissie voor Nederlandsche Indie (Commission for Natural Sciences of the East Indies). In 1820 Kuhl, in company with his friend and colleague Jan Coenraad van Hasselt (1797–1823) of Holland, travelled to the Dutch East Indies. Both had completed their studies at the University of Groningen. These young naturalists were active in collecting and describing their discoveries and paid special attention to the fishes. They travelled widely in western Java, collecting plants and animals, and made descriptions and drawings of many species. Both contracted tropical diseases and died in Java, Kuhl after only nine months and van Hasselt two years later. Their collections arrived intact at the Rijksmuseum and the ichthyological material was made available to Cuvier who, extravagant in his praise, referred to them again and again as 'unfortunate', 'zealous', 'indefatigable'.

Temminck, Schlegel and von Siebold

Cuvier maintained excellent relations with the zoologists of his day, particularly those of France, Germany, England and Holland. Conraad Jacob Temminck (1778–1858) was Director of Leiden's royal cabinet of natural history from 1820 to 1858 and was influential in the early years of the Rijksmuseum van Natuurlijke Historie and the Natuurkundige Commissie voor Nederlandsche Indie. He provided specimens for Cuvier from Leiden and Vienna, and the specimens and drawings sent by Kuhl and van Hasselt.

Philipp von Siebold (1796–1866) was born in Germany where he studied medicine. He moved to Holland and, as a medical officer with the Dutch East India Company, travelled to Batavia. He joined a trading expedition to Japan as a naturalist-physician and remained on the island of Deshima in Nagasaki Harbour for an extended period (1823–1829). At this time, Japan shunned contact with the west but the Dutch held a privileged position, being able to conduct limited trading from their island base. Von Siebold amassed remarkable collections of plants and animals which formed the basis of an ambitious work, *Fauna Japonica*. Heinrich Bürger (1806–1858), an assistant of von Siebold, continued to collect on his behalf, sending specimens back to Holland; von Siebold himself returned to a diplomatic post in Japan in 1859.

Temminck and his colleague Hermann Schlegel (1804–1848) studied von Siebold's collection of specimens and drawings of fishes to produce the *Pisces* volume of *Fauna Japonica* (1842–1850). This included the description of approximately 40 species found also in Australian waters, for instance the Starry Toadfish, *Tetraodon* [now *Arothron*] *firmamentum*; the Mirror Dory, *Zeus* [now *Zenopsis*] *nebulosus*; the Grey Seabream, *Dentex* [now *Gymnocranius*] *griseus*. Temminck and Schlegel had received a drawing of a distinctive fish from von Siebold's assistant, Heinrich Bürger, and they established the genus *Glaucosoma* for this fish. They did not assign it a specific name but John Richardson, at the time he studied the related Western Australian Jewfish, named the latter *Glaucosoma hebraicum* and Bürger's fish *Glaucosoma buergeri*.

Fish described from Australia in the *Histoire Naturelle des Poissons*

In spite of their vast energy and resources, Cuvier and Valenciennes were able to report relatively few species from Australia. Many of the fishes of northern Australia were described, but generally from specimens taken elsewhere in the Indo-Pacific, and frequently the French ichthyologists were unaware that the range of the species before them extended to Australia. Between them, Cuvier and Valenciennes described about 350 species which are now recorded from Australian waters.

The *Histoire Naturelle des Poissons* included descriptions of nearly 30 new species for which the authors gave Australian type specimens, and about half of these species are still regarded as valid. These were of particular interest as most were endemic to Australia. The majority were collected at Port Jackson and the Macquarie River in New South Wales; Western Port in Victoria; several coastal sites in Tasmania; and from King George Sound and Shark Bay in Western Australia. Their most useful collectors in Australia were the naturalists who accompanied Baudin (Péron and Lesueur), de Freycinet (Quoy and Gaimard's first voyage), d'Urville (Quoy and Gaimard's second voyage) and Duperrey (Lesson and Garnot). Among the southern Australian fishes described from specimens presented by Quoy and Gaimard, Cuvier described the Swallowtail, *Beryx* [now *Centroberyx*] *lineatus* from King George Sound, Western Australia.

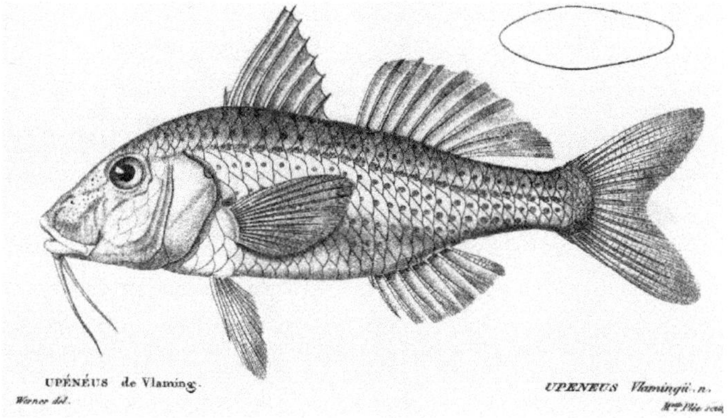

The Bluespotted Goatfish, *Upeneus* [now *Upeneichthys*] *vlamingii* collected by Quoy and Gaimard at King George Sound. From Cuvier and Valenciennes 1829.

The Blue Weed-whiting, *Odax semifasciatus* [now *Haletta semifasciata*] was collected by Péron, probably in Tasmania, and described by Valenciennes. From Cuvier and Valenciennes 1839.

The Southern Crested Weedfish, *Cristiceps australis* was collected by Péron and later by Quoy and Gaimard from 'Hobart-Town'. Valenciennes' illustration shows an adult giving birth to live young. From Cuvier and Valenciennes 1836.

Significance of the *Histoire Naturelle des Poissons*

This grand summary of ichthyology, the *Histoire Naturelle des Poissons*, although incomplete, was invaluable to workers in all parts of the world and provided a good foundation for classifying new species. It was the principal text used by zoologists who dealt with fish until it was gradually replaced by Günther's *Catalogue* as the latter was published over the course of the 1860s. Thus Cuvier, by means of the *Régne Animal* and the *Histoire*, guided naturalists working on fishes through the early and middle years of the 19th century, providing a logical framework for classification and the mechanism for identification.

The *Histoire* was important to Australian ichthyology for several reasons; its strengths and weaknesses deserve some comment. First and foremost it was the most modern treatment available and included descriptions of some fishes endemic to Australia and many species that were described from elsewhere in the Indo-Pacific but whose range included Australian seas. Its strengths included an attempt to describe in detail all of the fishes known to that time. Wherever possible the authors worked with well preserved fishes and attempted to verify the descriptions of earlier workers by reference to their own specimens. Cuvier encouraged dissection and the preparation of skeletons for study. He and Valenciennes were remarkably well informed about, and well supplied with, the earlier literature; and they kept abreast of recent publications and even those in preparation. Their definitions of species and higher groups were presented systematically.

The weaknesses of the work included some of those common to all museum workers of the day. The authors were usually obliged to work with small numbers of each species and frequently the specimens chosen for preservation by the collectors were of small size. The state of preservation was sometimes poor due to the inexperience of some of the collectors, the difficulties under which they were operating, and the rigours of transport by sea from remote localities. Lastly, the attempt to describe all known fishes was frustrated by the sheer volume of the task: Cuvier died in 1832, by which time only eight of the 22 volumes of the *Histoire* had been published.

Valenciennes had been left with an immense task. He followed the plan established by Cuvier but stopped publication with volume 22 in 1849. This left a number of groups still to be revised: the pipefish and seahorses (Syngnathiformes), toadfish, filefish, boxfish, sunfish (Tetraodontiformes), eels (Anguilliformes), cod (Gadiformes), whiptails (Macrouriformes), cuskfish (Ophidiiformes), flatfish (Pleuronectiformes) and clingfish (Gobiesociformes).

Cuvier and Valenciennes integrated the work of their predecessors in a critical fashion, the result being by far the most extensive and reliable summary of ichthyology that had been attempted to that time. As a comparative anatomist, Cuvier strove to establish the basis of the discipline now described as functional morphology. In so doing, he relied almost exclusively on observations made by himself or his co-worker, Valenciennes. For ichthyology in general, and Australian ichthyology in particular, John Richardson and Albert Günther were the most significant successors to the achievement of the *Histoire.* Richardson used its classificatory framework to place his discoveries but Günther also elaborated and modified the system.

Gray at the British Museum, and the Bennetts

Gray

At the British Museum in 1824 John Edward Gray (1800–1875) was appointed assistant to the keeper of the natural history collections. Gray, who had early shown an interest in natural history, aided his father in preparing an ambitious botanical work. He was assistant to Wilfred Leach, then Keeper of Zoology at the Museum, and to Leach's successor, John Children. Gray had studied medicine for six years but he had no interest in practising:

> I always felt that medicine could do little and that the larger part of the patients required better food and solace. (Gunther 1980)

Gray was an active, precise naturalist with a keen appreciation of the advances being made in his subject in France and Germany. He visited the important centres of learning on the continent and was always alert to opportunities for enlarging the collections at the museum. He accepted an enormous workload which included publishing catalogues of a variety of animal groups and many other papers from his own original work. Gray negotiated the purchase of collections, undertook much of the administrative work of the museum and dealt with the many local and foreign specialists who came to work with the collections.

Prior to Gray's ascendancy some Australian fish were represented in popular British publications. George Perry's *Arcana* (1810–1811) illustrated a few Australian fish, including the Common Seadragon, *Phyllopteryx taeniolatus*; little appears to be known about the author. Edward O. Donovan (1768–1837) was a member of the Linnean Society in London and a professional illustrator. He exhibited natural history specimens and published many illustrated books on the fauna of Britain and elsewhere. He dealt incidentally with fishes, generally those from British waters, but also illustrated a few Tasmanian fish. He described the Brown-striped Leatherjacket, *Balistes* [now *Meuschenia*] *australis* (Donovan, 1824).

Gray's publications on ichthyology spanned over 30 years, beginning with a brief account of the fishes collected by Captain Phillip Parker King five or six years earlier (Gray 1827). King had commanded surveying cruises to Australia's north-west coast in small vessels, the *Mermaid* and the *Bathurst* (1818–1822). His instructions included the usual injunctions to observe and collect in the cause of natural science. He complied and, since Allan Cunningham was aboard, the botanical results were substantial (Hordern 1997). King also collected animals, including some fishes, but the collections were random and their preservation inadequate.

The greater part of Gray's work in ichthyology was in the 1830s, much of it stimulated by the increasing influx of specimens from the Indo-Pacific. After this time, John Richardson (see page 46) became very active in ichthyology in Britain and Gray encouraged him to deal with most of the collections and publications on fishes.

In his early days as assistant keeper Gray's most effective informant was Major-General Thomas Hardwicke (1756–1835). During his 45 years with the East India Company, Hardwicke served zoology by recording his own observations, collecting specimens and engaging Indian artists to make coloured drawings of animals. Gray prepared and published *Illustrations of Indian Zoology* (1830–1835), a beautiful work with over 200 plates which exhibited some of Hardwicke's collection of drawings; about 30 of the plates illustrate fish from Indian waters. Included were new descriptions of a few fish found in tropical Australia, such as the Pipehorse, *Syngnathus* [now *Solegnathus*] *hardwickii* (Gray, 1830).

Ichthyology comprised a small fraction of Gray's work, and only a few of his papers were concerned with Australian fishes. He prepared notes on some of the fishes collected by Ronald Gunn (see page 107), who wrote:

> I have this season, from my residence being so near the sea, procured a few fish, but have not numbered them. They were all caught in a seine at Circular Head except one species called a 'Nurse'; I caught it at Western Port [Victoria]. Another species of shark, called here Sword Fish, is abundant; the peculiarly prolonged and armed snout [Saw Shark, *Pristiophorus* sp.] struck me as curious: 3 specimens are sent. A fish called Sea Hedge-hog or Porcupine [diodontid] is also very abundant. On being brought to land it inflates itself into a perfect sphere, the skin as tight as a foot-ball, and the spines erect and stiff in all directions from the body ... Parrot fish, so called from the shape of the head and mouth I suppose, as also perhaps from the various and beautiful colours. Of these I send twelve specimens of two or three species; one specimen, although much faded and altered, is still very beautiful ...(Gray 1838).

Gray was impressed by the 'Nurse' which proved to be a Port Jackson Shark, *Heterodontus portusjacksoni*; he thought that his new specimen might be the only one in European collections and gave a description. From this collection, he also described the 'Parrot fish' of Gunn, now usually known as Cowfish (Ostraciidae), and established a new subgenus (*Aracana*) for them. Gray thought that the 12 specimens represented several new species, being unaware of the remarkable differences in form and colour between the sexes in species of this genus.

A new lamprey from Australia was the occasion for Gray (1851a) to attempt a revision of the family; the fish had been taken in South Australia, probably in the Onkaparinga River. This was *Geotria australis*, a species which, unknown to Gray, he had also encountered in the form of its velasia larva sent from Chile and described by him as *Velasia chilensis* (Gray 1851b); he erected the genus *Velasia* in this publication. In the collections made by the medical officers of HMS *Herald* in Western Australia Gray found a new pipefish, for which he established the genus *Haliichthys* (Gray 1859).

Gray's fish descriptions were brief, generally superficial, and his use of existing genera was often inaccurate. However, his contribution to the ichthyology of Australia was considerable and much greater than would seem from this meagre list of papers. He was an efficient curator: he pursued collections, generally had them described by the appropriate specialists, and maintained them in good order for future study. Gray developed an efficient

The large Ribboned Pipehorse, *Haliichthys taeniophora* [now *H. taeniophorus*] from Shark Bay. From Gray 1859.

system for registering new acquisitions and his type specimens could usually be located by zoologists revisiting his work.

The Bennetts

The ichthyological Bennetts of the first half of the 19th century ought to be distinguished one from another as their names are cited as authors of a number of species whose range includes northern Australian waters. John Whitchurch Bennett (see page 68) described fishes from Sri Lanka (1828–1830). Frederick Debell Bennett (1806–1859) sailed on a whaler (1833–1836), describing the fishes collected during the expedition in his *Narrative of a Whaling Voyage ...* (1840). His brother, George Bennett, resided in Sydney and worked on the Australian Museum collections (see page 98).

Edward Turner Bennett (1797–1836), a surgeon and zoologist, known for his part in founding the Zoological Society of London, published many papers on fishes. He described the fishes of the collection of the Zoological Society and collections were sent to him from many parts of the world. Captain Frederick William Beechey, commanding the *Blossom*, investigated the Pacific and Behring's Strait (1825–1828). The voyage, as was usual, had many objectives, including a search for a north-west passage. The naturalists aboard were very active and collected large numbers of plants and animals. Gray and Richardson were involved in reporting the zoology from the voyage but an account of the fishes was prepared by Lay and Bennett. George Tradescant Lay (?–1841) was an officer and naturalist on the voyage and E.T. Bennett an experienced zoologist. In their report (Lay and Bennett 1839) they described fishes from the Pacific, particularly the Ryukyu Islands and Hawaii. This included a few species that extend to northern Australia, such as the Goby, *Valenciennea longipinnis*, a Snake Eel, *Leiuranus semicinctus* and the Red-ribbon Wrasse, *Thalassoma quinquevittatum*. The description of the latter was based upon a drawing made by Beechey.

Richardson

John Richardson (1787–1865), a British naturalist who was largely independent of the British Museum, contributed many works on the fishes of Australia and the Pacific. He was born in Dumfries, Scotland, the first of 12 children. Apprenticed at 13 to his uncle, a surgeon, Richardson studied in Edinburgh and obtained his licence to practise from the College of Surgeons in 1807. He was appointed assistant-surgeon in the Royal Navy, serving during the wars with Napoleon, and later as a surgeon of the Royal Marines in the war with the United States (1813–1814). He returned to the study of medicine and natural history, presented a thesis on yellow fever, and was granted his MD in 1816 (DNB; Stewart 1931).

Arctic experience

Richardson practised as a surgeon in Leith, the port of Edinburgh, for two years before being offered a position as surgeon-naturalist to John Franklin's first overland expedition. He was to collect and observe the objects of natural history interest while aiding Franklin in his search for a north-west passage into the Arctic Ocean. This part of the world had not previously been visited by a naturalist and it was in this capacity that Richardson was expected to make his major contribution.

This was Richardson's introduction to natural history in the field. In his journal he made a multitude of observations on the geology, plants and animals that he encountered. He showed himself to be a competent botanist and zoologist on this expedition, where he was also to play the parts of unsuspecting cannibal and reluctant executioner (Houston 1984). Franklin's expedition camped through winter in 1820 at Winter Lake and set off north in June of 1821. Richardson's journal provides a remarkable record of the journey: nearly a month on snowshoes from their winter quarters to the Coppermine River, three weeks by canoe down the river, six weeks by canoe along the shores of the Arctic Ocean, eight weeks of hardship and near starvation walking through the 'Barren Grounds' back to the winter quarters, and six weeks of further privation and anxiety as the remnants of the expedition gradually received relief. Richardson fulfilled his functions as a naturalist most conscientiously throughout the journey to the Arctic Ocean and along its shores, in a part now known as Coronation Gulf.

These experiences of the lakes, rivers and Arctic shore provided Richardson's first experience of ichthyology. He gave careful descriptions of over 20 species of fish, with notes on their importance to the Indian and Inuit people native to the region. In the appendix to Franklin's narrative of this journey Richardson presented his *Notice of the Fishes* (1823). He travelled with Franklin again a few years later, with better boats and organisation, and accomplished much more. In 1836 he published *Fauna Boreali-Americana*, the third part of which was devoted to fishes (Richardson 1836). Here he included original descriptions of some of the fishes that he had first encountered on his initial journey with Franklin, cooperating with Cuvier who remarked 'Recently Mr. Richardson was willing to show us fishes that were taken during the last voyage of Captain Franklin to North America' (Cuvier 1995). Richardson undertook a third Arctic journey (1848–1849), led by James Clark Ross, in search of Franklin when his old friend was lost on his third attempt to find a north-west passage.

John Richardson. From Fortuine 1969, courtesy of the Canadian Medical Assocation.

Haslar

The Royal Naval Hospital at Haslar, Portsmouth was established in 1753, at the order of King George II, to help cater for the large number of sick and wounded sailors, and enlarged in 1762 when it housed over 2000 patients; Haslar accepted patients from all of the major battles from that time (Revell 1984). Richardson was appointed as a physician at Haslar in 1838 and helped to establish a natural history museum on the premises. Here, he assembled a large collection of fishes. Richardson explained his involvement in the museum and his publications that were based to a large extent on specimens held there:

> The superintendence of the Museum, established at Haslar Hospital through the exertions of Sir William Burnett, Inspector-General, being part of my official duty, I have thought that by publishing descriptions and figures of rare species in the Transactions of Societies, or in periodical works devoted to Natural History, I might advance the interests of science, and make an appropriate acknowledgement to the naval officers and others whose voluntary contributions support the Museum. (Richardson 1843a)

Although the Haslar collection contained a variety of specimens, Richardson concentrated on the fish:

> In selecting from the Museum ichthyological subjects first for publication, I am influenced by the fact that they offer more novelty than the other parts of the collection, which arises partly from the great proportion of existing fish which are still undescribed, and partly from the facilities which naval officers have of procuring specimens in that department of Natural History. (op. cit.)

Among Richardson's young doctors at Haslar was Thomas Huxley, the talented naturalist who became a leading proponent of Darwin's evolutionary concepts. Huxley intended to have a career in natural history but was experiencing great difficulty in finding an opening. He approached Richardson who arranged a position for him as naturalist on the voyage of the *Rattlesnake*. Huxley studied invertebrates on the voyage and the quality of his work helped to launch him on his subsequent career as a biologist.

Richardson worked in many fields, most notably introducing treatment for patients with psychiatric disorders within the normal hospital system, wherever possible avoiding their incarceration in the notorious asylums of the time. He continued to work at Haslar until his retirement from the Navy in 1855 (Lines 1997).

Richardson the ichthyologist

Richardson's name occurs throughout the literature of ichthyology as he described fishes from expeditions to many parts of the world. The Australian literature is particularly coloured by his enterprise through his work on the British naval collections from the voyage of the *Erebus* and *Terror* (1844–1848) and, to a lesser degree, from the *Sulphur* (1844–1845), the *Samarang* (1848) and the *Herald* (1845–1851). Richardson's enthusiasm for natural history resulted in useful collections being made by British naval personnel; the material brought back to England was sent to the appropriate specialists, usually at the British Museum, and Richardson dealt with the fishes himself. John Gray, Keeper of the Zoology Department from 1840, recognised Richardson as the leading authority on fishes in Britain. Gray referred collections of fishes for Richardson to determine and publish, and they sometimes worked together on collections of fish and reptiles. Richardson acknowledged the help he received from Gray in the concluding paragraph of his introduction to his illustrated work on Australian fishes:

> I cannot conclude this brief notice ... without expressing my thanks to John E. Gray, Esq., of the British Museum, for the valuable aid which he has afforded me now and at former times, and without which, living as I do at a distance from London, I could not have ventured upon the task I have set myself. (Richardson 1843c)

Gray consulted the collections and reference works of the British Museum on Richardson's behalf.

The chief references upon which Richardson relied were the works of Cuvier and Valenciennes, and those of Johannes Müller. Cuvier had prepared the framework, presented in his *Régne Animal* of 1816 (and revised in 1821), for a comprehensive account of fishes. The greater part of the *Histoire Naturelle des Poissons* was available by the time Richardson began his work on the fishes sent from Tasmania in 1839. When the publication of the *Histoire* was suspended with the appearance of the 22nd volume in 1842, most of the teleost groups had been revised (but see page 42). Gray then had the British Museum commission a German zoologist, Johann Kaup of Darmstadt (see page 82), to revise the outstanding groups. He completed the pipefish (Kaup 1856a) and eels (Kaup 1856b); he also contributed work on cod (Kaup 1858) and some other groups.

As a surgeon who had been instructed in natural history, and being familiar with the methods of Cuvier and Müller, and the early work of Richard Owen, Richardson knew the techniques and value of dissection:

> That a correct knowledge of the structure and development of any group of animals is as necessary to the zoologist as an acquaintance with their physiognomy and habits, is so evident, that it seems unnecessary to affirm that no one can hope to become an accomplished ichthyologist unless he submits to the labour of patient anatomical investigation. (Richardson 1860)

Richardson dissected when practicable but many specimens reached him as skins, or in a badly preserved state, and sometimes he worked only with drawings; or a specimen might be one of a kind, and thus too valuable to damage. In all of these cases, he made the best of what was available to him, dissecting when the opportunity arose.

Richardson's experience with a wide variety of fishes from various parts of the world, together with his familiarity with the literature of the day and his attention to detail, generally allowed him to place his specimens logically within the existing system of classification. He did not try to modify the systems of the day and appeared to be a little disconcerted by Richard Owen's efforts in this cause, even enlisting Tennyson:

> Our little systems have their day;
> They have their day and cease to be;
> They are but broken lights of *Thee*,
> And thou, O Lord! art more than they.
> (quoted in Richardson 1860).

As fine a reference to the divine plan as could be had.

Aside from his work on collections from North America, Australia and the Pacific, Richardson contributed to the popular literature on fishes. He wrote the ichthyology section for the eighth edition of the *Encyclopaedia Britannica* (1856), praising (and borrowing from) Cuvier, Müller and Owen. He introduced the structure of fishes, with emphasis on their osteology, and recommended Richard Owen's *Lectures on the Comparative Anatomy and Physiology of the Vertebrate Animals* (1846) for more detail. In the *Encyclopaedia,* Richardson described some of the great diversity of structure and habit found among fishes, illustrating frequently with fishes that he had described from Australian seas. He presented a useful classification, based on that of Johannes Müller. The article concluded with a summary of fish biogeography, untroubled by evolutionary principles. Richardson, well aware of geological and palaeontological discoveries, speculated:

> ... when the ocean was in a condition to support ichthyic life it was peopled by the Great Creator, each species being placed in sufficient numbers in the district best suited to its structure and habits, and the purpose it was designed to serve, while at the same time the balance between reproduction and supplies of food was provided for by a sufficient number of the rapacious kinds. And,

> moreover, that new species were created when a change in the condition of the ocean and its boundaries led to the extinction of the older races, and rendered such an exercise of creative power necessary. (Richardson 1856)

The biogeography of fishes appears to have fascinated Richardson. He referred to aspects of this in various publications and summarised his observations in his introduction to a *Report on the Ichthyology of the Seas of China and Japan* (Richardson 1846b). At this stage his study of the collections of the *Erebus* and *Terror* was a work in progress and he had handled many collections from the Indo-Pacific and Australia. His observations include the bitemperate representation of some western Pacific genera, the close relationship between the fishes of Australia and those of New Zealand, and the possibilty of a barrier separating

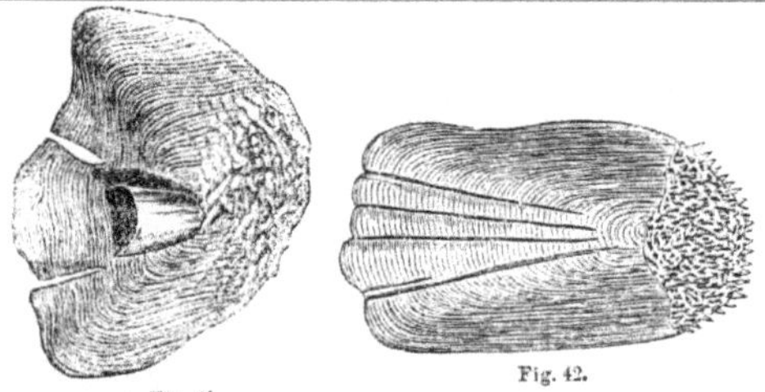

Fig. 42.

Fig. 41.

Fig. 41. A cycloid scale from the lateral line of *Saurus undosquamis*, one of the North Australian *Scopelinæ*.
... 42. Ctenoid scale of *Platycephalus cirrhonasus*, one of the *Solerogenidæ* of Botany Bay.

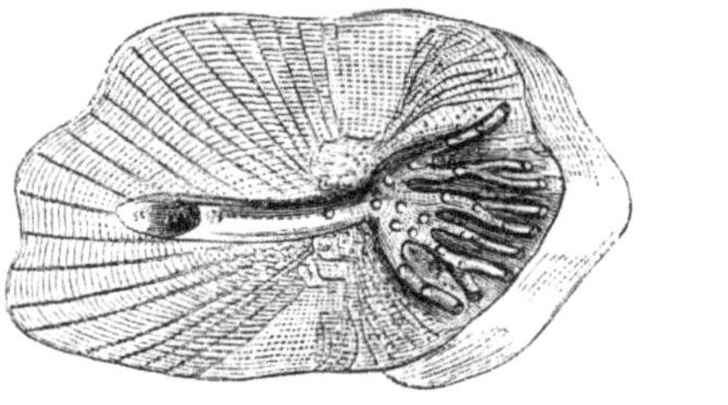

Fig. 38.
Cycloid scale from the lateral line of *Labrus laticlavius*, an Australian Cyclo-Labroid.

Fig. 35.
Placoid scales of *Aleuteres trossulus*, one of the *Balistidæ*, from Australia.

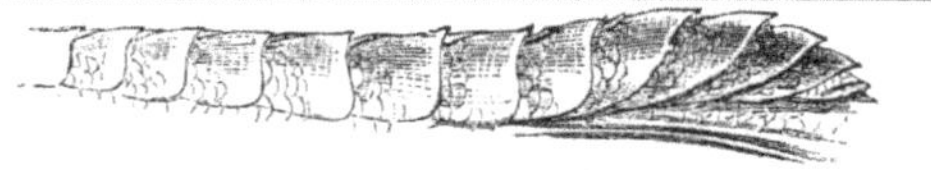

Fig. 37.
Ctenoid scales from the lateral line at the base of the caudal fin of *Caranx parapistes*, a Scomberoid from North Australia.

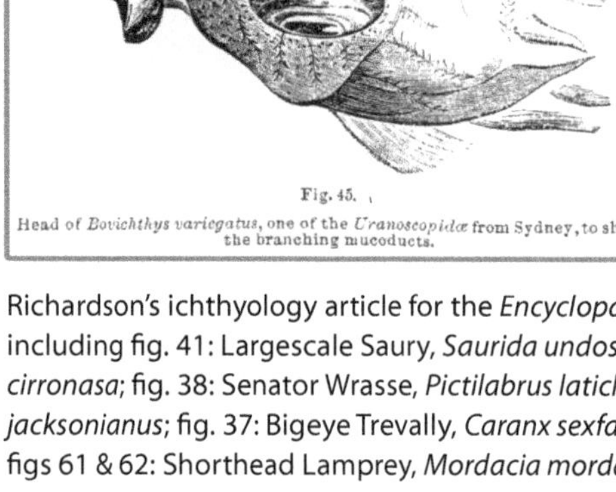

Fig. 45.
Head of *Bovichthys variegatus*, one of the *Uranoscopidæ* from Sydney, to show the branching mucoducts.

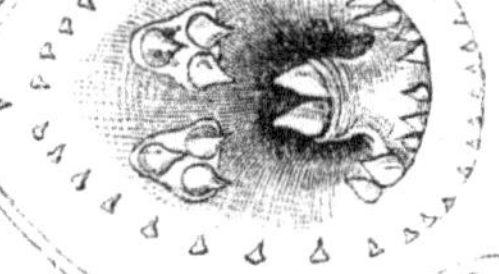

Fig. 61.
Labial teeth of *Petromyzon mordax*.

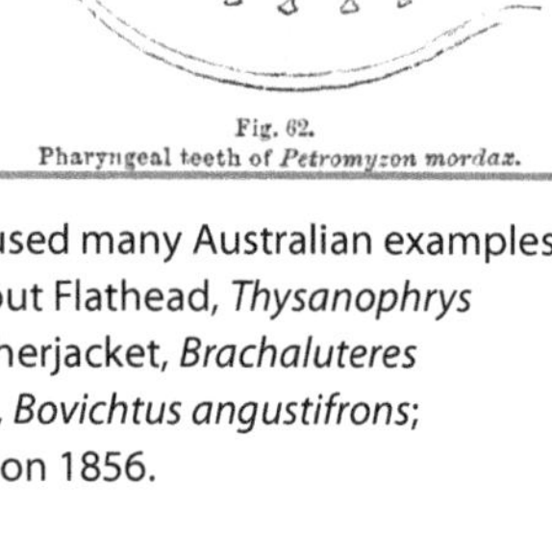

Fig. 62.
Pharyngeal teeth of *Petromyzon mordax*.

Richardson's ichthyology article for the *Encyclopaedia Britannica* for 1856 used many Australian examples, including fig. 41: Largescale Saury, *Saurida undosquamis*; fig. 42: Tassel-snout Flathead, *Thysanophrys cirronasa*; fig. 38: Senator Wrasse, *Pictilabrus laticlavius*; fig. 35: Pygmy Leatherjacket, *Brachaluteres jacksonianus*; fig. 37: Bigeye Trevally, *Caranx sexfasciatus*; fig. 45: Dragonet, *Bovichtus angustifrons*; figs 61 & 62: Shorthead Lamprey, *Mordacia mordax*. Extracts from Richardson 1856.

the Indo-west Pacific fishes from those of the East Pacific. With regard to the extent of the Indo-west Pacific fauna, Richardson said:

> ... we have from Africa eastward, within the warmer districts of the ocean, a continuous range through the Indian ocean and archipelago, the Malay archipelago and Polynesia, which embraces three-fourths of the circumference of the globe.

Richardson, aware of the partial nature of his data, referred to the '... desultory observations I have thrown out respecting the distribution of fish ...' but concluded that '... the ichthyology of the Australian seas has an Asiatic character as opposed to the Atlantic or South American assemblages of species.'

Later, Richardson (1860) contributed a slim *Fishes* volume to the *Museum of Natural History, Zoology* series of handbooks. On this occasion he used British fishes to illustrate most of the groups, but also inserted a few of his southern hemisphere favourites.

Use of drawings

Illustration is frequently an integral part of a taxonomist's work. Accurate drawings were, in the 19th century, sometimes regarded as a sufficient basis for the description of new species. For fishes, this was particularly the case as those brought back from prolonged voyages were often in a poor state and, even if well preserved, colours and markings were faded or lost.

Richardson frequently used drawings as a basis for the description of new species. For instance, he derived a large part of his *Report on the Ichthyology of the Seas of China and Japan* (1846b) from a collection of drawings made by Chinese artists and commissioned by Mr John Reeves.[3] In many cases Richardson described new species of fish from the drawings alone, but sometimes specimens preserved dry or in alcohol were also available.

Further examples of Richardson's generally judicious use of drawings are found in his *Icones Piscium* (1843c) and in the reports on the fishes of the *Erebus* and *Terror*, and of the *Sulphur* (mentioned below). He was aware of the limitations of drawings but his use of them inevitably led to taxonomic problems for later workers.

Richardson's Australian fishes

Between about 1839 and 1850 Richardson received Australian fishes from a variety of sources, including geographic areas that had not been explored by Cuvier's collectors. Most of the Australian specimens were dealt with in the *Proceedings of the Zoological Society of London* (1839, 1840, 1841a and 1850), in a paper published in nine parts in the *Annals and Magazine of Natural History* (1842–1843), in an account of the fishes of the *Erebus* and *Terror* expedition (1844–1848), in the *Transactions of the Zoological Society of London* (1841b) and in a report on the fishes of the *Beagle* surveys (1846a). Naval officers who acted as part-time naturalists collected, and sometimes drew, the fish; the specimens and drawings were sent to Haslar. John Gray forwarded to Richardson specimens that were sent to the British Museum.

3 John Reeves (1774–1856) arrived in Canton in 1812 to take up the position of Assistant Inspector of Teas for the East India Company and, with encouragement from his employer, proceeded to amass a remarkable series of natural history drawings. Reeves' son helped to expand the collection which ultimately included representations of nearly 500 fish (Whitehead 1969).

Richardson's publications on Australian and New Zealand fish are considered below according to the region from which the specimens were collected; his report on the *Erebus* and *Terror* is also considered as a whole.

Richardson's sources of, and publications on, Tasmanian fishes

Richardson's interest in Australian fishes was first aroused by collections from Tasmania, a region for which few fish were then known. His paper was introduced:

> Dr. Richardson read his account of an interesting collection of fish formed at Port Arthur in Van Diemen's Land, by T.J. Lemprière, Esq., Deputy Assistant Commissary General, by direction from His Excellency Sir John Franklin, K.C.B., Lieutenant Governor, and now deposited in the museum of the Royal Naval Hospital at Haslar. (Richardson 1839)

In this and two subsequent papers Richardson recorded over 30 species of fish from Tasmania (Richardson 1840, 1841a, 1841b) (see colour plate 5, lower). The work included the description of 25 new species, over half of which are still regarded as valid, and the establishment of several new genera. However, several of the new species descriptions, particularly those of the leatherjackets (Monacanthidae), were too brief to ensure recognition by later workers.

Although Richardson's descriptions were sometimes terse, they usually sufficed and occasionally included items of particular interest. In describing the Marblefish, *Aplodactylus arctidens* from Tasmania he recognised its affinity with a similar Chilean fish, *Aplodactylus punctatus*, described in the *Histoire Naturelle des Poissons*. Cuvier, throughout his work, encouraged dissection and close attention to dentition. Richardson followed his lead, quoting from the description in the *Histoire* by way of comparison, and wrote:

> In the Van Diemen's Land fish, the teeth stand in eight or nine crowded ranks in the upper jaw, and in five or six in the lower one, those of the interior rows being very much smaller in all their dimensions, but otherwise shaped exactly like the teeth of the exterior rows, which resemble those of *punctatus*. Their points show three small lobes, the middle lobe being largest and most prominent. The species further differs from *punctatus* in having four caeca, [around the pylorus of the stomach] but its food appears to be similar, the intestines having been found filled with large fragments of seaweed ... (Richardson 1839)

In the Tasmanian collection there were two small leatherjackets which Richardson described as new species: *Aleuteres maculosus* and *Aleuteres paragaudatus*. Of the latter, he remarked 'This handsome Aleuteres is named in allusion to the striped upper vestments of the Roman ladies.' Many leatherjackets exhibit marked sexual dimorphism; Richardson had the male (*paragaudatus*) and female (*maculosus*) of a single species, the Bridled Leatherjacket, which had been described as *Aleuteres* [now *Acanthaluteres*] *spilomelanurus* by Quoy and Gaimard (1824) from specimens taken at Port Jackson. Richardson noted a resemblance between the fish collected by Lemprière and the description of that from Port Jackson, but also noticed some differences. He refers to his use of microscopic examination:

> In both the Port Arthur *Aleuteres* the minute prickles of the skin, when examined by a good microscope, appear to be solitary and to spring from a globular base. (Richardson 1840)

Richardson was intrigued to find a large crustacean apparently in the gut of one of the small leatherjackets:

> One of our specimens [of *Aleuteres paragaudatus*] had the gut and the whole abdomen distended by a large *Idotea*, full of roe, not at all crushed, and apparently little digested: a portion of its tail fin protruded at the anus of the *Aleuteres*. (Richardson 1840)

He was unaware of the true nature of this '*Idotea*': a female isopod parasite (*Ourozeuktes bopyroides*) living within the body cavity, but not the gut, of the leatherjacket.

The same paper included the description of a fish thought to be a parrot fish:

> It was presented to the Museum of Haslar by Mr. Conway, formerly medical superintendent of a convict ship, and since deceased. (op. cit.)

The specimen was mounted, precluding dissection, and was said to have been caught in Hobart or Sydney. The jaws were fused, hence the belief that it was a parrot fish (Scaridae), but Richardson noticed several differences from that family. He described a new genus for the specimen, naming it *Oplegnathus conwaii*. This was the Knifejaw, which subsequently proved to have been collected in South Africa; much later, a southern Australian member of the genus was discovered (see page 257).

Coincident with Richardson's studies of the Tasmanian collection was the publication of papers on sharks and rays by Müller and Henle (1842). Among the fishes from Port Arthur was an Electric Ray but Richardson (1840) initially hesitated to describe it as new as he was aware of, but had no access to, Henle's earlier paper *On Narcine, a new genus of Electric Rays together with a Synopsis of Electric Rays* (1834). When he obtained the paper he described the Tasmanian Numbfish, *Narcine tasmaniensis* as a new species of Henle's genus (Richardson 1841).

Over the next few years Richardson continued to publish occasional descriptions of fishes from Tasmania in papers which were chiefly concerned with fishes from northern Australia. Lemprière, his efforts scrupulously acknowledged by Richardson, continued to send new fishes; among them were the Red Handfish, *Cheironectes* [now *Sympterichthys*] *politus* (Richardson, 1844) and the Thornback Skate, *Raia* [now *Dentiraja*] *lemprieri*:

> This ray has received its specific name in compliment to Deputy Assistant Commissary-General F.J. Lemprière, to whose exertions the ichthyology of Van Diemen's Land is much indebted. (Richardson 1845).

In 1842, Richardson received some illustrations of fishes 'made by a convict artist at Port Arthur, Van Diemen's Land, under the inspection of Dr. Lhotsky ...' (Richardson 1842).

Johann Lhotsky (1795–1866), from the Ukraine, was a botanist and collector who worked in South America and Australia; Governor John Franklin had employed him to collect specimens around Hobart. Richardson referred to several of these illustrations, deriving detailed descriptions of the fishes represented from the drawings alone.

In his account of the ichthyology from the voyage of the *Erebus* and *Terror* Richardson dealt with several other Tasmanian fish, including a spectacular species of Boarfish. The specimen was far from perfect:

> Of this fish I have received merely a mutilated head from Dr. W.P. Jones, Surgeon in the Royal Navy, to whom it was presented by Mr. Moriarty of Hobart Town ... The specimen was caught in a net in one of the inlets of Storm Bay, by some fishermen, and carried by them to Mr. Moriarty, as a fish which they had never seen before. (Richardson 1845)

The fish, a Longsnout Boarfish, *Histiopterus* [now *Pentaceropsis*] *recurvirostris* is distinctive and proved to be common in southern Australia (see page 81).

Richardson's sources of, and publications on, northern Australian fishes

Contributions to the Ichthyology of Australia (1842–43) was mostly involved with specimens received from northern Australia. Richardson's descriptions of fishes in this and subsequent papers were painstaking and much more detailed than those in his earlier work; their value to future ichthyologists was consequently greater. He appreciated the importance of good drawings and was aware that limited illustration was a defect in some of his early papers on Australian fish. This became even more important in his papers on northern Australian species, some of which he described from drawings alone.

In a typically heroic sentence, Richardson described his main sources for *Contributions to the Ichthyology of Australia*:

> Mr. Gould having had the kindness to place in my hands for examination a collection of fish, procured by his assistant Mr. Gilbert, at Port Essington, on the north coast of New Holland, I purpose in the following paper to give a summary of my observations thereon, together with a few remarks upon some drawings made by Lieutenant Emery, R.N., of fish captured on the north-west coast of the same country. (Richardson 1842)

Gilbert (see page 113), due to the nature of his wanderings, could usually collect only skins. These were adequate for his ornithological specimens but fish skins were much less satisfactory: the general shape of the creature was often hard to reconstruct, the colours faded, and parts important for identification were often lost. Richardson commented:

> Mr. Gilbert's specimens, numbered from 1 to 37, are all dried skins of one side of the fish. The colours of most are consequently altered, and in a few instances the vomerine and palatine bones have been cut away, but in other respects they are in excellent condition. (Richardson 1842)

The loss of bones from the roof of the mouth (vomer and palatines) could be particularly troublesome as knowledge of their dentition was important for identification. Lieutenant Emery's drawings presented different problems:

> Lieut. Emery not being an ichthyologist, has sometimes omitted to portray the minute serratures of the opercular pieces [gill covers], and has not always distinguished the spinous from the articulated rays [in the fins]. (Richardson 1842)

In this paper Richardson described 24 new species of northern Australian fish, less than half of which are still regarded as valid. These included some species endemic to the streams and rivers of northern Australia. Gilbert supplied field notes too, of which Richardson made full use:

> Mr. Gilbert informs us that this little fish [the Blackbanded Rainbowfish, *Atherina* (now *Melanotaenia*) *nigrans*] is a tolerably abundant inhabitant of the freshwater streams that flow into the harbour of Port Essington, and that it is very easily taken with a hook baited with flies or fresh meat. Yalgurnda is its native name. (Richardson 1842)

Most of the marine fishes described have extensive Indo-Pacific distributions and several had been previously named (particularly by Cuvier, Valenciennes or Bloch). However, a few of these wide ranging species had not before been encountered by ichthyologists; for instance, a tropical snapper known in Queensland as the Stripey Snapper, *Mesoprion* [now *Lutjanus*] *carponotatus*, and found from the Bay of Bengal and the Philippines to northern Australia. In describing this species Richardson quoted Gilbert again:

> The native inhabitants of the shores of Port Essington call this fish the 'Mungundju,' and it frequents the deep water in rocky places at the entrance of the harbour. (Richardson 1842)

A recent account of Gilbert's collecting activities in the Northern Territory includes an annotated list of his fish specimens with illustrations of some of his specimens that are now held in the Natural History Museum, London (Fisher and Calaby 2009).

The fishes that Richardson described in 1842–43 included some recorded by Solander and Parkinson on Cook's first voyage. Among them was the Sicklefish, *Drepane punctata* taken near 'Endeavour's Careening place' at the mouth of the Endeavour River in north Queensland (Richardson 1842). Richardson consulted a pencil sketch by Parkinson and notes left by Solander (who had named it *Chaetodon punctata*) but the fish later proved to have been described by Linnaeus in 1758. Another species, taken off Bustard Bay in May 1770, was named *Labrus iris* by Solander. Richardson could not identify the fish from Solander's notes but thought that it might prove to be new; he simply quoted Solander's description under the name given by that author, '... we quote [Solander's] description chiefly to excite the attention of collectors who may visit this part of Australia.' (Richardson 1843a) Solander had regarded the fish as a wrasse, and Richardson had insufficient grounds to doubt this. However, it proved to be the Paradise Threadfin Bream, *Pentapodus paradiseus* (Günther, 1859).

In 1843 Richardson also prepared the first of what he intended to be a series of papers illustrating rare fishes. This was the *Icones Piscium*, the first issue dealing largely with fishes from north-western Australia drawn by James Emery (Richardson 1843c) (see colour plates 6, 7 and 8). Unfortunately, the project was not continued. Richardson introduced the *Icones*, explaining that he could not publish as many figures as he wished in the periodicals he had been using. He explained his choice of Emery's drawings:

> The first fasciculus [of the *Icones*] contains exact copies of drawings from the portfolio of James B. Emery, Esq., late first lieutenant of the *Beagle* surveying vessel employed on the western coasts of Australia. These representations of the recent fish are valuable records of the true colours of the species, and were evidently executed with much care, but, like all drawings of fish not made under the eye of an ichthyologist, they are unavoidably deficient in the expression of certain generic characters. This very circumsta`nce renders their publication most desirable, for verbal descriptions of them may mislead the naturalist but the plate enables him to judge for himself, and will also guide any officer who may hereafter visit the same coast in his search for desirable specimens. (op. cit.)

Twelve species were described, all illustrated; two had been described previously by Richardson (1842) as new species, but not then figured. The other ten species were all described and named as new species in the *Icones*. Of these ten species, only three are still regarded as valid. Happily, these include a magnificent fish collected by Emery at Barrow Island, north-west Australia, which Richardson named for him. Emery's drawings bear notes in two different hands. The faintest writing gives the numbers of the drawings and includes some suggested names and is perhaps Emery's. The later hand is presumably Richardson's and gives lengths, names, sites of collection and indicates which drawings will be reproduced in the *Icones*.

Richardson's sources of, and publications on, temperate fishes of Western Australia

The *Icones* also included the 'Imperial Aulopus [*Latropiscis purpurissatus*] ... taken among the rocks of the Houtmans Abrolhos ...' Once again, the species was described wholly on the basis of the drawing. The waters of these islands are inhabited by fishes of both temperate and tropical origin, *Latropiscis purpurissatus* being found around the southern half of Australia.

John Gould donated to the British Museum some fish specimens that he had obtained while in Western Australia. Gray made them available to Richardson who described two as new species: the Western Blue Grouper, *Labrus* [now *Achoerodus*] *gouldii* and the Moonlighter, *Chaetodon* [now *Tilodon*] *sexfasciatus*. Both are common in southern Australian waters, but they had not come to the notice of ichthyologists. Gould saw the Blue Grouper in a fresh state and Richardson quoted his notes on the colours of this fish.

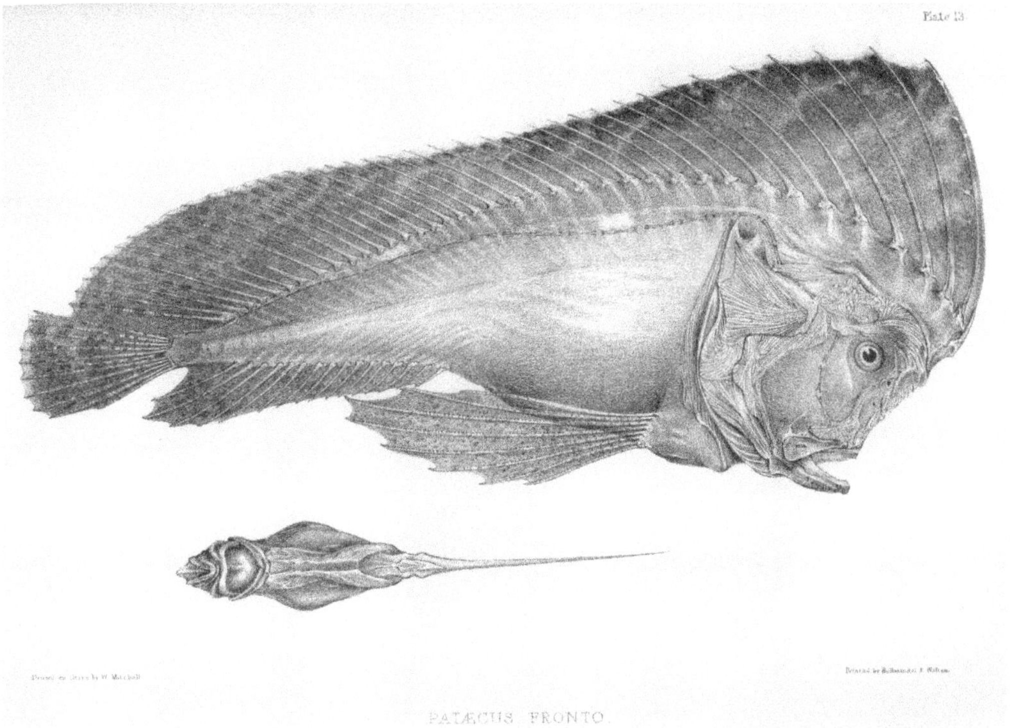

The Red Indian Fish, *Pataecus fronto,* described by John Richardson from a single dried specimen presented by Captain George Grey. From Richardson 1844.

Robert Neill at King George Sound

Richardson's best informant for temperate western Australian fish resided at Albany on King George Sound. This was Robert Neill (1801–1852), mistakenly referred to as 'J. Neill' in many publications, who was employed in the British Army Commissariat. Born in Edinburgh, Neill arrived in Hobart in 1826, where he became a member of the Van Diemen's Land Society, a scientific body. He served in various parts of Tasmania and moved to Albany in 1835, being promoted to the position of Deputy Assistant Commissary-General in 1837. Throughout this time he collected natural history specimens, sending them to contacts in Britain.

In 1840–41 Edward John Eyre, then resident magistrate of the Murray River district, undertook an overland expedition from Adelaide to King George Sound in Western Australia. He published an account of the expedition in 1845 for which Neill painted several illustrations. In an appendix entitled *Catalogue of Reptiles and Fish found at King George's Sound, by Deputy Assistant Commissary-General Neill* Richardson gave an account of Neill's fish drawings, supplying notes and identifications, many of the latter tentative, and in a couple of instances describing new species on the basis of the drawings alone.

Neill said that he had been prompted to study the fishes of the region at the suggestion of his friend, Captain George Grey, then (1839) the Resident Magistrate at King George Sound. Grey (see page 111), later Governor of South Australia, encouraged the study of natural productions and among his early contributions to Australian natural history was a specimen of the Red Indian Fish, *Pataecus fronto* which he sent to the British Museum from south-western Australia.

Neill, following Governor Grey's advice, sent his notes and drawings to John Gray at the British Museum. He was modest about his efforts and wrote to Gray from Albany:

> Nothing is assumed as to the execution of the drawings; in fact it often occurred when I set off in my little skiff, (especially in the outset) that seven or eight species were procured in the course of the excursion, which compelled me to make drawings of all when I came home tired in the evening; forwarding them to ensure, as far as possible, their colours before they became extinct – a sort of forced effort in respect to the execution has, therefore, only been effected. (Richardson 1845)

Neill went to some trouble to determine and record the aboriginal names for the fishes he drew; he also supplied the names used by the settlers and those of the sealers when he could learn them. This diversity of nomenclature at such an early date gives a good indication of the origin of much of the confusion about popular names of fishes in Australia. Neill enhanced the value of his drawings by pasting one or two scales from the subjects of his sketches onto the page. He said of his source of specimens, 'Most have been taken by my own hook; some by the native's spear, and some by the seine net.'

His comments on the aboriginal names were quoted by John Gray in his introduction to this paper:

> The younger natives generally [give] different names to those of the elder; but finding the fish named by the latter more descriptive, I have, of course, in most instances, adopted them ... I find, also, the old men are more minute in *species*; the younger often call very different fish by the same name ... (Richardson 1845)

Neill was proud of his drawings, and Gray quoted his account of an aboriginal friend's response to them:

> As regards the fidelity of the drawings, it may be worth while to mention a singular mistake made by my friend *Toolegetwalee*; one of the oldest and most friendly savages we have of the King George tribe; who, in looking over my collection to assist me in naming them, observed that the drawings were a little raised off the paper; and like a monkey, began to touch them with his long talons; of course, I flew to their rescue, and asked what he meant?
>
> '*Inilen* how make em? me *twank* skin put him on!' which literally means – 'Ah! I now see how you do it, you put the skin on!! From want of paper of uniform size, I was obliged to use any paper which came to hand, cut the figures out, and afterwards paste them on clean paper; which circumstances gave rise to the

> poor savage's mistake, and it was not until I actually cut one out before him, that he could be convinced that he was in error ... (Richardson 1845)

Although this demonstrates the condescending attitude common to settlers and officials, Neill clearly enjoyed the confidence of at least some of his aboriginal acquaintances.

The Deputy Assistant Commissary-General was a little diffident about his ichthyological exertions:

> The effort has afforded me much amusement, and it will be still more agreeable, if they will in any way contribute to a better knowledge of the subject. (Richardson 1845)

Recalled to Britain in 1848, Neill was appointed to the Windward and Leeward Islands in the West Indies where he was Assistant Commissary-General. In 1852 he, his wife and four children died at Barbados of yellow fever (Glover 2007).

Richardson's account of Neill's drawings

Gray sent Neill's drawings to '... my excellent friend, Dr. John Richardson of Haslar, one of the first Ichthyologists now living ...' There were 59 drawings in all. Richardson identified them as best he could, supplying scientific names where possible, and arranging them in systematic order. The original drawings were retained in the Zoology Department of the British Museum. He based two new species descriptions solely on these drawings. For instance, Richardson's description of *Girella zebra*, the Zebrafish, so-called by the sealers of southern Australia, was taken directly from one of Neill's drawings to which there was 'a single scale attached'. Richardson noted that Neill's drawing was 'nearly a foot long'. Neill had explained:

> The outline of nearly every specimen was taken from *actual profile*, by laying the fish upon the paper – in this way I defied error in outline – of course, afterwards carefully drawing and correcting various parts which required it, in a free or rough manner, time not admitting of much pains. (Richardson 1845)

Neill's version of the aboriginal names for the fish was '*Kgnmmul* or *Karraway*, The Striped.'

Richardson's catalogue of Neill's drawings generates most interest from the inclusion of Neill's remarks about the fish. For example:

- Red Mullet, *Upeneichthys vlamingii*: 'Native name ... Kgnarnuck (the bearded).'
- Snook, *Sphyraena novaehollandiae*: 'It comes into the shallow bays in summer; and being a sluggish fish, is easily speared by the natives, who esteem it to be excellent food. It will lay for a minute looking with indifference at its enemy, while he poises the fatal and unerring spear.'
- Mulloway, *Argyrosomus japonicus*: 'Grows to a great size; as I am informed by the natives, that they often spear individuals weighing sixty or seventy pounds.'

- Australian Salmon, *Arripis truttaceus*: 'It congregates in vast shoals, and pursues the fry of other fishes in shallow bays, but never enters fresh water.'
- Strongfish, *Dactylophora nigricans*: 'Inhabits rocky points of sandy bays, where they love to run in and root up the sand with their fleshy mouths. They are sluggish, and easily speared by the Aborigines, whose chief food it constitutes at certain seasons ... The *Toojenong* grows to a large size exceeding twenty pounds in weight.'
- Silver Trevally, *Pseudocaranx georgianus*: 'Very common in shallow sandy bays, and forming the staple food of the natives, who assemble in fine calm days, and drive shoals of this fish into weirs that they have constructed of shrubs and branches of trees.

The last of Neill's drawings as listed by Richardson is of a leatherjacket. After a brief description, Neill commented:

> Inhabits deep water in rocky places, and is very common. It is esteemed for food by the Aborigines; is much infested by an Isopode named *Nettong*, or *Toort*, by the natives. This insect inserts its whole body into a pocket by the side of the anus, separated from the gut by a thin membrane. The fish to which the insect adheres are yellow; those which are free from it are of a beautiful purple colour.[4] (Richardson 1845)

Neill observed that the parasite was external to the gut (c.f. Richardson's comment, page 53).

Gray later received dried skins prepared by Neill (some were of the fishes represented in his drawings) which provided an opportunity for Richardson (1850) to revisit the drawings. Of the 12 species from King George Sound described in this paper, Richardson described eight as new; all but one of these is still regarded as valid. Richardson reviewed the genus *Cheilodactylus* (in the broad sense in which it was then understood), discussing 13 species, for nine of which he had specimens.

Richardson's last contribution to the ichthyology of Australia was the description of another novelty from King George Sound (Richardson 1858): the Tubemouth, *Siphonognathus argyrophanes* for which Richardson created a new genus.

New Zealand ichthyology

Aside from the pioneering work of Cook's naturalists, the New Zealand fish fauna was poorly served in the literature prior to studies by Richardson and Gray. Richardson brought together the work that had been done, identified and described new finds and presented a *Report on the present state of ichthyology of New Zealand* to the British Association for the Advancement of Science (Richardson 1843b). He and John Gray published a checklist of New Zealand fishes with descriptions of new species collected by Ernest Dieffenbach, MD, who had been Naturalist to the New Zealand Company (Richardson and Gray 1843).

4 This refers to the Sixspine Leatherjacket, *Meuschenia freycineti* (Quoy and Gaimard, 1824). The fish parasitised usually have the colours of the female (yellow with broad dark stripes), presumably by the choice of the parasite or by its effect upon the host.

Apistes [now *Glyptauchen*] *panduratus* described from a drawing and dried skin sent by Neill from King George Sound. From Richardson 1850.

Dieffenbach had travelled extensively while in New Zealand, collecting natural history specimens, including fishes.

Richardson utilised Joseph Banks' collection in gathering materials for his *Contributions to the Ichthyology of Australia* (1842–43):

> In preparing the following notes, I have availed myself as often as my professional avocations and residence at a distance from London would allow, of the valuable collection of drawings made in Cook's first and second voyages by Parkinson and [Georg] Forster, now in the Banksian library.
> (Richardson 1842)

New Zealand fishes made up only a small part of the paper, about ten species in all, but included several previously undescribed species. He also consulted Solander's notes in the British Museum.

Among the new species established by Richardson from the records of Cook's naturalists was the Eastern Red Scorpionfish, *Scorpaena cardinalis*. The fish is common on the east coast of Australia and in northern New Zealand. It was collected on Cook's first voyage at Queen Charlotte Sound and described by Solander. Richardson transcribed Solander's manuscript description and added details from Parkinson's pencil sketch of the specimen (Richardson 1842).

The *Sulphur* and the *Samarang*

Among the collections from the voyage of the *Sulphur* (1836–1842) were fish from the Pacific; Richardson (1844–1845) published a report with descriptions of new species, including some that range into Australian seas. The *Samarang* (1843–1846) collections were also treated by Richardson (1848) and again some wide ranging species were described, including the Whitebelly Roguefish, *Apistes leucogaster* from the 'Sea of China.' The South African ichthyologist, J.L.B. Smith (1958), later created the genus *Richardsonichthys* for this fish.

The *Erebus* and *Terror*

Of the first importance to Australian ichthyology was Richardson's *Ichthyology of the Voyage of H.M.S. Erebus & Terror* (1844–1848). Captain James Clark Ross commanded these vessels on a cruise of the southern ocean, intended to explore the limits of Antarctica, from 1839 to 1843. The expedition left England in September 1839, passing via Cape Town to the Kerguelen Islands and arriving in Hobart in August 1840. They sailed south into Antarctic waters, returning to Hobart in April 1841, and arriving in Sydney in July. Next they sailed to New Zealand and thence, once again, into Antarctic waters. The last stage of the voyage saw them visiting the Falkland Islands and returning to England via Cape Town, Rio de Janeiro and the Azores. The natural history collections from this expedition were immense.

The botanist Joseph Hooker was naturalist on the voyage and Ross, who had an enthusiasm for zoology, cooperated to ensure that natural history collection remained a prime occupation throughout. Fishes were taken by nets, by hook and line, and even from the stomach contents of seals and seabirds. As was usual on voyages of this sort, collection was more effective than preservation and Richardson's introductory remarks to his account indicate the condition of the material with which he was obliged to work:

> In no expedition that ever sailed from Europe has more care been taken to collect the zoological productions of the sea, than in the pre-eminently scientific one of the Erebus and Terror. The commanding officer, an accomplished zoologist, had previously paid much attention to Ichthyology, and, under his fostering superintendence, ample collections of fish were made at Van Diemen's Land, Australia, Kerguelen's Land, Cape Horn, the Falklands, and wherever an opportunity offered of drawing the seine or trawl, or dropping a hook. The specimens thus obtained filled many casks, and numerous jars and bottles, and it were greatly to be wished that so much industry had met with the full measure of success that it deserved; but we have to regret that, during a voyage protracted for upwards of four years and a half, including every possible change of climate, and during which the ships were buffeted by many severe gales, and sustained innumerable shocks in forcing their way through the ice-packs of the Antarctic Seas, the specimens suffered very severe damage. Owing to the deterioration of the spirits in jars that were crowded with fish, and the long continued action of the brine, where that liquid was employed, very many specimens entirely perished, or merely fragments of skeletons could be rescued from the mass. (Richardson 1844)

There were many sources for the specimens upon which the *Ichthyology of the Voyage of H.M.S. Erebus & Terror* is based. The fishes captured on the voyage made up only the nucleus of the specimens used for this publication. Some fish taken by the officers of the *Beagle* on its surveying voyages were described by Richardson at this time. Robert Brown, the botanist who had travelled with Flinders over 40 years earlier, had charge of the drawings made by Ferdinand Bauer on the *Investigator.* He made the drawings available to Richardson who used them as the basis for describing and naming several species. Bauer's beautiful depictions of leatherjackets (see page 24) and the Rainbow Fish, *Scarus* [now *Heteroscarus*] *acroptilus* from southern Australia were the sole basis of Richardson's descriptions of these species. Some of the observations and drawings made by Robert Neill at King George Sound (see above) were also utilised.

Approximately 230 species are included in the *Erebus & Terror* Report, of which Australian species make up about one half. The remainder is from New Zealand (15 species), Subantarctic regions (15), a scattering through the Indo-Pacific and a few Atlantic species. In this work, Richardson devoted 25 pages to a review of eels, including several Australian species. A report on the fishes of the Antarctic and Subantarctic regions formed another major component of the report.

Eels had been omitted from the *Histoire Naturelle des Poissons* and Kaup's revision of the British Museum collections was still ten years in the future. For this part of his report, Richardson reviewed the British Museum's Moray Eels (Muraenidae), including a comparative account of them and a key. He also dealt with Snake Eels (family Ophichthyidae, all referred to *Ophisurus* by Richardson), Freshwater Eels (Anguillidae) and a variety of other families (all as *Congrus*).

The collections from high latitudes were hard won and Richardson was well aware of their significance, remarking 'Except what may be gleaned from the notes and drawings of the Forsters, who accompanied Cook on his second voyage, nothing is known of the fish inhabiting the seas beyond the fiftieth parallel of south latitude.' He had before him a variety of unfamiliar forms from Tierra del Fuego, the Falkland Islands, Kerguelen, the Auckland Islands and 'the Antarctic Glacial Ocean, off Victoria Land.' He established the genus *Notothenia* for *Notothenia coriiceps* from the Auckland Islands and Kerguelen and described a number of species that he attributed to the new genus (Richardson 1844).

The officers of the *Beagle* were fascinated by the natural history of Australia and proved to be effective collectors. John Lort Stokes (1811–1885) sailed in the *Beagle* on its three major surveying cruises: P.P. King's surveys in South America to the Straits of Magellan; Robert Fitzroy's voyage with Darwin; and lastly the intensive surveys of the Australian coast (1837–1843) initially commanded by John Wickham, and later by Stokes himself. Stokes became a close friend of Charles Darwin during the *Beagle* surveys in South American waters and was stimulated by Darwin's enthusiasm for natural history (Hordern 1989). In Australia James Barker Emery (c.1794–1889), first lieutenant on the *Beagle*, was among the oldest of the officers aboard and had seen service in the navy since 1808; as self-appointed naturalist and taxidermist he collected and preserved fish, drawing them with great care (see colour plates 6, 7 and 8), and thereby contributing most to ichthyology of all the ship's officers. Benjamin Bynoe (1804–1865) was surgeon and naturalist on the *Beagle* and his contributions, together with those of Emery and Stokes, were applauded by Richardson.

Datnia [now *Amniataba*] *caudavittata* from the Harvey River, Western Australia; specimens were collected by Surgeon Bynoe of the *Beagle*, and others supplied by John Gould. From Richardson 1845.

Richardson paid his naval collectors the courtesy of naming some of their finds after them. Thus there were *Scorpaena stokesii*, described in an appendix to Stokes' account of the voyage (Richardson 1846a) and *Scorpaena bynoensis*, described in the *Erebus & Terror* report, although they proved to represent a single species: the Painted Scorpionfish, *Parascorpaena picta* (Kuhl and Hasselt, 1829). Emery was celebrated in the *Icones Piscium* (see page 56). Richardson remarked of a small goby, *Gobius* [now *Amblygobius*] *bynoensis* that the '... specific name has been bestowed in honour of Benj. Bynoe, Esq., Surgeon in the Royal Navy, who, while serving in the *Beagle*, made extensive collections of animals and plants, which he presented to the Museum of Haslar Hospital.' Among Bynoe's finds was the Yellowtail Grunter, *Datnia* [now *Amniataba*] *caudavittata* caught in the Harvey River, Western Australia.

As Governor of Western Australia, George Grey sent Richardson fishes which he described in his account of the *Ichthyology of the Voyage of H.M.S. Erebus & Terror*. Richardson described the Beaked Salmon, *Rynchana* [now *Gonorynchus*] *greyi* and the Redbait, *Emmelichthys nitidus* from Grey's specimens.

The voyage itself was the largest source of Australian fish discussed in the report, 43 being described as new species. Of these, 6 were from freshwater, 15 from marine waters of northern Australia and 22 from southern Australia. Over half of these new species are still regarded as valid. Many of the new forms puzzled Richardson, particularly the River Blackfish, *Gadopsis marmoratus* and the Shore Eel, *Cheilobranchus* [now *Alabes*] *dorsalis*. Several of the new species represented families not previously found around Australia,

Shorthead Lamprey, *Petromyzon* [now *Mordacia*] *mordax* was described from a single small Tasmanian specimen. From Richardson 1846.

Tassel-snout Flathead, *Platycephalus cirronasus* [now *Thysanophrys cirronasa*] from a Botany Bay specimen. From Richardson 1848.

for instance: the Spiny Eel, *Notacanthus sexspinis*; Common Stinkfish, *Callionymus* [now *Foetorepus*] *calauropomus*; Banded Bellowsfish, *Centriscus* [now *Centriscops*] *humerosus*. Some of the more impressive Australian endemic fishes were among the collection presented by Ross: Western Australian Jewfish, *Glaucosoma hebraicum*; Callop, *Datnia* [now *Macquaria*] *ambigua*; Shorthead Lamprey, *Petromyzon* [now *Mordacia*] *mordax*; Tassel-snout Flathead, *Platycephalus cirronasus* [now *Thysanophrys cirronasa*].

The Eastern Frogfish, *Batrachomoeus dubius* had been described by White (1790) and Ross secured new specimens for Richardson.

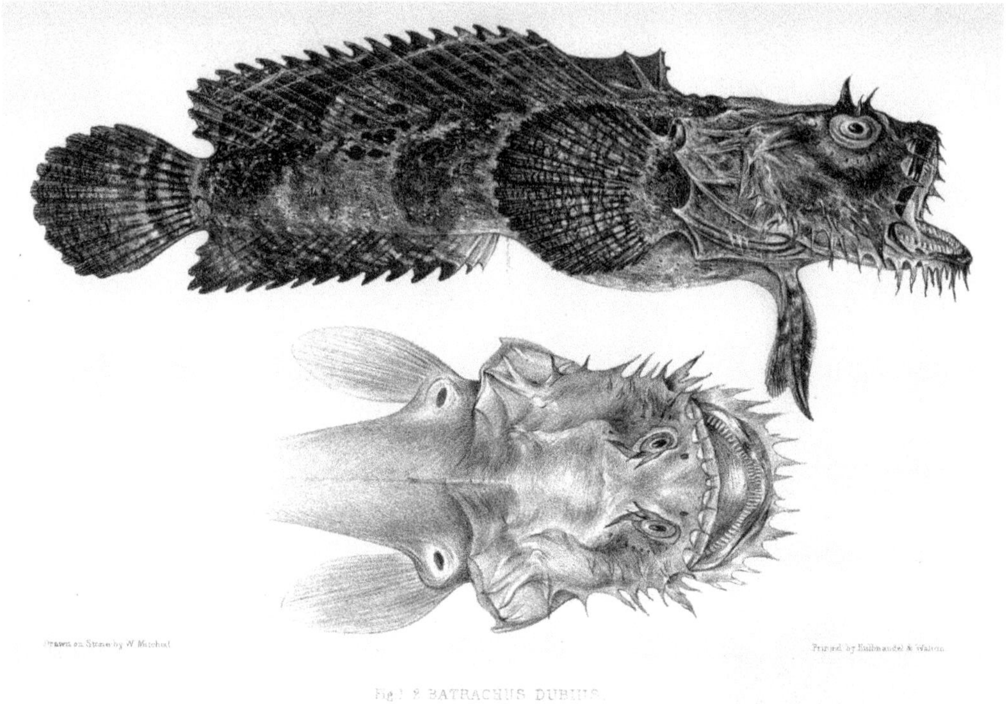

The Eastern Frogfish, *Batrachus* [now *Batrachomoeus*] *dubius* from Sydney. White (1790) described this fish as *Lophius Dubius* and Richardson refers his own Sydney specimens to this species, noting that White's 'figure which accompanies this notice is execrable (see fig. page 22), and almost useless as a means of determining the species.' From Richardson 1844.

Richardson's contribution

John Richardson's contribution to ichthyology was substantial. He is still recognised as the author of about 200 species of fish. He reported on the fishes of many countries and holds a particularly significant place in Australian ichthyology due to his efforts in obtaining collections of specimens and drawings, chiefly through his British naval contacts. His work was careful and, even where the descriptions were brief, most are still attributable to actual species. He wrote on the biogeography of fishes on a number of occasions but some of the theories that would have begun to explain his observations were still over a century in the future. He extended the work of Cuvier and gathered many of the specimens that Günther would use for his catalogue. Only after the publication of Richardson's *Ichthyology of the Voyage of H.M.S. Erebus & Terror* was it possible to see the unique nature of the southern Australian fish fauna.

Ichthyology and empire

The year of Queen Victoria's accession, 1837, was also the year that John Gray was given a formal appointment as assistant at the British Museum. The 64 years of Victoria's reign constituted a period of confidence and relative prosperity which stimulated the natural sciences in Britain. There were no direct military threats to Britain during this time. The British expressed their confidence in part through indulging in military expeditions that made large parts of the world readily accessible to British naturalists for the first time. Napoleon was a fading memory and the Kaiser loomed only dimly towards the end of the Victorian era. Explorations and voyages merged with the imperatives of empire and the natural science harvest was immense. Collections by naval personnel were handled by Richardson. Naturalists among the military, particularly army surgeons, made collections and gave accounts of the plants and animals of the countries in which they found themselves. Gray at the British Museum initially presided over the acquisitions made and work done in the Department of Zoology, and many of his best contributors were army surgeons or members of the civil service.

Playfair in Zanzibar; Buchanan, Hardwicke, Day, McClelland and others in India; and Andrew Smith in South Africa: all were soldiers of Britain. John Bennett had been an officer of the Royal Navy and joined the army in Sri Lanka. They all participated in the discovery and scientific description of the fishes of the Indo-Pacific. The geographic range of a considerable number of the species described by these naturalists includes northern Australia so their names are associated with Australian fishes through authorship or by courtesy of authors. These soldiers shared the same background as Grey, Mitchell and other soldier-naturalists whose field of endeavour included Australia.

The activities of the East India Company included investigations of the natural history of the countries in which it had commercial interests (Lawson 1993). Among its representatives who had a significant influence in ichthyology were Theodore Cantor and Francis Buchanan; an even earlier contribution made under the aegis of the Company was that of Patrick Russell MD.

Russell

Patrick Russell (1726–1805) was born in Edinburgh and studied medicine. He worked for 20 years in Aleppo, where he combined his work as a doctor with the study of natural history. He moved to the coastal city of Vizagapatam (now Visakhapatnam), India where he was employed by the East India Company to study natural history (Cunningham 1853). Russell (1803) published *Descriptions and figures of two hundred fishes; collected at Vizagapatam on the coast of Coromandel.* Many of Russell's fishes range into northern Australian waters.

Buchanan

Francis Buchanan (1762–1829) was also a medical officer in the pay of the East India Company. He was born in Perthshire, Scotland, took his medical degree in Edinburgh in 1783, and was appointed surgeon on a British man-of-war. He began his service with the East India Company in 1794 as a surgeon in Bengal and in 1800 was dispatched by Lord Wellesley through southern India, studying the religion and customs of the people, their methods of agriculture and commerce, and the natural history of the country. Buchanan investigated the economy of Bengal from 1806 to 1813, producing an enormous manuscript and eventually publishing a history of Bengal. He returned to England in 1815 and then retired to Scotland where he succeeded to his mother's estate, taking the additional name of Hamilton; thus his writings appear attributed to Buchanan, Hamilton or Hamilton-Buchanan.

Buchanan's contributions to natural history were mostly in botany and ichthyology. He was stationed at the mouth of the Brahmaputra River early in his career in India and studied the fishes of the region, in particular the fishes of the Ganges (Day 1875). In all, he made observations on the freshwater fishes of India over a period of 20 years. On returning to England, Buchanan found himself initially denied access to his drawings and original manuscripts but, after many trials, was able to publish his well-known text, *An Account of the Fishes found in the River Ganges and its branches* (1822). Albert Günther, an unforgiving critic of ichthyological endeavour, wrote that this work was 'distinguished by a greater accuracy of ... drawings ... than was ever attained before' (Günther 1880a). Buchanan's use of local talent in preparing the drawings is seen from an extract from one of his letters, written in 1897 and quoted by Hora (1927):

> I have given my old painter a gold mohur a month and have him employed on fishes. I am attempting to have him do the outlines with some degree of accuracy; when he succeeds in that I shall begin to colour.

Buchanan's account of the Ganges fishes included the description of over 250 new species, several of which have wide Indo-west Pacific distributions. The Largetooth Flounder, *Pleuronectes* [now *Pseudorhombus*] *arsius* ranges around Australia, being absent only from the cold temperate waters of Victoria and Tasmania; the Sevenspot Archerfish, *Coius* [now *Toxotes*] *chatareus* (Hamilton-Buchanan, 1822) is common in the coastal streams of tropical Australia.

J. W. Bennett

John Whitchurch Bennett arrived in Sri Lanka, the country of his ichthyological endeavours, in 1816. He published *A selection from the most remarkable and interesting fishes found on the coast of Ceylon from drawings made in the southern part of that island from living specimens* (1828–1830). Aside from his years in Sri Lanka, not very much seems to be known about Bennett. He had been an officer in the Royal Navy and joined the army in Sri Lanka as a lieutenant, became a Sitting Magistrate in 1826, and departed the following year (Information from dust jacket of facsimile edition of *Fishes of Ceylon*, 1994).

An accomplished artist, Bennett also had a good knowledge of the fishes of his region. He employed his spare time in drawing inshore reef fishes from living specimens. He was proud of his work but anticipated criticism of his drawings arising from the apparent exaggeration of the colour. He offered defence by anecdote:

> ... one of the senior Provincial Judges, a religious and hospitable character, who had resided twenty-four years in Ceylon, complimented the author on the 'fertility of his imagination', when the first drawing of the Hembili-Girawah, No. 14, met his view. His breakfast table, the next morning, displayed two living specimens of that splendid fish, in a vase of water, which had been caught within eight hundred yards of his own residence. With an exclamation, he did justice to the author; and in a most impressive manner expressed his admiration of, and ascribed all glory to the works of Him, to whom all things are possible as Nature's God!

In this work, Bennett illustrated 30 species, 26 of which he described and named, in the belief that they were new to science; a few of these are presently regarded as valid. He recorded his gratitude to 'General Hardwicke, of the Honorable East India Company's Bengal Artillery' (see page 44) for assisting him with the fishes, and named one of the new species, the Sixbar Wrasse, *Sparus* [now *Thalassoma*] *hardwicke* for him.

Day

Francis Day (1829–1889) entered the Madras medical service in 1852, served with the British Army through the second Burmese war, and travelled extensively through India. He was a keen naturalist with a special interest in fishes and was appointed Inspector-General of Fisheries in India. He wrote many papers and reports on medical and natural history subjects and left India with the rank of Deputy Surgeon-General in 1876.

From 1859 to 1862, while stationed at Cochin, he collected and preserved fishes in his spare time and published *The Fishes of Malabar* (1865). At this time there was some appreciation of the damage to fisheries due to weirs that had been built, for irrigation purposes, across the seven principal rivers of the Madras Presidency. Day visited these sites in southern India in 1867 and reported

> ... the fish which, prior to the erection of the weirs, had ascended the rivers during the season of the rains for the purpose of spawning, were not only prevented from proceeding up stream to spots suitable for the deposition of their ova, but were collected in such vast numbers immediately below these weirs, which they vainly attempted to pass, that the wholesale manner in which they were caught by the native fishermen almost amounted to extermination of the spawning fish of each season.

The significance of Day's findings was obvious. He advocated the collection of more information and this was given practical recognition by his appointment to the post of Inspector-General of Fisheries in India. From this time, much of Day's activity was directed towards accumulating materials for his major work on the marine and freshwater fishes of the whole of India. His publications in the scientific journals were not well received by Albert Günther, the doyen of ichthyology at the British Museum. Instead of offering guidance and encouragement in private to an aspiring amateur ichthyologist, Günther published harsh and derisive criticisms. This led to a protracted and scientifically unproductive battle in the literature over a number of years (Whitehead and Talwar 1976).

Day continued his study of ichthyology in retirement in England, where he had brought extensive collections of fishes from India. He published his major work: *The Fishes of India: being a Natural History of the Fishes known to inhabit the seas and fresh waters of India, Burma, and Ceylon* (1875–1888). His studies in London were hampered by continuing poor relations with the irascible Albert Günther, without whose authority Day was unable to utilise the relevant collections in the British Museum. The conflict was aggravated by Day securing the services of George Ford (see page 71), the best scientific illustrator of the day, whom Günther had intended to employ in illustrating his own work. Ford was in poor health by this time but he completed most of the plates for the first two parts of Day's work.

The latter part of the 19th century saw the publication of the first of the fairly comprehensive regional accounts of fish fauna in the Indo-Pacific. Day's *Fishes of India* remains among the most successful works of its kind. The layout of the book was well designed for ease of use, there being synopses of the genera found in each family prior to the generic descriptions, and synopses of the species of each genus prior to the specific descriptions. Detailed synonymies were presented and Day gave good colour descriptions (generally from his own observations), distributions, Indian names where known and offered limited notes on the biology of some species. The plates were mostly attractive and accurate.

Fishes of India has been made a little harder to evaluate and use than it might have been due to the scattering of type specimens in various collections around the world. To assist researchers, Whitehead and Talwar (1976) prepared a 'Table of new genera and species described by Francis Day, with registered numbers for possible type specimens in eleven institutions'; the institutions were located in Calcutta, London, Vienna, Berlin, Leiden, Florence, Paris, Leningrad, Chicago, Cambridge (Massachusetts) and Sydney.

Many of the marine fishes that Day described and illustrated are distributed widely in the Indo-west Pacific region, including the shores of tropical Australia. In the *Proceedings of the Zoological Society of London* and in *Fishes of India* he also provided the original descriptions of some species listed in the Australian fauna; for instance, the Radiant Rockcod, *Serranus* [now *Epinephelus*] *radiatus* (Day, 1868) was described from Madras, India and is found in north-western Australia.

Cantor

Theodore Cantor (1809–1860) of the Bengal Medical Service (East India Company) was a Danish born physician and naturalist who wrote notes on Indian fishes and was later based in Penang. Here he obtained specimens from the local fishermen and published a *Catalogue of Malayan Fishes* (1850). He has been criticised for his lack of field work and concentration on the larger, marine fishes of Penang Island. The misleadingly inclusive title of his work is the most valid general criticism. His collections became the property of the East India Company and were transferred to the British Museum in 1860.

Many of Cantor's Malaysian species are found in northern Australian waters. There are illustrations of some of these species in his *Catalogue of Malayan Fishes*; his new species descriptions include the Beady Pipefish, *Syngnathus* [now *Hippichthys*] *penicillus* and the Tripodfish, *Triacanthus* [now *Pseudotriacanthus*] *strigilifer*, both described from the 'Sea of Pinang' and both including northern Australia in their range.

Playfair and George Ford

Robert Lambert Playfair (1828–1899) collaborated with Albert Günther of the British Museum to produce *The Fishes of Zanzibar*, an account of the fishes of the western extremity of the great tropical Indo-West Pacific fauna. Born in St Andrews, Scotland, Playfair sought a career in the army and entered the Madras artillery in 1846. However, his abilities came to the attention of leaders of the Indian civil service and he was given work in diplomacy and administration. In 1854 he was in Aden as acting-governor and, from 1862 to 1866, as political agent in Zanzibar (DNB). Here, he interested himself in the magnificent fish fauna of the island and soon many casks of fishes were being dispatched from the British Residency to Günther at the British Museum.

In 1866, Playfair was on leave and distributed his time between Scotland and England, where he worked on the fishes for his book. He visited Peters (see page 81) in Berlin and examined the fishes which Peters had obtained from Mozambique. He worked with Günther in the British Museum and corresponded with him from Scotland and Germany. Playfair appears to have been in a cheerful commotion in completing the book; some of his correspondence with Günther is quoted in a reprint (Myers 1971), and extracts convey his mood:

> ... Sunday is always a dreary day in Scotland, but when one is in the Highlands and the rain is falling in continuous sheets, and the nearest church (Presbyterian!!) four miles off, the horror of the situation can hardly be described ...
>
> ... As for me I am sick of illness, and pine for the pleasant vaults of the British Museum and the fragrant pisci-alcoholic atmosphere which pervades the same ...

After the *Fishes of Zanzibar* (Playfair and Günther 1867) was published Playfair expected to resume his duties (and ichthyological pursuits) at the end of his leave. However, he was appointed consul-general in Algeria, and later in Tunis, where he had not the same opportunities for original descriptive work among the fishes, although he sent some specimens to Günther. He directed his energies to exploration and to historical researches which led to his producing valuable works on the history of the 'Barbary States'.

The Fishes of Zanzibar is a remarkable work, the production of an unusually effective collaboration:

> Playfair and Günther's *The Fishes of Zanzibar* was unique among regional works in that the naturalist who obtained the collections (Playfair), joined a metropolitan ichthyologist of the greatest renown (Günther) in authorship, while the principal illustrator (Ford) was the ablest of his time. (Myers 1971)

George Ford (1809–1876) spent his early life in South Africa where he began drawing for the adventurous Andrew Smith.[5] Ford was associated with the British Museum from shortly after his arrival from South Africa and his talents were employed by a number of well-known naturalists, including John Gray, Albert Günther and Charles Darwin. Francis Day had his services in illustrating his *Fishes of India* although by this time (1874) Ford had been working for the British Museum naturalists for many years.

Illustration by woodcut dates back nearly to the origin of printing in the 15th century and techniques of engraving on copper and steel permitted a great improvement in quality. However, only with the introduction of lithography[6] did illustration reach its peak of beauty and accuracy (Myers 1971). Ford used this method for the ichthyological publications which he illustrated. He died while working on Day's fishes in 1876.

Günther described some of Playfair's specimens in separate papers and a few in his *Catalogue*, but the greater number was reserved for the *Fishes of Zanzibar*. In this work, nearly 500 species were listed, 45 of which were thought to be new, and described accordingly. Most of these 'new' species, a little less than half of which are now regarded as valid, were illustrated by Ford. Among these, the Spectacled Leatherjacket, *Monacanthus* [now *Cantherhines*] *fronticinctus* has an extensive distribution in northern Australia.

5 Andrew Smith (1797–1872), an army surgeon who developed a passion for natural history, was stationed in South Africa where he established and directed the natural history museum. He produced an ambitious series of volumes: *Illustrations of the zoology of South Africa* (1838–47) and the work was illustrated by George Ford; when Andrew Smith returned to London in 1857, Ford accompanied him.

6 Lithography consists of drawing on polished lithographic stone with a grease pencil and immersing the stone so that it can absorb water. Printer's ink is then applied and adheres only to the grease pencil drawing, leaving the stone clear of ink. The paper for the plate is then applied to the drawing and the ink impressed by running a roller across it, following which the paper is carefully separated from the stone. The lithograph can then be hand painted with water colours (Myers 1971).

Continental ichthyology

Workers in Germany and Austria received fishes from Australia in the mid to late 1800s: Müller, Kner, Klunzinger, Steindachner, Peters and Kaup were the most prominent. Usually their descriptions were accurate and their publications valuable, based on a good appreciation of the variability of species and the importance of the early literature. In France, ichthyology after Cuvier and Valenciennes was served by Guichenot, Duméril and their associates.

Müller and Henle

Johannes Peter Müller (1801–1858) was born in Nuremberg, Germany and studied medicine and natural science. Professor of Physiology and Anatomy successively at Bonn, then Berlin, Müller contributed to many aspects of natural science. He rejected the speculative theories of his day, relying upon observation, dissection, microscopic examination and, when practicable, physiological experimentation (DSB). Among his most significant publications were studies of embryonic development, the function of the compound eye of arthropods, and the electrophysiology of nerve conduction.

The science of comparative anatomy was dominated by Cuvier when Müller began his researches in the 1830s. His work on fishes included reviews of the most primitive of vertebrates (lampreys), primitive fishes (ganoid fishes and lungfish), cartilaginous fishes (sharks and rays) and a revision of Cuvier's fish classification. He also showed the significance of the lancelet in relation to its vertebrate ancestry. Müller's classification of fishes affected Australian ichthyology since it was espoused by Günther, whose publications were used as a taxonomic framework by most Australian zoologists up to about 1900.

Johannes Müller. From Duke-Elder 1958.

Friedrich Gustav Jacob Henle (1809–1885) was the student and co-worker of Müller. He studied medicine and became an assistant to Müller at Berlin. Henle worked in the fields of comparative anatomy, histology, physiology and pathology (DSB). Among his early studies was a description of *Narcine*, a genus of electric rays; Henle produced a major revision of these animals (1834). In so doing, he overlooked the common electric ray (the Coffin Ray) of southern Australia, *Hypnos monopterygium*, which had been poorly described in an obscure publication (Shaw 1795) (see page 14).

Müller and Henle cooperated to produce major systematic works on sharks and rays, culminating in their *Systematische Beschreibung der Plagiostomen* (1838–1841). The work was based primarily on their own dissections and established the outline of elasmobranch taxonomy for the remainder of the 19th century. In this book, and a paper on the genera of elasmobranchs (Müller and Henle 1838), nearly 40 new genera were defined and most of these are retained today. In all, Müller and Henle described over 100 new species and about 60% of these are regarded as valid. Many of the pelagic sharks range throughout the warm seas of the world, their wide distributions commonly including Australian waters, and some (such as the Bull Shark, *Carcharhinus leucas* and the Common Blacktip Shark, *C. limbatus*) were first described in the *Plagiostomen*. The ranges of benthic sharks and rays are generally more limited but Müller and Henle had specimens from many parts of the Indo-Pacific and from these described a significant number of species found also in Australia. They also described the Common Stingaree, *Trygonoptera testacea*, an endemic Australian species. The ray had been drawn by Spöring on Cook's first voyage; in describing the species Müller and Henle published the drawing, reversed, with Solander's description.

The influence of the *Plagiostomen* could hardly be overstated. Elasmobranchs were omitted from the *Histoire Naturelle des Poissons* of Cuvier and Valenciennes. Günther's treatment of these animals in his *Catalogue* (1870) was based upon the work of Müller and Henle. The *Plagiostomen* remains the largest step that has been taken in understanding the relationships and classification of sharks and rays; later work has been built upon this foundation.

Kner

Rudolf Kner (1810–1869) studied medicine in Vienna and assisted Johann Heckel with fish collections. He became Professor of Zoology at the University of Vienna, specialising in ichthyology (Herzig-Straschil 1997), and reported on fishes collected by naturalists aboard the Austrian Frigate *Novara* (Kner 1865–67). Kner did not approve of the classification used by Günther in his *Catalogue* but employed it for his own work 'under repeated protest' (Günther 1865a) as it had gained wide acceptance. Günther, with heavy humour, took Kner to task over the introduction of new terms: 'It would also be well not to introduce terms like 'Pseudokieme' (pseudo-gill), as pseudo-ichthyologists might be induced to adopt them.' When the third and final part of Kner's *Novara* report was published, Günther was rather patronising:

> The Recorder has had occasion to carefully examine several portions of the work; and his confidence in the conscientiousness and correctness of Prof. Kner's determinations has been much strengthened during his study.

The *Novara* visited Sydney in 1858 and the zoologists aboard improved their collections by exchanges with the Australian Museum. They approached Lieutenant Governor William Denison, a supporter of natural science, who made it clear that international cooperation was desirable; he passed on the correspondence to the Australian Museum:

The Silver Sweep, *Scorpis lineolatus* [now *S. lineolata*], collected in Sydney. From Kner 1865. Courtesy Museum Board of South Australia.

The Bridled Goby, *Gobius* [now *Arenigobius*] *bifrenatus*, collected in Sydney on the *Novara* expedition, described and illustrated by Kner. Waite reproduced Kner's illustration, adding a black pupil to the image. From Waite 1923.

> The inclosed letters will explain the requirements of the Scientific Staff on board of the Austrian Frigate Novara and I should hope that we may be in a position to profit by the exchange thus offered to us. The subject had better be brought under the notice of the Trustees at a special meeting to be called for the purpose. (AMA)

The exchanges included Sydney fishes which were dealt with by Kner (1865 and 1867).

In Vienna, Kner had assistance from two young zoologists who were to become well-known ichthyologists: Steindachner and Canestrini (see below). In later publications, Kner described further new species from Sydney; approximately half of the species he described from eastern Australia are at present regarded as valid.

The Southern Maori Wrasse, *Ophthalmolepis lineolatus* (Valenciennes, 1839) from the Sydney collections of the *Novara*. From Kner 1865. Courtesy Museum Board of South Australia.

Klunzinger and von Mueller

Carl Benjamin Klunzinger (1834–1914) studied medicine but had a strong preference for natural history. He relinquished his hospital appointment to take up a post as a sanitary physician at a small Egyptian port on the Red Sea. Here he spent five years collecting marine animals, particularly fishes, for the Natural History Museum of Stuttgart. On his return, he worked on his fish collections and published *Synopsis der Fische des Rothen Meeres* (1871). Klunzinger spent several more years collecting fishes from the Red Sea, and published another major work on them in 1884 (Botros 1971). His work on the Red Sea fishes constituted Klunzinger's major contribution to ichthyology but he also received fishes from Australia, particularly from Ferdinand von Mueller.

Ferdinand von Mueller (1825–1896) obtained a qualification in pharmacy at the University of Kiel, Germany and migrated to Australia. He was appointed Government Botanist in Melbourne in 1853 and participated in the North Australian Exploring Expedition from 1855 to 1857. On his return to Melbourne, von Mueller was made Director of the Botanical Gardens (ADB). He corresponded with botanists and other specialists in natural history throughout Europe. From his own explorations and from collectors in many parts of Australia and New Zealand he received specimens, including collections of fish. He sent the fish to Klunzinger at Stuttgart and Steindachner at Vienna.

Von Mueller sent fishes to Stuttgart from Victoria (Port Phillip Bay and the 'Yarra Lagoon') and a few from the lower reaches of the Murray River in South Australia. Klunzinger (1872) published an account of these collections, naming over 20 as new species, although over half of these had already been described. Among the new species was the Blue-spotted Flathead, *Platycephalus speculator.*

Carl Benjamin Klunzinger. From Klunzinger 1964.

Although Klunzinger published on Australian fishes only twice his work was significant for Australian ichthyology. His second paper (Klunzinger 1879) was also based on von Mueller's collections housed at Stuttgart and included fishes from south-western Australia, South Australia, Victoria, north Queensland and the Northern Territory; a few New Zealand fishes were also described. The paper included a valuable survey of previous publications on the fishes of Australia and New Zealand and presented an annotated list of nearly 260 species of Australian fish. The synonymies, derived in part from Günther's *Catalogue*, were as complete as Klunzinger could make them and represented a real advance. He thanked Steindachner for assistance with identifications and access to some of the more obscure literature. In revising his earlier paper, he submerged in synonymy a few of the species there described as new; he also decided that some of the specimens which he had identified with known species were in fact new. In all, Klunzinger described 27 new species in this paper, of which about one third are now regarded as valid, illustrating most of them.

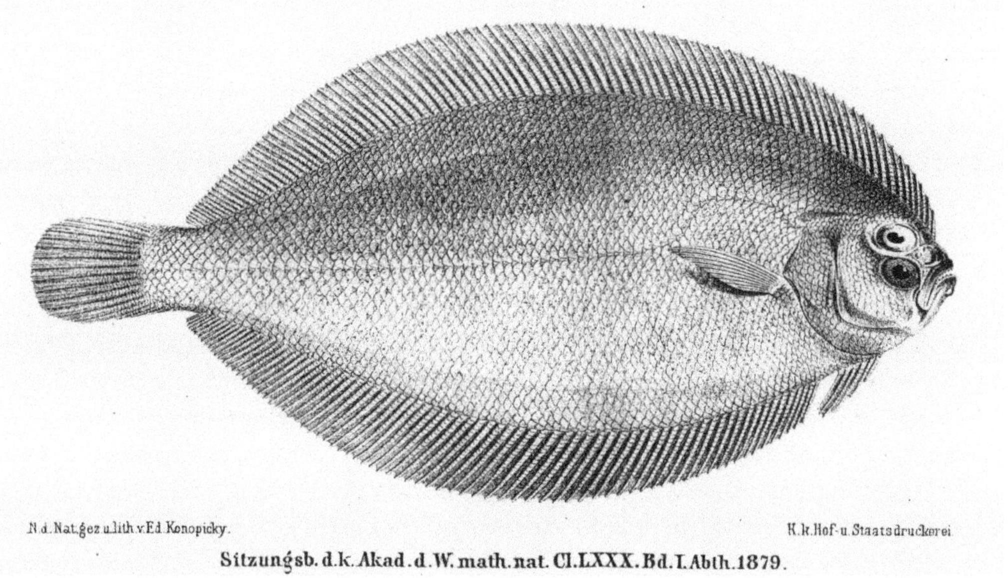

Mueller's Flounder, *Pseudorhombus Mülleri* [now *Arnoglossus muelleri*] from Hobson's Bay, sent by von Mueller. From Klunzinger 1872.

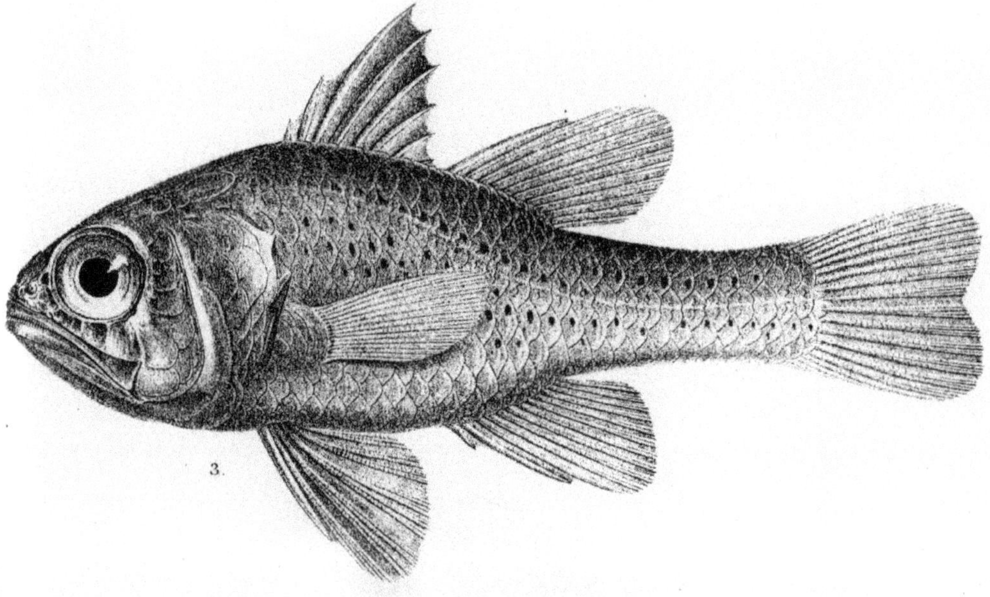

The Orange Cardinalfish, *Apogon punctatus* [now *Vincentia punctata*] sent by von Mueller, collected at King George Sound, Western Australia. From Klunzinger 1880.

The ease with which the work of different ichthyologists can be assimilated and assessed varies greatly. Klunzinger's work was substantial and localised to just two papers, easily assimilated into the whole through exact descriptions with accurate illustrations, and a clear advance through the use of extensive synonymies.

Steindachner

Franz Steindachner (1834–1919) studied natural science in Vienna and specialised in ichthyology. He worked on the fish collections of the Kaiserlich-Königliches Hof-Cabinet, enlarging them through collecting in Europe, the Canary Islands and Africa during the 1860s. He spent several years working with Louis Agassiz in Boston and accompanied Agassiz on a South American collecting expedition. Steindachner returned to zoological work in Vienna in 1874 and became Director of the zoological collections of the Kaiserlich-Königliches Naturhistorisches Hof-Museum in 1887. He directed deepsea expeditions in the Mediterranean between 1891 and 1893 and expeditions in the Red Sea between 1895 and 1898. His last collecting trip, at the age of 69, was to Brazil in 1903 (Herzig-Straschil 1997).

Steindachner received fish collections by gift, exchange and purchase, from all over the world, describing over 1000 species of fish. His most important Australian work was in the form of major papers describing collections sent to him at intervals. He dealt with fishes from Port Jackson (Steindachner 1866), listing 71 species and describing 21 of these as new, including the Eastern Sea Garfish, *Hyporhamphus australis*. Some of the specimens came from Gerard Krefft, others from von Mueller and from the collections of the anatomist Joseph Hyrtl. He also referred to Australian specimens donated by Dr.

Schütte and honoured him with a generic eponym in describing the Eastern Pomfred, *Schuettea scalaripinnis* from Port Jackson. The following year, Steindachner reported on collections from north Queensland (Cape York and the Fitzroy River near Rockhampton) and added a few species to his Port Jackson list (Steindachner 1867).

Franz Steindachner, courtesy of the South Australian Museum.

Von Mueller provided more specimens for Steindachner (1879), most from Victoria (Hobson's Bay, Melbourne) and Queensland (Townsville). From this paper it is clear that Steindachner had contact with Klunzinger and access to the Stuttgart Museum specimens.

From South Australia Wilhelm Haacke, while Director of the South Australian Museum, sent fish to Vienna for identification and description. Steindachner (1884) published the results of his study of these fish. This was an elaboration of an earlier report (1883); most of Steindachner's major papers on Australian fish were heralded by the publication of a quick report, or abstract, with short descriptions of new species, presumably to secure priority of authorship. The 1884 publication was a valuable addition to the knowledge of South Australian fishes. In this paper Steindachner described the Wavy Grubfish, *Percis* [now *Parapercis*] *haackei,* naming it after the donor.

Steindachner had corresponded with von Mueller and Australian zoologists, including Krefft, Ramsay, Haacke, Ogilby and Waite. He was still actively seeking Australian fishes as late as 1898 when he wrote to Edgar Waite at the Australian Museum (as translated by South Australian Museum staff):

> Esteemed Colleague,
> With today's mail I forward to you my latest ichthyological paper and take this opportunity to ask you to send me specimens of Australian fish, particularly those which you have described in recent years.
> For my part, I shall send you within the next few months a collection of pretty fish from the Red Sea, which would probably be absent from your museum.

Steindachner sent Waite copies of his publications and this note was enclosed with *Die Fische der Sammlung Plate* (Steindachner 1898). The paper held interest for Waite and others as many of the genera represented in the account of this collection, made by Dr. Plate from the coast of South America and Juan Fernandez, are also found in southern Australia and New Zealand. Another of Steindachner's papers with relevance to Australia was *Fische aus dem Stillen Ocean* (1901), an account of fishes sent to Steindachner by Professor Hugo

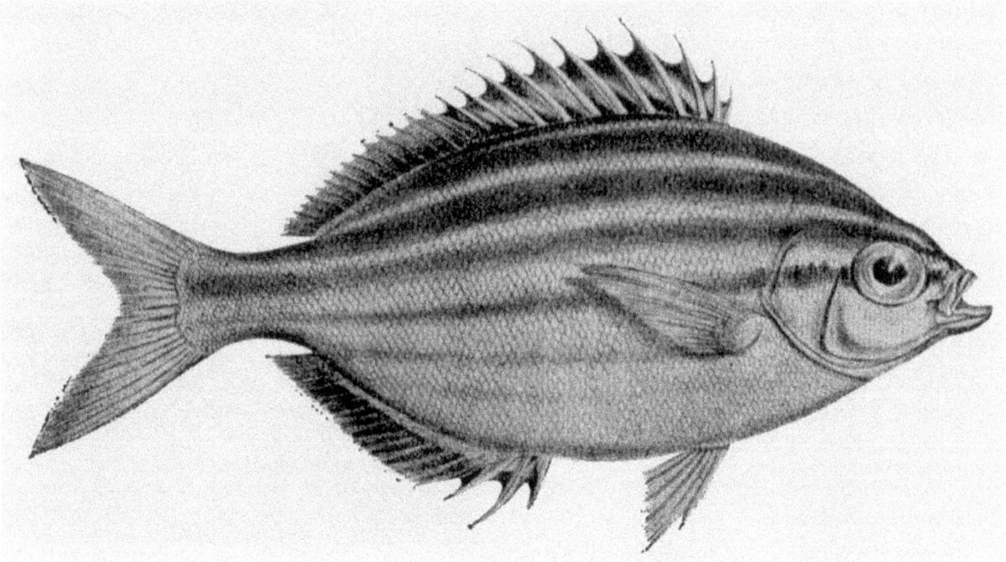

The Mado, *Atypichthys strigatus* (Günther, 1860) was illustrated by Steindachner from his Port Jackson collection. From Steindachner 1866.

Steindachner described the Southern Sole, *Solea haackeana* [now *Aserragodes haackeanus*] from Gulf St Vincent, South Australia in 1883 and illustrated it the following year. From Steindachner 1884.

Schauinsland of the Natural History Museum of Bremen. The collection was derived from a voyage in the Pacific in 1896–1897; most of the fishes were from Hawaii but a few were taken in New Zealand and South America.

Steindachner had a benevolent influence on Australian ichthyology, corresponding with practitioners of the science in Australia and producing accurate descriptions of the species that came to his attention. It was unfortunate that his publications were sometimes overlooked by Australian workers.

Canestrini

Giovanni Canestrini (1835–1900) was a zoologist with very wide interests in the natural sciences, best known now for having translated Darwin's *On the Origin of Species* into Italian. He studied at the University of Vienna and was recruited by Kner to work on the *Novara* collections. He published on fishes from 1860 to 1872, primarily on the fishes of Italy. Canestrini also handled specimens from other parts of the world, including at least one small collection from Australia. The latter consisted of temperate forms which he listed, discussing several in detail (Canestrini 1869).

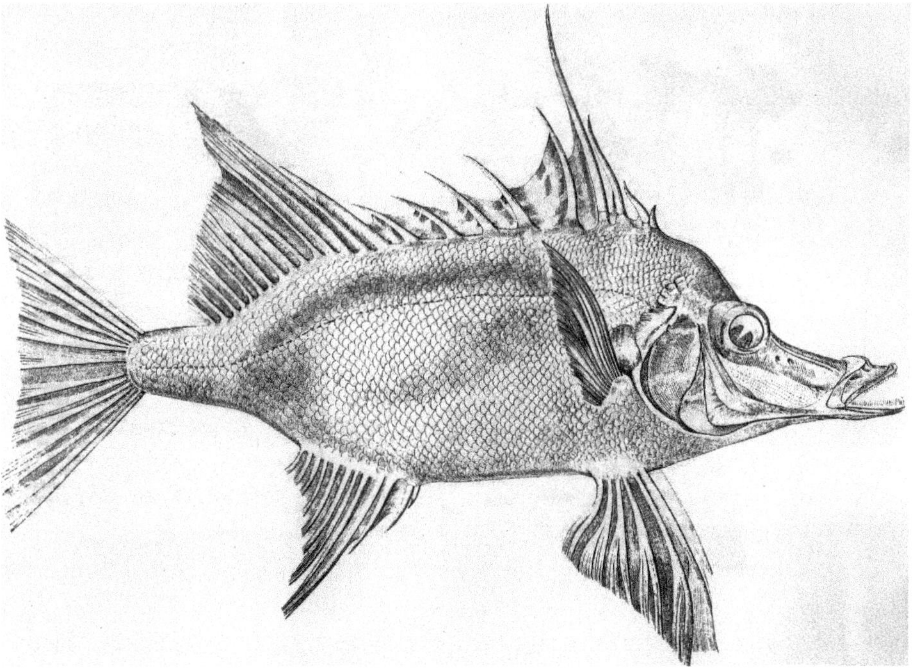

The Longsnout Boarfish, *Pentaceropsis recurvirostris* (Richardson, 1845), described originally from a head only, was re-described and illustrated by Canestrini. From Canestrini 1869.

Peters

Wilhelm Carl Peters (1815–1883) was born in a small town in Schleswig. He studied anatomy and zoology in Copenhagen, then medicine and natural science in Berlin, receiving his degree in medicine in 1838. Among his instructors was Johannes Müller. Peters carried out field work on marine invertebrates in the Mediterranean and later became an assistant to Müller in the Anatomy Museum in Berlin. At this time Berlin was home to an extraordinary assemblage of talented men dedicated to natural science, notably Johannes Müller, Jacob Henle and Alexander von Humboldt.

In 1842, Peters set out on a six year expedition through Mozambique, the Comores Islands and Zanzibar, travelling up the Zambezi River and returning to Berlin by way of India and Egypt. He collected, described and drew a wide variety of zoological specimens. He was appointed to a professorial post in Medicine in 1853 and became Assistant Director, and then Director, of the Zoologisches Museum, holding the latter position from 1857 to the time of his death in 1883 (DB).

An accomplished zoologist, Peters published frequently on a range of animal groups; his papers on fish collections from many parts of the world were published between 1844 and 1882. Berlin received specimens of reptiles and fish from a German settler, Richard Schomburgk, living a little north of Adelaide. Among the fish were several new species, including the small Cobbler Wobbegong, *Crossorhinus* [now *Sutorectus*] *tentaculatus*; Shorthead Seahorse, *Hippocampus breviceps*; and Blue Devil, *Plesiops* [now *Paraplesiops*] *meleagris* (Peters 1864a, 1864b and 1869). Peters also received a collection from Sydney through Gerard Krefft of the Australian Museum, describing some new species (Peters 1866). Peters reported on the fish collection of the German *Gazelle* expedition which had set out in 1874, shortly after the British *Challenger*, on a two year voyage around the world, conducting oceanographic research. From this collection Peters described the Flatface Seahorse, *Hippocampus planifrons* taken in north-western Australia (Peters 1877).

Kaup

Johann Jakob Kaup (1803–1873) was born in Darmstadt, Germany and worked in many branches of natural history, contributing most to palaeontology, herpetology and ichthyology (NDB). He had contact with some of the most important biologists of his time, in particular Owen, Agassiz and Cuvier. John Edward Gray (see page 43) engaged him to undertake revisions of some animal groups at the British Museum, resulting in his *Catalogue of the Apodal fish* [eels] *in the Collection of the British Museum* (1856) and *Catalogue of Lophobranchiate Fish* [pipefish and seahorses] *in the Collection of the British Museum* (1856). Kaup had the advantage of access to collections of pipefish from the curators of other European museums: Duméril (Paris), Temminck (Leyden), Heckel (Vienna), Rüppell (Frankfurt), Lichtenstein (Berlin) and Krauss (Stuttgart). He described from southern Australia the large and extraordinary Brushtail Pipefish, *Leptoichthys fistularius* Kaup, 1853 and the Seamoth, *Pegasus lancifer* Kaup, 1861.

Guichenot

Antoine Alphone Guichenot (1809–1876) was a zoologist of the Muséum National d'Histoire Naturelle in Paris. He followed Valenciennes in dealing with new fish collections at the museum, publishing for over 30 years up to about 1880. Castelnau sent him fishes from South Africa, south-east Asia, South America and Australia. Guichenot established a new genus for Australian Gurnard Perches, describing the Common Gurnard Perch, *Neosebastes scorpaenoides* from Melbourne specimens (Guichenot 1867).

Duméril

Auguste Henri André Duméril (1812–1870) was Professor of Herpetology and Ichthyology at the Muséum National d'Histoire Naturelle in Paris from 1857. His work on fish was marred by a penchant for describing new species with little justification. However, he published original descriptions of a few Australian species, including the Varied Carpetshark, *Hemiscyllium* [now *Parascyllium*] *variolatum*.

Bleeker

Progress in ichthyology has always been uneven, great advances in particular regions or subjects commonly being made by individuals over relatively short periods of time. The middle years of the 19th century saw an enormous advance in knowledge of the Indo-Pacific fish fauna. The focus of diversity in this region is within the seas of the Indo-Australian archipelago and a remarkable proportion of its species were revealed to European science by a Dutch army surgeon.

Pieter Bleeker (1819–1878) was born in Zaandam, Netherlands and initially trained as an apothecary in Amsterdam. Although attracted to the natural sciences, he qualified as a surgeon in 1840. He failed to obtain a position with the Natural History Museum at Leiden and travelled to Paris for further study (Weber and de Beaufort 1964). In 1841 he received a commission as a surgeon in the army of the Dutch East India Company. From 1842 to 1860 Bleeker worked mostly in Batavia (now Jakarta) and, during this time, assembled a collection of approximately 2500 fish species, describing many of them as new. Most of his fishes were sent to him by army surgeons and civil service officials who were posted in different islands throughout the Dutch East Indies.

At the same time, Bleeker performed his medical duties and organised and supported scientific societies. He also established and edited scientific journals, beginning with *Koninklijke Natuurkundige Vereenigung in Nederlandsch Indie* in 1844. He corresponded with ichthyologists and other scientists throughout the world and wrote on a wide variety of subjects in the biological sciences, medicine, geology and anthropology; in zoology, he published on many vertebrate and invertebrate groups, both marine and terrestrial.

The sophistication of Bleeker's publications improved rapidly. In ichthyology, he began with species descriptions and regional lists, but progressed to include group revisions. By 1845 he was working towards a comprehensive account of the fishes of the East Indian archipelago, a scheme that involved frequent exchanges of fish collections with his many correspondents.

Bleeker returned to the Netherlands in 1860 with his vast fish collections, retired from military service in 1863, and concentrated on the preparation and publication of the text and illustrations for his *Atlas Ichthyologique des Indes Orientales Néêrlandaises*. During this period he visited

Pieter Bleeker. From Weber and de Beaufort 1964.

the main natural history institutions of Europe; a list of his contacts at that time reads like a Who's Who of mid 19th century ichthyology:

> With each issue of the Atlas, Bleeker's fame increased, and among his correspondents were men like F. Day (Cheltenham), the Dumérils (Paris), Th. Gill (Washington), J.E. Gray and A. Günther (London), J.J. Kaup (Darmstadt), R. Kner (Vienna), A. Kölliker (Würzburg), J. von Liebig (Munich), Chr. Lütken (Copenhagen), W. Peters (Berlin), W.P.E.S. Rüppell (Bonn), H.E. Sauvage (Paris), C.T.E. von Siebold (Munich), J.J.S. Steenstrup (Copenhagen), and L. Vaillant (Paris). (Boeseman 1983).

Günther (1864c) reported in the *Zoological Record*:

> Dr. Bleeker has commenced collecting the results of his numerous labours on the Ichthyology of the East Indian archipelago in a splendid work published under the auspices of the Dutch Colonial Government. All the species observed by the author during nearly twenty years of ichthyological study are described and figured. The descriptions are more or less republications of those formerly given. The plates are coloured drawings made in India, and very accurate, and therefore quite sufficient for scientific purposes ... On the whole, the work is invaluable and almost indispensable for the determination of the fishes of that fauna ...

Günther continued with a brief grumble about '... the most trivial characters being used for the distinction of genera and groups ...' Bleeker in part financed his work by selling specimens to the British Museum, where Gray and Günther were anxious to add his species to their collection. He responded to Günther's complaints about the condition of the specimens provided to the British Museum: 'If it is true that I did not always send the largest or most beautiful specimen of the requested species, I hope that you will take into account that, if one possess for instance only two specimens, one does not preferably dispose of the best example.' Bleeker responded also to Günther's repeated disparagement of the published work, commenting that he expected criticism but resented the manner in which it was delivered, claiming 'always to have had the conviction that errors committed by a simple autodidact would be forgiven by his colleagues and that criticisms of a too personal kind would prove detrimental only to their author.' (both passages quoted in Boeseman 1983).

The text and illustrations for the *Atlas Ichthyologique* were completed, but publication ceased with Bleeker's death. Nine of an intended 14 or 15 volumes were published between 1862 and 1878 and the plates for the remaining volumes were published recently (1983). His type specimens either remained in Leyden or were sold to the British Museum to help finance the publication of the *Atlas*. After his death, the collections were auctioned and a small portion was eventually purchased by the National Museum of Victoria, Melbourne.

In 1874, Bleeker commented on the part he played in making known the fishes of his region and also on the origins of his other collections:

> I have had the privilege to enrich science with more than 1300 new species of fishes. More than 200 of these species, however, occur outside the Indian Archipelago, in Japan, China, Siam, Bengal, Ceylon, Madagascar, Guinea, Surinam, etc., so that there remain over 1100 species for the East Indies ...'
> (Boeseman 1983).

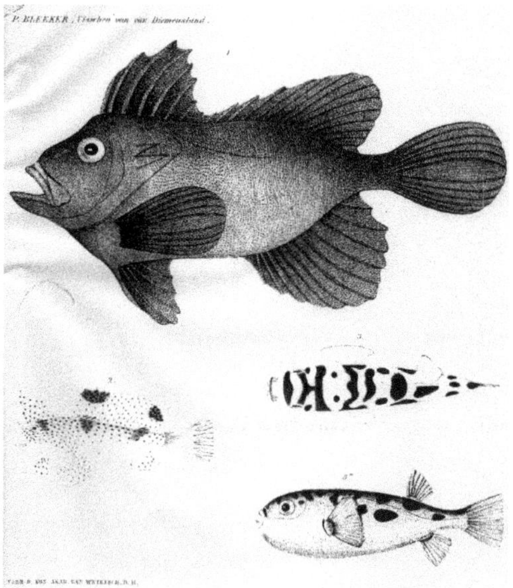

Pieter Bleeker's fishes from Tasmania: The Red Velvetfish, *Gnathanacanthus goetzeei* Bleeker, 1855; The Spotted Handfish, *Brachionichthys hirsutus* (Lacépède, 1804); The Barred Toadfish, *Gastrophysus* [now *Contusus*] *richei* (Fréminville, 1813). From Bleeker 1855.

The 'etc.' included Australia.

Several collections were sent to Bleeker from Australia. He published descriptions of a few new species from the mouth of the Derwent River in Tasmania, from Port Jackson and from King George Sound. His friend, J.W. Goetzee, sent him a collection from Hobart from which he described several new fish. In this paper Bleeker (1855) presented a biogeographical table, showing the distribution of fishes from mainland Australia (Nieuw-Holland), New Zealand, Norfolk Island, Tasmania, and the Auckland Islands. His report on a collection from Port Jackson (Bleeker 1863) included a description of the Green Moray, *Gymnothorax jacksoniensis* [synonymous with *Gymnothorax prasinus* (Richardson, 1848)] and, on two separate occasions, he published new descriptions of fishes endemic to the east coast of Australia: Günther's Wrasse, *Pseudolabrus guentheri* (Bleeker, 1862) and the Mossback Velvetfish, *Paraploactis trachyderma* (Bleeker, 1865) .

Bleeker published the initial descriptions of many northern Australian fish as they were among the collections made for him in Indonesia (see colour plate 6, lower). He received a particularly good representation of the larger fishes from coral reefs of the archipelago and this is reflected in his authorship of a significant proportion of these fishes. For instance, he appears as the author of about 15% of the species of wrasse (Labridae) found in tropical Australia. In all, Bleeker described nearly 300 species which are found in Australian waters, the great majority inhabiting the tropics. Among these is a tiny mimic, the Blacksaddle Leatherjacket, *Alutarius* [now *Paraluteres*] *prionurus* (Bleeker, 1851).

Gilbert Whitley (1964) suggested that Bleeker was 'probably the world's greatest ichthyologist' and a bibliographic index to his fish publications (Weber and de Beaufort 1964) contains 500 entries. Although he handled relatively few specimens from Australia, Bleeker's importance to Australian ichthyology is evident from the frequency with which he is cited in taxonomic works on Australian fish and the proportion of Australian fish species for which he prepared the original description.

Günther

Albert Günther (1830–1914), from Tubingen in Germany, came to the British Museum to work on reptiles and fishes. He was appointed Keeper of the Zoological Department in 1875.

Günther's Catalogue and collections at the British Museum

Günther published about 170 papers on fishes, aside from his *Catalogue of the Acanthopterygian Fishes in the British Museum*, the first volume of which was issued in 1859; John Gray (see page 43) wrote the preface to each volume, the work being concluded with the eighth volume in 1870. Günther expressed his view of the work:

> The collection of the British Museum, together with its Library, have enabled me to introduce original diagnoses of most of the species, and to compile the synonymy and revise that stated by Cuvier, thus raising this work above the point of a mere compilation. (Günther 1859)

Günther had excellent access to the early literature and received new papers shortly after, or even before, they were published. He listed 330 publications which he cited in the first volume of his work, the first 30 or so dating from Aristotle to the tenth edition of Linnaeus' *Systema Naturae* (1758). He also listed the collections to which he had access. In succeeding volumes he incorporated the literature as it appeared and listed further collections as they came to hand.

In the *Catalogue*, Günther used the system of classification developed by Johannes Müller as the framework but modified the criteria used to define some of the families. This represented an improvement over Cuvier's system, incorporating definitions of ganoid fishes (as made by Louis Agassiz) and other refinements. Günther offered much more than a simple catalogue, describing the fishes in detail, including many new species and deriving his own family definitions as he proceeded. He responded briefly to criticism of the scheme he used:

> ... frequent as are the objections against Müller's modifications of Cuvier's system, no one has proposed any arrangement which would give a more satisfactory result, if put to the test of carrying it out to a detailed subdivision. (op. cit.)

Günther constructed keys to assist identification, but only on a limited basis, generally to facilitate the separation of large genera into groups of species. He was very conservative with regard to genera, frequently resisting their subdivision when this was proposed by other workers.

Collections of fishes came to the British Museum from all over the world: Patagonia, the Amazon, Ecuador, Nepal, China, Zanzibar, Morocco, St Helena, the Arctic ... Günther wrote reports on the fishes and reptiles he received over the course of more than 50 years (1857 to 1910). From Australia, he acquired the largest number of specimens between 1861 and 1878, and published over a dozen papers on Australian fishes during this period. These collections were among his main resources for studying the fishes of Australia. However, in the introductory pages to the various volumes of the *Catalogue* he listed few collections from Australia:

- Volume 4 (1862): 'A Collection of Marine Fishes from Sydney, presented by G. Krefft, Esq.'
- Volume 5 (1864): 'Further collections of Fishes from Australia and New Zealand, sent by Mr. G. Krefft.'
- Volume 6 (1865): 'Further Collections of Marine and Freshwater Fishes from Australia, sent by G. Krefft, Esq.'
- Volume 8 (1870): 'A Collection of Marine Fishes from Tasmania. Purchased.'

Thus Günther's main source of fishes from Australia, while he was preparing his *Catalogue*, was Gerard Krefft (see page 102), the Curator of the Australian Museum. This source was lost with Krefft's expulsion from the Museum in 1874 and most Australian specimens thereafter were referred to Castelnau, or to the museums of continental Europe.

The voyages of the *Rattlesnake* (1846–1850) and the *Herald* (1852–1861) provided specimens from northern Australia and waters immediately to the north. John Gould (see pages 54 and 56) presented the British Museum with collections of fish, many taken by Gilbert around Port Essington in the Northern Territory. George Bennett (see page 98), among a few other residents in Australia, is acknowledged by Günther for sending Australian fishes.

The collections contemporary with the writing of the *Catalogue* were supported by older material. The Haslar Museum specimens of John Richardson had been moved to the British Museum and there were even older collections. In particular, the collections of the Zoological Society and the East India Company were held by the British Museum. Edward Smith Stanley, the 13th Earl of Derby (1775–1850), had been an enthusiastic naturalist and a president in turn of the Linnean Society and of the Zoological Society. He was known for maintaining an extensive menagerie and also kept a zoological museum from which Günther received specimens, including Australian fish.

For the *Catalogue*, Günther worked his way through the British Museum collections in a systematic order. Thus he began volume I with two small families, the Gasterosteidae (Sticklebacks) and the Berycidae (fish related to Nannygai); then the Percidae (the perch-like fishes), a large family which he covered in about 200 pages. He wrote careful diagnoses of the families, genera and species in much the same way as Cuvier and Valenciennes had done before him. At the end of each volume he brought the earlier volumes up to date with addenda. As he proceeded, Günther used short papers to describe collections or small numbers of species, and included new species described in these papers in the volumes of the major work. For instance, he described a seadragon from South Australia in the *Proceedings of the Zoological Society of London* (1865b) and incorporated it in volume VIII of the *Catalogue* (1870). Günther's short papers included descriptions of fishes whose groups had already been dealt with in his *Catalogue;* rather than include them in the addenda of subsequent volumes, he contemplated a major addendum to be published after the *Catalogue* had been completed, but this addendum was never published.

In the last volume of the *Catalogue*, Günther (1870) discussed the state of knowledge of the number and diversity of fishes. He explained his criteria for listing a new species:

> I consider a species to be well established only when it is founded on characters which, from an examination of numerous examples, are found to be permanent, not subject to *gradual* variation, and not dependent on season, sex, or age – or which are known to be so from the examination of allied forms ... By this principle I have been guided throughout the work ...

He gave his reasons for requiring multiple specimens of each species:

> A species to which no particular interest is attached ought to be represented by at least three specimens, as a certain number of examples are necessary to fix specific characters. But there are a number of other fishes which cannot be perfectly understood without a much greater number of examples. Such are those which undergo with age changes so considerable that the stages of development have been described as different genera, or those which exhibit most extraordinary sexual differences, or are so subject to variation as to have given rise to the creation of numerous nominal species, or those which have a wide geographical distribution.

He acknowledged the '... many imperfections of this work' but implied that the greatest source of error lay outside his control. Thus, 'In the present work 6843 species are regarded as well established and described, whilst 1682 others are doubtful and referred to by name only.' The doubtful ones were those described by other authors! Günther noted in the *Zoological Record* (1872a) that only about 5% of the species he received from Australia at this time were undescribed. He proceeded to describe these but, at the present day, less than half of his 'new' Australian species described in the *Catalogue* or elsewhere are regarded as valid.

The inadequacy of new collections, in the absence of Richardson's influence, was a source of anxiety:

> ... the National Collection has been dependent on the booty of private collectors. One of the chief sources, the navy, which assisted Sir J. Richardson in bringing together the magnificent collection of Haslar Hospital, now transferred to the British Museum, has failed entirely since the great English ichthyologist withdrew from active life.

Günther went on to draw attention to the energy with which fish collections were being amassed by the expeditions of Austria, Prussia, Russia and the United States; and by the collectors working for 'Messrs Godeffroy of Hamburgh' (see page 94) and Pieter Bleeker (see page 83) '... who has made us acquainted with the astonishing variety of fishes in the East-Indian archipelago ...' Günther also bemoaned the lack of teaching in ichthyology, contrasting British apathy in this area with the universities of Europe and the United States and saying, reasonably enough, 'Of course this state of things is in immediate connexion with the defective system of scientific education.'

Short papers

Günther's short papers on fishes from Australia, mostly published in the *Annals and Magazine of Natural History* and the *Proceedings of the Zoological Society of London* between 1861 and 1878, were based on fishes sent from Australia to the British Museum. Most of the specimens forming the subject of these papers were collected in Queensland (particularly Cape York, Bowen and Mackay, with a few from Moreton Bay), New South Wales (most from Port Jackson), Victoria (around Hobson's Bay), Tasmania (around Hobart), South Australia (around Adelaide and a few from Port Lincoln) and Western Australia (Fremantle, and some from the north-west). A summary of Günther's Australian collections and collectors follows, with some examples of the fishes they found.

Queensland

Günther (1864a) published the description of a remarkable fish which had been taken in the Burdekin River, Queensland and preserved as a dried skin in the Australian Museum. Krefft sent photographs of the fish and Günther later located a specimen in the British Museum: 'It had been sent by the unfortunate Mr. Gilbert (see page 113) as a specimen collected by Leichardt.' This was the first record of the ancient lineage of 'Bonytongues' (Osteoglossidae) from Australia: the Southern Saratoga, *Scleropages leichhardti*.

Krefft contributed specimens that he received at the Australian Museum from Queensland and New South Wales. For instance, Günther (1864b) reported the Barred Grunter, *Therapon* [now *Amniataba*] *percoides* from 'the Fitzroy River, near Rockhampton, in Queensland.'

Among the small collections Günther received from north Queensland was one '... brought by Hr. Dämel from Cape York ...' from which the curious Spotted Eel Blenny, *Notograptus guttatus* was described (Günther 1867b). This elegant little fish belongs to a genus endemic to northern Australia and New Guinea. The spectacular Harlequin Tuskfish, *Xiphochilus* [now *Choerodon*] *fasciatus* was also described from Cape York (Günther 1867a) (see colour plate 9, upper).

The *Alert* was surveying north Queensland waters in 1881 for the British Government and an assessment of the zoology of the region was one purpose of the cruise. The fishes were sent to Günther who published his findings in 1884, including the description of several 'new' species, of which only the Swallowtail Dart, *Trachynotus* [now *Trachinotus*] *coppingeri* is now regarded as valid. This fish was named for R.W. Coppinger, MD, the navy surgeon of the *Alert* who had, among his duties, that of zoologist:

> ... the following pages are by themselves a lasting testimony to the great service rendered by Dr. Coppinger to the National Museum and to the cause of science. I must not allow this opportunity to pass without duly acknowledging the energy and skill with which he performed his work. The collections were

> made with singular judgment, the specimens (many of them most fragile and delicate) preserved, labelled, and packed with the greatest care; and, beside, full lists were prepared by him giving additional, and in many cases most valuable, information. (Günther 1884).

Günther's praise was not disinterested: he wanted naturalists to take part in all such navy ventures.

New South Wales

Gerard Krefft was Günther's best source of fishes from New South Wales. They had been on good terms for years and sustained a friendly correspondence. Ichthyology was peripheral to Krefft's own interests and, as it was not one of his areas of expertise, he sent a great number of specimens to Günther. Many were described in the *Catalogue*, others in the short papers.

Some fishes collected by Eduard Daemel (see page 95) around Port Jackson came to Günther's notice when they were received from the Godeffroy Museum. Among these he found and described the familiar Black Rockcod, *Serranus Dämelii* [now *Epinephelus daemelii*] of Sydney (Günther 1876a).

In the New South Wales Court of the Fisheries Exhibition of 1883 was a large stuffed fish from Port Jackson which E.P. Ramsay had identified as *Pachymetopon grande* (a South African endemic sparid described by Günther from specimens taken at the Cape of Good Hope). Günther examined Ramsay's specimen after the exhibition and described it later (Günther 1886) as the Silver Drummer, *Pimelepterus* [now *Kyphosus*] *sydneyanus.*

Victoria

'A collection of fishes from Victoria, sent to the International Exhibition, and procured for the British Museum' was described by Günther (1863). He was hampered by the mode of preservation: all were stuffed skins. However, he proceeded to describe several new species from this collection, including the Estuary Perch, *Lates* [now *Macquaria*] *colonorum,* Bearded Rock Cod, *Pseudophycis barbata* and a demoiselle known locally as Scalyfin, *Glyphidodon* [now *Parma*] *victoriae.*

Tasmania

Morton Allport (see page 110) was Günther's most active Tasmanian donor. Günther (1872b) described the Longfin Pike, *Lanioperca mordax* [synonymous with *Dinolestes lewini*] (see page 158) and the Banded Morwong, *Chilodactylus allporti* [synonym of *Cheilodactylus spectabilis* Hutton, 1872]; a few years later he described further species from Allport: Rosy Perch, *Callanthias allporti*; Red Velvetfish, *Holoxenus cutaneus* [synonym of *Gnathanacanthus goetzeei* Bleeker, 1855] (see page 85); Barred Grubfish, *Percis* [now *Parapercis*] *allporti*; and Shorthead Worm Eel, *Muraenichthys* [now *Scolecenchelys*] *breviceps* (Günther 1876a).

Using specimens from Allport, Günther (1876b) published a study of the urogenital papilla of a Weedfish (clinid), concluding 'It is not improbable that this fish is viviparous, and, consequently, that copulation is necessary for the fecundation of the ova ...' He refers to the fish as *Clinus despicillatus*, giving its length as 14.5 inches (36 cm); judging from the

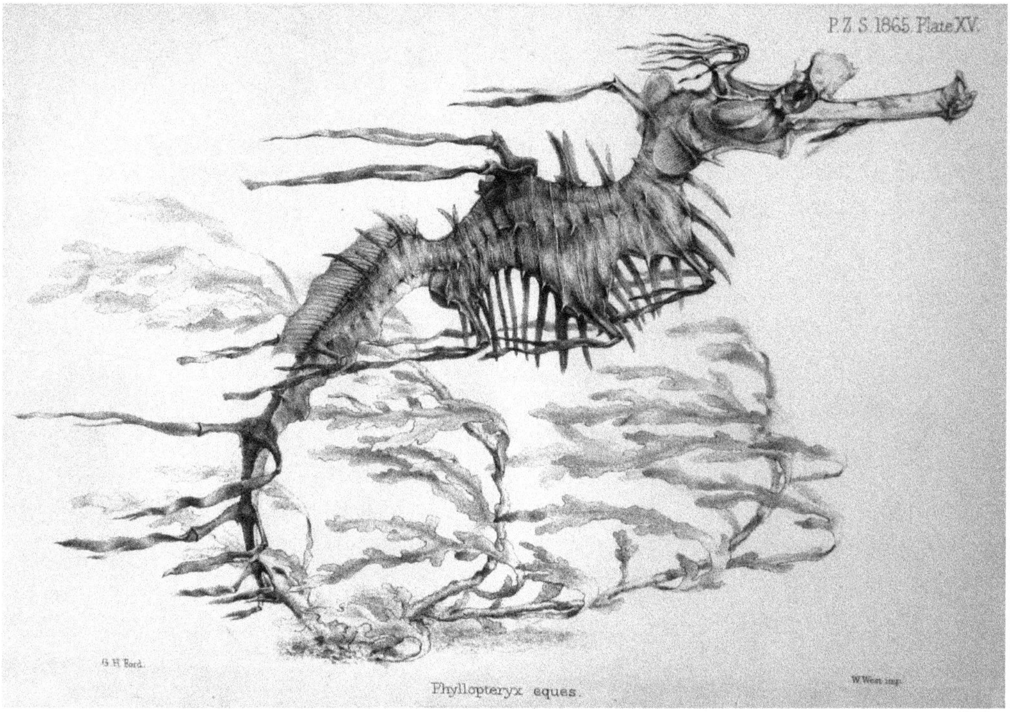

The Leafy Seadragon, *Phyllopteryx* [now *Phycodurus*] *eques*, drawn by Ford in its algal habitat. From Günther 1865, courtesy of the Museum Board of South Australia.

size, he may have had *Heteroclinus johnstoni*, a large clinid that was not formally described until 1886 (see page 410). Valenciennes (1836) had drawn attention to viviparity in a Tasmanian clinid (see page 41) but Günther apparently overlooked this.

South Australia

Among the earliest specimens to reach Günther from South Australia was a magnificent Leafy Seadragon, *Phycodurus eques* found at Port Lincoln and sent to the British Museum by George French Angas (see page 100). Günther's description (1865b) was illustrated by G.H. Ford who provided beautiful drawings of this fish, and the related Common Seadragon, *Phyllopteryx taeniolatus* (see colour plate 10, upper).

In 1886, Günther noted that the British Museum had received a collection of fish from Adelaide. This included two species of berycid fishes, one of which, the Bight Redfish, Günther described as *Beryx* [now *Centroberyx*] *gerrardi*.

Western Australia

Günther had few western Australian sources but received a small collection from Mr Duboulay (possibly Mr Francis du Boulay, a beetle collector and natural history artist) who later sent specimens from New South Wales to Castelnau. This collection included a few fish from the north-west, and Günther (1867b) named the Scribbled Angelfish, *Holacanthus* [now *Chaetodontoplus*] *duboulayi*.

The Australian Lungfish, *Ceratodus* [now *Neoceratodus*] *forsteri* from Günther's description of a specimen sent by Krefft. From Günther 1871. Courtesy Museum Board of South Australia.

The Australian Lungfish

Gerard Krefft had described the Australian Lungfish in 1870 (see page 102). He communicated with Günther who introduced his own paper:

> At the beginning of last year my valued correspondent, Mr. Gerard Krefft, Curator of the Australian Museum, Sydney, informed me of the discovery in Queensland of a large *Lepidosiren*-like animal[7], of which he enclosed a photograph, and which he assigned to *Ceratodus*, a genus of fishes known from fossil teeth only, naming the species after its discoverer, the Hon. William Forster, *Ceratodus forsteri*. (Günther 1871a)

Günther carried out a detailed study and wrote a long report, describing the anatomy of the fish and discussing its placement within the classification of the day. In the same year (Günther 1871b) he described a supposed second species of Lungfish (*Ceratodus miolepis*) from the Dawson River in Queensland, but he had underestimated the natural variation within the species.

Günther's *Study of Fishes*

Günther had been asked to provide a section on ichthyology for the *Encyclopaedia Britannica* to replace that of Richardson (1856). He obliged, eventually expanding this essay into his *Introduction to the Study of Fishes* (1880). The latter work began with a history of ichthyology, providing an interesting summary that was largely derived from the much larger section in the first volume of Cuvier's *Histoire Naturelle des Poissons* (1828). This was followed by chapters on the anatomy, physiology and biogeography of fishes. The 'systematic or descriptive part', comprising the last 400 pages, was an account

7 *Lepidosiren paradoxa* is the South American Lungfish, described in 1837 by Leopold Fitzinger, an Austrian zoologist.

of the families and genera of fishes with definitions and some small illustrations, many of the latter taken directly from Richardson (1856).

Pungent criticisms of fellow zoologists were part of Günther's attempt to dominate the ichthyology of his day. Sometimes he was met with ineffectual grumbles of the kind Castelnau (1872a) published; but he had more effective adversaries. His early criticism of the work of Francis Day provoked a long-standing battle which included Günther imposing restrictions on Day's access to the British Museum collections; Günther intended that his *Catalogue* would be illustrated by the leading scientific artist of the day, George Ford (see page 72), who had illustrated his earlier works (Whitehead and Talwar 1976). Day forestalled him by engaging Ford to illustrate his *Fishes of India.* Bleeker also came in for some criticism:

> His desire of giving a new name to every individual, to every small assemblage of species wherever practicable, or of changing an old name, detracts not a little from the satisfaction with which his works would be used otherwise. (Günther 1880a).

Günther reserved some of his most vigorous criticism for Theodore Gill who had said that the eight volume catalogue reflected moderately well the state of the science at that period but that the *Introduction to the Study of Fishes* was out of date and inaccurate: 'The errors commence on the first page and flow in an almost uninterrupted, but varying, stream to the end of the work.' (Gill 1881). Günther criticised Gill's work in the *Zoological Record* (see below) and elsewhere.

The biogeography section of the *Study of Fishes* included the first useful analysis of the distribution of Australian fishes. In the account of the shore fishes of the 'South Australian District' of the 'Southern Temperate Zone' were some observations on the character of the fauna. Günther took this district to include New Zealand, Tasmania and the east coast as far as Sydney and noted the '... influx of tropical forms on the east coast', together with the close relationship between the fishes of the southern half of Australia and those of New Zealand. He remarked on the 'extraordinary development of *Monacanthus*' (Leatherjackets) on the coast of South Australia (intending southern Australia) and the greater presence of 'gadoids' (cod-like fish) in New Zealand. Günther recognised that this work was limited by the paucity of collections, particularly to the west: '... the western half of the south coast of Australia is still almost a *terra incognita.*' He did not realise that large numbers of species remained undiscovered in more populous regions, referring to 'the thorough manner in which the ichthyology of New South Wales and New Zealand has been explored.'

Zoological Record

Industrious and accurate, Günther was essentially a museum worker, with not much apparent appreciation of the importance of observations in the field. This led him into many conflicts with ichthyologists working in remote areas. He was a great organiser and his establishment in 1865 of the *Record of zoological literature*, continued by the Zoological Society after 1870 as the *Zoological Record,* was a contribution to taxonomic order almost comparable with that of Linnaeus. However, although this annual publication was intended to record the current literature rather than to assess it, Günther used it for 'preaching scientific taxonomy from his newly created pulpit' (Whitehead and Talwar 1976). He referred to himself as 'the Recorder' and variously praised, patronised or sniped at his colleagues. Theodore Gill (see page 157) published *Synopsis of the Pleuronectoids of the Eastern Coast of North America* (1864) and Günther commented '... Mr. Gill would much advance Ichthyology by giving us serviceable descriptions, instead of limiting himself to synoptical tables with minute pseudo-generic subdivisions.' (Günther 1864).

The mode of formation of scientific names caused Günther to protest from time to time. J.G. Cooper created the genus *Gillichthys* for a Goby in *On new genera and species of Californian fishes* (Cooper 1864) and Günther commented 'We hope Dr. Cooper will not continue to use this barbarous name, which is not worthy of the author or of the man whom he intended to honour by it, but substitute for it a form more in accordance with the rules of nomenclature, like Gillia.' The comment was ignored and *Gillichthys* stands.

Museum Godeffroy, Amalie Dietrich, Andrew Garrett, Eduard Daemel

The Godeffroy family had been Huguenots for a century when the religious intolerance of Louis XIV caused them to flee from France to the Hansa city of Hamburg in 1737 (Spoehr 1963). The Godeffroys established a trading firm that thrived for several generations. Johann Cesar Godeffroy (1813–1885), the sixth of that name, owned a large Hamburg shipyard. He developed a great interest in natural history, spurred by the curiosities brought back through his overseas trade. In the 1850s his ships were trading in the Pacific and their captains had instructions to collect natural history objects. Their collections were brought to Hamburg where Godeffroy commissioned a scientist from Zurich, Dr E. Graeffe, to establish a natural history museum in about 1860. He employed a curator, J. Schmelz, who held the position for 20 years, and welcomed scientists from Germany and elsewhere who wished to study his specimens.

Godeffroy financed collectors, some of whom stayed for years in the Pacific amassing specimens. Dr Graeffe, one of several who had contracts with Godeffroy, spent ten years collecting in Samoa and Tahiti. The period of intensive collecting activity was approximately 1860 to 1875; thereafter, the disturbances of the Franco-Prussian war caused economic problems and resulted in the bankruptcy of the firm.

Among Godeffroy's most productive collectors was Amalie Dietrich (1821–1891), a German naturalist, who collected in the then sparsely settled Queensland from 1863 to 1872. She travelled by horse and cart, with all her collecting gear, preserving a wide range of flora, fauna and aboriginal artefacts (and skeletons). Dietrich worked from Moreton Bay to Bowen, shipping specimens to Hamburg where Godeffroy had experts assess and report on them (Sumner 1993; Lowie 1968). Günther published descriptions of new species from these collections in the *Journal of the Museum Godeffroy* in 1873.

Andrew Garrett (1823–1887) was a self-taught naturalist and artist from New York who specialised in molluscs and fishes. He spent many years collecting and painting fishes in various parts of the Pacific; he collected for Godeffroy from 1873 to 1875, sending both specimens and drawings to Hamburg. Günther provided descriptions of the species in the collection in *Andrew Garrett's Fische der Südsee* (1873–1910) and the work was illustrated by plates taken from Garrett's water colours. Garrett had described a few of these fishes himself in the *Proceedings of the Californian Academy of Science* in the 1860s.

Eduard Daemel (1821–1885) was an entomologist and collector of natural history specimens. He operated in eastern Australia from 1867 to 1874, sending specimens to Australian buyers and to the Godeffroy Museum. Some fishes sent by Daemel from Cape York were received at the British Museum and duly described by Günther.

Challenger expedition

The corvette *Challenger* set out on a cruise of oceanographic and biological discovery in 1872. Professor Wyville Thomson was leader of the team of scientists; naturalists aboard were Henry Moseley and John Murray. The expedition arrived back in England in 1876 and Günther studied the fish collections. He described a few new species from the *Challenger* collections in 1878, including some deep sea Dragonfish (Stomiidae), and began publishing the main *Challenger* work with his *Report on the Shore Fishes collected by H.M.S. Challenger during the years 1873–1876* (Günther 1880b). Subsequently were published his *Report on the Deep-Sea Fishes* (1887) and *Report on the Pelagic Fishes* (1889).

Much of the deepsea work of the *Challenger* was relevant to the Australian fauna although little of it was performed within Australian waters. The report included accounts and excellent illustrations of many deepsea fishes which inhabit Australian seas. The *Challenger* collected shore fishes in New South Wales at Twofold Bay and Sydney, including several previously undescribed Australian species. In the *Report on the Pelagic Fishes* Günther described a Flyingfish, *Cypselurus* [now *Exocoetus*] *naresii* that was collected off Cape York, Queensland; he named it for Captain George Nares, R.N., F.R.S. who commanded the *Challenger*.

A party from the *Challenger*, attracted primarily by the possibility of discovering primitive fishes, conducted a brief excursion to the Mary River. Wyville Thomson explained (Günther 1880b):

> Those [rivers] in which *Ceratodus* [Australian Lungfish] had lately been discovered had the greatest interest for us, for we hoped that, besides getting a good supply of *Ceratodus* in all stages, we might by effective netting and other means find some additional forms of the Dipnoi [Lungfish].

While their vessel remained in Sydney a party consisting of Lieutenant Aldrich, John Murray, Wyville Thomson 'and a couple of bluejackets' sailed to Brisbane with 'a stock of trammel nets, lines, and other fishing appliances, a box of dynamite cartridges, fowling pieces, and collecting gear of all kinds'. They saw the sights of Brisbane, took a coastal steamer to Maryborough, and travelled 'in a couple of buggies' to a site on the Mary River 'about twenty miles above Maryburgh.' Here they settled down to collect in earnest:

> For about ten days the river was fished day and night with net and rod, and fishes of several species were taken in large numbers, but we found no trace of *Ceratodus* or of any allied form. We had taken with us a number of powerful dynamite cartridges, and these were thrown, with a few feet of Bickford's fuse

The Crested Flounder, *Lophonectes gallus* was trawled off Port Jackson. From Günther 1880.

Fish trawled off Twofold Bay, New South Wales: Peacock Skate, *Raja* [now *Pavoraja*] *nitida* described from this young male, just 20 cm in length; Spiny pipehorse, *Solenognathus fasciatus* [synonym of *Solegnathus spinosissimus*]. From Günther 1880.

> attached, into the deeper pools, and in a minute or so a shock like a blow from a heavy wooden mallet was felt on the bottom of the boat, one could see a slight rise on the surface of the water, and perhaps a hundred fishes of different sorts and sizes rose to the surface and floated on their backs or sides.

No Lungfish were found among the bombed fish but the Lieutenant caught one on a baited line and two other specimens were donated by aborigines. The fishes secured on this expedition, and some sent by a cooperative resident at a later date, allowed Günther to give a brief account of the fishes of the Mary River including the description of a new Glassfish, *Ambassis marianus* Günther, 1880.

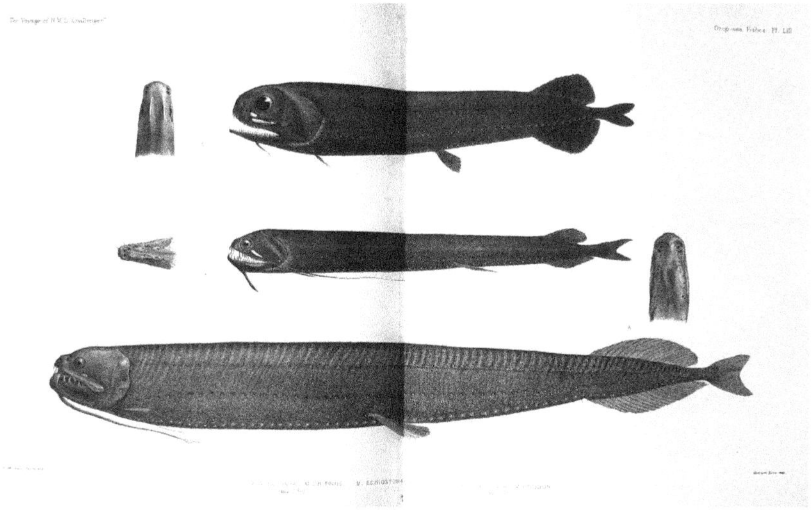

Deepsea stomiid fishes described by Günther (1878) in his preliminary *Challenger* report and illustrated in the final report: *Opostomias micripnus* (upper, from south of Australia); *Echiostoma barbatum* (middle, from Madeira and since found in Australian seas); *Pachystomias microdon* (lower, from north-west Australia). From Günther 1887.

The biologists from *Challenger* sampled the ocean between Australia and New Guinea, taking deep water forms in the Arafura Sea and some fishes characteristic of northern Australian waters. They also used a seine at Albany Island in Torres Strait, adding some shore fishes of the region to their collections.

Günther's contribution

The standard of Günther's own work was very high although the sheer volume of specimens with which he dealt compromised that standard at times. At one time or another he made most of the errors that he condemned in other workers: some of his descriptions were insufficiently detailed; inadequate appreciation of variation with growth and sex led him to describe the same species on more than one occasion; his familiarity with the literature was extensive but he was not infrequently at fault. Günther described a significant proportion of the Australian fish fauna: approximately 230 species, about one third of which were described from Australian specimens, and over half of these are still valid.

Günther's publications were of great importance to the development of ichthyology in Australia, dealing with collections and describing new species, offering criticism of the standard of work produced by Australian zoologists and providing publications that were highly valued by those workers. The *Catalogue* and the *Study of Fishes* were the chief tools available to the ichthyologists of the mid to late 1800s. However, Günther's resistance to new concepts in the classification of fishes marred his otherwise benign influence on the work of the ichthyologists of his time. The *Zoological Record* was perhaps the most important part of Günther's legacy, making known as it did (and does) the zoological works of all authors.

Fish collection and ichthyology in Australia (1800 to 1870)

The overland expeditions of the first half of the 19th century sought geographical knowledge, with an emphasis on finding waterways and assessing pastoral and mining possibilities. However, there was often a subsidiary interest in natural history and useful collections were made by some explorers. There were also professional collectors, such as John Gilbert and Amalie Dietrich, who were paid to collect a wide range of biological specimens. In the absence of effective Australian institutions, most specimens prior to about 1870 were sent directly to England and Europe. 'Natural history specimens and Aboriginal objects from expeditions by Grey, Eyre and Mitchell were sent directly to the British Museum for analysis and description.' (Jones 1996). With the gradual development of Australian museums, specimens were accumulated, exhibited and studied in these institutions. However, the British Museum remained the benchmark for 'recruitment, publication and even scientific authority' for many years (Kohlstedt 1983). The activities of collectors and zoologists are considered below state by state.

New South Wales

A 'Colonial Museum' was begun in Sydney in about 1828 at the instigation of Governor Darling and through the efforts of the Colonial Secretary, Alexander Macleay, and others. William Holmes had arrived in Sydney in 1827 and was commissioned to collect and arrange zoological specimens for a proposed museum; he was appointed to the newly-created position of 'Colonial Zoologist' and, aided by the Reverend Charles Wilton, specimens were assembled to create the museum. Holmes died at Moreton Bay in 1831. The museum was housed in various unsuitable rooms around Sydney until a new building, dedicated as a museum, was completed in 1849 (Rainbow 1922; Strahan 1979; Etheridge 1916 and 1919).

Bennett, Angas and Wall

The Australian Museum was first known by its present name in 1836, when George Bennett (1804–1893) was appointed zoologist; he was later known variously as the director, superintendent or curator. Bennett was born in Plymouth, England and took the first of his many sea voyages at the age of 15. He studied medicine and obtained membership of the Royal College of Surgeons in 1828. Bennett and Richard Owen had been fellow students at the Royal College of Surgeons, and Owen had become the most influential comparative anatomist in England at this time.

Bennett is known mostly for his accounts of natural history: *The Wanderings in New South Wales, Batavia, Pedir Coast, Singapore and China ...* (1834) and *The Gatherings of a Naturalist in Australasia ...* (1860) were popular texts in their day and contained a fund of information about the Australian environment. Bennett's natural history notes now appear

George Bennett. Courtesy of the Australian Museum Archives.

a little naive although he appreciated the plants and animals that he observed and generally limited his scientific publications to subjects with which he was most familiar.

Bennett introduced the *Gatherings* with his account of some pelagic animals he had encountered, including the description of a supposedly new species of Flyingfish. He named the fish *Exocoetus solitarius* 'from not being seen in large flocks like the others' but it has been synonymised with *Cheilopogon cyanopterus* (Valenciennes, 1847). He also reported 'A fish which excited my very great surprise when I first captured it in the towing-net...' This was a tiny, luminous shark, 'perfectly black' and just '5½ inches' [14 cm] long. He thought it '... not a little singular that my brother, the late Mr. F.D. Bennett (see page 45), obtained a specimen of this fish almost in the same latitude ...' His brother had named the shark *Squalus fulgens* in 1840; it is synonymous with *Isistius brasiliensis* (Quoy and Gaimard, 1824).

In 1829, Bennett made his first visit to Australia and he returned in 1832, travelling and collecting extensively in New South Wales. He made a third trip to Australia in 1835, in the same year writing a *Report on the Epidemic Catarrh or Influenza Prevailing among the Sheep in the Colony of New South Wales in the Year 1835*. He established a successful medical practice in Sydney and began his long association with the Australian Museum. He wrote and sent specimens of animals and fossils to Richard Owen, and some to John Gould. The inequalities of Bennett's relationship with Owen have been the subject of a study (Newland 1991) demonstrating the dependent status of early scientists in Australia. Such relationships were open to abuse such that Owen, 'the London based curator [of the Hunterian Museum at the Royal College of Surgeons] could use Bennett's material and writings to enhance his own scientific reputation.' (op. cit.) Bennett, who was expected by Owen to function primarily as a discriminating collector, kept Owen informed of his activities: 'I have just sent the bird stuffer of the Museum to collect fish and shells with the soft parts about Botany Bay and Port Jackson, the result I have no doubt will be some valuable specimens.' (Bennett to Owen, 1834, quoted by Whitley, 1963) The 'bird stuffer' was John Roach, a convict, of whom Bennett also wrote 'The present collector and bird stuffer I sent with Major Mitchell and hope he will not get speared by the natives ...' (Bennett to Owen, 1836, quoted by Whitley 1963)

Bennett (1837) prepared a *Catalogue of the specimens of Natural History and Miscellaneous Curiosities deposited in the Australian Museum*. He was functioning as the secretary and zoologist of the Museum, the catalogue being prefaced:

> The Committee of the Australian Museum wish it to be understood that the arrangement of the following Catalogue is wholly due to the labours of the Secretary, who alone is responsible for the names and synonyms attached to the specimens.

The fish section was brief and included the 'snout of the Saw-fish', a 'Skull of the Sword-fish' and 'An Enormous Head of the River Cod-fish of the Colonists.'

During his time in charge of the museum Bennett continued to work as a government surgeon; Whitley (1963) mentioned that his duties at the museum were interspersed with attendances at public hangings, of which there were apparently about six per week.

George French Angas (1822–1886), English naturalist and artist, travelled in Australia and New Zealand between 1843 and 1845; he returned to Australia in 1850 and was secretary at the Australian Museum from 1853 to 1860. Angas published many articles on natural history, particularly on molluscs, and returned to London in 1863 (ADB). George Bennett published notes and a drawing of a juvenile Damselfish given to him, together with a specimen in spirits, by Angas. The Damselfish was spectacularly coloured and both Angas and Bennett thought it an undescribed species. Bennett consulted Cuvier's *Histoire Naturelle des Poissons* in London and decided that the fish was *Glyphisodon biocellatus* [now *Chrysiptera biocellata*]. He made it the subject of a presentation to the Zoological Society of London (1859), recommending it as a potential aquarium fish. Bennett quoted Angas:

> This brilliant and elegant little fish is found in the pools amongst rocks at low spring tides, both on the outer coast and in several localities inside the harbour of Port Jackson ... The extreme brilliancy of the colours, gold and azure, as the little creatures dart in and out amongst the cavities of the rocks, reminds one of jewels flashing in the sunlight ... They generally make their appearance on the coast about November, and remain till May; during the winter months I have looked for them in vain. The usual size varies from 1 to 2 inches in length ... it is probable that 4 inches is the largest size they acquire.

George French Angas. Courtesy of the Australian Museum archives.

Angas was a meticulous artist, schooled in scientific illustration: the fish he represented appears to be a juvenile White Ear, *Parma microlepis* (see colour plate 9, lower).

Among his other writings, Bennett produced short papers on some fishes and sharks of New South Wales. In the *Medical Gazette*, Bennett (1871) made a clinical report on a case of poisoning by the Toadfish, *Tetractenos hamiltoni*: 'Some public interest having been excited by the recent death

(January 18th, 1871) of two boys from imprudently eating the 'Toad Fish,' I consider a few observations upon it, together with a description and figure of the fish, may be of interest ...' He described the signs and symptoms of the poisoning and advised the use of an emetic.

Bennett seemed to be the authority on fishes during his residence in Sydney. He informed himself on those aspects of natural history that came to his attention, including the fishes of Port Jackson.

William Sheridan Wall (1814–1876) had studied anatomy in Dublin and, with his brother Thomas, migrated to Sydney in about 1840. He became the 'preserver' at the museum and was curator from about 1842 to 1858 (Etheridge 1916). Around 1853 he wrote articles for the *Illustrated Sydney News*, including drawings and descriptions of some of the more spectacular fish of Port Jackson (Whitley 1963). Thomas Wall also contributed specimens to the museum; he died on Kennedy's overland expedition to Cape York.

Mitchell

The Surveyor-General of New South Wales, Major Thomas Mitchell (1792–1855), led expeditions through the inland of eastern Australia and published an account of his experiences in 1838. The fate of the natural history specimens derived from these journeys is of special interest. The mammals and reptiles were described by 'Mr Ogilby' (father of J.D. Ogilby) of the British Museum; the fossils were sent to Richard Owen, then Director of the Natural History Department at the British Museum; but Mitchell himself described the plants and fishes. The latter were clearly illustrated in the original publication and the descriptions, although brief, refer unmistakably to distinct species and are still accepted as the original descriptions.

Mitchell makes it clear, in his entry for 12 December 1831, that the large 'cod' which he describes from the Peel River is already known to 'the colonists':

> At an early hour this morning, one of our men caught a fish, which weighed eighteen pounds; but, according to the natives, this was no uncommon size. These fishes are most erroneously called cod by the colonists, although they certainly very much resemble cod in taste. The flakes are firmer than sea cod, and equally white, the fish affording a very light and palatable food.

This was the Murray Cod, *Acerina* [now *Maccullochella*] *peelii*, the largest fish of Australian fresh waters. Mitchell cautioned that his fish might prove to be the same species as the Trout Cod (*Maccullochella macquariensis*) described by Cuvier and Valenciennes in 1829. For many years it was thought that there was only one species but the nomenclature has been clarified of recent times and Mitchell's species stands.

Five weeks later, fishing in a deep lagoon that Mitchell decided was supplied by overflow from the Namoi River, his men took another interesting species:

> We caught some small fish, two of them being of a rather singular kind, resembling an eel in the head and shape of tail, although as short in proportion to their thickness as most other kinds of fish ... On my return to the camp in

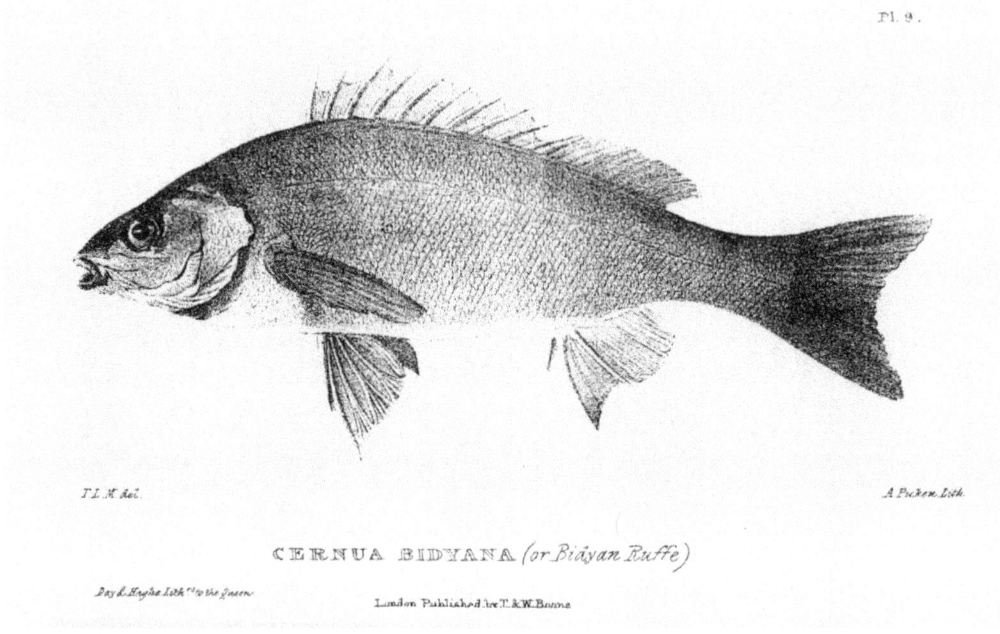

Silver Perch, *Cernua* [now *Bidyanus*] *bidyanus*, one of several common freshwater fishes described by Thomas Mitchell. From Mitchell 1838.

> the evening, I made a drawing of the eel-fish, which we had caught earlier in the day.

In the formal description of this fish he mentions that the aboriginal name is 'tandan' and he latinises this as tandanus in the specific epithet. This is the Freshwater Catfish, *Plotosus* [now *Tandanus*] *tandanus.*

In the entry for 24 January 1832 Mitchell mentions a third species, '... a firm but coarse-tasted fish, having strong scales; this made a groaning noise when on the hook.' Once again he gives the aboriginal name, 'bidyan', and uses it in the specific name which he gives the fish. He suggests 'growler' as the common name. The fish was *Bidyanus bidyanus,* of the family Terapontidae, a group well known for their ability to make sounds through sonic muscles attached to the swim-bladder. Those from Australian fresh waters are generally known as 'grunters' and a salt water relative is the 'trumpeter'.

Krefft

Johann Ludwig Gerard Krefft (1830–1881) was born in the Duchy of Brunswick. He attended college in Brunswick up to 1845 and then worked in a mercantile firm. In 1850 he went to the United States and came to the Victorian goldfields in 1852. He travelled with Wilhelm Blandowski (see page 106) on an inland expedition, exploring the Lower Murray and Darling Rivers, and together they made a large natural history collection (ADB; Paxton and McGrouther 1997). He worked on the collection, cataloguing it for the National Museum of Victoria.

Gerard Krefft. Courtesy of the Australian Museum Archives.

In 1860, Krefft was appointed Assistant Curator of the Australian Museum, then Curator in 1864. He was an enthusiastic collector and published his discoveries promptly.

Scientific isolation was one of the marked features of many of the early zoologists in Australia but Krefft kept in touch with recent advances through the literature and through correspondence with such figures as Charles Darwin, Albert Günther, Louis Agassiz and Richard Owen. Unlike many zoologists in Australia and elsewhere he accepted Darwin's evolutionary theory. Among Krefft's more important publications were *Snakes of Australia* (1869) and *Mammals of Australia* (1871).

When an enormous Manta Ray was found at Sydney in 1868, Krefft reported and named it in the *Illustrated Sydney News* as *Ceratoptera alfredi*; it proved to be a synonym of *Manta birostris* (Walbaum, 1792). Krefft sent most of his marine fish specimens for identification to Albert Günther at the British Museum (see page 87). Günther wrote to Krefft in May 1861:

> ... If you know or ascertain that the Trustees or Curators of your Museum are desirous, of having named your species of Reptiles and Fishes, you may inform them that I am quite willing to do it, if they will send them in convenient lots. This may be done in various ways: either I return the whole of the specimens after having named them, or you send me such specimens as you can spare from your collection, each with a certain number, & I return the list of names corresponding to the numbers. In this case, those specimens only could be procured for the British Museum, which either are desiderata, or come from a new locality ...
>
> Finding that the Australian Fish-fauna is by no means worked out, I am very anxious to collect as many marine & freshwater specimens as possible. You would greatly advance the object of my work if you could direct a part of your attention to this subject. I prefer so much specimens in spirit to those which are dried or skinned ... double price is paid for the former by the British Museum. I could at least take three specimens of each species, & should like very much to hear from you in your next letter whether you or one of your acquaintances could undertake to collect fishes. If separate copies of my papers are of any use to you I shall be glad to send them. (AMA).

Clearly Günther was keen to secure a source of Australian fish and Krefft responded well, supplying specimens for the whole period of his employment at the Museum. Their correspondence was affable though chauvinistic, Krefft writing to Günther in March 1869:

> ... I think the marriage epidemic like the tidal waves has gone round the world and all the wise fellows bordering on the age of forty have been hooked at last; I am glad to hear that I stand not alone in this age of folly and that such an old fox as Günther has been caught at last.
>
> I confess I have gone and done it ... nobody found me out for a good while as I was supposed to be the only person living in the Museum. (AMA).

His experiences with Blandowski may have encouraged Krefft to undertake the description of some freshwater fishes. He named four new species of gudgeon from the streams of New South Wales (Krefft 1864) and a few other freshwater fishes from New South Wales and South Australia (Krefft 1868a). All four of Krefft's Gudgeons are still regarded as valid; the fishes of the 1868 paper had previously been described. In listing the vertebrates of Tasmania, Krefft (1868b) gave a brief account of Tasmanian freshwater fishes; he listed only four species: the Eel, Grayling (*Prototroctes maraena*) and two species of *Galaxias*. He gave no account of marine fishes.

Krefft's many articles for the general public, published in the *Sydney Mail* and the *Sydney Morning Herald*, included the original description of the Australian Lungfish in 1870. This was the greatest of Krefft's discoveries. The presence of a large, unidentified fish in south-east Queensland, in the Mary and Burnett Rivers, had been known to local settlers for at least 20 years prior to its formal description. It was caught and eaten, and known as the Burnett Salmon, due to its pink flesh. In 1870 the Minister for Lands in New South Wales, Mr William Forster, brought the fish to Krefft's attention.

Fossilised teeth of a peculiar 'coxcomb' form were well known from Triassic deposits in Europe; they were thought to be remains of an extinct shark and had been named *Ceratodus* by Louis Agassiz (1838). Krefft found that the Burnett fish possessed these teeth, noted its lung and its large thick scales, and identified it as a species of *Ceratodus*.

Krefft's formal description of the fish as *Ceratodus forsteri*, named for William Forster, appeared in the *Proceedings of the Zoological Society of London* (1870b) but he notified the local populace of his discovery in a letter to the editor of the *Sydney Morning Herald*:

> ... Fancy a triton or newt three feet in length, covered with huge scales the size of a crown piece, with four scaly flappers or fins instead of legs, with four plates of horny teeth, two in each jaw, which resemble segments of a cog-wheel in a horizontal position – with a tail like a Siluroid or Catfish, with small eyes and a partly cartlilaginous skeleton, and the modern *Ceratodus* is complete.
> (Krefft, 1870a).

This letter was published some ten months earlier than the London journal and so is cited as the original description of the fish.

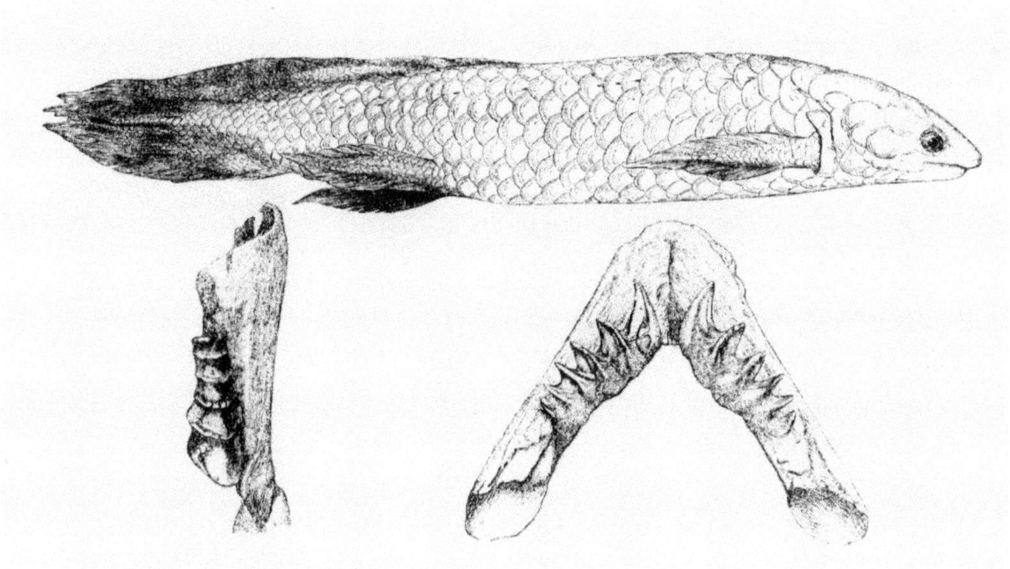

The Australian Lungfish, *Neoceratodus forsteri*. From Krefft 1870.

Krefft obtained extra specimens of Lungfish for his correspondents in London:

> The Principal Librarian
> British Museum
> London
>
> Sir
> I have the honour to inform you by direction of the Board of Trustees of the Australian Museum that a case containing 5 Ceratodus Forsteri and a collection of other fishes from the Burnet River in Queensland has been dispatched to you by the outgoing mail steamer. It is the wish of the Trustees that the specimens of Ceratodus should be handed over to Dr Günther who will select such specimens as he may require for the British Museum Collection and will exchange the remainder for the benefit of this institution. The case contains also a Myrmecobius [the Numbat, *Myrmecobius fasciatus*, a Western Australian marsupial] in spirits which you will be good enough to hand to Professor Owen who has applied to the Trustees for a specimen of this rare animal.
>
> I have the honour to be Sir
> Your most obedient servant
> Gerard Krefft
> Curator

Günther acknowledged receipt of the fish (AMA); he had already received, dissected and published a description of an earlier specimen.

From this time, there was an active trade in Australian Lungfish. They initially fetched about ten pounds apiece and were eagerly sought by museum curators in Australia and overseas, as purchases and through exchanges. Very large numbers of the fish were caught for sale and the price fell: Whitley (1940) quotes the Australian Museum minutes for 6 October 1876:

> The following telegram was received from J.S. Ramsay, Esq., Maryborough: 'Have nine Ceratodus alive. Will Museum purchase and what price you paying all expenses to Sydney. Reply immediately'. Mr. Macleay stated that owing to the number of Ceratodus sent to Europe, the price had fallen considerably and he thought that they would not be worth more than 10/- each in England or Germany ... It was decided to offer 10/- for each specimen delivered alive in Sydney.

Gerard Krefft was dedicated to the best interests of the Museum and may have upset the trustees by resisting the practice of enhancing personal collections at the expense of the Museum. It seems that he was less than diplomatic in exchanges with these gentlemen and they dismissed him in 1874. He denied their power to discharge him and barricaded himself inside the Museum, but was finally ejected. Sadly, this altercation marked the end of his career in natural history and he died in 1881 (ADB).

Victoria

Victoria's separation from New South Wales was accomplished in 1851 and, within a few years, there developed support for the establishment of a natural history museum in Melbourne. The assembling of a collection (which formed the nucleus of the National Museum of Victoria, now Museum Victoria) seems to have begun in about 1854, with the appointment of a Government Zoologist. Although the total accomplishment in ichthyology in 19th century Victoria was small, some zoologists paid attention to the fishes, in particular, Wilhelm Blandowski and Frederick McCoy (see page 152).

Blandowski

Wilhelm Blandowski (1822–?) was born in Upper Silesia. His father was a lieutenant-colonel of the Prussian Medical Corps and Wilhelm spent two years in the Royal Prussian Cadets (ADB). He had some education in the natural sciences and arrived in Australia in 1849, intending to study the geology and natural history of the country.

Blandowski had some success at Castelmaine on the Victorian goldfields and made a trip to Cape York, Queensland, collecting natural history specimens. When the museum of natural history was established in Melbourne in 1854, he was appointed Government Zoologist and in the same year helped to found the Philosophical Society of Victoria. He was soon active in the field and, in 1856, led an expedition, in which Gerard Krefft participated, exploring the Lower Murray and Darling Rivers. They brought back a large number of specimens including many fish from which Blandowski described 19 species. He had been offended by the treatment he received from some members of the Philosophical Society and they, in turn, found cause for offence in his naming some of the fish specimens after them: they took his descriptions, '... low forehead, big belly and sharp spine' and '... slimy, slippery fish. Lives in the mud' to refer to themselves. He denied any intention to offend but would not alter the paper (Blandowski 1858). The affronted councillors rejected this part of his article so these pages were never published. They survive in draft form for the *Transactions of the Philosophical Institute of Victoria* for 1858 in the archives of Museum Victoria (MV). Blandowski later took most of the drawings and specimens from this trip with him when he returned to Germany in 1859.

The illustrations prepared by Blandowski for publication had been drawn with care and generally allow identification of the species. The accompanying text does not provide much help with identification but the number of rays and spines in the fins is indicated on most of the drawings. Most impressive is the fact that Blandowski's party, apparently with enthusiastic assistance from the aborigines (Humphries 2009), collected the whole of the known freshwater fish fauna of the lower Murray with the exception of a couple of small species of gudgeon.

In describing the fishes from his expedition, Blandowski made the common error of describing several different growth stages of the Silver Perch, *Bidyanus bidyanus* as different species. He also failed to notice that some of his supposed new species had already been described [Callop, *Macquaria ambigua* (Richardson, 1845) and River Blackfish, *Gadopsis marmoratus* Richardson, 1848]. It seems that, had the dignitaries of the Philosophical Society seen fit to leave his paper uncensored, Blandowski would appear as the author of eight species of fish from the Murray.

Tasmania

With such early settlement, beginning in 1803, it is not surprising that Tasmanian zoologists produced very early work on fishes. The enthusiasm for natural history in Tasmania generally took the form of observations and collections that were sent to overseas specialists. This was especially so with plant collections but zoology frequently benefited from the activities of these early naturalists as well.

Gunn

Ronald Campbell Gunn (1808–1881) arrived in the colony in Van Diemen's Land in 1830 and became assistant superintendent of convicts, and later a police magistrate, in Launceston. Gunn's interest was stimulated most effectively by the Professor of Botany at Glasgow University, William Hooker. Hooker provided advice and books and Gunn sent him specimens of plants, many of which were undescribed. Gunn also communicated with zoologists but generally found them less cooperative than Hooker, and his best efforts were always in botany.

Gunn sent fishes to Hooker, who passed them on to John Gray at the British Museum. The covering letter was published (Gray 1838) and it is clear that fish collecting had a very low priority. Gunn remarked 'I have this season, from my residence being so near the sea, procured a few fish, but have not numbered them ...'. He went on to discuss briefly his latest offerings, and concluded:

> I cannot and do not devote much time to the pursuit, but I plainly see that much might be done. To a sincere lover of natural history, possessed of knowledge and the necessary means, a finer field than Van Diemen's Land could scarcely be found. Crabs are very various and curious; fish also. And indeed a good cask of spirits might be filled with sundries highly interesting to a scientific person.

This looks very much like a plea for encouragement and perhaps for some remuneration. Gunn provided many zoological specimens for Gray, not least a live Tasmanian Tiger (Baulch 1961), but the British Museum of the day was miserly and Gray apparently gave Gunn little support (Finney 1993).

Gray (1838) reported on Gunn's fish collection, which included a Port Jackson Shark, *Heterodontus portusjacksoni* and several Cowfish, which Gunn referred to as 'parrot fish'. The latter were here briefly described by Gray as three new species. He was unaware of the sexual dimorphism characteristic of this genus, describing *Aracana ornata* from a male and naming the female as a separate species, *Aracana flavigaster*.

The Franklins

In 1837 a new governor, John Franklin, arrived in Hobart. He had previously cruised Australian shores as a midshipman on the *Investigator* with Matthew Flinders, and was known for his naval career and as an Arctic explorer. As a fellow of the Royal Society, and friend of Robert Brown and Joseph Banks, Franklin had a particular interest in supporting natural history investigation in the colony (Finney 1993). His companion in Arctic exploration had been John Richardson (see page 46), at this time the most active British ichthyologist, with a great interest in the fishes of Australia.

John Franklin and his wife Jane were both keen to assist the study of natural history in Tasmania. The Governor became the patron of the 'Mechanics' Institute' and supported the efforts of local naturalists, including Ronald Gunn, Thomas Lemprière and Edmund Hobson. He recommended government funding for the Institute and the appointment of a colonial naturalist. The funding was slow to be approved and niggardly, and the proposal for a naturalist's position was rejected by the Secretary of State for the Colonies. The Franklins helped to found the Tasmanian Natural History Society, predecessor of the Royal Society of Tasmania, and encouraged the publication of its journal (Parrot 2005). Franklin actively encouraged Lemprière to collect fishes for John Richardson.

In 1841 Jane Franklin's private museum was opened in Kangaroo Valley on the outskirts of Hobart, a miniature example of classical architecture at the foot of Mount Wellington (Stanbury 1987). It was intended to house works of art and nature for the enlightenment of the colonists but it did not capture their interest. The project may have failed because the Franklins left Hobart in 1843, unpopular with a powerful faction who resented their efforts to improve education and bring a wider perspective to the colonists. A quaint structure, the museum remains as a monument to Jane Franklin's optimism. Not until 1863 was a structure built specifically to house the collections of the Royal Society of Tasmania, which had been accumulating for over 20 years.

Hobson

Edmund Charles Hobson (1814–1848), a physician, was the colonial naturalist proposed by Sir John Franklin. He was born in Paramatta, New South Wales and moved with his family to Hobart as a child. He studied medicine with a Hobart surgeon and completed his medical degree in Germany. On his return to Tasmania he became associated with the natural history community that was supported in part by the Franklins' patronage. Hobson helped to establish the Tasmanian Society and published the results of his natural history investigations in the *Tasmanian Journal of Natural Science*. He moved to Port Phillip in 1840 but continued his association with the Tasmanian naturalists, still publishing in the Society's Journal (Finney 1993).

The first publication of the Tasmanian Journal led with Hobson's paper *On the Callorynchus Australis* (Hobson 1842), an exposition of the anatomy of the Chimaera, now known as the Elephant Fish, *Callorhinchus milii*. The subject of his study, '... this beautiful and singularly formed fish ...' taken in the Derwent River at Hobart, was an adult male. Hobson mentioned that Elephant Fish entered the Derwent in November in numbers and

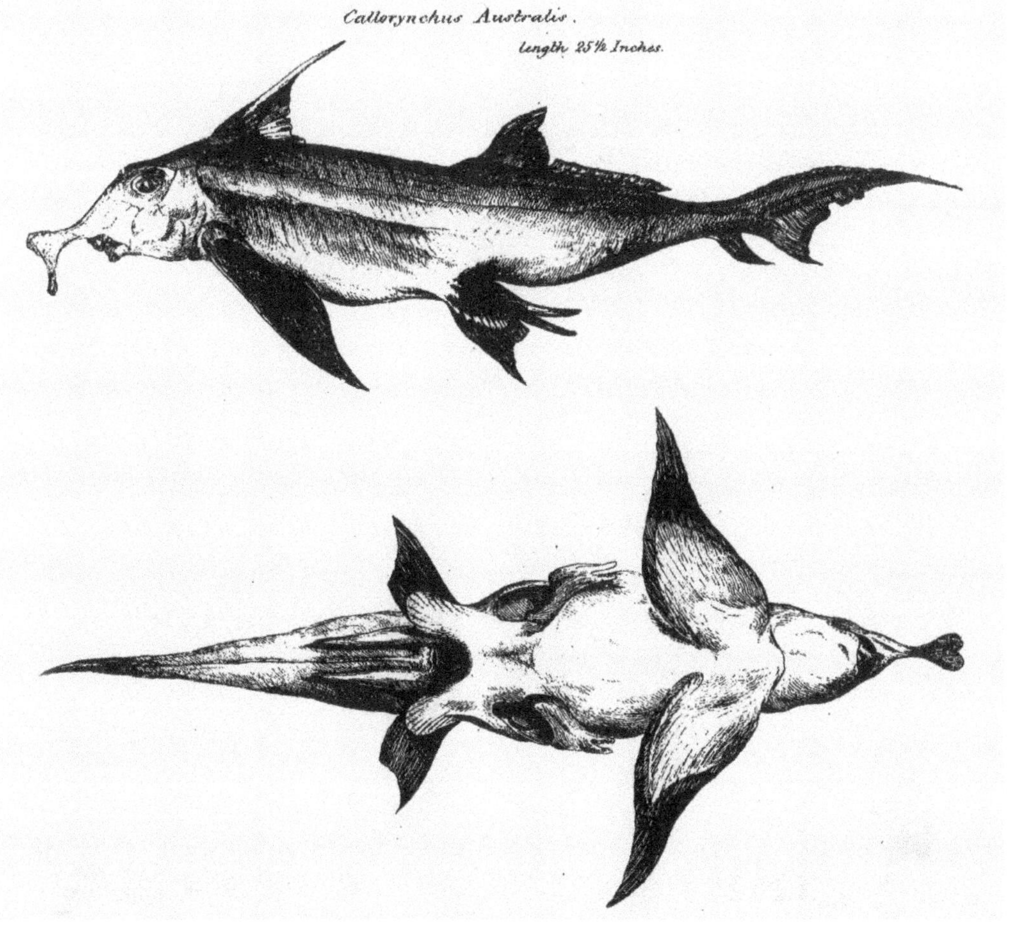

The Elephant Fish, *Callorhinchus milii* (Bory de Saint-Vincent, 1823). From Hobson 1842.

that '... my friend Mr. Ronald Gunn ... has taken them in the Tamar.' The dissection was described and some attempt made to relate form and function. Aside from the faulty identification of an 'accessory organ of respiration', Hobson's study was reasonably objective and accurate. This was the first ichthyological paper by an Australian-born naturalist (Whitley 1964).

Lemprière

Thomas James Lemprière (1796–1852) came to Tasmania in 1822. He was at first a settler and trader, and later a military official. As Deputy Assistant Commissary-General at Port Arthur in 1837, he was drawing Tasmanian birds and communicating with British naturalists. An active naturalist and an accomplished artist, Lemprière's activities were important to the scientific community in Tasmania. He established a museum at Port Arthur, and recorded in his diary for 24 December 1837 the loss of the head of a swordfish which he had hoped to procure for his museum; the Superintendent of Convicts forestalled him (Whitley 1966). Curiously, Lemprière has been maligned in a recent novel (Flanagan 2001) in which he has been fancifully misrepresented as physically and morally degraded.

Robert Johnston (1883) (see below) mentions that '... we are indebted to Mr T.J. Lemprière, who was the first Tasmanian naturalist who formed an extensive collection of fishes.' This collection was apparently assembled at the suggestion of Governor Franklin who forwarded them to his old colleague, John Richardson at the Royal Naval Hospital at Haslar. The fishes, comprising about 30 species, were deposited in the museum of this establishment and described by Richardson (see page 52).

Bicheno

James Ebenezer Bicheno (1785–1851), a barrister and social reformer, came from England in 1842 when appointed Colonial Secretary of Van Diemen's Land during Franklin's governorship. He had published studies in botany and ornithology, shared other interests with the Franklins, and was respected for the manner in which he carried out his official duties (ADB). Bicheno seized an ichthyological opportunity when '… an old fisherman brought me a fine specimen of a saw-fish, caught in the Derwent.' The fish, '… measuring full 5 feet from the extremity of the snout to the end of the tail', was an adult female:

> The reproductive organs were in a favourable state for examination; for there was found in her two young ones, – the largest 8 inches, and the smallest 6 inches in length, – together with 12 eggs, the largest being about the size of a pullet's egg, – a tolerable proof of the species being ovo-viviparous. (Bicheno 1851)

Bicheno regretted that the Saw Shark, *Pristiophorus cirratus* '… did not fall into the hands of some more experienced physiologist.'

Allport

Morton Allport (1830–1878) was one year old when the family moved from Staffordshire, England to Hobart. He was admitted to the bar in 1852 and established a legal practice.

However, natural history was his great interest and he maintained a wide correspondence with members of the learned societies to which he belonged in Tasmania, Victoria, London, New Zealand and Belgium (ADB).

Allport was one of many Tasmanian naturalists who viewed the introduction of northern hemisphere salmonid fishes as the most significant event in Tasmanian ichthyology. However, he also made extensive collections of native fishes and compiled a manuscript catalogue of native species which was useful to Robert Johnston when he came to assemble his checklist (Johnston 1883). Allport sent fish to Albert Günther at the British Museum and several of his specimens were recorded as new records for Tasmania or described as new species.

South Australia

The colony of South Australia was proclaimed in 1836 but its natural history was not much studied by residents until the establishment of the South Australian Museum 30 years later.

Grey

Among the earliest natural history collections from South Australia were those sent to the British Museum by Captain George Grey (1812–1898). Grey's was a restless intellect, modified by military discipline and a Victorian sense of responsibility for imperial affairs. He exemplified the high official whose interest in natural history was both prescribed and personal. For 30 years he was used as a trouble-shooter for the imperial interests of Britain in the southern hemisphere, being Governor of South Australia (1841–45), Governor of New Zealand (1845–53), High Commissioner in South Africa (1854–61), and Governor of New Zealand again (1861–68) (Henderson 1907).

While soldiering in Ireland Grey had conceived a plan for overland exploration in the north-west of Australia. His scheme was approved by the Secretary of State for the Colonies and he sailed in the *Beagle* in 1837. He led two hazardous expeditions: overland from Hanover Bay to the Glenelg River in 1837–38 and in the Shark Bay region in 1839; the latter journey concluded with an overland course of 500 kilometres, after shipwreck, from Gantheaume Bay to Perth. These explorations yielded useful geographic information and among his orders were the usual instructions to assess the resources of the country and to collect natural history specimens. The zoological discoveries from this work included a specimen of the Thorny Devil, a curious lizard, which Grey sent to John Gray of the British Museum; the description of this species as *Moloch horridus* appeared in an appendix to Grey's account of his overland journeys in 1841. Among other specimens he forwarded was the Red Indian Fish, *Pataecus fronto*, described by John Richardson (1844). Grey was appointed Resident Magistrate at King George Sound in 1839, where he was responsible for developing and administering plans for practical interaction with the aborigines; during his year in this position he compiled a useful vocabulary of the aboriginal language of the region and advanced schemes for education and employment.

In South Australia, although his main objectives were to improve relations between the aboriginal communities and the white settlers, and to rescue the state from bankruptcy, he found time to support the manoeuvres that preceded the establishment of a state museum (Hale 1956). Grey maintained his personal interest in the South Australian Museum, sending specimens from his later posts in New Zealand and South Africa.

The Natural History Society of South Australia was formed in 1838 and the desirability of establishing a museum was expressed; however, it was not until the Adelaide Philosophical Society became active that campaigning for a state museum was effective. The South Australian Institute, comprising a public museum and library, became a reality in 1861, by which time a variety of mineralogical, anthropological, botanical and zoological specimens had been assembled.

Waterhouse

The first curator of the South Australian Museum was Frederick George Waterhouse (1815–1898), a keen naturalist who had arrived in Australia in 1852; he presided over the opening of the new museum to the public in 1862. Waterhouse instituted a far-flung system of exchanges, exporting many specimens of the local fauna and of aboriginal artefacts in order to obtain collections of exotic zoological and anthropological material. The policy was short-sighted in that so much of the South Australian material rapidly became unobtainable due to the high rate of extinction of terrestrial mammals and the abrupt degradation of aboriginal culture (Hale 1956).

During the late 1860s and early 1870s Waterhouse made collections off the east coast of Eyre Peninsula and elsewhere (Glover and Olsen 1985) but ichthyology clearly had a rather low priority as, by the end of 1863, there were still only 20 specimens of fish in the museum (Hale 1956). Waterhouse sent specimens overseas to various specialists, including Kaup at Darmstadt and Peters in Berlin.

Waterhouse became enthusiastic about trawling as a means of collecting and his expeditions were noticed in the local papers:

> ... Mr Waterhouse, the enthusiastic curator of the South Australian Institute Museum, went on another expedition in the Gulf in search of rare fish, and was amply repaid for his trouble. The trawl used was very small compared with those of the British Channel boats, but it has brought up many fish supposed to have been hitherto entirely unknown. On a former occasion several varieties were taken, and at once forwarded to Count Castelnau, an eminent French naturalist at present prosecuting his researches in Victoria ... The section of St. Vincent's Gulf lying between Point Malcolm and the entrance of the Port River has been fully overhauled, and some beautiful as well as curious fish captured. The curator now intends making an extended cruise to Port Gawler to inspect that locality ... (*The Register*, 1872)

Such experiments were not without their problems:

> With praiseworthy zeal Mr. F.G. Waterhouse, the Curator of the Adelaide Museum, sailed again on Friday with a view of testing the fishing-ground near Port Gawler. The night was spent in the Gulf, and on Saturday morning two or

> three rich hauls quite repaid the Curator for his trouble. A strong breeze sprang up and the head of the cutter's mast was carried away, consequently he was compelled to return to the jetty to repair the damages.

These were the first serious attempts to sample the fishes of the South Australian Gulfs. From them, Castelnau described a number of new species (see page 122).

Queensland

With the separation of Queensland from New South Wales in 1859, Brisbane gradually became the natural centre for explorations into the natural history of the region. The Queensland Philosophical Society was established in 1859 and, as usual with similar societies, began to accumulate specimens. These assumed the appearance of a museum during the 1860s but it was not until the collection was moved to a separate building, and Karl Staiger was appointed as an assistant, that much attention was paid to fishes.

Staiger

Karl Theodor Staiger (1833–1888), from Germany, was appointed as government analyst and custodian of the Queensland Museum in 1872. Mineral assays took up much of his time but he performed the curatorial functions to the satisfaction of the museum trustees. He continued to donate specimens to the museum, even after William Haswell was appointed in his stead (Mather 1986). Staiger organised the existing collections and amassed additions. He took an active interest in the fishes and relied upon Castelnau for identifications, with some curious results (see page 125).

Leichhardt

The explorations of Ludwig Leichhardt (1813–1848) were made more useful by his broad interest in natural history. He included notes on fishes in his journals but his most spectacular contribution to ichthyology was the collection of *Scleropages leichhardti* from the Burdekin River, Queensland (see page 89).

Gilbert

John Gilbert (1810–1845), came to Australia with John Gould (1804–1881), and the latter's wife and illustrator, Elizabeth, in 1838. Gould's chief interest was ornithology but he functioned as an entrepreneur, willing and able to deal in natural history specimens of any kind. Gilbert was employed by Gould as a zoological collector and, when Gould returned to England in 1840, Gilbert remained in Australia. Gilbert's activities were wide-ranging: he collected in Tasmania, near the 'Swan River settlement', in the north-west and near Port Essington (near the present site of Darwin). He took his specimens back to London but, still in Gould's pay, returned to Western Australia to make more collections in 1842 and 1843.

Gilbert's contributions were enhanced by his field notes. Richardson quotes his observations of a Frogfish, *Halophryne diemensis*:

> this fish is an inhabitant of the mud at the head of the harbour of Port Essington, where it may be frequently seen creeping over the surface when the tide has left. It is very difficult to capture, for on the slightest appearance of danger it plunges down instantaneously. (Richardson 1843a)

At Gould's suggestion, Gilbert moved to Moreton Bay in Queensland. He seized an opportunity to join Leichhardt's exploration to the Gulf of Carpentaria but on this expedition, collecting again in the vicinity of Port Essington in 1844, he was speared and killed by aborigines (Moyal 1986).

With the continuing development of Australian museums and universities, individuals and eventually full-scale expeditions were sent out with specific directions to collect in a more scientific manner. As the scientific institutions of Australia became more organised and effective the custom of sending specimens to overseas institutions was gradually abandoned.

PART THREE
(1870 to 1930)

Prior to 1870 the study of Australia's fishes, as distinct from the collection, preservation and dispatch of specimens, was largely undertaken outside Australia. From about 1870, much of this study was conducted within Australia. At first, most of this work was independent of institutions, later almost exclusively within Australian museums. The quality of the work done was quite uneven but towards the end of this time the essential characteristics of the fish fauna had been delineated and a good tradition of careful work had been established. Four figures – Castelnau, Ogilby, Waite and McCulloch – stand out among the students of fish in these years, for their enthusiasm and for the sheer volume of the work they accomplished. Their work and its merits, together with that of less significant students of fish, are discussed here.

Castelnau

Count Castelnau (1810–1880) or, as he was sometimes known, François Laporte, came to Melbourne in 1862 and was soon absorbed in the ichthyology of his new home. There can be no mistaking the authenticity of his ardour for natural science: he had pursued the study of zoology throughout the whole of his adult life in many countries and his enthusiasm for his work enlivens his published reports. Castelnau had studied with Cuvier, a most precise taxonomist, and had a wide experience of fishes in different parts of the world. In spite of this his work contained many errors, oversights and even absurdities. Essentially he worked in isolation and was generally subject to criticism only through the published literature; as such, the criticism was delayed and not constructive in effect, whatever the intention. The sheer volume of Castelnau's published work on Australian fishes has meant that it must be considered in detail as virtually all of his successors have discovered, often to their enduring frustration.

Castelnau's career prior to his arrival in Australia

In Paris, Castelnau learned from Georges Cuvier and other Parisian zoologists, including Duméril. Entomology was his primary zoological interest and by 1840 he had published about 45 papers, mostly on entomological topics. His was an adventurous career: as diplomat and zoologist he journeyed widely, leading successful expeditions and holding responsible positions in several countries. Between 1837 and 1841 he travelled in Canada, studying the fauna of the Canadian Lakes and enquiring into the political system in that country. At Cuvier's request he collected specimens of fish which he sent to Paris.

Castelnau's most ambitious exploit was leading an expedition in South America for the French government. Over a period of nearly four years, he and his companions travelled from Rio de Janeiro, exploring the Mato Grosso and the Paraguay River; they passed through Bolivia to Lima in Peru, then to the source of the Amazon from which they returned to Para. The party studied the flora, fauna, geology and anthropology of these regions, making extensive observations and collections. Much of this material was lost, and one of the explorers killed, in a skirmish with Indians. However, a monumental work was derived from the remaining records: *Expédition dans les parties centrales de l'Amérique du Sud, de Rio de Janeiro à Lima, et de Lima au Para: exécuté par ordre du gouvernement Français pendant les années 1843 à 1847*. Publication was organised by Castelnau and comprised many volumes, written by members of the expedition and by specialists in Paris. With the assistance of Guichenot and Duméril at the Jardin des Plantes in Paris Castelnau prepared the section on fishes and had it magnificently illustrated by 50 handcoloured lithographs. The specimens were preserved in the Paris collections. This work included the description of 117 species of fish that Castelnau believed to be new to science; about one in three of these is at present accepted as valid.

The work on South American fishes appeared in 1855. Meanwhile Castelnau had been appointed French Consul at Bahia, Brazil, where he continued to pursue his zoological interests. The delay in publication meant that he was able to include some of the fish he found in these latter journeys in northern Brazil in the volume on South American fishes. From 1856 to 1858 he was French Consul at the Cape of Good Hope. From this base he travelled around southern Africa, making zoological collections and paying particular attention to the fishes. He published two papers on the fishes of South Africa, including the description of 73 supposedly new species, about one in five of which is now regarded as valid. An examination of his publications on the fishes of South America and South Africa show that his concept of a species was imprecise, even for his day; he had little appreciation of the variation that may occur within a species. His observations were sometimes superficial and his species descriptions insufficient in relevant detail and marred by inaccuracies. For these reasons, Castelnau's publications attracted criticism, particularly from Albert Günther of the British Museum.

Castelnau's diplomatic career entailed another move in 1858: he was instructed to establish a French Consulate in Thailand. From his base in Bangkok he roamed and collected in parts of India, Sri Lanka, Vietnam, Singapore and Java, and communicated with the leading ichthyologist of the Indo-Pacific region, Pieter Bleeker. On occasion, Castelnau painted watercolours of fishes that he regarded as new species; Bleeker (1860) based descriptions of some Cardinalfish (Apogonidae) and a Goby on Castelnau's annotated paintings of fishes he collected at Singapore (Russell *et al.* 2010).

The diversity of Castelnau's interests is reflected in his many publications: he wrote on entomology, ornithology, palaeontology, anthropology; on the domestication of the llama, alpaca and vicuna, suggesting their introduction to Algeria as domestic animals; on the distribution of South American reptiles; on the abundance of tigers in Singapore. However, his main interests were insects (chiefly beetles) and fishes.

After the Asian interlude, Castelnau returned to Paris, where he suffered a protracted illness. He refers to this and to the frustration of his work on Asian fishes in the introduction to a later publication (Castelnau 1872a):

> In India, I continued my ichthyological labours. At Bangkock I collected the sorts of the great Mainam River; at Saigon, those of the Meklong; and, during a more or less lengthened stay at Malacca, Sumatra, Java, Ceylon, and Singapore, I described and sketched from nature over 750 sorts. On my return to Europe, I began to put in order my voluminous notes, but having been obliged, on account of sickess, to interrupt my work, I was, on my recovery, struck with a most disagreeable surprise, in discovering that my servant had, for more than one month, used the sheets of paper on which I had bestowed so much time and labour to light the fires, and other parts of my learned lucubrations were discovered in the last place in the world where an author would be proud of finding his works.

This work he abandoned and fish taxonomists can surely sympathise with the Count, while perhaps deriving some selfish satisfaction from the non-appearance of what promised to be a plethora of potential synonyms.

In 1862 Castelnau arrived in Australia and in 1864 was appointed Consul General for France, a position he retained until 1877. In Melbourne he resumed his study of the Coleoptera, while amassing a formidable collection of fishes. From 1872 to his death in 1880 Castelnau published almost exclusively on Australian fishes.

Castelnau dispatched collections of fishes from many parts of the world, including the type specimens of his new Australian species, to the Muséum National d'Histoire Naturelle in Paris. Valenciennes recorded some of these donations in the *Histoire Naturelle des Poissons.*

Australian ichthyology on Castelnau's arrival

The greatest contributions to Australian ichthyology to this time were from collections made by the early French and British naturalists. Richardson attempted to correlate their work with his own descriptions of collections made by, or at the behest of, British naval personnel. The great encyclopaedic work of Cuvier and Valenciennes had been the standard text but that of Günther was complete by the time Castelnau began publishing on Australian fish.

In New South Wales, George Bennett had published some observations on fishes and Krefft had studied a few of the freshwater fishes, sending the marine fishes that came to his attention to Günther. In Victoria, Blandowski's expedition had made known some freshwater fishes and Frederick McCoy had written a little on the marine fishes. Morton Allport had assembled a collection of marine fishes in Tasmania and made a manuscript list. Frederick Waterhouse in Adelaide had collected some of the marine fishes, sending them to Kaup and Peters.

Thus there was no ichthyologist at work in Australia in the 1860s: there was a place to be filled. Museum and other collections were still sent overseas. Günther (London), Peters (Berlin), Klunzinger (Stuttgart), Steindachner (Vienna) and Kaup (Darmstadt) were all recipients of collections from Australia in the 1860s and 1870s and published accounts of many species new to science. Generally, their works were of a higher standard and their descriptions more detailed and accurate than those published by Castelnau.

Summary of Castelnau's published Australian work

Castelnau's first publications in Australian ichthyology were undertaken in 1872–73. He had been gathering specimens from the Melbourne fish markets and from a variety of other sources for several years before this. The papers entitled *Contributions to Australian Ichthyology* were in nine parts, arranged according to the regions from which the specimens were derived.

For the Victorian Exhibition of 1872–73 Castelnau (1873a) wrote a brief essay on the *Edible Fishes of Victoria*. He continued to receive collections from different parts of Australia and reported on these in 1875 (*Researches of the Fishes of Australia*), 1878 (*Australian Fishes*) and in some smaller subsequent papers.

The fishes of New South Wales formed only a small part of Castelnau's work until he published an *Essay on the Ichthyology of Port Jackson* (1879). In this paper, as usual, he included descriptions of many fishes that he believed to be previously undescribed.

To give an impression of some of the problems Castelnau's publications have caused taxonomists, the account below includes assessments of the fate of the names he applied to Australian fishes. In particular, the numbers of his new species still regarded as valid, and those of his new genera presently in use.

Contributions to Australian Ichthyology: introduction

The nine parts that comprised this series of papers were published in the journal of the Royal Zoological Acclimatization Society of Victoria (1872–73). In the first, Castelnau gave an account of the fishes of Victoria, generally as represented in the Melbourne fish market, but including a few of the small fishes collected from the intertidal region of Port Phillip Bay. This was a major paper, 242 pages in length, including the description of many new species; it was systematically arranged, using Günther's scheme and Castelnau here reproduced Günther's definitions of the families of fishes.

The other parts of *Contributions to Australian Ichthyology* were not so long, dealing with smaller collections from South Australia (parts 2 and 4), Western Australia (part 8) and northern Australia (parts 5 and 6). He issued supplements to the Victorian list (parts 3 and 9) and reported on a small collection from New Caledonia (part 7).

Contributions to Australian Ichthyology: Castelnau on Günther

In the introduction to the first paper, Castelnau included a lengthy and querulous defence of his earlier work, provoked by Albert Günther's criticism. This passage, several pages in length, showed the naivety of some of Castelnau's views. He considered zoological classification to be largely a filing system and so criticised Günther:

> Many of Dr. Günther's superior divisions are established only on anatomical characters, and I think this most objectionable, as it would exclude from the study of science all those who would not have at their entire disposition one of the large museums of Europe, and even the fortunate zoologists who are so situated can only, in comparatively few instances, sacrifice valuable specimens. I think that anatomical characters ought only to be used to confirm zoological ones. (Castelnau 1872a)

Castelnau confused the construction of a rational classification with a scheme for identification.

Castelnau offered some rather more valid criticisms of Günther's work. He considered Günther to have the faults of many zoologists who did not participate in field work and tended to overlook the significance of first-hand observations:

> Dr. Günther, in most cases, prefers describing the colours from the discoloured specimens he has at his disposition sooner than adopting the

> description of travellers who have seen the sorts alive, and in many cases have made drawings from specimens so taken. For instance, *Phractocephalus Hemiliopterus* is described as a fish of obscure tinges, when he had my plate under his eyes, showing its beautiful colours, drawn from the living specimen. (Castelnau 1872a) (See colour plate 10, lower.)

Reliance on the specimens' colours in preservative is a conservative and sometimes misleading approach but often still used when first hand observation is not possible. Günther frequently used reliable colour descriptions, as seen from many of his publications, but he was without confidence in Castelnau's work as a whole.

Günther's fastidious approach to the structure of scientific names offended Castelnau:

> He [Günther] changes the names, derived from Greek or Latin, that he considers badly composed, and this I think useless, as it only increases synonymy unnecessarily. (Castelnau 1872a)

Reasonable criticism, but there is irony in Castelnau complaining about unnecessary synonymy.

Of Castelnau's many criticisms of Günther's work, that made with perhaps most justice concerned the latter's concentration on type specimens held in the British Museum:

> … Dr. Günther appears not to have examined the Parisian collection, which is certainly the most important in the world, on account of the immense quantity of typical specimens it contains. (Castelnau 1872a)

Günther's defence is implicit in the title of his work, *Catalogue of the Fishes in the British Museum*, although it was clearly intended as a definitive summary of knowledge of the world's fishes, so Castelnau's criticism had some validity.

Many ichthyologists suffered from Günther's energetic and often unnecessarily personal criticisms. Castelnau's response to such provocation appeared rather ineffectual:

> It is also to be regretted that the learned doctor does not add to his qualities a little more indulgence towards other naturalists, often his predecessors in the science. He has placed himself, by his works, quite high enough not to be quite so bitter towards those who have not, like himself, the privilege of not being mistaken … It is evident that Dr. Günther's greatest delight is to find fault with everyone and with everything. When he mentions a plate, he must almost always add one of the following epithets:- 'Not good.' 'Bad.' 'Very bad.' And in many cases he has never seen the fish in question, and the drawing is due to one of those artists who, having devoted the labours of their lives to zoology, have become naturalists of no small merit themselves. (Castelnau 1872a)

Castelnau contrasted Günther's abrasive style with Cuvier's more generous approach:

> If Cuvier was obliged to rectify a traveller or a little known naturalist, he would do it in such an indulgent way as to encourage him to pursue his labours … (Castelnau 1872a)

Günther spoke of himself as 'The Recorder' and in the (royal) plural appraised Castelnau's 1872 *Contributions* (*Zoological Record* 1872):

> The paper is entirely based upon the Recorder's Catalogue; and we observe with pleasure that a great improvement of Count de Castelnau's descriptions has taken place when compared with those of his former publications. With regard to the distinction of the species we cannot enter here into their critical examination, but will only mention here that the collections received from the same seas at the British Museum contain about 5% undescribed species, whilst Castelnau claims the finding of more than 30%. We cannot help thinking that so great a difference is simply due to a different treatment of the matter. The introductory remarks of the author are chiefly directed against the Recorder's former treatment of ichthyological questions (that of the scientific nature of Count de Castelnau's works included), and will be read with pleasure by amateurs, and judged by zoologists.

Günther's comments were largely justified but they did not serve to restrain the Count.

Contributions to Australian Ichthyology: Victoria

In this, Castelnau's most ambitious paper, he dealt with the marine and freshwater fishes that had come to his notice in Victoria to this time. A few sharks and rays were mentioned but their treatment was cursory.

Most of his fishes for this paper came from the Melbourne fish market but a few were donated by friends and acquaintances. Castelnau here listed 148 species for the Victorian fauna but many were misidentified. He named 60 new species, few of which are at present regarded as valid; and 9 genera, 2 of which are still in use.

Some species were described as new more than once in this paper, Castelnau being misled by variations in form and coloration with growth and between the sexes. The Common Galaxias, *Galaxias maculatus* was awarded four new names; the Bluethroat Wrasse, *Notolabrus tetricus* three new names. The Bastard Trumpeter, *Latridopsis forsteri*, first named by Castelnau in this paper, was described under three different names, *forsteri* having page priority. The Macquarie Perch, *Macquaria australasica*, certainly a species which exhibits a good deal of natural variation, was named five times in three different genera, two of which Castelnau created here for this species.

Unbeknown to Castelnau, a few of his new species were described earlier in the same year (1872) by Klunzinger, who had been sent specimens from New South Wales. For instance, Castelnau's Southern Cardinalfish, *Apogon guntheri* was synonymous with *Vincentia conspersa* (Klunzinger, 1872) and Castelnau's Yarra Pygmy Perch, *Microperca yarrae* with *Nannoperca obscura* (Klunzinger, 1872).

The following year Castelnau published two supplements to his account of Victorian fishes (1873b). These consisted of notes on 38 species, including descriptions of 12 new species. None of these supposed new species is now regarded as valid.

Contributions to Australian Ichthyology: South Australia

The fishes of South Australia had received little attention prior to Castelnau's paper (*II.- Note on some South Australian Fishes*):

> I have received from Mr. Waterhouse, the able and indefatigable Curator of the South Australian Museum, a most interesting series of small fishes, collected by himself in the St. Vincent's Gulf, by means of a trawl. (Castelnau 1872b)

Waterhouse (see page 112) initially sent fish to Europe for identification and description, but found Castelnau's responses to his specimens prompt and encouraging. Presumably Castelnau was anxious to establish authorship of the novelties from this collection as he gave very brief descriptions of 15 new species (of which 6 are now valid) and 5 new genera (of which 4 are presently in use). The next year he produced a second paper with more complete information on what he regarded as 'those sorts deserving a particular notice': he expanded his diagnoses of the new species and genera, and added a couple of new species (Castelnau 1873b).

Fishes from the vicinity of Adelaide had not been thoroughly sampled and Waterhouse had himself assembled a collection, by the use of a small trawl in Gulf St Vincent, which included a high proportion of undescribed species. Among the most remarkable forms was the Tasselled Anglerfish, *Rhycherus filamentosus*. This was fully described, unlike a strange Prowfish, *Neopataecus waterhousii*, which Castelnau described briefly in the 1872 synopsis, and apparently overlooked in the 1873 paper.

Waterhouse contributed South Australian fishes again a year later and Castelnau described them in 1875 (see below).

Contributions to Australian Ichthyology: Northern Australia

Two papers in this series dealt with fishes from northern Australia: *Notes on Fishes from North Australia* and *Notes on Fishes from Knob Island*. The former was based largely on a collection from Port Darwin, donated by Frederick Waterhouse of the South Australian Museum and included notes on, or descriptions of, 15 species; this included descriptions of 9 new species (2 now recognised) and 2 new genera (neither now in use). The Knob Island paper consisted of Castelnau's description of a jar of fish. Castelnau explained:

> The following species were contained in a small bottle purchased from a sailor, who said they came from Nob or Knob Island, in the north of Australia [1 km east of Cape Grenville, Queensland]. I am unable to find out where this is situated, but I am inclined to believe that it may be one of the numerous islands of Torres Straits, as the sorts have a decided tropical appearance. (Castelnau 1873b)

There were notes on, or descriptions of, 11 species; this included descriptions of 7 new species; 2 are now recognised, including the beautiful coral-dwelling Flame Hawkfish, *Neocirrhites armatus*.

Contributions to Australian Ichthyology: **Western Australia**

Most of the specimens for this paper had been collected by Reverend Bostock of the 'Swan River settlement': a variety of interesting freshwater fishes were among the specimens. Notes on, or descriptions of, 47 species were given; this included descriptions of 19 new species. Among the few still recognised is the Nightfish, named *Bostockia porosa* after the Reverend Bostock, one of Castelnau's most helpful donors. Also described in this paper was the Southern Fiddler, *Rhinobatus* [now *Trygonorrhina*] *dumerilii*, 'named in honour of my late friend, August Duméril.'

Contributions to Australian Ichthyology: **New Caledonia**

Castelnau had a collection made by a French merchant, based in Melbourne, during a few months he spent in Noumea. The report consisted of notes on, or descriptions of, 15 species; this included descriptions of 5 new species (1 now recognised) and 1 new genus (not now in use).

Researches on the Fishes of Australia

This rather grand title heralded a series of notes and descriptions of everything ichthyological that had come into Castelnau's hands over the preceding couple of years. Many of the donors were experienced and dedicated naturalists so their collections contained some interesting specimens. The newly appointed custodian of the Brisbane Museum, Karl Staiger (Mather 1986), sent collections from Queensland (Cape York and the Brisbane region); Reverend Bostock, who returned to England at about this time, sent fish from Perth, plus a few from the Dampier Archipelago and the north-west coast; a Mr Petard sent a few fish from the Richmond River, New South Wales; and the curator of the South Australian Museum, Frederick Waterhouse, sent another collection from the vicinity of Adelaide.

Nearly 130 species are considered, including descriptions of 75 new species (of which about one in seven is now recognised) and a host of new genera (few of which are now in use). The freshwater Pinkeye Mullet, *Mugil* [now *Trachystoma*] *petardi* from the Richmond River and the Breaksea Cod, *Serranus* [now *Epinephelides*] *armatus* from the Swan estuary were among the new species in the *Researches*.

In this paper Castelnau contrived to present as a great mystery a drawing that Staiger had sent him of a large fish, captured in a lagoon near Brisbane. He described its features from the drawing and named the fish *Brisbania staigeri*, creating a new genus for it in the process. This was of some significance as Staiger was eventually to send a specimen of this fish, and also because Staiger sent another drawing to Castelnau, not long after this, which resulted in Castelnau making a colossal blunder.

Australian Fishes: New or Little Known Species

Castelnau had shown an increasing interest in the fishes of New South Wales. He visited Sydney in 1876, met William Macleay and Edward Ramsay, and corresponded with them thereafter. Shortly after this he moved to Sydney and began publishing in the *Proceedings of the Linnean Society of New South Wales*.

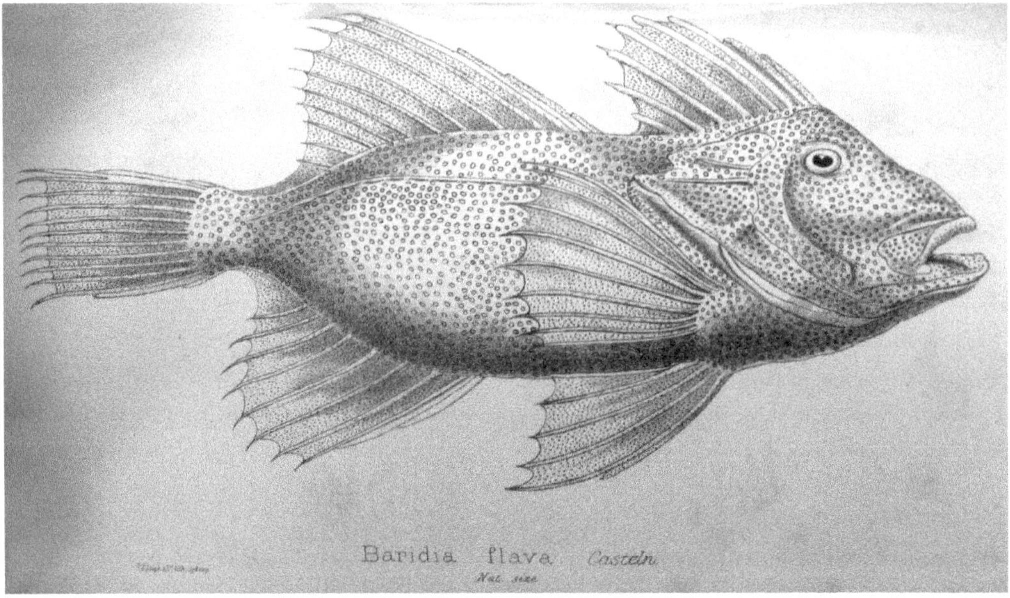

Beridia flava, spelled *Baridia* in error on the plate, proved to be *Gnathanacanthus goetzeei* Bleeker, 1855. From Castelnau 1878.

Castelnau's *Brisbania staigeri* proved to be *Megalops cyprinoides* (Broussonet, 1782). From Castelnau 1878.

In 1876 Castelnau visited Brisbane and obtained fishes from a variety of sources in Queensland, including Moreton Bay and the Brisbane River; Rockhampton; the Norman, Dawson and Fitzroy Rivers; and Cardwell. In this paper, he described fish from Queensland and a few from New South Wales, Victoria and South Australia. He provided notes on, or descriptions of, 31 species; there were descriptions of 16 new species (2 now recognised) and 2 new genera (1 now in use).

The paper is unusual among Castelnau's Australian publications in featuring two illustrations. One of these was of his mystery fish, *Brisbania staigeri*. While in Brisbane, Castelnau had Staiger organise fishermen to try to capture a specimen from the lagoon from which it had been reported. This failed but while he was still in Brisbane a fisherman brought him a specimen of the 'so-much desired fish', caught in the Brisbane River. Castelnau published the drawing and a careful description but failed to recognise the fish as belonging to the genus *Megalops*, well described and illustrated by Cuvier and Valenciennes and in many other works. He also named and figured *Beridia flava*, a beautiful and distinctive little fish sent to him from Portland, Victoria; this species had been described and illustrated by Bleeker in 1855 as *Gnathanacanthus goetzeei*.

One of Castelnau's donors, a Mr Gulliver, sent specimens from the Norman River, Queensland. For him Castelnau named the Nursery Fish, *Kurtus gulliveri*, the male of which develops a hook-like protuberance from the forehead on which he carries the eggs.

Castelnau's short papers

Castelnau derived further collections from Mr Gulliver who resided for a couple of years at a settlement on the Norman River about 20 miles from its mouth on the Gulf of Carpentaria. Castelnau (1878b) listed 25 species, including descriptions of 13 new species (3 now recognised) and 2 new genera (1 now in use).

He also published a brief account (Castelnau 1878c) of a few more Queensland fishes that came to his attention.

On a New Ganoid Fish from Queensland

Another drawing sent by Karl Staiger formed the basis for this paper; Castelnau quoted Staiger:

> It is only found in a single waterhole in the Burnett River, living together with *Ceratodus* [Australian Lungfish]; and when in August, 1872, I was in Gayndah, I got it on the breakfast table, brought in by blacks from a distance of about eight to ten miles. I had the fish for breakfast, remarked its curious shape, and asked the then Road Inspector to draw it for me, which he did ....

Castelnau concluded that the sketch represented a ganoid fish of an unknown type. He incautiously described it as a new species on the basis of the drawing and speculated on its relationships, naming it *Ompax spatuloïdes*:

> In our present knowledge of this singular fish, some inconvenience might arise from giving it a significant name; and I think it is preferable to design it under the mysterious historical one of *Ompax*.

The word ompax was apparently associated with the religious rites of Demeter, themselves shrouded in mystery: the name was not entirely inappropriate.

Whitley (1940) quotes an article in a Sydney paper, *The Bulletin*, in which the mystery fish is explained as a practical joke played on Staiger by the hands on Gayndah station.

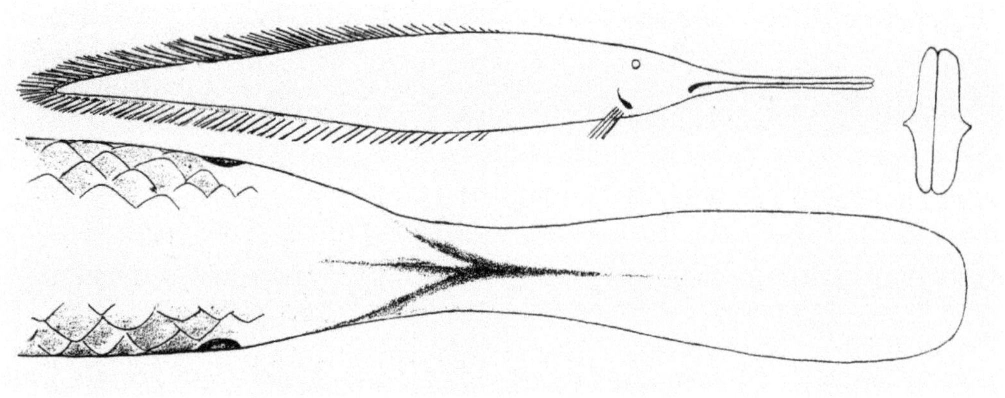

Ompax spatuloïdes, a hoax. From Castelnau 1878.

Whitley suggests that the 'fish' may have been constructed from portions of an eel, a mullet, and perhaps the beak of a platypus.

Ompax crops up in the scientific literature, generally with appropriate scepticism. For instance, Macleay (1881):

> I think it probable that no such fish was ever found, but I give it a place in my Catalogue, in order to direct attention to the subject, as the likelihood of finding Ganoid Fishes in Queensland waters is asserted by several very distinguished Naturalists.

McCulloch (1930) dismisses it with a curt 'probably mythical.' Eschmeyer (1998) unfortunately includes the generic name *Ompax*, with the comment 'Perhaps mythical.' The recent Australian catalogue (Hoese *et al.* 2006) sensibly omits the name altogether.

Ichthyology of Port Jackson

There had been little published specifically on the fishes of Port Jackson up to 1879, with the exception of a paper by Franz Steindachner (1866). Castelnau cited this publication twice but, in some ways, seemed scarcely aware of it. For instance, he overlooked the distinctive Pomfred, *Schuettea scalaripinnis* and some other species described and illustrated by Steindachner in this paper. This was typical of his approach: he quoted the early voyagers, Cuvier and Günther but tended to overlook contemporary publications. In this way he virtually ignored Steindachner (1866) and Klunzinger (1872). The latter received collections of Australian fish from Ferdinand von Mueller, the celebrated Melbourne botanist, who sent them to Stuttgart rather than handing them over to Castelnau.

Castelnau's paper on the fishes of Port Jackson presented lists comparing the fish fauna of Victoria with that around Sydney. This attempt at a biogeographical analysis was compromised by taxonomic problems, many of which were created by Castelnau himself; and, of course, by the incomplete sampling of the two regions. These lists were followed by notes on, or descriptions of, 87 species; 16 were supposedly new species (few are now recognised) and 3 new genera (none now in use). The valid new species included the Yellow

Crested Weedfish, *Cristiceps aurantiacus* and the Tiger Pipefish, *Syngnathus* [now *Filicampus*] *tigris*, both from Sydney.

William Macleay, well known for encouraging the study of natural history and hospitable to visiting naturalists, gave Castelnau some help, for which he received formal recognition:

> Mr. W. Macleay owns a splendid collection of Australian zoology, and has most liberally lent me specimens and valuable books. (Castelnau 1879).

Macleay was then collaborating with Doctor Alleyne (see page 134) and beginning to publish short papers on Sydney fishes. E.P. Ramsay, appointed Director of the Australian Museum in 1874, was efficient in amassing collections and had developed an interest in the fishes of Port Jackson; however, he published nothing of significance to ichthyology during Castelnau's lifetime. Castelnau also mentioned Ramsay's assistance:

> Mr. Ramsay, of the Australian museum, has also done all in his power to assist me by lending me any of his own books I had not in my own library. (Castelnau 1879).

However, Castelnau's relationship with Sydney zoologists seemed a little fragile. He was demanding, as shown by extracts from some of his brief letters to Ramsay as Director of the Australian Museum. (ML). His handwriting was abominable but, although there are irritating gaps, an impression of the man emerges:

[Castelnau (visiting Sydney) to Ramsay. July 6 (presumably 1876)]

> Dear Sir,
> I have been several times to the Museum without finding.
> I want to examine some of your fish, but I cannot do it through the [?] glasses.
> Let me know if I could call, in the morning and get the [?] glass removed
> You will oblige me,
> Yours [?],
> F. de Castelnau

[Castelnau (from Melbourne) to Ramsay. Nov. 21, 1876]

> ... I have neither heard nor obtained an answer from Mr. Macleay or Dr. Cox. I am led to believe that Sydney people, following the example of the Lords of the middle ages, neglect terribly the great art of writing ...

[Castelnau (from Melbourne) to Ramsay. Dec. 5 1876]

> ... It is with pleasure that I learn that you have two Ichthiologists at Sydney, I did not know you had one. Fishes are very difficult animals to study and require much time to make out.
> My papers on Australian fishes seem to have been very much overlooked and I do not see any mention of Bleeker's papers or those of New Guinea. I should

> say the same of the works of most of the French Expeditions.
> I know it often takes some days even to be certain what family a fish belongs.

[Castelnau (living in Sydney) to Ramsay. Nov. or Dec. 30, 1878]

> I am very much annoyed at my paper on Australian fishes not having been [?] separately. It seems to me that a certain number of the [part?] it is contained in might be torn up and divided amongst the different authors. Please, in all cases see that I have a few.
> The fish you showed me yesterday as being called the Blue fish is my Ruppelia prolongata. It belongs to the family of Nandidae. [This was the Southern Blue Devil, *Paraplesiops meleagris*]

[Castelnau (from Mayfield House, Mordialloc, Melbourne) to Ramsay. March 6, 1879]

> ... My eyes have been bad lately and have much prevented me from working.
> I sent my work to Mr. Macleay but, according to Sydney Fashion, he has never answered me. [?] wrote to me and by his letter I expect he has got my work.
> Pray beg him to return it when he has seen it sufficiently as it is the only copy there is of it in Australia and I often want to consult it.

[Castelnau (from Mayfield) to Ramsay. June 19, 1879]

> ... Mr Macleay has just sent me the last number of the Proceedings and my copy of the paper of the fishes of Sydney. [?] plate has been left out. I do not know for what reason as it might have been at least as interesting as to give a dozen figures of the common shark that has been figured often already and much better.

Castelnau thought he had been shabbily treated by his Sydney acquaintances but he clearly gave them a bad time in return. '... It is with pleasure that I learn that you have two Ichthiologists at Sydney, I did not know you had one. Fishes are very difficult animals to study and require much time to make out.' (see extract above) This acerbic comment refers to Macleay and Alleyne who had just published a large paper on fishes. He makes his views even clearer in his essay on Port Jackson fishes:

> The study of ichthyology is attended with much more difficulty than that of any other branch of zoology; not only are the species very numerous and often very difficult to distinguish one from the other, but they also present difficulties from their size and mode of preservation ... The rules of most of the Australian Museums do not allow specimens to be removed from their buildings; and whatever may be the kindness of the curators, they cannot break these rules.
> On the other hand, numerous books are required for the study of fishes; and those are not generally found in the scientific establishments of the country.
> These reasons explain why certain species from Sydney contained in the Australian Museum are not included or described in this paper.

> The Brisbane and Adelaide Museums have adopted a different system, and have sent me all their fishes, and thus their Ichthyological collections, &c., are completely named, while those of the other museums are the reverse. (Castelnau 1879)

Castelnau's work in Sydney probably helped to confirm Macleay and others in their views that Australian animals ought to be studied in Australia. This encouraged the development of zoological science within Australia, but led to burgeoning synonymies created by workers who, like Castelnau, did not have sufficient expertise as taxonomists.

Castelnau's contribution

Castelnau's work was not seen in perspective by his contemporaries and early successors within Australia. Macleay, de Vis, Johnston and Lucas generally regarded him as a reliable authority, and the standard of their own work is probably derived in part from Castelnau's example. Even as late as 1890 Arthur Lucas, introducing his own list of Victorian fishes, wrote 'The Count's labours have made the work of those who follow him mainly of a supplementary nature.'

Australia's ichthyologist of the 1870s was industrious and enthusiastic; he handled a significant proportion of the fish species found in southern Australia but his publications have engendered confusion and some dismay in succeeding generations of fish taxonomists. It was left to another generation of workers, specifically Ogilby, Waite and McCulloch, to understand the problems that Castelnau had caused and to set about unravelling the tangles he had left in Australian fish taxonomy.

Interregnum

François Laporte de Castelnau died in Melbourne in 1880 and, for the next five years, research on Australian ichthyology was not dominated by any one person. The succession was not clear. After 1880, the work was performed increasingly within, or in close association with, Australian museums. Castelnau's publications had stimulated greater interest in fishes from naturalists in Australia and their study proceeded in several centres: Sydney (Edward Ramsay and William Macleay), Brisbane (Charles de Vis), and Hobart (Robert Johnston). Their publications were often interesting, particularly since they were dealing with such an imperfectly known fauna, but their work on Australian fishes generally lacked rigour. Most of their publications on fishes appeared between 1880 and 1885 and in that latter year James Douglas Ogilby arrived in Australia, bringing a new talent to the study of Australian fishes, and more besides.

In summary:

- Ramsay published independently (later with Ogilby) chiefly1880–1886.
- Macleay's main publications on Australian fishes 1879–1886.
- de Vis' main fish publications 1882–1885.
- Johnston's main fish publications 1882–1885.

In this section are also inserted brief accounts of the activities of a number of other authors whose publications at about this time included some ichthyology: Haswell, Miklouho-Maclay and Tenison-Woods in Sydney; McCoy in Melbourne; Haacke in Adelaide.

Ramsay

Edward Pierson Ramsay (1842–1916) was born in Sydney. From childhood he was fascinated by natural history, studied medicine at the University of Sydney for two years, and was as a young man most interested in entomology and ornithology. The latter was his favourite pursuit among the natural sciences. His association with the Australian Museum began in about 1860 but the Director, Gerard Krefft, was unhelpful. Ramsay was appointed Curator in 1874, after Krefft was discharged, and held the position until his retirement through ill health 20 years later (ADB; Paxton and McGrouther 1997).

Ramsay was an active and competent curator, and greatly increased the museum's collections; he also encouraged, and participated in, natural history explorations. He published many papers in scientific journals, mostly concerning birds, mammals and fish, and remained available to the Australian Museum as a consultant ornithologist until 1909.

The Australian Lungfish, made known to science in 1870, provided the subject for Ramsay's first ichthyological publications, brief notes that were published in the *Proceedings of the Zoological Society of London* (Ramsay 1871 and 1876). Thereafter, his fish papers appeared in the *Proceedings of the Linnean Society of New South Wales*, generally dealing with new records of fishes from Port Jackson. His interest in fishes was known and he served on the 1880 Royal Commission enquiring into '... the actual state and prospects of the Fisheries of this Colony' and later became a member of the Fisheries Commission, on which he served for 13 years.

Edward Pierson Ramsay. From Etheridge 1917.

At the Melbourne Exhibition of 1881 Ramsay exhibited fishes for the Australian Museum (Ramsay and Ogilby 1886) and in 1883 he visited London, representing New South Wales at the Great International Fisheries Exhibition. Only New South Wales and Tasmania of the Australian states presented exhibits and the New South Wales 'Court' was a great success. This was the occasion of Ramsay meeting Francis Day (see page 69) and negotiating the purchase, for £200, of an important collection of Indian fishes for the Australian Museum. The purchase consisted of about 2000 specimens, representing nearly 800 species, and proved a valuable addition to the Australian Museum collection (Whitehead and Talwar 1976). After the exhibition, Ramsay visited several natural history museums in Europe. He met museum administrators and zoologists and studied the composition and display of their exhibits.

Ramsay corresponded with museum curators in Australia and Europe. For instance, he appealed to Albert Günther at the British Museum for assistance in identifying a Sunfish; and there survives correspondence with Enrico Giglioli (1845–1909), Director of the Royal Zoological Museum at Florence, which appears to have begun prior to the exhibition in 1883 and continued at least until 1890 (AMA). Some exchanges of fishes and other specimens were effected in due course, between Florence and Sydney, but Giglioli promised more than he performed and Ramsay eventually found the correspondence irritating. There seemed to be no end to Giglioli's requests: aboriginal artefacts, seals, platypus, dugong and detailed lists of fishes. In 1883 he asked for '... as complete a set as possible both stuffed and in spirit ...' and proceeded with a list of about 50 genera. These appeared to represent the fishes which were taken to London for the exhibition and Giglioli hoped for more: '... I do not know whether you can dispose of any of the Tasmanian exhibits, if so I should like very much to secure a set.' (Giglioli to Ramsay, 30 September 1883, ML).

At Giglioli's representation, Ramsay received a quaint honour: Cavaliere of the Crown of Italy from King Humbert (Etheridge 1917). Another of his honours was the receipt in 1886 of the honorary degree of Doctor of Laws from the University of St Andrews

(Scotland). Thereafter he could be, and often was, addressed as 'doctor', a circumstance which was sometimes treated with mockery (WD).

Ramsay's interest in the fishes of New South Wales was consistent: he prepared an annotated list, unfinished and unpublished, which dealt with many species. Some of the accounts in his notebooks, entitled 'Fishes – Rough Notes 1880–1886', are expanded as though he contemplated producing a semi-popular work on fishes (AMA). Under each species heading there is a lengthy description, weighted heavily towards details of colour markings, with some notes on size, weight and where captured or collected. He included notes on biological information and behaviour. For instance, under the heading 'Cossyphus gouldii [*Achoerodus viridis*] ... Blue Groper' he appends the note 'feeds on Echini'; under 'Trigla poliomata [*Pterygotrigla polyommata,* a Gurnard]', he notes 'a large specimen examined last week had roe, but not ready for spawning'; and under 'Therapon cuvieri [*Pelates sexlineatus*] ... Trumpeter Perch', he says 'It has received its name from the peculiar sound emitted when first caught, resembling the words "toot", "toot", "toot" at a second or half second interval until it dies.'

The fish collection at the Australian Museum was expanded greatly through Ramsay's exchanges with, and purchases from, other institutions both within and outside Australia. Local species were secured by purchase from the Sydney fish market, donations by individuals of fish captured or stranded, and by personal collecting. In his notes for April 1882, he reported 'There is a small Seine trawl at work in the harbour towed before the wind by two boats.' He inspected the catch and 'took photos of 1 Flounder, 2 Sole, 3 Taraglin, 4 White trevally.' Over many years he conducted his own trawling excursions in Sydney Harbour, inviting friends, colleagues and museum staff to participate. In November 1882 he mentions 'Trawling between jetty near the Torpedo Station and large sailing mark near Vaucluse Point' in connection with the capture of a flounder which he thought represented a new species. Ramsay also bought specimens from professional collectors and sent museum staff on collecting expeditions. Alexander Morton (later of the Tasmanian Museum) provided a dozen species of 'Fishes from the Burdekin' and these were duly described by William Macleay (1884a).

Among Ramsay's pet projects was the retrieval of stranded Sunfish (*Mola*, but then referred to the genus *Orthagoriscus*). He kept notes on no less than eight of these impressive animals that found their way into Port Jackson during his time at the museum. Most of the specimens were enormous but Ramsay was undaunted. He numbered his specimens and in November 1882 refers to 'Orthagoriscus sp. No. 2' which was '5 ft. 8.5 In. [170 cm] from the vent to the base of the dorsal fin', found at Manly and 'Bought of Skinner for the sum of £5.0.0.' Later the circumstances of the collection of 'Orthagoriscus sp. No. 3' are described:

> A very large specimen was found aground near Chadwick's Mills in Darling Harbour. It was seen among the shipping and for some unknown reason forced itself up between the vessels and the shallows into black mud. Boats prevented its escape and finally a chain was passed under it, and it was hauled bodily up alive by a crane on to the timber skids of the wharf and placed on a truck ... This fish measured 8 feet [240 cm] in length and eleven feet [330 cm] from tip to tip of the fins ... It weighed 1 ton 3 cwt. [about 1170 kilograms].

As a stuffed skin, this fish formed part of Ramsay's display at the International Fisheries Exhibition in London, following which it was donated to the British Museum. Giglioli described the fish as *Orthagoriscus ramsayi* [*Mola ramsayi*] (Giglioli 1883). 'Orthagoriscus sp. No. 4' was bought by the museum from another Manly fisherman:

> This specimen weighed one ton 6 cwt [about 1330 kilograms], we had great difficulty in bringing it home having to tow it from Manly a distance of 5 miles or more, across the heads where as there was a Sea on we nearly lost it. Length 8 ft 4 Inches [250 cm], width across the fins from tip to tip 12 feet [360 cm].

He reports with manifest satisfaction 'This specimen is mounted and slung in the central hall 1st. floor of the New Wing.' Its installation perhaps made up for 'the very large one destroyed by Krefft.' This was an allusion to Ramsay's predecessor who had acquired 'No. 1' in 1871.

William Macleay (see page 134) was among the most enthusiastic supporters of the Australian Museum, despite being in competition with it for specimens for his own enormous collections. Ramsay noted (14 September 1882, AMA) that 'The Hon. William Macleay brought today a fine specimen of a new Cheilodactylus of which the following is a description taken from the type itself ...' The description, however, consisted merely of the number of spines and rays in the fins and a careful account of the fish's colours. A photograph of the fish is inserted and Ramsay added a later note: 'This he has called C. mulhallii.' Macleay named the fish for the fisheries inspector who brought it to him. Macleay and Ramsay were often misled by the published opinions of Castelnau, in this case by the Count's insistence that this species has a complex colour pattern (Castelnau 1872a). Castelnau, with attractively patterned juvenile specimens before him, wrote of the Strongfish, *Cheilodactylus* [now *Dactylophora*] *nigricans*, 'It is a mistake that the authors [Günther] describe this fish as of an uniform blackish grey'; Macleay, following this advice, created *Cheilodactylus mulhallii* for a dull-coloured adult specimen.

Ramsay published 10 short papers on fishes between 1880 and 1884, noting new records for Port Jackson and describing about 20 species as new. Of particular interest was *Description of a New Species of Hemerocoetes from Port Jackson* (Ramsay 1881a). The specimens were dredged from sand near the entrance to Sydney Harbour and named *Hemerocoetes* [now *Creedia*] *haswelli* for William Haswell, biologist, whose dredging expeditions provided the museum with a variety of creatures. Ramsay (1881b) also described a coral reef Pipefish from the Solomon Islands as *Syngnatus* [now *Corythoichthys*] *intestinalis*, under the misapprehension that it was a parasite dwelling in the gut of sea cucumbers; it is common on Australian coral reefs. From George Town at the mouth of Tasmania's Tamar River he was sent a tiny fish found 'clinging to stones at low water' (Ramsay 1882) and named the Broad Clingfish *Gobiesox* [now *Creocele*] *cardinalis*.

Until 1890, museum researchers in Sydney were obliged to publish primarily in overseas journals. Ramsay, at the urging of some of his staff, suggested to the museum trustees that the Australian Museum should establish its own publication. Thus in 1890 the

Records of the Australian Museum first appeared, providing thereafter a more or less effective instrument in which the staff published the results of their researches.

With the appointment of Douglas Ogilby as Scientific Assistant in 1885, Ramsay's work on fishes virtually stopped. From 1886 to 1888 Ramsay and Ogilby published 18 papers jointly, with Ramsay as senior author, although it is clear from both style and content that the work was primarily attributable to Ogilby. Ogilby's familiarity with the wider fish literature, interest in synonymies and attention to the more taxonomically significant characters are the most obvious signs of his contribution. The paucity of references cited by Ramsay is particularly noticeable: he appeared to rely principally on Günther and Castelnau. After 1888, Ramsay's fish work at the museum was in Ogilby's hands. However, Ramsay had local knowledge, expertise at obtaining specimens, a very useful personal library and contacts with overseas curators. Ramsay's most significant contributions to Australian ichthyology lay in placing these tools at the disposal of the zoologists he appointed. Two of these appointments were highly significant to Australian ichthyology: Ogilby and, in 1893, Edgar Waite.

Macleay

William John Macleay (1820–1891) is best remembered for his pastoral and political activities but he was also a patron of natural science and published a number of zoological papers. He was born in Caithness, Scotland and entered the medical school of the University of Edinburgh in 1834. He and his two younger brothers were orphaned in 1837 and his uncle, Alexander Macleay, advised him to emigrate: he came to Sydney in 1839. He was already attracted to medical and scientific studies and his relatives in Sydney encouraged his interest in natural history. Alexander Macleay, and his sons George and William Sharp Macleay, were all enthusiastic about the opportunities for discovery in the natural sciences in Australia.

In 1841 William Macleay was appointed as a magistrate and in 1848 he took over a large property on the Murrumbidgee. He became a prominent politician and continued to increase his pastoral holdings. His active career in natural history began around 1861 when he became a Trustee of the Australian Museum and the following year he helped to found the Entomological Society of New South Wales. He leased Elizabeth Bay House from his cousin, George Macleay, in 1865 and inherited his cousin's collections which were housed there. He greatly augmented these collections to make a significant natural history museum.

The Linnean Society of New South Wales was formed in 1874 and William Macleay was its first president. He associated with doctors Alleyne and Cox, both of whom had an enthusiasm for the fishes of Sydney Harbour. Haynes Gibbes Alleyne (1815–1882), a medical practitioner who became medical adviser to the government and president of the Medical Board in Sydney, appears to have been an ichthyological mentor to Macleay, being senior co-author of Macleay's first fish papers. James Charles Cox MD (1834–1912) participated in collecting excursions, donating fish (and other) specimens to the Australian Museum or sometimes directly to Macleay. Cox was the author of a short pamphlet on fishes of New South Wales: *An Alphabetical List of the Fishes Protected under the Fisheries Act of 1902, with Remarks on each Species* (1905).

The HMS *Challenger* set out from Portsmouth in 1872 on a three year voyage of oceanographic exploration. The vessel called at Sydney in 1874 and Macleay entertained the scientific staff at his home, showed them his museum and arranged field trips on Sydney Harbour. His own interest in marine life appears to have commenced at about this time; in anticipation of the *Challenger* visit, he had begun collecting the local fishes. Biologists from the *Challenger* introduced Macleay to dredging, an occupation which he embraced with his usual enthusiasm: he hired a steam yacht, purchased a dredge, and was soon conducting expeditions in and around Port Jackson. By the end of 1874 Macleay had begun to plan an ambitious collecting trip to the tropical north.

William Macleay. From Fletcher 1893.

In 1875 he bought and fitted out the barque *Chevert* for a scientific expedition to Torres Strait and New Guinea. Public expectations for this expedition were unduly high, thanks to the press, who represented it as an exploration of New Guinea. Macleay was satisfied with the natural history collections made but the enterprise was widely perceived as a failure. He later referred to it as 'my fishing trip to New Guinea.' The *Chevert* expedition lasted about 4 months. Significant collections of fishes were made at Cape Grenville in North Queensland, the islands of Torres Strait (particularly Darnley Island) and Hall Sound in New Guinea.

Macleay published over 70 zoological papers, mostly on entomology and ichthyology. From 1877 to 1886, he authored over 30 papers on fishes, published exclusively in the *Proceedings of the Linnean Society of New South Wales.* His work consisted of an account (with Alleyne) of the fishes collected on his Torres Strait expedition; many short papers describing new records or new species; description of a collection from Darwin; four papers on the fishes of New Guinea (Macleay 1882d, 1882e, 1883a, 1883b); summaries of the Australian members of a few fish families: Clupeidae (1879b), Mugilidae (1880) and Pleuronectidae (1882a); his *Descriptive Catalogue of Australian Fishes*; and some papers on elasmobranchs (with Miklouho-Maclay).

In *Ichthyology of the Chevert Expedition* (Alleyne and Macleay 1877) 145 species of marine fishes were listed, 47 of which were supposedly new although few of these are still regarded as valid; the sharks and rays were reserved for a later paper (which did not eventuate). Most of the species listed were taken in far north Queensland waters, particularly Torres Strait, and a few were from the adjacent coast of New Guinea. The report showed

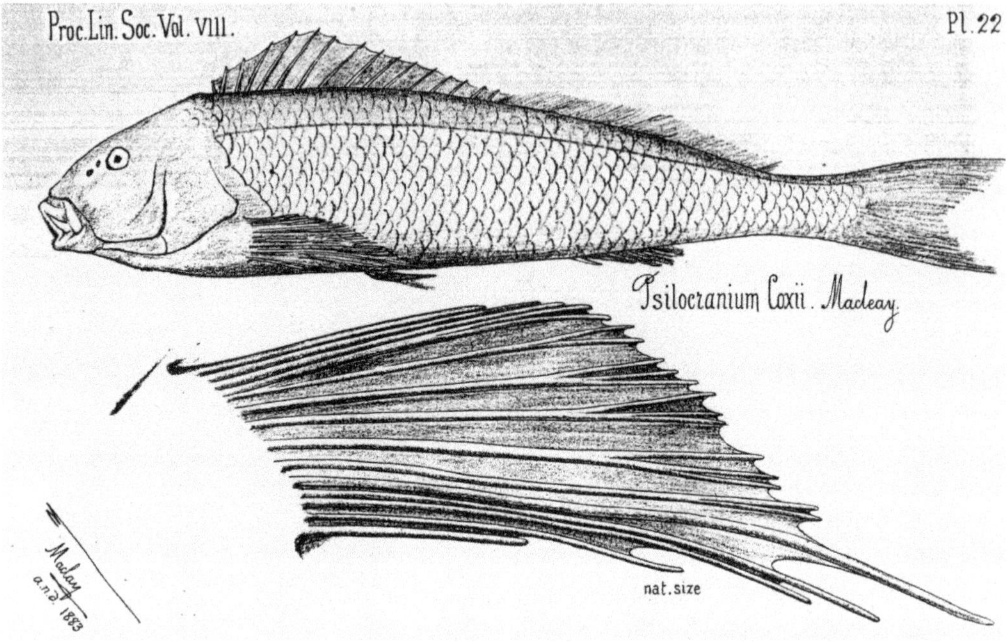

Macleay's *Psilocranium coxii*, a synonym of *Dactylophora nigricans* (Richardson, 1850), was drawn by his Russian friend, Baron Miklouho-Maclay. From Macleay 1884.

much more enthusiasm than insight. Apparently Dr Alleyne was familiar with the species of Port Jackson but he was lost in the taxonomy of northern Australian fishes. However, the work had its virtues: all of the 'new' species were illustrated and, wherever possible, a number of each species was preserved and carefully labelled. Many of these specimens still survive, tucked away in the Macleay Museum at Sydney University or in the Australian Museum; they have been the subject of extensive study by fish taxonomists over the years. Among the new fish described (Alleyne and Macleay 1877) were the Bluespotted Trevally, *Caranx bucculentus* from Cape Grenville, Queensland and the Brassy Trevally, *Caranx papuensis* from Hall Sound, New Guinea.

Among Macleay's short papers were several accounts of individual fishes and collections, most from the vicinity of Sydney. New species were described but some of these publications provided Sydney fishes with unwanted synonyms. In dealing with a specimen of a female Rainbow Fish, *Heteroscarus acroptilus* (Richardson, 1846), Macleay described it anew as *Heteroscarus castelnaui*, celebrating '... the founder of the genus [*Heteroscarus*], and the most distinguished of our naturalists - the Count de Castelnau.' (Macleay 1879a) This was appropriate in that the Count had already endowed this species with two superfluous names. In the same paper, Macleay described the well-known Spotted Whiting, *Sillaginodes punctata* (Cuvier, 1829) as a new species, creating a new genus for it into the bargain. The Strongfish, *Dactylophora nigricans* was brought to his attention by the Fisheries Inspector, Thomas Mulhall, and was duly described as *Chilodactylus mulhallii* (Macleay 1882c); in 1884, the same species was brought to him, once again by Mulhall, and Macleay described it again as a new species, *Psilocranium coxii*, creating a new genus for it as well. In fairness, it should be known that '... the taxidermist of the Institution [Australian Museum] had already skinned the fish, and thrown away the body.' (Macleay 1884b).

Macleay also received fish from further afield. Some of these collections were from inland waters: the Reverend Tenison-Woods presented him with a small collection from the Palmer River (Macleay 1882b); Alexander Morton, then collecting for the Australian Museum, provided fishes from the Burdekin River in north Queensland (Macleay 1884a); and a friend from Yass sent him fishes from the upper Murrumbidgee (Macleay 1885). A report on a collection of fishes from Darwin (Macleay 1878) included the description of over 20 new species, of which a few are still valid. Among the freshwater fish presented by Tenison-Woods was the Freshwater Sole, *Synaptura* [now *Brachirus*] *selheimi* collected by Dr Selheim[8] from the Palmer River using 'a very small hook baited with a fragment of a blade of grass.' Morton's efforts on the Burdekin with '... nets of various kinds, lines, dynamite and every appliance for catching or trapping fish that could be procured ...' produced a wide variety, including the Sooty Grunter, *Therapon* [now *Hephaestus*] *fuliginosus*.

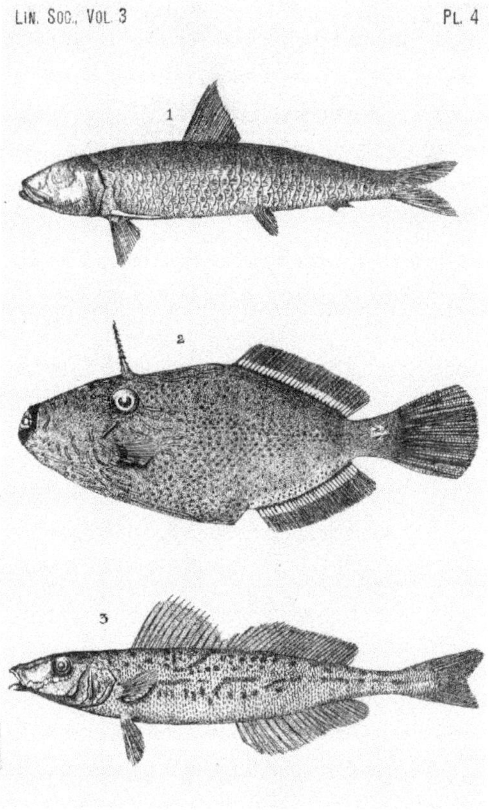

Etrumeus jacksoniensis [synonym of *E. teres*] from Port Jackson; *Monacanthus guttulatus* [synonym of *Acanthaluteres brownii*] from King George Sound; *Isosillago maculata* [synonym of *Sillaginodes punctata*] from King George Sound.
From Macleay 1879.

In 1882 and 1883 Macleay published a series of four papers on fishes from New Guinea: 'Mr. Andrew Goldie, the well-known New Guinea Explorer and Naturalist, has from time to time for the last year or more, sent me collections of Fishes taken by him at Port Moresby, and Cuppa Cuppa.' Goldie (1840–1891), who had settled in New Guinea in the late 1870s, ran a trading post and explored the country; he collected zoological specimens for the Australian Museum and plant specimens for Ferdinand von Mueller. In these four papers a total of 405 species were listed, 56 of which were described as new, although most had been described by earlier authors.

The Fishery Commissioners of New South Wales organised a trial with a beam trawl, run from the *S.S. Dove*, between Jervis Bay and Port Jackson in 1883. The results were poor but Macleay gave an account of the few fishes taken (Macleay 1884c) and described two new

8 Philip Selheim (1832–1879) was a German pastoralist with an interest in natural history. He emigrated to Queensland in 1855 and, in 1874, was a government mining official, warden of the Palmer River goldfields.

species, a gurnard and a skate, from about 40 fathoms. A few months later, he reported two more new species from trawling outside Port Jackson (perhaps from further examination of the catch of the *Dove*), a flathead and a stingaree (Macleay 1884e). All four of the new trawled species are still recognised as valid. Macleay was optimistic about the commercial possibilities of trawling in New South Wales, drawing attention to the '... existence of a true skate so near us and in such apparent quantity...' This referred to the Sydney Skate, *Raja* [now *Dipturus*] *australis* (Macleay 1884e). He criticised the conservatism of local fishermen who fished only with lines and seine nets: '... beyond these miserable appliances our fishermen seem incapable of going.'

In 1881 Macleay was president of the Linnean Society of New South Wales. The Society published his catalogue of Australian fish in four parts in 1881–1882, and issued his supplement to this in 1884.

> The following catalogue is compiled with the view of obviating or lessening the difficulty which the student of Ichthyology in this Country has to encounter from having to refer for the history and identification of species, to numberless publications in a variety of languages, which are only to be found in very large and well-stocked libraries. (Macleay 1881–1882)

Macleay listed 1133 species in his *Catalogue*, but he was quite uncritical of much of the material which he included and sparing with references, so his catalogue was much less complete and less reliable than he anticipated. A major weakness lay in Macleay having accepted Günther's definitions and descriptions without any attempt to verify them, and in naively copying Günther's errors. Worse, he accepted Castelnau as an authority for many species.

The *Catalogue* included descriptions of 88 new species, duly named. The quality of the descriptions was poor and this time illustrations were not presented. Most have simply encumbered the synonymy of known species although about one in five is at present recognised as valid. From these descriptions it can be seen that Macleay had access to specimens from many parts of Australia although most of his 'new' fishes were from Port Jackson (34) and north Queensland (21). Some remarkably common fishes were described from Port Jackson for the first time in the *Catalogue*: the Eagle Ray, *Myliobatis australis*; Rock Blackfish, *Girella elevata*; Spotted Dragonet, *Callionymus* [now *Repomucenus*] *calcaratus*; and from notes and a single large specimen supplied by E.P. Ramsay, the Pearl Perch, *Glaucosoma scapulare*.

A supplement to the *Catalogue* (Macleay 1884d) added another 157 species to Macleay's list of Australian fishes. Many of the additions were derived from Johnston's (see page 150) recent publications on Tasmanian fishes; also, Klunzinger's (see page 76) work on Australian fishes had at last come to Macleay's notice: '... I am enabled by the kindness of Dr. Klunzinger to include now some fishes described by him some years ago in the publications of German Societies, but not seen by me.' Macleay believed that Australia's animals should be described in Australian publications by zoologists resident in Australia:

> Without questioning the superiority in Biological Science of the Germans, it still seems unnecessary, as it is degrading, for the simple purpose of the identification of species and the compilation of a classified Catalogue of the animal life of the country to send, as South Australia has done, all her collections to Vienna, to be recorded in a foreign tongue and in foreign publications. (Macleay 1884d)

This view is easily understood: the German publications were not easily obtained in Australia and they had to be translated. The problem with Macleay's sentiment is best illustrated by his own *Catalogue*. He did not have sufficient knowledge of ichthyology to deal with a complex fauna that had already attracted a considerable literature.

Macleay's *Catalogue* was intended to provide a basis from which the study of Australian fishes could proceed but the work was premature and caused confusion. It was not possible to use the *Catalogue* to identify, with confidence, even common species. There were no keys, no illustrations, the descriptions were inadequate and often inaccurate, many species appeared under two or more names, and synonymies were not presented. Suffice it to say that this well-intentioned publication encumbered the literature with a plethora of new names, most of which have served to swell the synonymies of previously described species.

The Macleay Museum stands in the grounds of Sydney University, a quaint and rather equivocal symbol of Macleay's influence. His collections had been housed adjacent to his residence at Elizabeth Bay until Macleay presented them to the University in 1890. The gift was made on condition that a building be erected for the purpose, that his own curator (George Masters[9]) should continue to have the care of the collections, and that they be accessible to students of natural history and members of the Linnean Society of New South Wales. Macleay provided a generous endowment for a curator's salary and the government provided funds for the Macleay Museum to be built (Fletcher 1893). However, the museum gradually lost the function intended by its founder as the University departments encroached on its space and modified it to suit their purposes (Stanbury and Holland 1988). Macleay would certainly be obliged to hunt carefully to locate his specimens within the structure today, confined as most are to lofty and inaccessible niches.

Most of Macleay's zoological work was purely descriptive and, since he had no training in zoology, it is not surprising that it was imperfect. His enthusiasm for describing new species was unfortunate as he had not the skills to assess his specimens critically. He did leave a legacy of large collections and at least his type material was lodged in Australia. He gave generous support to the Australian Museum and the Linnean Society of New South Wales, donated his collections (and their curator!) to Sydney University, and endowed scholarships for the study of natural science at that institution. Perhaps his most significant contribution to ichthyology lay in the practical encouragement he lavished on natural science in Australia.

9 George Masters (1837–1912) was an entomologist and natural history collector who emigrated from England to Melbourne in 1856. He travelled throughout Australia as a collector and was assistant curator at the Australian Museum throughout the tenure of Gerard Krefft. He collected extensively for Macleay and accompanied him on the *Chevert* expedition. Although insects were his main interest, Masters also collected other groups, including fish. He was appointed curator of Macleay's Museum at Sydney University in 1874.

Miklouho-Maclay

Nicolai Nicolaevitch Miklouho-Maclay. From Greenop 1944.

Nikolai Miklouho-Maclay (1846–1888), a Russian with Scottish ancestry, was a friend and student of Ernst Haeckel (1834–1919), a celebrated German naturalist. Miklouho-Maclay spent several years in the 1870s living on the coast of northern New Guinea where he carried out ethnological studies in the region and investigated its natural history. His interest in the native people he encountered moved him to try to prevent their exploitation by European settlers (Greenop 1944).

Miklouho-Maclay visited Sydney in 1878 and was befriended by William Macleay with whom he collaborated on several projects. He stayed initially at Elizabeth Bay House, where Macleay's collections were housed, and worked on them. He wished to study the marine life of Port Jackson and promoted the concept of a biological research station on Sydney Harbour. Through Macleay's influence, the station was built at Watson's Bay in 1881 and Miklouho-Maclay worked there when he was not travelling. The gentlemen of the Linnean Society established six rules for the use of the station, ensuring that access would be granted to '... all Naturalists of the male sex, without distinction of nationality ...', that '... a fee of Five shillings a week' would be charged and that '[n]o tenant shall disturb any other by singing, or whistling, or any other unnecessary noise.' (anon. 1879).

Marine Biological Station, Watson's Bay. From Greenop 1944.

Three papers on elasmobranchs were published in the *Proceedings of the Linnean Society of New South Wales*, jointly authored by Macleay and Miklouho-Maclay. These papers were intended to be the beginning of a much larger study of the 'Plagiostomata [sharks and rays] of the Pacific' but this was pursued no further. Castelnau was tetchily indignant to find that one of his own illustrations had been omitted to make room for the plates associated with this paper (see Castelnau to Ramsay, June 1879, p. 128). His view was not without some substance as the publication (Miklouho-Maclay and Macleay 1879) contained little new information.

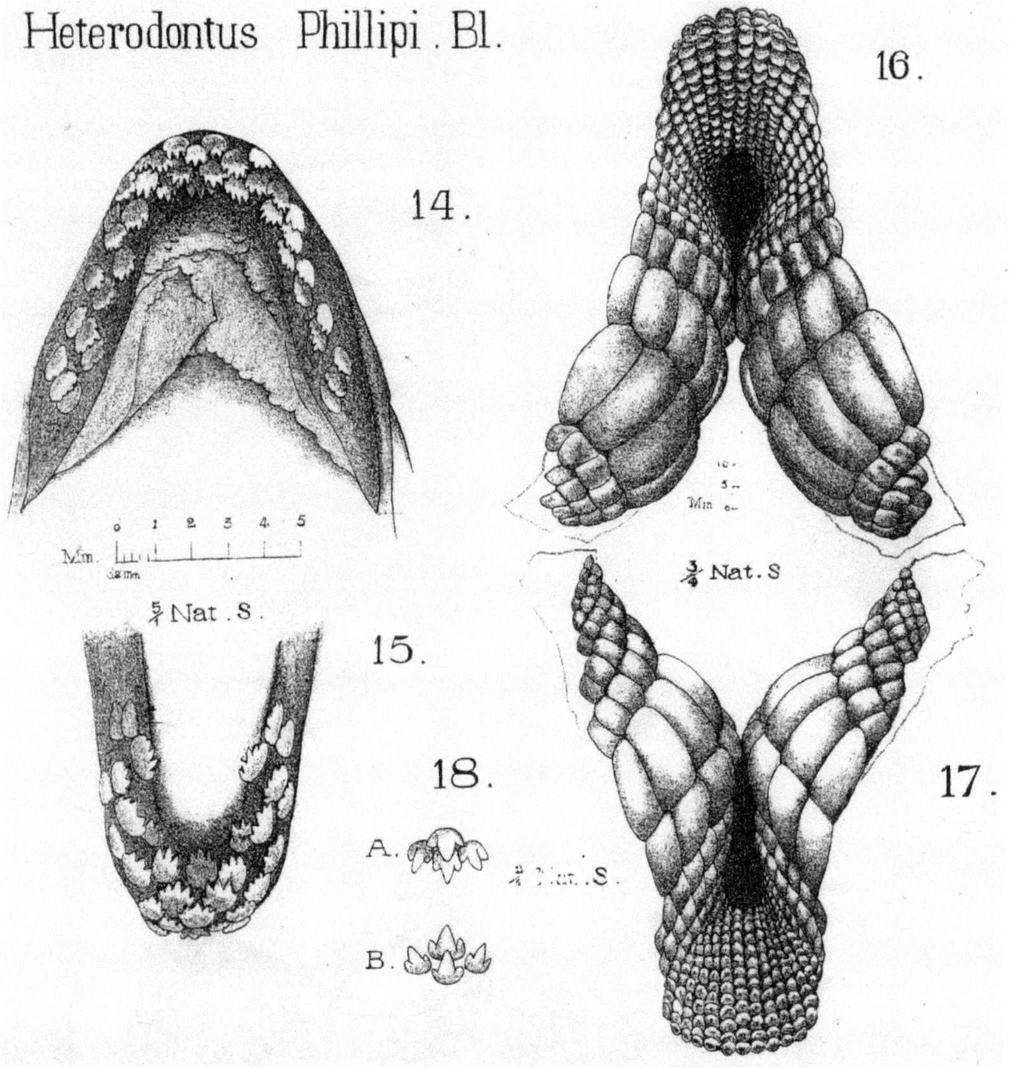

Heterodontus portusjacksoni dentition. From Miklouho-Maclay and Macleay 1879.

On Miklouho-Maclay's early death in 1888 his Australian wife, Margaret, donated part of his collection to the Macleay Museum; the greater part was sent to the Miklouho-Maclay Institute of Ethnology and Anthropology in Moscow.

Haswell

William Aitcheson Haswell (1854–1925) was born in Edinburgh and studied science at the University of Edinburgh. Thomas Huxley and Charles Wyville Thomson (leader of the *Challenger* expedition) were among his teachers and guided him towards the study of marine zoology.

Haswell came to Sydney in 1878 and began working on marine animals in the small laboratory at Watson's Bay on Port Jackson. His earliest zoological studies were conducted on marine collections from Macleay's *Chevert* expedition, polychaete worms providing the subject of his first publication. At this time the trustees of the Queensland Museum were looking for a curator; Haswell, who already had a promising reputation as a young zoologist, was offered the post. He accepted but the salary was poor and, within a year, he became aware of better opportunities in Sydney (ADB). Haswell was also aboard the British Survey ship *Alert* while it was in Queensland waters and wrote to a friend at the Queensland Museum:

William Aitcheson Haswell. From Walkom 1925.

> I have taken advantage of an exceptional opportunity for a zoologising excursion with the 'Alert,' and have seen a great deal that is new to me and obtained a good many specimens. (QMA)

He was a demonstrator in the zoology department of the University of Sydney in 1882, assistant to E.P. Ramsay at the Australian Museum in 1883, and lecturer in comparative anatomy and zoology at the University in 1884. In 1890, Haswell was appointed to the chair of biology at the University of Sydney and he remained an active member of the scientific community for many years. His successor at the university was S.J. Johnston, a friend of American ichthyologist, David Jordan; Johnston is known to Australian ichthyology through publishing a practical guide to the dissection of the Stingaree, *Trygonoptera testacea* for his students (Johnston undated).

Haswell became known for his part in writing a *Text-book of Zoology* with Professor T. Jeffrey Parker of the University of Otago, New Zealand. This book was first published in 1898 and ran through many editions. However, Haswell's largest contributions related to the marine invertebrates, particularly crustaceans. His *Catalogue of the Australian Stalk- and Sessile-Eyed Crustacea* (1882a), although largely a compilation, was an essential text for subsequent researchers.

Ichthyology was incidental to the main thrust of Haswell's work but he published a scattering of studies, mostly dealing with developmental or anatomical features of elasmobranchs. His *Studies on the Elasmobranch Skeleton* (1884a) comprised descriptions of fresh skeletons, and a few skulls, of southern forms and he commented on apparent relationships. A *Note on the young of the Saw-Fish Shark, Pristiophorus cirratus* (1884b) described embryos, noting the presence of a rudimentary shell around early embryos and

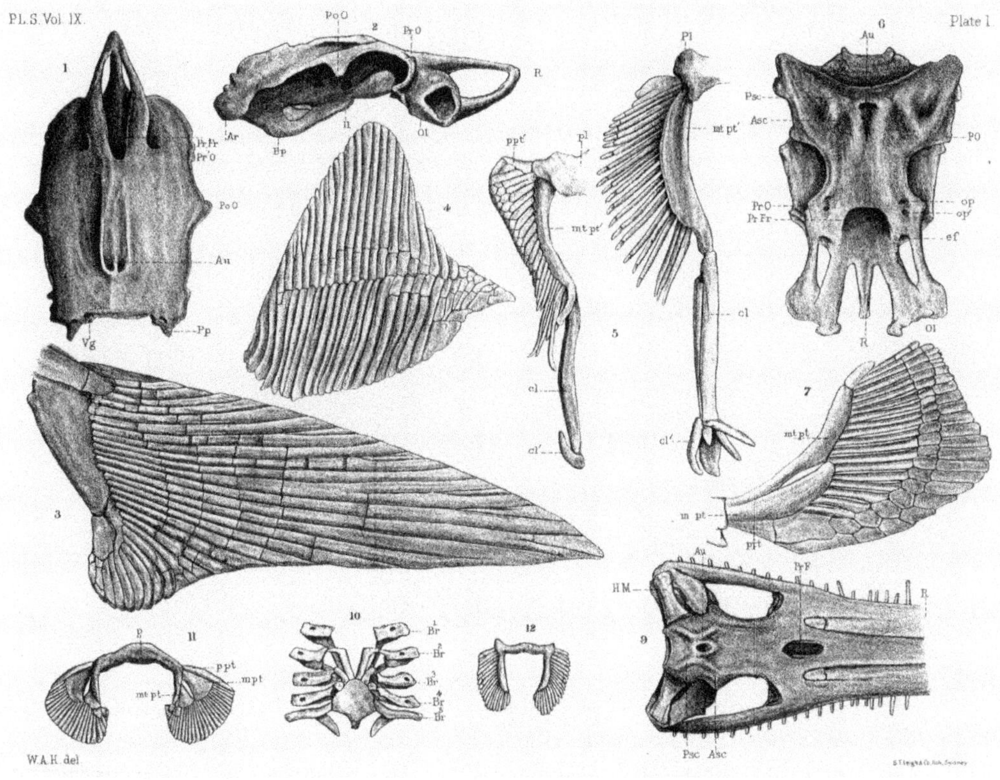

The skulls, fins and fin supports of several shark species. From Haswell 1884.

the presence of vascular uterine villi. *On the Development of Heterodontus (Cestracion) phillipi* [now *H. portusjacksoni*] (1897) exploited his discovery of eggcases containing developing embryos at Jervis Bay, south of Sydney and consisted of a thorough description of the histology of early stages of development in the Port Jackson Shark. He used the anatomy of the paired fins of the Australian Lungfish to modify the 'general theory of the vertebrate limb' (Haswell 1882b) and gave an account (Haswell 1890) of the histology of early stages of development in the Shore Eel, *Alabes dorsalis*.

Tenison-Woods

Julian Edmund Tenison Woods (1832–1889) was born in London and developed an interest in natural history while living on Jersey as a teenager. In spite of ill-health and some family opposition he became a catholic priest. He came to Tasmania in 1855 and was ordained in 1857 after time spent at the Jesuit seminary at Sevenhill, South Australia. He is remembered for his part in founding, with Mary MacKillop, the Sisters of St Joseph whose purpose was teaching the children of poor families, and for his original and advanced ideas as an educator (Press 1979).

Father Woods spent ten years in the south-east of South Australia where his work involved a great deal of travel on horseback but he found time to study the natural history of the region, to correspond with scientists in England, and to write many papers on South

Australian geology. His books, *Geological Observations in South Australia* (1862) and a *History of the Discovery and Exploration of Australia* (1865), were ambitious, well-informed works. He returned to Adelaide in 1867 and was made Director-General of Catholic Education. He adopted the form 'Tenison-Woods' for his works to prevent confusion when he discovered that there were two other clergymen-scientists publishing as 'Woods'.

Julian Tenison Woods. From Walkom 1925.

For the greater part of his life, excepting only his busiest time in Adelaide, Woods travelled through the Australian countryside; he continued to study natural history wherever he found himself. In Tasmania, his studies were intensified and commonly involved seashore life, particularly molluscs. He sent specimens to Professor McCoy in Melbourne, gave talks on natural history and presented papers to the Royal Society of Tasmania. During these few years he published about 20 papers, most on living and fossil molluscs, and on geology.

In 1876 he once again left Tasmania, this time to resume his travels through New South Wales and Queensland. He had entered on his most prolific period of natural history study and his work ranged over many subjects: geology, palaeontology, botany and zoology. Most of his zoological publications were on marine organisms, particularly molluscs, echinoderms and corals. During this time he sent specimens of freshwater fish to William Macleay.

In *Fish and Fisheries of New South Wales* (1882) Woods aimed 'to give a popular account of our useful fishes to the colonists'. The book inevitably contains errors and many omissions but it accomplished the stated purpose and remains a readable account of great historical interest. He refers to the '... wholesale destruction within the harbour [Port Jackson] caused by stake nets and seines with meshes almost small enough for a naturalist's hand-net ...' and to the problems in bringing a catch to market:

> Many freights of valuable fish are thrown away or abandoned and left to rot on the beaches about Broken Bay during the summer months, and not unfrequently in the colder months. This happens when the fisherman coming down the river with his boat fully laden finds a strong southerly wind blowing in the offing. He has no means of taking his fish to market overland, although the distance from the head of Pittwater to Manly cannot be more than about twelve miles. Accordingly, unless the head wind takes off or veers to a favourable quarter within twenty-four hours at most, he cannot possibly expect to save his freight, unless he have the good luck to catch the steamer which makes occasional visits to Broken Bay.

This book was prepared at the request of the New South Wales Fisheries Department, for the London Fisheries Exhibition of 1883. For this he won the Royal Medal, awarded by the King of the Netherlands for the best publication of the exhibition.

Fish and Fisheries was illustrated by photographs and a few of Frederick McCoy's plates. Woods was realistic as to the value of the catalogues of the day as a practical means for the identification of Australian fish and thought that his guide would help:

> Instead of taking an unknown fish and trying to make it out by the aid of a catalogue, such as Günther's or Macleay's, let the student compare a fish, the scientific name of which he is sure of, with the description given of it in these works. When this has been done in the case of ten or a dozen species in different families, the work of identification will be comparatively easy. All who are in reach of the Museum in Sydney, or any of our colonial collections, will find themselves so aided by the named species, that instead of a difficult study the whole process of learning will be a delightful recreation. (Tenison-Woods 1882).

He makes it sound so pleasant, and his book was certainly more user-friendly than the catalogues to which he refers, although only a small proportion of the fish fauna was represented.

Much of the information offered in *Fish and Fisheries*, and carefully acknowledged, was extracted from a series of newspaper articles by Mr E.H. Hill in 1874; from the *Report of the Royal Commission on the Fisheries* (1880); and from Günther's *Study of Fishes* (1880).

The book was a good example of Woods the Educator at work. He set out to inform but even his introductory synopsis of ichthyology was written with a light touch. He included some pleasant distractions to avoid making the account of fishes and fishing too tedious. The reader learns of his excursion in Tasmania's D'Entrecasteaux Channel with the barracouta fishermen; hears that the young of a common blenny '... are admirable objects for seeing the circulation of the blood under the microscope'; and Woods gives an engaging account of 'The Phyllopteryx' (Leafy Seadragon, *Phycodurus eques*):

Murray Cod, *Oligorus macquariensis*, a name referring to the Trout Cod. However, the plate shows the true Murray Cod, *Maccullochella peelii* (Mitchell, 1839). From Tenison-Woods 1882.

PLATE XXXI.

Cossyphus gouldii.—RICHARDSON.
THE BLUE GROPER.

Blue Groper, *Cossyphus gouldii*, a name referring to the Western Blue Grouper. The plate represents the Eastern Blue Grouper, *Achoerodus viridis* (Steindachner, 1866). From Tenison-Woods 1882.

> But of all the curious fishes that ever were seen *Phyllopteryx* is the most extraordinary. It is the ghost of a seahorse, with its winding sheet all in ribbons around it; and even as a ghost it seems in the very last stage of emaciation, literally all skin and grief. The process of development by which this fish attained to such a state must be the most miserable chapter in the history of 'natural selection'. If this be the 'survival of the fittest', it is easy to understand what has become of the rest ... The odd thing about these strange fishes is that their tattered cerements are like in shape and colour to the sea-weed they frequent, so that they hide and feed with safety. Thus the long ends of ribs which seem to poke through the skin to excite our compassion are really 'protective resemblances,' and serve to allure the prey more effectually within reach of these awful ghouls. The *Phyllopteryx* is therefore, in spite of his rags and emaciation, an impostor, and like many a sturdy human beggar puts on the aspect of misery more effectually to ply his trade. The appendages to the spines are well worth a study. Just as the leaf-insect is imitative of a leaf, and the staff insect of a twig, so here is a fish like a bunch of sea-weed. If this is development, it stopped here only just in time; one step more and it would have been a bunch of kelp. (Tenison-Woods 1882) (See figure on page 91.)

Woods accepted an invitation from an old friend, Sir Frederick Weld, then the Governor of the Straits Settlements, to come to Malaya for the purpose of reporting on the mining prospects of certain regions. From 1883 to 1886 he travelled through Malaya, Java, the Philippines, China and Japan. He was able to collect many geological, botanical and zoological specimens and presented reports on the mining prospects in several parts of Malaya. Among his last works was *Fisheries of the Oriental Region* (1888). It is a rather rambling account which ranges through fish poisons and isinglass, pirates and the aerial respiration of fishes (among many other topics) and was apparently one of a number of papers that he dictated from his bed, at the end of his life.

de Vis

Charles Walter de Vis (1829–1915) was born in Birmingham, England. He took his B.A. at Cambridge in 1849 and commenced a career in the church. He became a deacon in 1852 and a rector in Somerset in 1855. However, he had been interested in science for many years and left the church to take positions that allowed him to study natural history. He organised natural history collections and became curator in the Queens Park Museum, Manchester (Mather 1986).

De Vis was 41 when he arrived in Queensland intending to study Australian natural history. Twelve years later, in 1882, he was appointed curator of the Queensland Museum, supposedly on the recommendation of the Reverend Tenison-Woods (Mack 1956; Mather 1986). However, Tenison-Woods' 'recommendation' was guarded, at best. He stated that he knew de Vis only through his having written on 'geological matters' in a local paper and through inspecting his fossil collection 'which had evidently been gathered with much trouble and care' although the fossils were not classified. Woods continued,

> With regard to Mr Devis' zoological knowledge I have never had any means of framing an opinion & not knowing his wishes with regard to the appointment of Curator I did not make any enquiry & therefore cannot speak of his fitness or qualifications. He appeared to be a gentleman of education & beyond this I can offer no opinion. (Tenison-Woods to Commissioner of Public Works, 24.11.81, QMA)

A fine neutral reference, but de Vis was appointed in spite of this.

De Vis left Rockhampton in 1882 but some of his fellow naturalists continued to send him specimens, including fish. William Jaggard, a young schoolteacher at Rockhampton, was among his most enthusiastic donors, sending drums of fish in exchange for having the specimens named. Jaggard wrote repeatedly in 1884, arranging transport of a large barramundi and some fish collections from Rockhampton, Masthead Island, and elsewhere. He also sent a 'Giant Perch' from Rockhampton to Brisbane which, to Jaggard's dismay, required '5 cwts' (250 kilograms) of ice. (Jaggard to de Vis, various dates in 1884, QMA)

Charles Walter de Vis. From Johnston 1916.

His staff was small but de Vis was an active and conscientious worker in many fields and the greater part of his publications during this time related to vertebrates. In her history of the Queensland Museum Patricia Mather (1986) heads a section 'The Desperate Years, 1893–1910'. This was the time of the economic depression when the museum staff was reduced to only five and very little money was available for salaries, financing collectors or even routine maintenance. This must have been a frustrating climate for de Vis, for whom the best interests of the museum were clearly so important. He retired, reluctantly, at the age of 76 (in 1905), but remained on the consultant staff of the museum until 1912.

De Vis published many papers on fishes between 1882 and 1885, and a few thereafter. Generally this work consisted of brief accounts of new species, or new records for Queensland. From 1881 his papers appeared in the *Proceedings of the Linnean Society of New South Wales*. De Vis was a founding member of the Royal Society of Queensland, established in 1884, and he used their Proceedings when that journal was launched in the same year. The *Annals of the Queensland Museum* was first published in 1891 and, although the early issues appeared irregularly due to the poor financial position of the museum, de Vis used this medium for reports based on the museum collections (Mack 1956); few of these related to fishes.

De Vis published over 100 papers on natural history during his time in Australia and fishes formed the subject of about one fifth of these. His preoccupation with fishes was greatest during his first five years as Curator, during which he described nearly 200 new species. However, he had no real appreciation of the complexities of fish taxonomy; at present, only about one in five of his 'new' species is regarded as valid. His descriptions were often brief and inadequate, his knowledge of the literature was rudimentary, and he had no informed colleague to offer criticism of his work. He made many of the same errors as Castelnau and Macleay, lacking an appreciation of variations in form and colour with sex, growth and environment. This frequently led him to describe the same species under various names. For instance, the Blue Tuskfish, *Choerodon cyanodus* (Richardson, 1843) is described under three names; Günther's Wrasse, *Pseudolabrus guentheri* Bleeker, 1862, under four names; a Grouper, *Epinephelus fasciatus* (Forsskål, 1775) under three names. Each of these fish had already been described but de Vis overlooked the earlier work. He also provided the Scat, *Scatophagus argus* (Linnaeus, 1766) with two unnecessary names and commented that 'Two other species of Scatophagus are in the Queensland collection ... but both being quite young, I forbear to notice them further.' (de Vis 1884b) A small mercy.

In 1884–1885 de Vis published a numbered series of five papers, *New Australian Fishes in the Queensland Museum* (de Vis 1884a, 1884b, 1884c, 1884d, 1885), largely concerning fishes collected in Queensland. For a year or two both before and after this he issued shorter papers on fishes that he had obtained for the Museum. His role in acquiring fishes was not at all passive: he understood that the Queensland fish fauna was imperfectly known and he was successful in obtaining specimens. He established useful contacts with interested residents and sent out collectors, particularly to North Queensland waters.

The specimens that de Vis obtained came from many regions within Queensland but the majority were marine species from Moreton Bay and the lower reaches of the Brisbane

River. His next largest collections were from Cardwell and Cape York in North Queensland. He did not seem to appreciate the full importance of specifying the locality from which his specimens came, often citing 'Queensland', 'Queensland coast', 'Great Barrier Reef'. This may have been in part because his collectors gave insufficient information, perhaps due to poor briefing by de Vis. He also received a few small fishes from South Australia, his published work on which (de Vis 1883a and de Vis 1885) was brief and without merit.

Some of the fishes de Vis described were from elsewhere in the Pacific, in particular the New Hebrides (Vanuatu); and from the Bismarck Archipelago, north-east of New Guinea. In *Fishes from South Sea Islands* (de Vis 1884e) he listed 30 fish, with brief descriptions; for half of these he specified localities, the rest being from 'South Sea Islands'. He explained the origin of the specimens:

> To Government agents and captains employed in hiring hands for the plantations, I am indebted for several opportunities of examining fish from the prolific waters around the islands from which the labour supply is derived. (de Vis 1884e)

It appears that he numbered the notorious 'blackbirders' among his donors.

De Vis was alert to opportunities for expanding the fish collections: 'Hearing that a graving dock was being pumped dry, I sent for anything there might be left in it.' (de Vis 1882a) This produced some anchovies. The most reliable of the professional collectors employed by the Queensland Museum was Kendall Broadbent (1837–1911). He was initially engaged by Haswell during his brief period in charge of the Queensland Museum and the contract was renewed by de Vis. Broadbent collected for the Museum between 1880 and 1893, prior to which he had sent North Queensland fishes to Count Castelnau (1873) and to E.P. Ramsay at the Australian Museum (1877–1879). He sent animals, including fishes, from many parts of Queensland and Torres Strait (Mather 1986).

Broadbent, reasonably enough, liked de Vis to be aware of his trials and tribulations. He wrote from Tully, North Queensland in October 1882:

> I have a touch of the fever again. We have had a great deal of rain on the Tully River, and the scrubs are very unhealthy, we have to go into such nasty places. (QMA)

He spent time with Frank Jardine, who had settled at Somerset on Cape York. Jardine (1841–1919) had travelled overland with stock nearly 2000 kilometres from Rockhampton and was known for assisting natural history collectors and explorers. From Somerset, Broadbent wrote:

> The fish here are nearly all different to Cardwell. But I am at a great disadvantage, neither dynamite or net ... only line. Mr Jardine has a net here but it is no good, all broke ... I shall want it [dynamite] more on the reefs than I do here. (February 1884, QMA)

Again in need of supplies, Broadbent wrote before leaving Thursday Island:

> I shall want more spirits before I can go. Please send me my Drums full by return steamer … I shall also want Dynamite and caps, I cannot get these here. Please send me a good strong pair of scissors also a good skinning knife … (April 1884, QMA)

And after the trip to Cape York:

> I have been out to the reefs and a rough time I had of it, blowing strong south east winds all the time. If you had sent me plenty of spirits, I would have got you a fine collection of fish … I am getting very thin and weak, and shall not be able to go inland this year. Those wet scrubs at the Cape have done me no good and this last trip roughing it in small boats has nearly finished me up. You will please allow me to come home about November. (June 1884, QMA)

Some of de Vis' most interesting specimens came from Moreton Bay, Brisbane. These included the remarkable Pineapple Fish, *Cleidopus gloriamaris* for which he established a new genus (de Vis 1882b); the Venus Tuskfish, *Choerops* [now *Choerodon*] *venustus* (de Vis 1884f); and the Ornate Wobbegong, *Crossorhinus* [now *Orectolobus*] *ornatus* (de Vis 1883a). From Cardwell came the Shark Mullet, *Mugil* [now *Rhinomugil*] *nasutus*, unmistakably a mullet, but with the head modified to a remarkable degree (de Vis 1883b); and from a little further north at the Murray River came the Eastern Eelgoby, *Leme* [now *Taenioides*] *mordax* (de Vis 1883a).

When J.D. Ogilby began publishing in Sydney in 1885, de Vis virtually suspended his own work on fishes. He may have had an inkling of the inadequacy of many of his own descriptions and, when de Vis appointed Ogilby as his assistant in 1901, ichthyology formed the greatest part of Ogilby's work. De Vis' disappointment when he had first-hand experience of Ogilby's unreliability was profound but there was no opportunity for him to procure another assistant. For the remainder of his tenure as Curator de Vis granted Ogilby only intermittent and limited access to the museum collections and library.

Fortunately, de Vis' influence on ichthyology was largely confined to the fishes of Queensland waters and to the brief period of the interregnum, but he created a significant amount of taxonomic confusion. Boulenger of the British Museum was scathing in his criticism of de Vis' taxonomy, making no allowance for the difficulties under which he laboured (Ingram 1990). To de Vis' credit, he accumulated an extensive zoological library and large collections of the Queensland fauna. His successors in Queensland ichthyology benefited from these but were hampered by his undiscriminating approach to the description of new species.

Johnston

Born in Scotland, Robert M. Johnston (1843–1918) attended a village school in a small farming community and began working on a local farm. He left home at the age of 16 and worked to pay for an education through evening classes at a Glasgow University. He inclined towards natural history and attended classes in botany, geology and chemistry. He migrated to Tasmania in 1870 where he worked for the railways and later in the government Audit

Department. He had a talent for statistics and became government statistician in 1880 with responsibility, among other things, as registrar of births, deaths and marriages. His insights into administrative procedures, economics and social science were translated into effective organisation in government within Tasmania.

Whenever possible, Johnston pursued his natural history interests in the field and he contributed papers on palaeontology, geology, zoology and botany. His published account of Tasmanian geology proved to be an effective basis for geological studies in Tasmania (ADB).

Johnston began publishing on Tasmanian fishes with a paper describing two new species from the estuary of the Derwent (Johnston 1881). In 1883 he published *General and Critical Observations on the Fishes of Tasmania with a Classified Catalogue of all the known Species*: a summary of the state of knowledge of Tasmanian fishes at that time (Johnston 1883). Although useful, this little book had many faults which, considering Johnston's background and scientific isolation, is hardly surprising. However, it was a significant advance in the knowledge of Tasmanian fishes. In this paper Johnston described the Derwent Smelt, now known as Tasmanian Whitebait, *Haplochiton* [now *Lovettia*] *sealii*. The precise affinities of this unique little salmoniform fish, now placed among the galaxiids (Nelson 2006), has caused some controversy. Johnston noted 'It appears in the upper waters of the Derwent, near New Norfolk, in large shoals during the months of October and November.' They spawn at this time, having migrated into the river from the sea; the adults die and the young, after hatching, are carried down to the sea. Johnston noted that this fish and the now threatened Australian Grayling, *Prototroctes maraena* were both preyed upon by the introduced salmonids. Of the Grayling, he remarked 'Unfortunately, the introduction of the European Brown Trout into many of our rivers ... threatens the extinction of this most valuable native fish.' (Johnston 1883).

In the introduction to his *General and Critical Observations* (1883) Johnston paid tribute to Morton Allport (see page 110) and to Thomas Lemprière (see page 110). He explained 'My own labours in connection with the fishes of Tasmania only extend over the last six years, but during that period I have devoted much of my time to their investigation' and concludes '... these observations are respectfully submitted for the consideration of those eminent in the science.' As a compilation, the *Classified Catalogue* was relatively conservative although Johnston perpetuated the errors of Castelnau, McCoy, Macleay and Günther, among others. He explained something of his methods:

> Of the 188 species known to exist in Tasmanian waters I have personally examined the general characters of about 145 species. Indeed, I have been in the habit of making drawings and recording particulars of all individual fishes which came into my possession, whether rare or common. Of the more common fishes I have recorded characteristic particulars of several hundred distinct individuals. I mention this for the purpose of showing that the opinions herein advanced by me in respect of matters related to classification are based upon the observations of many individuals of the same species taken in different seasons and in various stages of development. (Johnston 1883).

In spite of this, about three quarters of the 20 new species he described proved to be synonymous with existing species. The fish fauna of Tasmania is much smaller than that with which Macleay, de Vis or Castelnau were engaged and, as Johnston gave detailed descriptions, it is usually easy to determine which species he had before him.

Over the course of the next 20 years Johnston published 15 papers on Tasmanian fishes, including additions to the checklist, accounts of attempts to introduce Atlantic Salmon (*Salmo salar*) and observations on the developing fishing industry. He is the author of several endemic Tasmanian fish including the Tasmanian Whitebait and the Pencil Weed Whiting, *Siphonognathus beddomei*. One of his virtues lay in confining his attentions to the fauna at hand, all of his 'new' species being taken in Tasmania.

McCoy

Frederick McCoy (1817–1899) was born in Dublin. He studied medicine initially but his main interests lay with natural history; he spent several years studying and classifying fossils and published significant works on palaeontology. He participated in the geological survey of Ireland and was invited to Cambridge by Adam Sedgwick, who praised his work in palaeontology. From 1849 to 1854 McCoy was professor of geology and mineralogy and curator of the museum at Queen's College, Belfast. He had devoted 16 years to palaeontology and geology by the time he was chosen for the position of professor of natural science at the new University of Melbourne, and he had produced a remarkable number of valuable publications (ADB).

The Melbourne University opened in 1855 and McCoy lectured in chemistry, mineralogy, botany, comparative anatomy and physiology of animals, systematic zoology and geology including some palaeontology. He was essentially a museum worker and appeared not to recognise the importance of field work, other than specimen collection, in zoology. His students were few and those few were dissatisfied. A museum collection had been started but, at the time of McCoy's arrival, it was in storage. He negotiated for the museum specimens to be housed in the university but was opposed by various factions in the lay and scientific communities. He resolved the matter by carrying off the collections to rooms in the university, so becoming the de facto Director (ADB; Wilkinson 1996). His position as Director of the Museum of Natural and Applied Sciences was made official in 1857 and he held this office, unpaid, until his death in 1899.

McCoy was vocal in his opposition to the evolutionary theory of Charles Darwin, in part perhaps due to his rather confined attitude to the natural sciences, resulting from intensive laboratory studies unsupported by field work. Perhaps for similar reasons he was an active member of the Acclimatization Society of Victoria, founded in 1861 to introduce 'suitable' exotic species of mammals, birds and fish. He apparently rejoiced in the proliferation of the rabbit and English song-birds.

The *Prodromus of the Palaeontology of Victoria* (1874–1882) and the *Prodromus of the Zoology of Victoria* (1878–1890) were McCoy's most ambitious projects. The latter work included descriptions and illustrations, many in colour, of 60 Victorian fishes (see colour plates 11 and 12). The philosophy behind the structure (or lack of structure) of this work is made clear in the introductory remarks. The publication was issued in 'decades'; that is, each issue (decade) comprised ten plates, with the accompanying descriptions and discussion. He seems to have seen these illustrations as an extension of the museum in as much as he intended to '... have a large number of drawings made, as opportunity arose, from the living or fresh examples of many species of reptiles, fish, and the lower animals, which lose their natural appearance shortly after death ...' (McCoy 1879). He planned '... to illustrate as many different genera as possible, and will deal first usually with species of some special interest, and of which good figures do not exist, or are not easily accessible.'

Frederick McCoy. Courtesy State Library of Victoria. Photographer: Johnstone, O'Shannassy and Co.

The stated intention was later to publish 'the final systematic volume' for each group of animals, by re-organising the *Prodromus* along systematic lines. He wrote in this vein when the task was in its early stages, and he was 62. Presumably he did not envisage completing this work personally or perhaps the expression of the plan was meant to give some rationale to his project. He produced 20 decades in two volumes over a period of 13 years, a considerable feat considering his other activities. Whatever the motivation, the work was interesting and useful although it represented a fragmentary approach to the zoology of Victoria. Many more illustrations were prepared than were published: this was, of course, an uncompleted work.

McCoy described a dozen 'new' species of fish but only the Southern Hulafish, *Trachinops caudimaculatus* is now regarded as valid. As McCoy stated elsewhere, he dealt with animals as they came to his notice:

> This little fish [*Trachinops*] created a great sensation by appearing in large numbers about the middle of October, 1884, at the piers at Williamstown, in Hobson's Bay, and, being reported to the Commissioner of Customs as the young of the Californian Salmon, were sent to me as an important matter to be determined ... The Inspectors of Fisheries and others dealing officially with the fishes of our waters are greatly retarded in their business for want of recognisable figures of most of the native sorts, many of which, like the present species, have never been figured. The illustrations of the natural colours of the living fishes which I expect to present in these Decades will, I hope, diminish the difficulty of recognising them in future, and enable observations on habits, migrations, and times and places of breeding of the different sorts ... (McCoy 1890)

A meticulous zoologist, McCoy's descriptions were detailed and reliable. He frequently disagreed with Castelnau's work, as with the Kingfish, *Seriola lalandi*:

> It is certain also that Count Castelnau's *S. grandis* of the Melbourne fish shops is the same species, the difference he gives in number of the rays of the dorsal and anal fins being simply inexplicable. The numbers of the fin rays and spines, as I give them, have been tested on great numbers of specimens.

Although saying that he would concentrate on the more common animals, McCoy could not resist a really spectacular fish and produced figures of Oarfish, *Regalecus Banksi* [now *R. glesne*], Ray's Bream, *Brama rayi* [now *B. brama*] and other rare visitors. Of Ray's Bream, he said:

> This remarkable fish appeared at Portland in considerable numbers in April 1884, being previously unknown to the fishermen ... This is considered one of the most delicious fishes for the table, of the Mediterranean, and the readiness with which the gourmets of Victoria recognise this quality in so unusual a visitor to our shores is rather amusing, most pressing letters requiring the name of this god-send arriving by every post for some time afterwards.

McCoy's earliest publication on Australian ichthyology (McCoy 1865) comprised an attempt to unravel the synonymy of the Australian members of the genus *Arripis* which includes the Australian Salmon (*Arripis trutta* and *A. truttaceus*) and the Tommy Rough

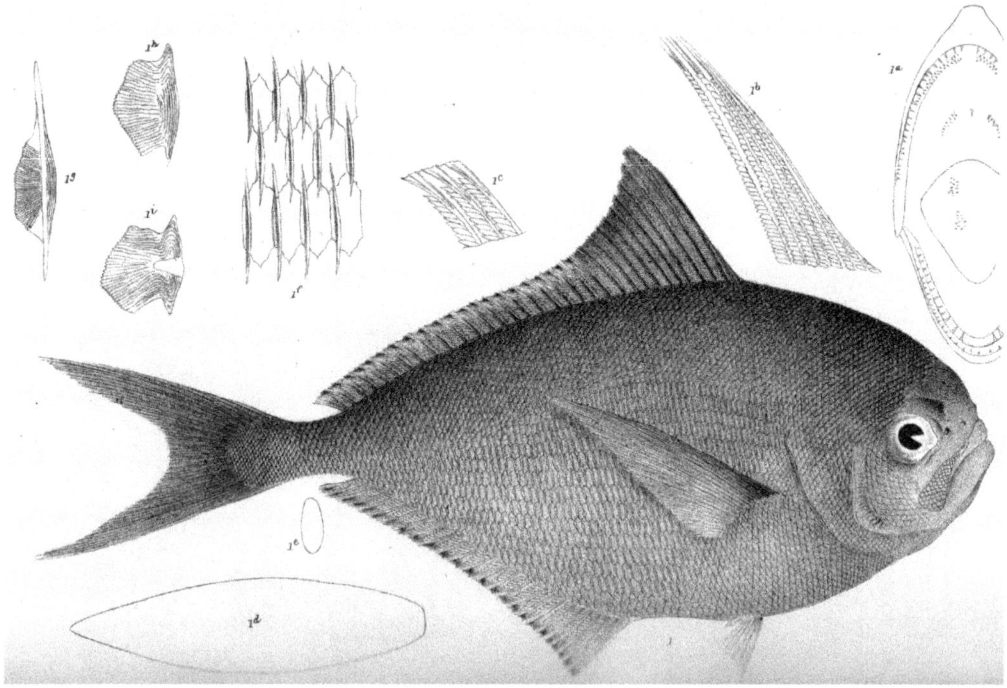

Ray's Bream, *Brama brama* (Bonnaterre, 1788) from Portland, Victoria. From McCoy 1887.

(*Arripis georgianus*). He concluded that only a single species was present but partially corrected the error in the second decade of his *Prodromus* (McCoy 1878). McCoy's descriptions of new species were occasionally published separately in the *Annals of the Magazine of Natural History*. He also wrote an essay *On the Recent Zoology and Palaeontology of Victoria* for the Intercolonial Exhibition held in Melbourne in 1866–1867 (McCoy 1867). The *Pisces* section of the essay was a summary of the fishes taken commercially around Melbourne.

Many early collections are still held at the museum in Melbourne where intermittently throughout his long stewardship McCoy conducted purchases and exchanges. He appears to have been disposed to seek specimens from interstate and overseas institutions rather than those of local waters, but continued to represent the local fishes in his *Prodromus*. In the 1880s and 1890s McCoy exchanged fish and other specimens with the museums in Adelaide, Sydney and Brisbane. In the 1890s he also managed exchanges with the Zoological Station in Naples and the U.S. National Museum. He made several small purchases, such as that from Count Castelnau, consisting of a small collection of fishes from Australia and New Caledonia. He received an extensive catalogue from Robert Damon of Weymouth, England who offered fishes from many Pacific localities, including the north-east coast of Australia (MV). These specimens were derived from the Museum Godeffroy. Bowen, Rockhampton and 'Sidney' were prominent among the localities and presumably represented collections by Amalie Dietrich and Eduard Daemel (see page 94).

New Guinea collections were also taken by McCoy, who appears to have had some from the Australian Museum and others directly from Andrew Goldie (see page 137) at Port Moresby. Goldie sent most of his fishes to the Australian Museum and to William Macleay but wrote to McCoy:

> I beg to advise two Casks of Fishes. This collection is composed mostly of species you did not get in your last ... (Goldie to McCoy, 5.3.81, MV)

He enclosed lists of fishes, including a 'catalogue of Fish collected in and about Port Moresby', listing 147 species (278 specimens) by their native names, and giving brief colour descriptions.

McCoy frequently misidentified fish specimens and redescribed known species. He had a limited appreciation of the variation in form normal to fishes, a fault common among workers of the time, and he was not sufficiently familiar with the literature of ichthyology. However, the parts of his *Prodromus* that were published helped to define the fishes of his region and its magnificent illustrations were nearly unique among Australian fish literature and still make it a pleasure to handle.

Haacke

When Frederick Waterhouse (see page 112) retired from the South Australian Museum Dr Wilhelm Haacke (1855–1912) was appointed. Haacke had studied zoology in Jena under Dr Ernst Haeckel (NDB). He travelled in New Zealand in 1881 and was appointed Curator of the South Australian Museum in 1882, then Director until he resigned in 1884. Haacke collected specimens of fishes, reptiles and various invertebrate groups. He drew attention to the lack of appropriate literature for identifying specimens and set about improving the museum library through subscription to research journals and the purchase of recent publications (Hale 1956). The lack of up to date references was, in part, his justification for sending a large number of zoological specimens to European researchers. He sent fish to Franz Steindachner in Vienna who named several species for Haacke. Haacke published his own description of the Shovelnose Ray, *Aptychotrema vincentiana* from Gulf St Vincent, South Australia (Haacke 1885a).

Wilhelm Haacke. From Hale 1956.

Haacke had an interest in orthogenesis, a form of evolutionary theory which he had supported and helped his friend, Eimer, to develop. The theory suggested that evolution was 'directed', and that selection operated gradually, altering anatomical characteristics in a predetermined manner to improve a creature's adaptation to its environment. Haacke had helped to form this theory of unnatural selection but was willing to consider facts that appeared to falsify it. In a study of a South Australian fish, locally known as the Trumpeter, *Pelates octolineatus*, he considered the place of 'Eimer's Theory' in the development of its colour pattern (Haacke 1885b). He had named this fish *Helotes scotus*, in the mistaken belief that it was undescribed, but made acute observations on the alteration in the colour pattern with age. On settling in seagrass beds the juveniles are uniformly yellowish brown in colour but they soon develop vertical bars. After they have reached about 25 millimetres in length they take on a pattern of longitudinal stripes. Between these states, they exhibit both patterns, assuming a reticulated appearance. Haacke interpreted the little 'rectangles' thus formed on the fish's side as contradictory to the orthogenetic theory: Eimer had stated that the ontogenetic change from a barred to a striped state would occur without the formation of an overlapping pattern.

On leaving Adelaide, Haacke returned to New Zealand and later became Director of the Frankfurt Zoological Gardens. His influence on ichthyology in Australia was slight but he helped to organise the collections and procedures in the South Australian Museum, and his successor, Zietz (see page 290), was appointed on his recommendation.

Overseas contributions

Initially, work on fishes within Australia was primarily in the form of collection, the fruits of this work being born in the form of description and some analysis in overseas publications. With the formation of Australian scientific bodies that produced their own journals, there was a powerful impetus towards local publication, bypassing the overseas experts. This had the good effect of encouraging the development of Australian science but Australian workers operated largely in isolation, sometimes with dubious results. The influence of overseas workers during this period is considered below.

The greatest contributions to Australian ichthyology were made by North American and British zoologists, the European influence gradually lessening, although the importance of type specimens held in European institutions has never waned.

North American Ichthyology: contribution to Australia

The fishes of North America were first studied by zoologists in Britain and Europe, the inhabitants of the United States being preoccupied with expansion and commerce largely at the expense of scientific pursuits. However, between 1840 and 1860 several American zoologists were collecting and naming their own fishes. They were dealing with several distinct faunas: Atlantic marine, east Pacific marine, and freshwater; few of the fishes they encountered were shared with Australia.

To the mid 1880s the development of Australian ichthyology was almost independent of that in the United States. Australian zoologists used the works of Cuvier, Richardson and Günther for guidance, most often simply inserting their discoveries into Günther's scheme. However, research in North America and elsewhere was showing weaknesses in Günther's work and much of the progress in ichthyology in the 1880s and 1890s was due directly to the efforts of American zoologists. Their influence on Australian workers was appreciable during this period, and steadily increased thereafter. It is outlined below and instances are integrated with the accounts of the work of Australia's three principal ichthyologists of the late 19th and early 20th centuries: Ogilby, Waite and McCulloch.

Cope

Edward Drinker Cope (1840–1897) of Philadelphia was best known for his studies of vertebrate palaeontology. He also examined the osteology of the principal groups of bony fishes in a series of studies which were extended by Theodore Gill. Cope described hundreds of new species of fish from North and South America and a few from further afield, including two leatherjackets (Monacanthidae) supposedly from Australia (see page 336).

Gill

Theodore Nicholas Gill (1837–1914), born in New York, was a zoologist with a special interest in fishes, mammals and molluscs. He began studying the fishes of New York and published a paper on this topic in 1856. From then until 1896 Gill published the results of his studies of many groups of fishes. His most valuable work took the form of reviews of genera or families of fishes, wherein by precise definitions he attempted to establish logical limits to these groups. Where possible, he based the work on osteological studies, partly from his own preparations and partly from a large series of fish skeletons purchased from Joseph Hyrtl. Hyrtl (1810–1894) was an eminent Austrian anatomist whose work included preparations of fish skeletons many of which, on his retirement from the university in Vienna, he sold to institutions in the United States.

David Jordan, the doyen of American ichthyology, praised Gill's work although with qualifications:

> Giving the greater part of his time to the study of fishes, or rather to what others had written of fishes, he was also a high authority on mollusks and mammals. Specimens he did not care to handle except in the form of dry and clean skeletons ... But he had an unprecedented mastery over the literature of science and a keener appreciation of the meaning of structure in classification and in evolution than that shown by any other naturalist. Nearly all of the few misconceptions in his work come from trusting to other writers in regard to statements inadequately verified by them and not at all by him. (Jordan 1922).

Many of Gill's studies had particular relevance to the Australian fish fauna and his work was utilised by Ogilby, Waite and McCulloch.

Gill became aware of the efforts of Australian ichthyologists and played a significant part in unravelling some of their taxonomic twists. For instance, he compared the original descriptions of Bleeker's Red Velvetfish, *Gnathanacanthus goetzeei* with Günther's *Holoxenus cutaneus* and Castelnau's *Beridia flava* and found them to be based upon a single species (Gill 1891a). In this paper, Gill reasoned that *Gnathanacanthus* was allied to some other southern hemisphere endemics (*Pataecus* and *Congiopodus*) and hoped that 'skeletons

Longfin Pike, *Esox Lewini* [now *Dinolestes lewini*] as illustrated by Griffith in the 1834 version of Cuvier's *Animal Kingdom*. From Griffith 1834.

or specimens to be skeletonised of the several types may be sent to the U. S. National Museum for examination.' Another of Gill's brief contributions was to identify the original description of the Longfin Pike, *Dinolestes lewini*, which was presented by Griffith in the tenth edition of Cuvier's *Animal Kingdom* (1834). Gill quoted from Griffith: 'Our figure of *Esox lewini* is from a drawing by Mr. Lewin, made in New Holland, of a species not hitherto noticed.' No description accompanied the drawing. In 1872, this fish was described as a new species independently in separate papers by Klunzinger, Castelnau and Günther. Gill recognised it from the work of 40 years earlier, even though the drawing lacked the first dorsal fin (Gill 1874).

Among the groups that Gill revised were a few of special significance to Australian ichthyologists. His study of the cirrhitoids[10] (Gill 1862a) is a good example: he split Günther's single family into four subfamilies, supplying unambiguous definitions and keys. This paper provided a practical guide to the taxonomy of the group.

Gill's review of Australian temperate wrasse (Labridae) was characteristic of his work. He established simple generic differences and attempted to encourage his Australian colleagues to work on the species:

> Only an exact comparative study of specimens of all, or most at least, will enable a correct judgment to be formed on the subject. Such a study is one of the greatest desiderata of Australian ichthyology, and it is to be hoped that some one of the able naturalists of the southern hemisphere will undertake it and express his conclusions *in antithetical terms*, whereby the relative degrees of affinity as well as distinctive characteristics may be understood. (Gill 1891b)

Gill was a precise author but exhibited occasional imaginative flights, as in describing the fin membranes of the handsome southern Australian Senator Wrasse, *Pictilabrus laticlavius*:

> ... the membrane behind the spines of the dorsalis and analis is not produced. The name *Pictilabrus* may be used to denote it. *Pictilabrus* is not only peculiar in the absence of pencils or penicillia [pencil-like protrusions of the fin membrane] behind the dorsal and anal spines, for the head is also smaller, and more abbreviated than in other species. If the meaning of the name *Pictilabrus* is demanded, imagination may play that the painting of the brilliantly colored fish has been completed and that the painter's brushes and pencils have disappeared. (op cit).

Gill sent reprints of his papers to Australian ichthyologists, providing a resource that helped to modify Günther's conservative influence. After he had virtually finished his taxonomic work, Gill prepared some semi-popular papers on a few groups of fishes; they were clearly intended to be entertaining as well as informative: *The Life History of the Angler* (Gill 1905) and *The Archer-Fish and its Feats* (Gill 1909) can still be read with pleasure.

10 Gill established separate subfamilies for groups now given familial status: Hawkfishes (Cirrhitidae), Marblefishes (Aplodactylidae), Kelpfishes (Chironemidae); he combined the Trumpeters (Latridae) and Morwongs (Cheilodactylidae) in a single subfamily.

Jordan

David Starr Jordan in 1891. From Jordan 1922.

David Starr Jordan (1851–1931) spent his early years on a farm near Gainesville, New York. He was first attracted to botany, but studied widely in natural history and, by the age of 23, had begun publishing keys to vertebrate groups of the north-eastern United States. He taught science and later biology in high schools and universities, accepting the position of President of the newly opened Leland Stanford Junior University in California in 1891. He remained with Stanford for the rest of his life (Jordan 1922).

Jordan was aware that fishes were the order of vertebrates about which least was known and this, together with attendance at a Summer School of Science for teachers and students of zoology run by Louis Agassiz,[11] induced him to focus seriously on ichthyology.

Breadth of knowledge and breadth of ambition marked David Jordan's endeavours. He was an effective and innovative teacher who introduced new approaches to higher education in the United States and was invited to lecture on education in many parts of the world. As a powerful advocate for international cooperation he travelled widely to promote the peaceful settlement of differences between nations. In ichthyology, he was an inspirational teacher and a researcher for whom no task appeared to be too great.

Many highly motivated young biologists were attracted by Jordan's teaching and their influence eventually extended into the taxonomy of fishes throughout the world. Early in his career as an ichthyologist Jordan manifested allergies to the chemicals used to preserve fishes so the majority of his research papers were the result of collaborations with his own students. They concentrated on the fishes of the United States, Central and South America, parts of the Pacific (particularly Hawaii, the Philippines and Japan) (Jordan 1922; Brittan 1997). Jordan's achievement included the naming of some two and a half thousand new species and the establishment of over one thousand genera. Later in his career he tended to split existing families and genera on the basis of relatively small differences and this work has been revised and simplified by subsequent workers (Myers 1963).

His grasp of ichthyology enabled Jordan to produce effective texts in both general ichthyology (*A Guide to the Study of Fishes*, 1905) and the popular aspects of the science (*Fishes*, 1907). He revised the latter (Jordan 1925), adding a daunting 'prefatory note' to its

11 Louis Agassiz (1807–1873), a Swiss who spent the first half of his life in Europe and his latter years in the United States, took a professorship at Harvard University in 1847. He collected his own observations in the field and published an enormous volume of descriptive natural history. His *Recherches sur les poissons fossiles* (1833–44) was written with precision in the tradition of Cuvier. However, Agassiz regarded the diversity and distributions of plants and animals as manifestations of special creation and, in common with many other naturalists of his day, opposed Darwin's views on evolution (DSB). Agassiz conducted his classes on the island of Penikese, near Cape Cod. Here, in 1876, Jordan began to study marine algae but was encouraged by Agassiz to turn his attention to fishes.

700 pages that began: 'In this volume the writer has tried to compress all that an educated man is likely to know, or to care to know, about fishes.'

Jordan had a great deal to offer his fellow ichthyologists, distant correspondents as well as his collaborators. Australian workers benefited from him directly and indirectly. He maintained an active correspondence with Ogilby, Waite and McCulloch, answering queries on taxonomic procedures and dispensing reasoned taxonomic opinions. He was actively engaged in helping to form, and reform, the rules of zoological nomenclature and kept his Australian correspondents up to date with these significant developments. Jordan also encouraged his associates to correspond with his Australian friends and there was an active exchange of collections and published work.

On two occasions Jordan visited Australia and he referred to the Australian fish workers with approval; in an account of his 1907 visit he said:

> The museum naturalists of Australia form an interesting and active group, the fact that the animals and plants of the region are still incompletely recorded adding zest to their study. Naturally I saw a good deal of the ichthyologists, even though I had no time for original investigation myself. Among them are three of excellent training and ability: Allan R. McCulloch, curator in the Australian Museum, a man of charming personality and a careful, accurate observer whose judgment in technical matters I value highly; Edgar R. Waite, his esteemed co-worker, then at Christchurch but soon after called to the Museum at Adelaide; and J. Douglas Ogilby, the accomplished curator of fishes in the Queensland Museum at Brisbane, and author of numerous papers of importance. (Jordan 1922)

On this visit, as a favour to Ogilby, Jordan found time to speak to the members of the Amateur Fishermen's Association of Queensland (see page 209).

The Genera of Fishes (Jordan 1917–1920) was published in four parts and concluded by *A Classification of Fishes* (1923). Jordan sought advice on problems associated with the preparation of this work and invited contributions and criticism from his fellow ichthyologists. Allan McCulloch was particularly responsive and his help was acknowledged:

> In preparing this paper I have been under special obligation to Allan R. McCulloch of the Australian Museum; to my former student, Carl L. Hubbs of the University of Michigan; and to Theodore D. A. Cockerell of the University of Colorado. (Jordan 1923).

These publications were of immense value to fish taxonomists and, in spite of their inevitable shortcomings, made fish taxonomy less onerous. Jordan thus contributed to the stability of the nomenclature of fishes, following the new 'International Rules of Zoological Nomenclature'.

Jordan with his students and associates revised many fish groups, some on a worldwide basis. An example is his review of the Silversides (Atherinidae), as then defined (several groups have since been excluded and make up separate families), that provided a key to the living genera (Jordan and Hubbs 1919). In this paper the authors named for McCulloch *Atherion maccullochi*, collected by him at Lord Howe Island and forwarded to Jordan. Some Australian fishes included in this review were later placed in separate families: the Pygmy Perch, *Nannatherina balstoni* Regan, 1906 and the Rainbowfish, *Melanotaenia nigrans* (Richardson, 1843).

A similar review of Anchovies (Engraulidae), based on the 'specimens represented in the Museum of Comparative Zoology' and the literature, was undertaken by Jordan and Seale (1926), wherein they named a Queensland Anchovy for Edgar Waite (*Stolephorus waitei* Jordan and Seale, 1926). The nature of Jordan's collaborations is shown in this paper. Being unable to handle the preserved specimens due to his allergies to the chemicals Jordan remarked:

> All the descriptions are the work of the junior author. The generic sequence and adjustments are chiefly the work of the senior author.

The regional studies of Jordan and his associates provided descriptions of many fishes from the Pacific. Some of this work was pertinent to Australian ichthyology, approximately 180 fishes now known to inhabit Australian waters having been first described from elsewhere in the Pacific by Jordan and his associates prior to 1930. This accounted for about 4% of the Australian ichthyofauna, mostly shore fishes of tropical and, to a lesser extent, warm temperate waters.

Jordan's work in the central Pacific, as distinct from his earlier and continuing work in the East Pacific, was started in about 1900 in Japan. He made several lengthy visits there, overseeing collections and interspersing his work with lectures and diplomatic contacts. He was full of praise for the Japanese people and their country and made training positions available to their aspiring biologists at Stanford. Jordan and his colleagues (particularly Snyder and Starks) published many papers on the fishes of Japan, mostly between 1900 and 1915. Shigeho Tanaka was largely trained in ichthyology by Jordan and produced a compilation of their work as a *Catalogue of the Fishes of Japan* (Jordan, Tanaka and Snyder 1913). Between 1908 and 1930, Tanaka described nearly 200 new species from Japanese waters.

Few Japanese fishes were found in southern Australian waters although Waite recognised a Goblin Shark, caught in a gill net at Goolwa, from Jordan's description of the animal. A Japanese specimen had been sent to Jordan by a Captain Owston, who dealt in natural history specimens (Jordan 1898).

Goblin Shark, *Mitsukurina owstoni*. Waite used the illustration from Jordan & Fowler 1903. From Waite 1921.

The waters of tropical Australia share many species with Japan. Among those first described from Japan are such familiar reef dwellers as the Neon Damsel, *Pomacentrus coelestis* Jordan and Starks, 1901 and the Schooling Bannerfish, *Heniochus diphreutes* Jordan, 1903.

A systematic examination of Hawaiian fishes was initiated in the late 1890s by Jordan and his colleagues. From 1897 to 1905 they collected and described the fauna (a remarkable proportion of it from the Honolulu fish market) and produced a series of papers, culminating in a book, *The Shore Fishes of Hawaii* (Jordan and Evermann 1905). The Hawaiian fish fauna exhibits significant endemism but is essentially an outlier of the Indo-west Pacific faunal region. About 25 of the species discovered by the American ichthyologists of Jordan's group have since been found to inhabit Australian waters; most are reef fishes but a few were sampled from deep water.

Jordan boasted to Waite that he had secured unprecedented collections from the reefs at Samoa through the use of explosives and poisons (Jordan to Waite, 17.8.1902, SAMA). As a result, the majority of new forms found there were small reef fish, particularly Cardinalfish, Demoiselles, Blennies and Gobies. Jordan and Seale described about 20 species that are also found in Australia.

The Philippines saw a great concentration of effort from Jordan's school, beginning in 1904 and continuing for nearly 40 years. The publications derived from collections in this archipelago spanned this time and were variously authored: Evermann and Seale 1904; Jordan and Seale 1905 and 1907; Seale and Bean 1907; Jordan and Richardson 1908; Seale 1910; Smith and Radcliffe 1911, 1912, 1913; Fowler 1918; Gilbert and Hubbs 1920; Herre 1927; Fowler and Bean 1928. Thereafter, to about 1945, Herre and Fowler continued the work. Each of the publications mentioned includes fish that range to northern Australia, in excess of 60 species in all.

Jordan sent separate copies of these papers to his contacts around the world, including his correspondents in Australian museums. Edgar Waite's collection of papers from Jordan and his associates numbered several hundred.

Garman

When David Jordan attended Agassiz' summer school in 1873, he met Samuel Garman (1846–1927), who functioned as Agassiz' helper and was an Assistant in the Museum of Comparative Zoology at Harvard University. Garman accompanied Alexander Agassiz, son of Louis Agassiz and Director of the Museum of Comparative Zoology, on several collecting expeditions, including two months of deep sea trawling in the East Pacific in 1891. The work was conducted from the U.S. Fish Commission steamer *Albatross*; collections were made from the Gulf of California and off Panama and Ecuador, including the vicinity of the Galapagos Islands. Garman's publication (1899) from this expedition included descriptions of deepsea fish. Early in his career he became intrigued by the structure of sharks and chimaeras and he published many papers on these groups over nearly 40 years, beginning in 1874.

Garman sent copies of some of his publications to Australian ichthyologists, including his lavish works on sharks and chimaeras, but he had little direct influence on Australian ichthyology. However, he described the Dwarf Sawfish, *Pristis clavata* of northern Australia (Garman 1906) and gave an account of fishes collected on an expedition led by Alexander Agassiz 'to the Islands and Coral Reefs of Fiji and to the Great Barrier Reefs of Eastern

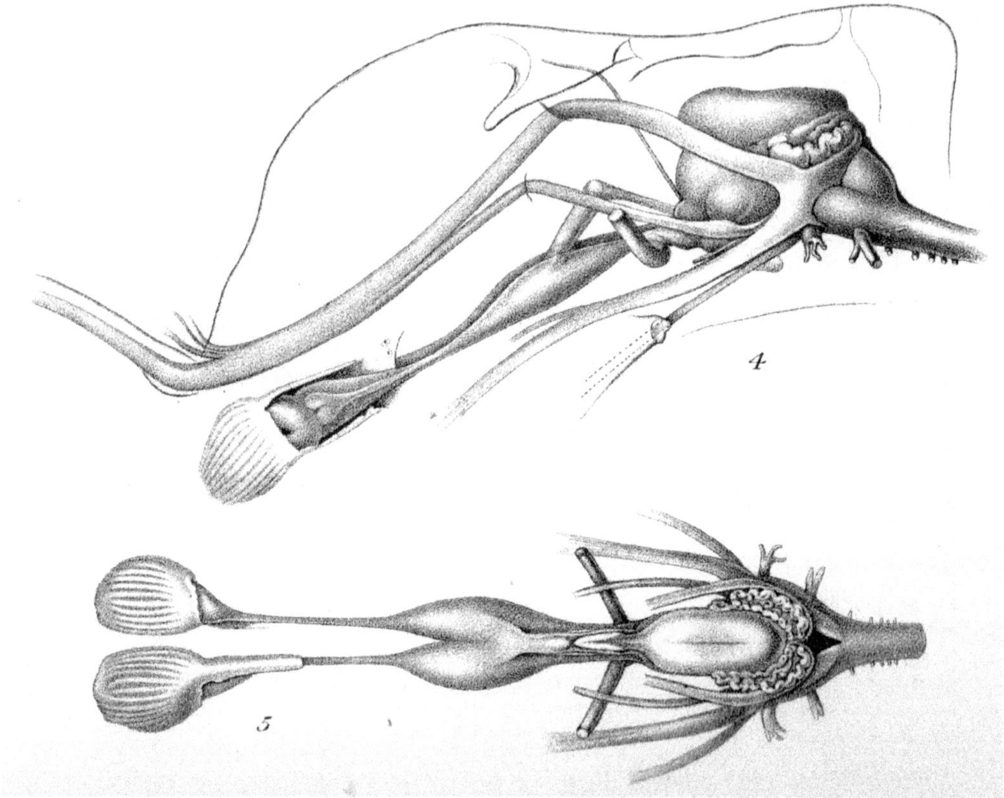

The brain of the Elephant Fish, *Callorhinchus milii* was studied by Samuel Garman. He noted that the specimen was '... sent by Mr W. Robertson from Hobart-Town.' From Garman 1904.

Gobius waitii (Garman, 1903) proved to be synonymous with *Paragobiodon xanthosoma* (Bleeker, 1852). From Garman 1903.

Australia.' The paper, misleadingly entitled *Some Fishes from Australasia* (Garman 1903) deals mainly with fishes of Fiji. Nevertheless, a couple of Queensland fishes were included, one of which Garman believed to be new and named for Edgar Waite.

Fowler

Henry Weed Fowler (1878–1965) studied under David Jordan at Stanford University and worked as a zoologist in the Academy of Natural Sciences in Philadelphia from 1903 to 1965. His main interest was ichthyology on which subject he produced an enormous number of publications. Fowler's *Fishes of Oceania* (1928) included many of the tropical Indo-Pacific fishes found in northern Australian waters and was followed by voluminous supplements in 1931, 1934 and 1949. This work on wide ranging Indo-Pacific fishes became essential to students of Australian ichthyology but it is generally outside the scope of this account.

Fowler reported on small collections of Australian fishes on several occasions. In 1908 he described new species from a small collection sent to the Academy of Natural Sciences of Philadelphia by a Melbourne conchologist, Mrs Agnes Kenyon, although only two are at present recognised as valid. He complimented a number of workers whom he saw as of special significance to Australian ichthyology in naming new species in this paper (Richardson, Lesueur, Bleeker, de Vis, Ramsay, Ogilby and Waite) but most disappeared into synonymy, only Bleeker and Lesueur surviving in the form of *Lesueurina platycephala* and a Seahorse, *Hippocampus bleekeri*.

Many years later a collection that had been made by members of the United States Exploring Expedition in the summer of 1839 in New South Wales was presented to Fowler and it provoked a short paper that was published in a Queensland journal (1953). The author noted: 'My thought of offering this paper is largely with reference to its historical value.' Fowler quoted from the journal kept by Mr Charles Pickering, an enthusiastic

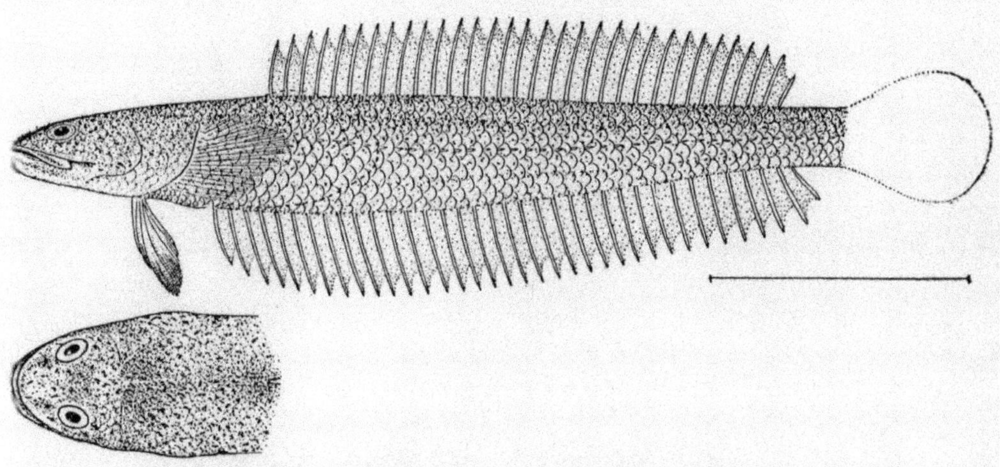

Fig. 9.—*Lesueurina platycephala* Fowler. (Type.)

The Flathead Sandfish, *Lesueurina platycephala*. Fowler established a new genus for this fish, named in recognition of Charles Lesueur's work in studying the fishes of the collection of the Academy of Natural Sciences of Philadelphia. From Fowler 1908.

conchologist, who formed part of the scientific staff of the expedition. Of the Garfish (*Hyporhamphus regularis*), which he called the 'Guard fish', he wrote:

> Several were seen swimming about and gathering objects floating on the surface, for which purpose the long under jaw was now found to be an exceedingly convenient instrument. Passing beneath it directs the object to the mouth, which receives it much more surely and without the disturbance caused by the attempts of ordinary fishes. The operation is much the same as that of the *Rhynchops or scissor-bill.* The fish in fact skims the surface below, as the bird from above.

The expedition remained in Sydney for a few months but their ichthyological results were negligible.

Further American contributions

Among the many useful publications by North American ichthyologists were *Oceanic Ichthyology* (Goode and Bean 1896) and *A Bibliography of Fishes* (Dean 1916–1923). The former was essential to workers who handled oceanic specimens anywhere in the world, the latter an invaluable key to the literature of ichthyology. These splendid publications were among those sent by their authors to their co-workers in Australia. Bashford Dean published a single paper on an Australian fish: he observed specimens of the Australian Lungfish in the Zoological Society's aquariums in London in 1904, summarised what had been previously published on its habits and added some observations of his own (Dean 1906).

Small contributions were also made by some American ichthyologists who participated in expeditions or studied the collections made by these expeditions. For instance, George Washington Vanderbilt the third (1914–1961) was a New York yachtsman and naturalist who owned several vessels with which he made worldwide collections of marine organisms. His voyages included that of the yacht *Ara* in the Mediterranean, Atlantic and Pacific (1926–1928); the *Alva* in the Mediterranean, Indian Ocean and South Pacific (1931–1932); the schooner *Cressida* in the South Pacific (1934); and the schooner *Pioneer* in the Caribbean and East Pacific (1941). Vanderbilt established a museum on Long Island, New York and an institute for the study of marine life. His fishes were studied by N.A. Borodin who reported a few fish collected from Queensland coastal waters and the Great Barrier Reef. This included a description of the Mossback Velvetfish, *Erosa australiensis* (Borodin 1932), although it has proven to be synonymous with *Paraploactis trachyderma* Bleeker, 1865. Borodin (1933) described the Spiny Eel Blenny, *Congrogadoides* [now *Congrogadus*] *spinifer* from a specimen collected at Broome by the American zoologist, Hubert Lyman Clark, known for his elucidation of the echinoderm fauna of Australia.

John Treadwell Nichols (1883–1958), well known to ichthyologists and herpetologists as the founder of the journal *Copeia* in 1913, was curator of fishes at the American Museum of Natural History. Harry C. Raven, also of that institution, conducted a number of scientific explorations, including a collecting expedition in North Queensland in 1921. From this collection, Nichols and Raven (1928) described *Rhadinocentrus* [now *Cairnsichthys*]

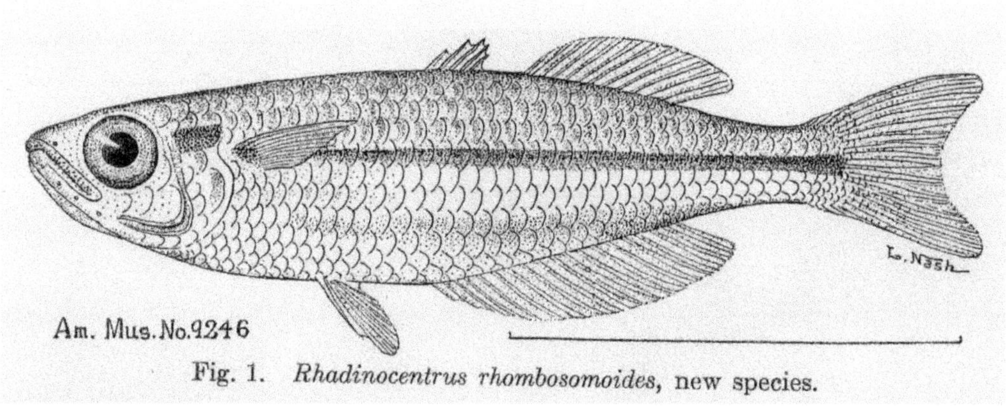

The Cairns Rainbowfish, *Cairnsichthys rhombosomoides*, collected from Babinda Creek, north-east Queensland. From Nichols and Raven 1928.

rhombosomoides. They later published a brief paper on the origin of Australian freshwater fishes (Nichols and Raven 1932); only a small fraction of the fauna was considered but they noted the close relationship between the freshwater fish fauna of the Cape York Peninsula and that of New Guinea. Nichols revisited this topic in publishing on the fishes collected during the expeditions to New Guinea and Cape York financed and led by the American zoologist and philanthropist, Richard Archbold.

A minor contribution to Australian ichthyology was made by Theodore Dru Allison Cockerell (1866–1948), Professor of Zoology at the University of Colorado. Cockerell was an Englishman who moved to Colorado at the age of 21, hoping that the warmer climate would promote resolution of his tuberculosis. His main interest was entomology, particularly bees, but he published on an enormous variety of subjects, including entomology, botany and palaeontology. He was essentially a Victorian naturalist, without college degrees, who published several thousand scientific papers, works of social comment, and books of verse and plays (Weber 2000).

The morphology of fish scales was among Cockerell's minor passions, and he published descriptions of the 'lepidology' of fishes from different parts of the world. Ronald Hamlyn-Harris of the Queensland Museum (see page 217) supplied him with specimens of Australian fish scales. Cockerell described these in detail, constructing some keys based on scale characters alone, and concluded that 'with rare exceptions the Southern species have scales closely resembling their northern congeners, when such exist' (Cockerell 1915). He published his descriptions of Australian fish scales in the *Memoirs of the Queensland Museum* (Cockerell 1913, 1915 and 1916).

Albert Günther's British successors: their relevance to Australian ichthyology

Ichthyologists working at the British Museum contributed to the understanding of the relationships between groups of fishes. Enormous cataloguing tasks had been performed by Günther and were initially continued by his assistant, George Boulenger. Taxonomy retained a high priority but other aspects of ichthyology gradually assumed greater importance.

Boulenger

George Albert Boulenger (1858–1937) succeeded Günther at the British Museum. He was born and educated in Brussels and began his career as a zoologist at the University and Museum of that city. Günther, then keeper of zoology at the British Museum, recruited Boulenger for that institution in 1882 and he worked there in the Department of Zoology for nearly 40 years. His published work was mostly in the fields of herpetology and ichthyology.

Boulenger began a revision of Günther's *Catalogue of Fishes in the British Museum* and published the first volume (Boulenger 1895), dealing with some important percoid families. This work was a great advance on the first edition, the author having resolved many taxonomic problems by detailed osteological studies, and determined the synonymies of many poorly defined species. However, the project then lapsed, no further

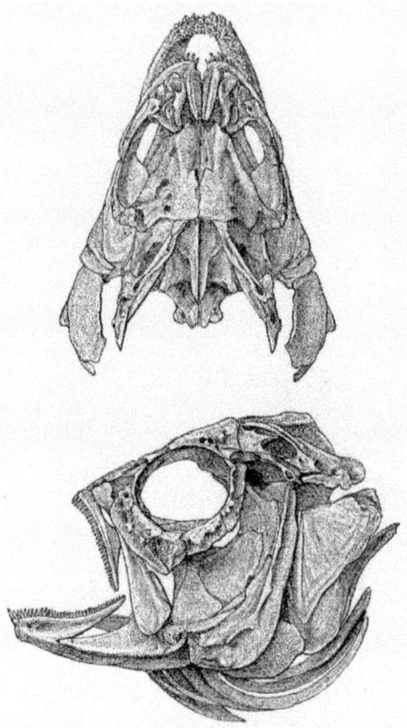

Skull of the Southern Blue Devil, *Paraplesiops meleagris* (Peters, 1869), part of the osteological studies Boulenger carried out in revising the classification of perciform fishes. From Boulenger 1895.

Günther described the Rosy Perch, *Callanthias allporti* in 1876 without illustration. Boulenger re-described and illustrated the fish from the original specimens sent to Günther from Tasmania by Morton Allport. From Boulenger 1895.

volumes being produced. Boulenger focussed his efforts elsewhere, particularly on the freshwater fishes of Africa.

His relevance to Australian ichthyology lay mainly in the single revised volume of the *Catalogue*, preparation of the ichthyological part of the *Zoological Record* from 1880 to 1904, and publication of a summary of the systematics of teleost fishes in the *Cambridge Natural History* (Boulenger 1904). He replied to queries from Ogilby and Waite, but the correspondence seems to have been sparse.

Regan

Charles Tate Regan (1878–1943) studied zoology at Cambridge and was appointed assistant keeper in the British Museum in 1901. He worked first as Boulenger's assistant, later having sole charge of the curation of fishes, and becoming keeper of the department of zoology in 1921. Regan became the Museum's Director in 1927 and retired from this position in 1938.

Regan undertook the systematic study of many fish families, of the sort carried out by Theodore Gill, basing his work on careful, accurate dissections and osteological studies with comparative exercises which elucidated the relationships of a number of difficult groups. His work enabled him to improve the knowledge of evolutionary trends and he summarised his conclusions in a series of papers and in articles in the 14th edition of the *Encyclopaedia Britannica* (1929). His work included studies of the reproductive organs of some viviparous freshwater fishes and preliminary studies of the species flocks of cichlid fishes in the African lakes. He also discovered the parasitic dwarf males found in Ceratioid Anglerfish of the deep sea (Regan 1925).

Regan named a few Australian fishes in the course of revising the groups to which they belonged, in particular in his revisions of the 'Suborder Plectognathi' (now order Tetraodontiformes) (Regan 1903) and the Galaxiidae (Regan 1906a). He also dealt with a very small collection from the King River, Western Australia (Regan 1906b), from which he described a Pygmy Perch, *Nannatherina balstoni*, named for the donor.

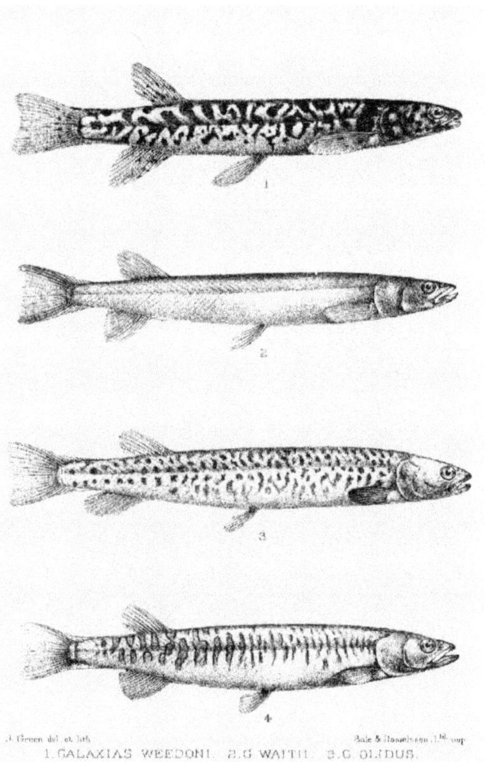

One of four plates showing '*Galaxias weedoni* Johnston' [synonymn of *G. brevipinnis* Günther, 1866]; '*Galaxias waitii* n. sp.' [synonym of *G. rostratus* Klunzinger, 1872]; '*G. olidus* Günther'; '*G. occidentalis* Ogilby'. From Regan 1906.

Regan appears to have supported a lively if intermittent and, on his part, sometimes rather didactic correspondence with the Australian ichthyologists of his day, a little of which survives in the archives of

Australian museums. He answered queries from Ogilby, Waite and McCulloch at various times but occasionally appeared to be in competition with them. His paper on Wobbegongs (Orectolobidae) (Regan, September 1908) was published slightly in advance of that of Ogilby and McCulloch (December 1908), and Regan (1909) then startled them somewhat by rapidly publishing the description of a new species of Wobbegong based solely upon a description and figure from their paper: this was *Orectolobus ogilbyi*, later synonymised with *Eucrossorhinus dasypogon* (Bleeker, 1867). In his diary, Edgar Waite noted that he was in competition with Regan in preparing his chief paper on Antarctic fishes.

Norman

John Roxborough Norman (1898–1944) performed military service in the First World War and was appointed to the British Museum in 1921, assistant to C.T. Regan. He was best known for his series of studies on 'coast fishes' (Norman 1935–38), dealing with the little known ichthyofauna of Antarctica (1938) and the adjacent region of Patagonia (1937a). He also wrote *A History of Fishes* (Norman 1931), a popular and readable work on ichthyology in the tradition of David Starr Jordan's *Fishes* (1907). His posthumously published *Draft Synopsis of the Orders, Families and Genera of Recent Fishes and Fish-like Vertebrates* (1966) was an attempt in manuscript to summarise the genera of fishes, distinguish them with keys, and present the relevant references. A few major groups were omitted but his work had wide circulation long before it was published.

Norman contributed to the descriptive work on some Australian collections, in particular on the flatfish from the *Endeavour* trawls (Norman 1926). He named eight new species from Australian specimens in this work, most of which are still recognised. For this he had the cooperation of Allan McCulloch and the latter's assistant, Gilbert Whitley, of the Australian Museum (see page 398). Norman's major systematic work included a revision of the flatfish on a worldwide basis (*A Systematic monograph of the Flatfishes (Heterosomata) Volume 1*, 1934) but this was incomplete. A great number of Australian forms were dealt with in this study which, together with the *Endeavour* publication, simplified later work on this difficult group.

Later Norman handled the fish collections from the British, Australian and New Zealand Antarctic Research Expedition 1929–1931 that had been led by Douglas

FLATFISHES.—NORMAN. 263

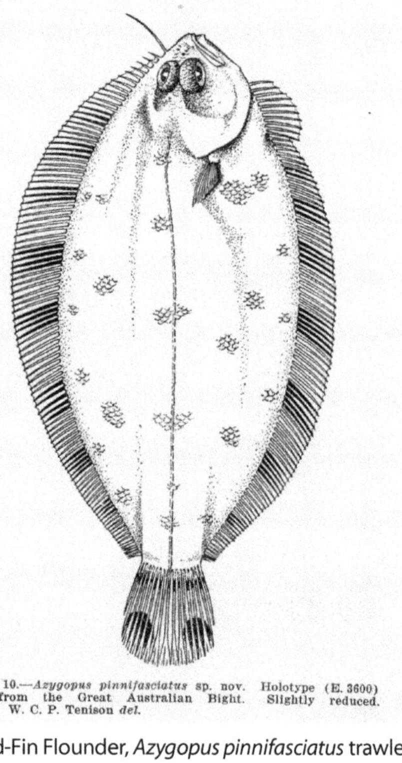

Fig. 10.—*Azygopus pinnifasciatus* sp. nov. Holotype (E. 3600) from the Great Australian Bight. Slightly reduced. W. C. P. Tenison *del.*

Banded-Fin Flounder, *Azygopus pinnifasciatus* trawled in the Great Australian Bight. From Norman 1926.

Mawson. The report began with notes on a small collection from southern Australia, including the description (Norman 1937b) of the Tasmanian Perch, *Lepidoperca tasmanica*. This paper also included fishes taken from the Antarctic and Macquarie Island.

The British contribution to Australian ichthyology from 1870 to 1930 was substantial, but less pervasive than that made by ichthyologists in the United States. Regan and Norman helped their Australian counterparts but there was no British equivalent of David Starr Jordan.

Scandinavian contributions

Dr Eric Mjöberg led a Swedish expedition (1910–1913) in the Kimberley region of Western Australia and in Cape York Peninsula. Large collections were made, mostly of zoological material, but the expedition has recently become notorious as Mjöberg also removed aboriginal relics and skeletons.

Mjöberg's fish were studied by Carl Hialmar Rendahl (1891–1969) in Sweden. At this time, Günther's *Catalogue* was still the basic reference used for identification (Åhlander *et al.* 1997), apparently updated only by Boulenger's volume on certain perciforms (Rendahl 1921). Rendahl's account of the fishes included several new species trawled off Cape Jaubert, north-western Australia (Rendahl 1921).

Dr Knut Dahl (1871–1951), a Norwegian zoologist, collected freshwater fishes in Arnhem Land and some marine fishes from Roebuck Bay in the Kimberley district in 1894–1895. The animals were deposited in the Royal Zoological Museum in Christiania and eventually sent to Rendahl for assessment. He described the collections and established several new species (Rendahl 1922).

Johannes Schmidt (1877–1933), a Danish biologist, carried out extensive studies on freshwater eels (species of *Anguilla*). He is best known for having located the spawning grounds of the European eel, *Anguilla anguilla* in the Sargasso Sea. He extended his studies of eels worldwide, including publications on the eels of New Zealand and those of Australia (Schmidt 1928). Schmidt secured, and acknowledged, the assistance of Australian zoologists, museum directors and fisheries departments who had specimens collected for him. Schmidt's paper on the Australian freshwater eels clearly distinguished the four species that are recognised today.

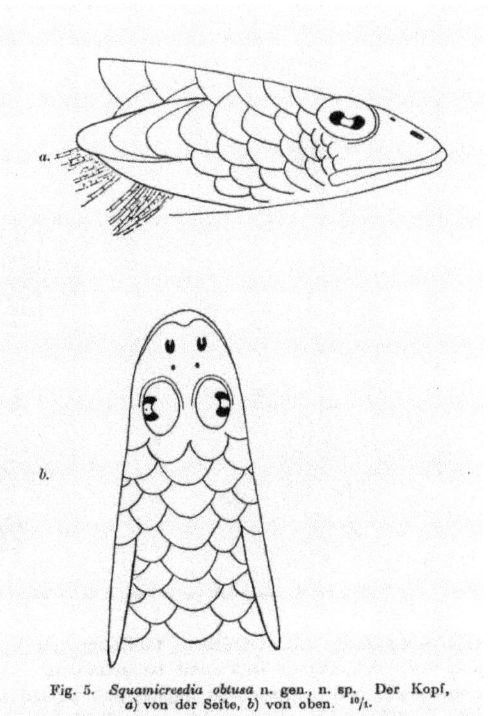

The Obtuse Duckbill, *Squamicreedia obtusa* Rendahl, 1921 from Cape Jaubert, Western Australia. From Rendahl 1921.

Dutch contributions

Max Weber (1852–1937), a comparative anatomist with a special interest in ichthyology, was for many years professor of zoology in Amsterdam. He carried out field work in Africa and among the islands north of Australia and reported on the collections made by Richard Semon (see page 211) who led an expedition to Australia and New Guinea in 1891–93. Among Semon's collections was a small fish from the Burnett River, Queensland. This was the Australian Smelt, *Prototroctes* [now *Retropinna*] *semoni* which Weber (1895) described.

Weber himself led a Dutch expedition to examine the marine life of the 'Malayan Archipelago' in 1899–1900. The Steamship *Siboga* was provided by the government of the Dutch East Indies and specimens were gathered by dredging, trawling, line fishing, barter and purchase. Enormous collections were amassed and the reports on these gradually appeared over many years, the fish being dealt with by Weber himself. His reports, particularly *Süsswasserfische von Neu-Guinea* (1907) and *Neue Fische aus Niederländisch Süd-Neu-Guinea* (1910), provided original descriptions of many species collected from the freshwaters of southern New Guinea that also occur in the streams of northern Australia. These included Gobies (Gobiidae), Gudgeons (Eleotridae), Terapon Perch (Terapontidae), Rainbowfish (Melanotaeniidae), Catfish (Plotosidae and Ariidae), Soles (Soleidae and Cynoglossidae) and Garfish (Hemiramphidae). In the preliminary (Weber 1909) and principle (Weber 1913) reports on the *Siboga* fishes Weber described many marine fish that are found in tropical Australia.

Bleeker's work was augmented by the *Siboga* report and both incorporated in the 11-volume *Fishes of the Indo-Australian Archipelago* (Weber and de Beaufort 1911–1962). This series, planned by Weber and in which he was an active collaborator in producing the first seven volumes, included a large proportion of the fishes of northern Australia.

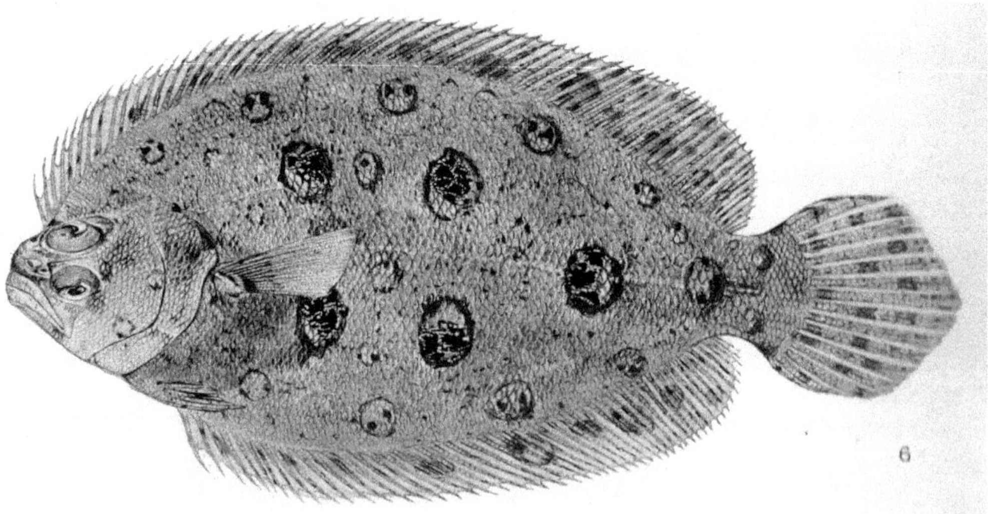

The Peacock Flounder, *Pseudorhombus argus* Weber, 1913 was taken by the *Siboga* in New Guinea. Its range includes northern Australia. From Weber 1913.

Ogilby

J.D. Ogilby was of great significance to Australian ichthyology. His work in Australia is readily divided into his time in Sydney (1885 to 1900) and in Brisbane (1901 to 1925).

Irish Reminiscences

Born in Belfast, in 1853, James Douglas Ogilby was educated at Winchester College, England (1866–69) and Trinity College, Dublin (1871–72). While residing in Brisbane late in life, Ogilby began experimenting with autobiography: fragments of this appear as odd pages and as limited entries in notebooks over a number of years. He called his reminiscences 'The Early Years of a Naturalist' and the most complete surviving manuscript, of about 150 pages, is held by the Oxley Library in Brisbane. He mentioned his mother, Adelaide Douglas, of the Earl of Moreton's family and William Ogilby, his father. He had an elder brother, Claud, and five sisters: Adelaide (Ada), the eldest; Isabella, 'My Lady Disdainful'; Beatrice; Louisa, who died at 18; and Edith, the youngest. The family lived in Northern Ireland at Altrachnee (sometimes spelled Altnachree) Castle, built by William Ogilby in 1840.

In his reminiscences, Ogilby frequently referred to his fascination with plants and animals, with bird-nesting and fishing. He recalled collecting sticklebacks in a lemonade bottle and keeping them in a glass globe aquarium as a child, and later he had Yarrell's books on British birds and fishes 'four precious volumes, which that dear illogical logician, the Dad, with misdirected liberality, had presented to me as a Christmas gift, in order to inculcate a love of Nature ...' His father, William I. Ogilby (1808–1873), was a well-known zoologist and Honorary Secretary of the Royal Zoological Society (1839–1846). At about the age of 15 Douglas (he preferred his middle name) was introduced to sea fishing at Portrush, on the coast of Northern Ireland. This 'helped in no uncertain way to fashion my future career' and he described, in convoluted style, the vivid impression made by a 'horn-eel':

> the tremendous leaps of which in its endeavour to get rid of the hooks greatly astonished me, as also, for although acquainted with it from the woodcut in Yarrell's British Fishes, I had never previously seen one in the flesh, did its personality, if one may use such an expression about a fish, with its long cylindrical body & elongated jaws with their rows of sharp & cruel-looking teeth; altogether it was an interesting revelation to me ...

He fished for trout and for eels, but it was the early encounters with marine fish that caught his imagination; at Portrush on another occasion:

> ... a most beautiful fish, the like of which I had never seen before & so it necessitated a search through the ever handy Yarrell, from which I discovered that it was a male 'Gemmeous Dragonet', a splendid fellow ...

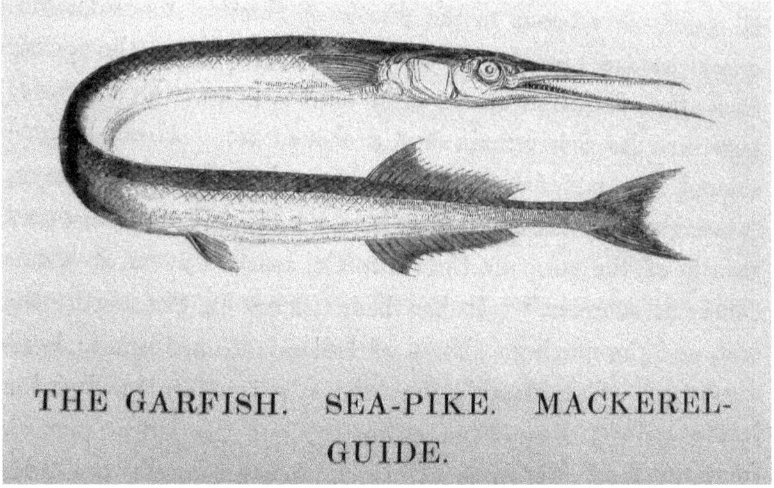

The Garfish of the British Isles, *Belone belone* (Linnaeus, 1761), was known locally as the Horn Eel. From Yarrell 1835–36. Courtesy Museum Board of South Australia.

The male Dragonet, *Callionymus lyra* Linnaeus, 1758, was known as the Gemmeous Dragonet. From Yarrell 1835–36. Courtesy Museum Board of South Australia.

Coming to Australia and prior experience in zoology

The Australian Museum advertised for a zoologist in 1884 and Ogilby applied, including what in retrospect might seem a suspiciously brief reference from his cousin, the Marquis of Hamilton: 'Mr Ogilby is a gentleman of high scientific attainments and of a persevering disposition' (AMA). He married Mary Jane Jameson in County Tyrone in November 1884 and they came to Australia in 1885, where he began work in February as an Assistant in Zoology at the Australian Museum.

When he arrived in Sydney, Ogilby was 32 and had considerable experience of natural history. He had been active in a naturalists' club in Ireland, where he had collected, prepared and identified specimens. He had fished and spent time on trawlers off the Irish coast and had consistently taken a particular interest in fishes. Ogilby's notebooks from the 1870s (SAMA) include lengthy notes on British fishes and some essays that may have been intended to form part of a book on fishes. He first published in 1874, in the *Zoologist (London)*, on a bat; in 1876 he had nine papers in the *Zoologist*, six of them on Irish fish. He travelled in North America in his late 20s, collecting birds, and published an 80 page catalogue of the birds of Texas (Ogilby 1883). He also worked briefly, and apparently in an unofficial capacity, at the British Museum.

Australian ichthyology as Ogilby found it

Ichthyology in Australia was at a critical stage in the 1880s. Count Castelnau, the Melbourne-based Consul of France who worked on the fishes of Australia in the 1870s, had confounded the taxonomic literature of Australian ichthyology with an almost unprecedented number of ill-conceived descriptions and misidentifications. In Brisbane, in the early 1880s, Charles de Vis had published a large number of confusing, inadequately researched papers on fishes, further muddying their taxonomic history. McCoy in Melbourne and Johnston in Tasmania were making occasional contributions, but neither was schooled in taxonomy. In Sydney, William John Macleay published many papers on fishes between 1877 and 1884; his *Descriptive Catalogue of Australian Fishes* was published in 1881 and his last contributions appeared in the *Proceedings of the Linnean Society of New South Wales* just after Ogilby's arrival in Sydney. Ramsay, Director of the Australian Museum and Ogilby's new employer had, in the early 1880s, published a few papers on fishes in the *Proceedings of the Linnean Society of New South Wales.*

Ogilby's experience of fishes was largely restricted to the fauna of Ireland but he soon perceived the opportunities afforded by the fish fauna around Sydney. He was aware of the importance of studying the animals themselves as well as their literature, soon came to appreciate recent advances in fish taxonomy and appeared to have the potential to apply a disciplined approach to the study of ichthyology. His publications almost certainly had the effect of inhibiting the effusions of at least de Vis and Macleay.

Resources at the Australian Museum

The Director of the Australian Museum, Edward Pierson Ramsay (see page 130), had been in charge of the Museum for 11 years when Ogilby arrived in 1885. An effective curator, Ramsay had amassed large, useful collections and established cooperative relationships with his counterparts in British and European institutions.

Although Ramsay's main interest was ornithology his activity in the collection and study of fishes prior to his association with Ogilby was extensive. He was a protégé of William J. Macleay, to whom he owed his position as Curator, and they both took a great personal interest in trawling in and around Sydney Harbour. Both men also commissioned collectors further afield, particularly in New South Wales and Queensland. Ramsay was intrigued by the fish fauna of Port Jackson and had made some preparation for a book on the subject, although neither he nor his mentor, Macleay, appreciated the complexities of fish taxonomy. Ramsay made his manuscript notes on Australian fish available to Ogilby, who recorded his gratitude (Ogilby 1887a).

At the Australian Museum, thanks largely to Ramsay, considerable resources were at Ogilby's disposal: large collections of fish were held and the Museum Library contained a respectable representation of ichthyological texts. When he purchased Francis Day's fishes in London, Ramsay also negotiated a number of exchanges of fishes with the 'Commissions' from Canada, China, Norway and the 'Straits Settlements' (now part of Malaysia). After the London Exhibition (see page 131), Ramsay visited some of the museums of Europe and arranged further exchanges with the Imperial Museum of Vienna, and with the natural history museums of Paris and Florence. Collectors sent Ramsay animals from New Guinea, the Solomon Islands, North Queensland, Port Phillip and Tasmania; he generally instructed them to pay attention to fishes, so all of these regions were represented in the Museum collection. Robert Etheridge (1917) said 'Dr. Ramsay encouraged numerous collectors, both amateur and professional, and the earlier Museum Registers include many entries of zoological collecting from all parts of Australasia, gathered together for him by local Missionaries and the officers of vessels trading in the South Seas.' He was also active in arranging exchanges with his fellow curators within Australia, and there are registrations indicating transactions with the National Museum of Victoria, the Tasmanian Museum, and the museums of Adelaide and Brisbane. When staff members were travelling 'on service' for the Museum, they also were expected to make collections of fishes. These collections, purchases and exchanges provided Ogilby with a remarkably diverse hoard of specimens upon which to draw, and he wasted no time in beginning to study them.

The gentlemen of the Linnean Society of New South Wales expected the Australian Museum scientific staff to participate in their society; Ogilby was soon exhibiting specimens and presenting papers at their meetings, becoming a member of the society in 1887. Macleay was still influential in the Society and in the Museum, and his own enormous natural history collections, including great numbers of fish, were at this time still lodged at Elizabeth Bay House in Sydney. Ogilby had access to these collections and made reference to them in his earliest publications.

Obtaining specimens

Ogilby used and extended Ramsay's far-flung network of donors, and undertook collecting excursions himself whenever possible, usually around Sydney. In the autumn of 1886 he was in Brisbane, where he accompanied Henry Tryon, assistant curator of the Queensland Museum, on a three day trip collecting and trawling in Moreton Bay (Ogilby 1903).

The fish registers at the Australian Museum, which show the accession of specimens, present a faithful record of many of Ogilby's activities. One of his first jobs at the Museum was to register specimens from the 'old collection', and he began with 860 of the fishes that Ramsay had purchased from Francis Day. These registrations are then mixed with the day to day acquisitions, sometimes single specimens that Ogilby bought at the markets, sometimes donations from residents around Sydney, but also some larger collections, such as those from New Guinea and Lord Howe Island; and exchanges with museums of Europe and North America. His own collecting excursions around Sydney are also evident from the registrations.

Ogilby often showed considerable ingenuity and persistence in obtaining specimens, seen in his use of the markets and fish-shops, the Bondi Aquarium, and amateur fishermen's associations of Sydney and later Brisbane. He conducted field trips whenever possible and corresponded regularly and at length with individual donors.

Among Ogilby's correspondents was a missionary, Reverend H.P. Schlenker. When about to return to New Guinea he wrote to Ogilby, asking for advice on the collection and preservation of animals. This elicited from Ogilby a closely written five and a half page letter of encouragement and advice; and it was very good advice, such as a routine for preparing labels, using soft pencil, always noting the locality and date, keeping a notebook to 'jot down any peculiarities, no matter how trivial, which may strike you in the habits of the animals ...' He inserted a paragraph on catfishes which, he said, 'will be included in the first part of my work on the 'Fishes of Australasia'.' (AMA)

Ogilby's early publications in Australia (1885 to 1890)

Alone or with Ramsay, Ogilby published nearly 40 papers in his first 4 years in Sydney. At first, in 1885, he appeared as sole author; from 1886 to 1888, the papers bore Ramsay's name as senior author. Most were published by the Linnean Society of New South Wales but, in 1887, Ogilby began contributing to the *Proceedings of the Zoological Society of London* as well, and these latter papers all indicated Ogilby as the sole author. He was also a regular contributor to the fledgling journal, *Records of the Australian Museum*, until he lost his status as a member of the staff in 1893; he then resumed publishing in the *Linnean Society of New South Wales Proceedings*.

Ogilby's early papers were often simple records of new occurrences, usually from Sydney Harbour, or descriptions of new, or supposedly new, species. However, he carried out some revisions of fish groups and attempted to reconcile the inadequate descriptions of previous authors with the species he was handling. In these early papers Ogilby's own descriptions were usually adequate, or at least an improvement on those of his immediate predecessors, though not usually as detailed as his descriptions in later years. Nearly three quarters of the species Ogilby described as new during this period are still considered valid.

From his Moreton Bay excursion Ogilby described several species, including a Pipefish, *Ichthyocampus* [now *Campichthys*] *tryoni* (Ogilby 1890a). Several new species were described from Port Jackson, such as the Blackspot Boarfish, *Histiopterus* [now *Zanclistius*] *elevatus* (Ramsay and Ogilby 1888c). A collection from the Strickland River in New Guinea was reported and new species named, including the Smallmouth Catfish, *Arius* [now *Cinetodus*] *froggatti*, named for the collector[12] (Ramsay and Ogilby 1886a).

After being in Australia for little more than 12 months, Ogilby published a *Catalogue of the Fishes of New South Wales with their Principal Synonyms* (1886). This included some notes,

12 Walter Froggatt (1858–1937) was a Melbourne born entomologist who participated in an expedition to New Guinea in 1885 during which he made a small collection of fishes. He was later engaged by William Macleay, principally as a collector of insects, and wrote extensively on Australian entomology.

generally limited to common names, maximum length, edible qualities and distribution. It was based on Macleay's *Catalogue*, with the synonymies in part drawn from Günther's multi-volume work. Although Ogilby's *Catalogue* was essentially a compilation from these sources, the author showed an appreciation of the complexities that had been introduced to Australian fish taxonomy by the publications of the previous decade. He provided a framework within which to include the new records published around that time by himself and Ramsay. The association with Ramsay was productive although it may have galled Ogilby to publish papers, some of which appeared to be his work alone, as the junior author.

Most of Ogilby's publications were of high quality, but a few absurdities were perpetrated. For example, Ogilby celebrated Macleay's birthday by doing memorable violence to the rules of nomenclature:

> ... we obtained in the trawl up Middle Harbour a magnificent *Ammotretis* [flounder], a genus of which very few individual examples are known; ... It is probable that the species described by Mr. Macleay as *A. zonatus* is identical with our fish, the banded appearance, which induced him to give it the specific name, having been caused by contact, when dying, with some foreign substance, and as the name '*zonatus*' is unsuitable and misleading, I would suggest that this fine species should be re-named *Macleayi* in honour of our distinguished Secretary [of the Linnean Society of New South Wales], on whose fete-day it was obtained, and who originally described it. (Ogilby 1885).

A. zonatus proved to be a synonym of the Longsnout Flounder, *Ammotretis rostratus* Gunther, 1862.

In some of the joint papers of Ramsay and Ogilby, the latter's influence is particularly clear. For instance, the occurrence of the extraordinary Hairtail Blenny, *Xiphasia setifer* in Sydney Harbour prompted a review of the available literature, a tentative synonymy and re-description '[i]n view of the meagre descriptions of this curious form which are available to us ...' (Ramsay and Ogilby 1886b). Similarly in 1888 a joint paper appeared, almost certainly attributable solely to Ogilby, regarding a pelagic fish known now as a Squaretail (Ramsay and Ogilby 1888b). Two years earlier, a seven inch specimen had been found stranded at Lord Howe Island and the visiting magistrate, Mr. H. T. Wilkinson, sent the fish to Macleay. Macleay had been intrigued by its unusual form, and described it as a new species, creating a genus for it. Macleay also illustrated *Ctenodax wilkinsoni*, as he called the 'very extraordinary Fish', with sketches of the general form, the scales and dentition (Macleay 1886a). A few months later Ramsay contacted Macleay, having identified the fish as belonging to the genus *Tetragonurus*, and Macleay promptly acknowledged this (Macleay 1886b). The contribution by Ogilby and his Director in 1888 was prompted by the finding of another specimen of this fish on Lord Howe Island. Ogilby wrote to Ramsay's contact 'Professor Giglioli, of Florence, for an example of the Mediterranean form, which with his usual promptitude he forwarded ...' Macleay donated his original specimen to the Australian Museum collection and with these three specimens to examine, and taking into account previous descriptions of similar fish from Madeira and the Mediterranean, Ogilby gave a thorough description and drew reasoned conclusions regarding the existence of different species within the genus. The nature and execution of this paper were quite beyond the ambition and capacity of Ramsay (Ramsay and Ogilby 1888b).

The aquarium at Bondi Beach was a popular attraction in the 1880s. Ogilby took an interest in their collection and wrote an annotated catalogue of the specimens, a 'popular' work. This pamphlet, a *Catalogue of Fishes and Other Exhibits at The Royal Aquarium, Bondi* (1887b) dealt with nearly 100 species of fish and presented notes on life histories, habitat, edible and angling qualities, and maximum size. It was intended to accompany an inspection of the exhibits and is surely one of the strangest contributions to ichthyological literature, being interleaved with pages of advertisements and offering incidental information: 'The Palace Emporium, Haymarket, is the place to buy a silk dress', 'Anthony Hordern & Sons sell Silks, Velvets, and Plushes.' In the same year Ogilby published a description of a Snakeskin Wrasse from Rose Bay in Port Jackson that he had spotted among the Bondi exhibits. He named the fish *Eupetrichthys angustipes*, establishing a new genus for it, and devoted the better part of a page to a detailed description of its colours. He acknowledged the donation, 'For the pleasure of describing this most beautiful fish we are indebted to the commendable liberality of the proprietors of the Royal Aquarium, Bondi, who handed to us the living fish on its being pointed out that it was of a species unknown to us.' (Ramsay and Ogilby 1888a).

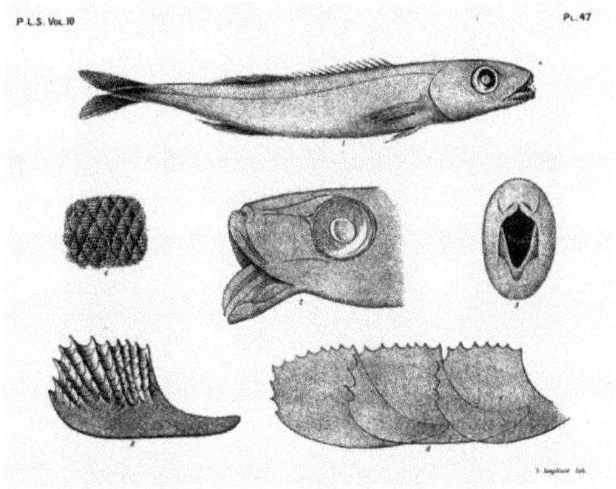

Macleay described the Bigeye Squaretail as *Ctenodax wilkinsoni* but this proved to be a synonym of *Tetragonurus atlanticus* Lowe, 1839. From Macleay 1886.

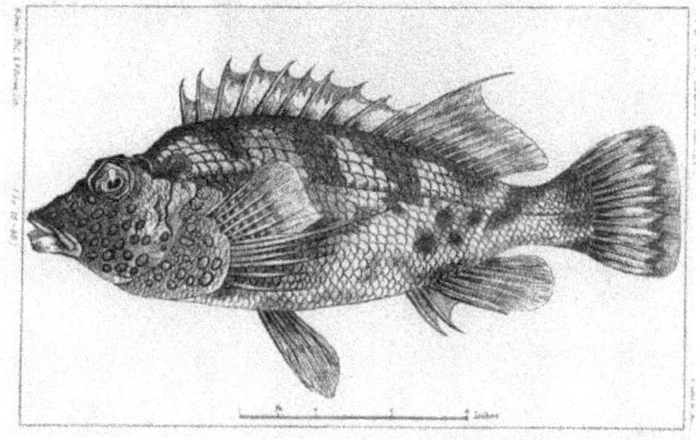

The Splendid Hawkfish, *Cirrhitichthys* [now *Notocirrhitus*] *splendens* was described by Ogilby in 1889 from a single specimen sent from Lord Howe Island. Six of the new species in this paper were drawn by H. Irwin. From Ogilby 1898.

The Reptiles and Fishes of Lord Howe Island (1889), published as an Australian Museum Memoir, included Ogilby's most ambitious ichthyological work to that date. It resulted from an Australian Museum expedition to the island, during which many fish specimens were taken. He also had access to Macleay's Lord Howe Island specimens at Elizabeth Bay House and a number of others purchased from collectors or donated by residents. The result was a paper recording 88 species, 14 of which were described as new. It is a tribute to the quality of this early work that the majority of these species are still recognised as valid. The descriptions were careful and generally detailed, although a few were not up to the standard of his later work. Six of the new species in this paper were illustrated by H. Irwin but some of the drawings were disappointing. Five times as many fishes are now recorded from Lord Howe Island, but from this limited sample of the fish fauna Ogilby drew some conclusions that were essentially accurate: the Australian fauna is strongly represented, there is a significant relation with New Zealand and the rate of endemism is low.

Previous experience with matters ichthyological is often evident in Ogilby's work. In 1888 the diagnosis of a mass mortality of fishes in the National Park south of Sydney apparently presented no problems, 'a very cursory examination sufficed to satisfy me that death was due to the attacks of a species of *Saprolegnia*: being cognisant from personal experience of the ravages caused by the same or a similar fungus among the fresh-water fishes of the British Isles ...' (Ogilby 1888b). He went to the site and found an eel that was affected. His colleague, Thomas Whitelegge, made some slides of the fungus taken from the eel and these were exhibited at a Linnean Society meeting. Ogilby attributed the infection to stagnancy of the creek caused by drought and suggested that the weir gates should be opened to allow the creek to dry up altogether.

The *Australian Museum Records*, intended as a medium for the Museum staff to publish their work, first appeared in 1890 with Ramsay as editor (Etheridge 1917). Up to this time, the Museum publications had generally taken the form of catalogues of the specimens. Ogilby (1888a) completed a catalogue of the *Recent Palaeichthyan Fishes* (sharks, rays and some primitive fishes) and was expected to prepare similar guides to the mammals and reptiles.

Loss of Museum appointment

By 1890, Ramsay was ill and frequently absent and Ogilby had been effectively in charge of Mammals, Reptiles and Fishes for some time. However, he frequently drank spirits to excess and bouts of this unacceptable behaviour had by this time caused repeated embarrassments at the Museum. In 1890 he was dismissed for being repeatedly intoxicated on duty. The Trust Minutes for October 1890 are terse, 'The Curator having reported verbally that Mr. Ogilby had again not been as steady as was desirable ... it was resolved that Mr. Ogilby's services be dispensed with after a week's notice.' (AMA) His hand writing in the specimen register, which had been continuous from 1885 to 1890, no longer appears after August 1890. However, he was retained on a contract basis and continued to have access to the Museum resources for the purposes of research on fishes and to complete the long overdue catalogue of mammals.

In 1891 Ogilby was able to report to the trustees, 'Having now concluded the ms of the *Descriptive Hand-list of Australian Mammals* to the satisfaction of Dr. Ramsay, I would ask you to entrust me with a similar, and similarly needed work on the Australian Reptiles and Batrachians.' (AMA) The *Catalogue of Australian Mammals*, a well-known and useful contribution to Australian zoology, was finally ready for the printer in 1892.

Arrival of Edgar Waite at the Australian Museum

In July 1892 Edgar Waite of Leeds answered an advertisement that appeared in the journal *Nature* for the position of Assistant in Zoology at the Australian Museum; a replacement

Edgar Waite, courtesy of Duncan Waite.

for Ogilby was required. Waite was 13 years junior to Ogilby in age and infinitely his junior in experience in ichthyology. Like Ogilby, Waite had expertise in a wide range of animal groups but had until then concentrated on working with birds and insects. The advertisement implied that experience with fish was necessary but Waite, who had almost none, thought that he could soon obtain it on the job. Waite arrived in Sydney in April 1893 eager to take up his duties but was disconcerted by Ramsay's conduct. Ramsay was ill and in the last year of his curatorship; Waite was keen to begin work but initially there was little guidance and no supervision, and Ramsay seemed to resent his presence. Both Waite and Ogilby had cause to complain: it was obvious at once to Ogilby that his replacement was ignorant of ichthyology; and Waite was initially perplexed by Ogilby who, although without official standing at the Museum, nevertheless attended more or less regularly and was active in ichthyological research.

In April 1893 Waite noted in his diary:

> My predecessor, J.D. Ogilby had charge of the Mammals, Reptiles and Fishes and is now engaged in writing catalogues but is not on the staff. He is an excellent man at his work but had to be discharged on account of drunkenness. I am told that he was borne with time after time & was even offered the chance of being placed in a curative hospital but would not accept. drink with him had become a disease and he says that he cannot resist it. he had many and many a time had to be carried to some place of secret in the Museum. His tipple is Museum spirit & it has to be kept out of his way. He takes fresh spirit if he can get it but as that is almost impossible he drinks that in which all kinds of things have been preserved & leaves the specimens dry. (WD, 11.4.93)

Waite had first hand experience of this a month later:

> My late week in the Museum. In the morning Ogilby went into the spirit room with me & took away 2 type Opisthognathus [Jawfish] for examination, when I went back at 2.0 he was helpless & neither of the specimens could be found.

Two days later the jars, with their desiccated specimens, were discovered behind some books in the museum library.

Catalogues of Fishes and supervision of Ogilby's work

Ogilby told Waite that he was engaged on a work on fishes for the government. A complex form of agreement had been drawn up and the Principal Under Secretary wrote to Etheridge, then effectively the chief administrator at the museum, in August 1893:

> Dear Sir,
> The Chief Secretary having given authority for the employment of Mr. J.D. Ogilby in connection with the revision of the list of Fishes of Australia at the rate of £3. per week with the assurance that the work will be completed within twelve months from the date of commencement, to be carried out under your supervision, and in order to comply with these conditions may I request you to be kind enough to certify to the due performance of the work in order that monthly progress payments may be made to Mr. Ogilby through you. (AMA)

This implied that Ogilby was to bring Macleay's *Descriptive Catalogue of Australian Fishes* up to date. The Catalogue had been published in 1882 and Macleay made some additions in 1884, but the better part of a decade working with Australian ichthyology had shown Ogilby that the Catalogue needed complete revision and that this could only be done effectively by a thorough examination of the already extensive literature, and of the fishes from which it had been derived. Macleay had trusted to Günther's judgment in the basic list, thereby perpetuating Günther's errors. He had then added a large number of fish species from his own work and from that of Castelnau and de Vis. Ogilby certainly knew that he could not sort out this taxonomic labyrinth in 12 months. He must have assumed that his contract would be extended once he was seen to be making progress.

In July 1893 Ogilby completed *Edible Fishes and Crustaceans of New South Wales*, published for the World's Columbian Exposition, held in Chicago in 1893. This was 200 pages of information, and some speculation, about the fishes of New South Wales (the crustaceans occupy only a few pages) which he introduced with a caution: 'The present volume must not be considered otherwise than as an instalment of a much larger and more important work, which it is intended, shall include within its scope the entire fishing industries of the Colony, and which it is hoped will be completed within a few years.' By 'the Colony' he implied the whole of Australia: his work in progress was intended as nothing less than a definitive Fishes of Australia.

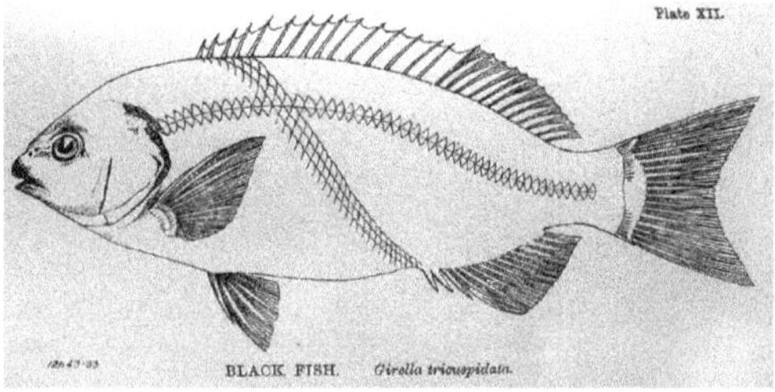

Irwin's drawing of the Luderick, *Girella tricuspidata* used in Ogilby's *Edible Fishes and Crustaceans of New South Wales*. From Ogilby 1893.

The *Edible Fishes* indicated Ogilby's ambitions in ichthyology and something of the sort of information he considered proper for a book on Australian fishes. It also demonstrated how much information he had collected since he produced his preliminary catalogue seven years earlier. There was certainly no one else in Australia at the time who would have thought it necessary or indeed possible to assemble the detailed life history data he sought. In the introduction, he drew attention to the importance of understanding life histories and provided as much information as possible, some drawn from Tenison-Woods' *Fish and Fisheries of New South Wales* (1882). Whenever possible, Ogilby prepared his descriptions from fresh specimens procured from the Sydney markets. The standard of illustration in the *Edible Fishes* was less than he aspired to, consisting of 48 outline drawings by H. Irwin; these showed fin rays, some head detail, and some indicative scale rows, but were neither scientifically adequate nor aesthetically pleasing.

Robert Etheridge, geologist and palaeontologist, became Curator of the Australian Museum in 1895 on Ramsay's retirement. From Dun 1926.

Robert Etheridge Junior (1847–1920) became Curator of the Australian Museum in 1895 although, due to Ramsay's ill-health, he had been performing most of the curatorial tasks for some time prior to this. He was placed in an unenviable position between Ogilby and the government. He inspected Ogilby's work at intervals but could hardly be expected to judge it effectively. Ogilby had officially estimated that the book would take 12 months to complete but after 14 months, in October 1894, he needed Etheridge's support to negotiate an extension of the time allowed. The Curator wrote to the Colonial Secretary's office 'I have seen the M.S. from time to time, as payments have been made to Mr. Ogilby, and I consider that the work has been carried out in a satisfactory manner.' And a month later he gave an unqualified recommendation:

> Such a list as Mr. Ogilby has been engaged upon is all important to science, not only in Australia, but throughout the world, and is of much practical importance to the development of the fisheries of this colony, as our present catalogues are extremely inaccurate and often deceptive.
>
> Under these circumstances, I venture to suggest that the authorization formerly given should be extended for a further period of twelve months, which should be sufficient to complete the work.
>
> If this is not done, the work already performed will be of little practical value, and I know of no one better qualified to execute it than Mr. Ogilby. (AMA)

This was generous of Etheridge who had little enough reason to rely on Ogilby.

Developments in relationship between Ogilby and Waite

Waite and Ogilby met frequently at the Museum and in the early mornings at the fish markets where they scanned the catches for interesting specimens. Waite was familiarising himself with the species and Ogilby was on the lookout for specimens to use for descriptions for his work. They exchanged books and on at least one occasion Ogilby and his wife came to the Waites for dinner. At this time Ogilby was still writing papers for the Linnean Society and several times had Waite read the papers for him at the Society's meetings. They also participated in local collecting trips, often with other Museum personnel. In June 1893, Ogilby was very distressed and said that his wife was ill; she died a few days before Christmas, 1893. Waite and Ogilby's other museum colleagues were sympathetic but could not understand Ogilby's retreat into his addiction to alcohol.

Ogilby had given Waite a copy of his *Catalogue of the Fishes of NSW* ('& then borrowed 5/- from me') (WD, 26.8.93) and Waite records passing an evening interleaving and binding this book, which he used as a basis for some of his own work on fishes. Ogilby's influence on Waite can hardly be overestimated. Waite was well aware of the older man's expertise and, perhaps rather grudgingly, benefited from it. Ogilby gave him his papers and articles, helped with identifications, and there are plain indications that he was a harsh but useful critic of Waite's early work on fishes.

Waite's family included many more or less talented artists and he had learnt to draw and paint at an early age. In due course he had adapted his talent for the purposes of scientific illustration. His skills were appreciated by his colleagues at the Museum, who took advantage of him to some extent. However, Waite found the drawings time-consuming and presumably felt that he was being exploited. When listing the drawings that he had done up to 1899, he wrote that he would no longer illustrate for his colleagues, only for his own papers. In May 1894 Ogilby asked Waite if he would prepare the illustrations for his work on the fishes of Australia. At first, Waite refused, as he says in his diary 'No! for many reasons.' He did not wish to become known as Ogilby's illustrator and shied away from any formal professional relationship with such an unstable character. He also stated in his diary that he did not want to take the work from the other illustrators who were 'dependent on their pencils for a livelihood' (WD, 22.5.94). Five days later Ogilby, examining a snake that he regarded as an undescribed species, called it after Waite; and a week later, Ogilby renewed his request, saying that the other illustrators did not draw fish well enough. Waite records '150 fish would have to be drawn in perhaps 2 years. Ogilby mentioned 22/6 each drawing.' Three days later Ogilby bought a fish from the Sydney market, a Dart, and Waite began to draw it that evening. Over the next 12 months, Waite continued drawing for Ogilby, but these drawings did not illustrate Ogilby's works; presumably payment was not made as Waite retained the drawings (see pages 244–5).

Ogilby's life and work seen through his diary for 1895-96

At about this time Ogilby apparently stopped drinking alcohol. He entered on a turbulent but professionally productive phase of his life, publishing 15 papers between 1896 and 1899 and the quality of this work was generally high. Much of his effort was directed to research for the revision of Macleay's Catalogue but there were many distractions.

A detailed account of Ogilby's life from October 1895 to April 1896 can be gleaned from a fragment of his own diary,[13] which exists in the form of an exercise book that survived among the papers that Waite carried from Sydney to New Zealand, and then to Adelaide. Its inclusion among Waite's papers is almost certainly due to Waite having redeemed Ogilby's books from a pawnbroker in Sydney after Ogilby had departed ignominiously for Brisbane.

Ogilby was boarding at Glebe and working hard, but living in most unusual circumstances which could hardly have been conducive to the pursuit of ichthyology. He documented his day to day activities, his thoughts and feelings, and each fragment of work as it was done. He had a number of projects running concurrently. His work habits and methods of research, perhaps more fragmented at this time than usual, are of particular interest. His ichthyological activities were all more or less aimed at completing the great work on Australian fishes, but it is easy for the reader of his diary to overlook this when confronted with him delving at random into a barrel of fishes sent to him from Tasmania; strolling off to the Macleay Museum, by this time established in its own building at Sydney University, for the purposes of unravelling some of the confusing descriptions published some 30 years earlier; setting out in the early hours of the morning to the fish markets to purchase specimens (and table fare); wandering into the Australian Museum for an hour or so to consult their specimens and library; and organising fishing and collecting trips to the coast. He was preparing, for his own use, indices to fish references in various journals, particularly the *Proceedings of the Linnean Society of New South Wales*, and also writing a complete account of the sharks and rays of Australia. Throughout this period he persevered with his long-standing correspondence with overseas ichthyologists.

Once again, there was correspondence over the revision of the Fish Catalogue. Frank Farnell, Director of Fisheries wrote to the Chief Secretary in May 1895 on Ogilby's behalf; and a little later one of the museum's trustees, Dr James Cox, an enthusiastic collector of fishes, also wrote in support. In a postscript he asked that Ogilby be given a return railway pass to Cooma as 'The headwaters of the Snowy River have never been satisfactorily examined for fish and Mr. Ogilby is most anxious to go up himself and do so.' (AMA) Well, he wanted a holiday and he had one. He received his pass and had an energetic time collecting birds, mammals, spiders and fish in July 1895. He sent a box of specimens back to the Museum, with a covering letter; Waite received the box and added a dismissive note, 'The box has been looted by Rats which have more or less eaten everything ...' (AMA)

The boarding house which was Ogilby's residence at this time was only two or three miles west of the Australian Museum, and even closer to the Macleay Museum and to the fish markets. He was apparently a paying guest of Ada (perhaps a relative?) and her semi-estranged husband, Frank. Ada's domestic strife and moods upset the sentimental Ogilby remarkably: he could not sleep after harsh words, was enormously distressed at Ada's tribulations, and made extraordinary efforts to keep her cheerful. Some of his typical days are worth considering.

13 Ogilby's diary, from which most of the information in the following nine paragraphs is derived, was consulted at the South Australian Museum library in 2001.

A 'drum' of fish specimens arrived at the Australian Museum from Alexander Morton of the Tasmanian Museum, Hobart and was delivered to Ogilby at his home on 5th October, 1895. He dipped into the drum, starting with the fish at the top and working his way down day by day. His sense of discovery and sheer delight is evident time and again amid the domestic preoccupations described in his diary, for instance:

> Sunday 13th October 1895
> ... I having shelled the peas for Ada's dinner, dipped into the drum again and brought out a prize in the shape of two magnificent specimens of Gnathanacanthus goetzi ... the diagnosis of these fishes took me till dinner time to draw up, and it is abundantly evident that Johnston's H. guntheri cannot be defended [he refers to *Holoxenus guntheri*, a synonym of the Red Velvetfish, *Gnathanacanthus goetzeei*]; the differences are merely individual, or at most sexual ... After dinner I lay down and read for an hour and a half when I prepared some coffee and fruit for Girlie [Ada], and then did a bit of fishing, undrumming this cast a Callanthias allporti [Rosy Perch] which I more than half described before tea time ... and I to my room to write to Morton ...

The image of Ogilby is memorable: casting for fish to 'undrum', settling down to describe them, and before bed writing to his friend, Morton, to thank him and report his latest catches from the drum.

Another typical account a couple of weeks later:

> Friday 25th October 1895
> I could not sleep all night so got up at 4 o'c. and worked ... I left the house at 9.15 and walked to the University where I did an hour's work and then took the tram into town ... after going to the Museum and getting Etheridge's address, leaving a note in at the Legislative Assembly Rooms, and calling at the Col. Sec's about my voucher, I came home ... [letter] from Morton; also two parcels from him, the one containing two pairs of Anchovies, the other an Antennarius which I am not acquainted with. I spent the afternoon making a rough description of the latter ... After tea I went on with my own work, which is getting on nicely now, but I do not see yet how I am going to get the illustrations done for it.

By way of explanation of the above: the day before this Ada had been particularly unreasonable, hence the lack of sleep. He went to Sydney University to consult the Macleay Museum's fish collections, then to the Australian Museum. The 'voucher' was Etheridge's certificate to state that Ogilby had made certain progress in his work on fishes and should be paid at the Colonial Secretary's Office. The letter referred to was from Alexander Morton (see page 409), the curator of the Tasmanian Museum, Hobart, who sent many specimens to Ogilby. The 'Antennarius' was a small Anglerfish which Ogilby described in great detail and sent back to Morton, together with the description, which Morton duly published, naming the fish *Antennarius mitchelli* [now *Trichophryne mitchellii*] (Morton, 1897). Finally, Ogilby was worried about the illustrations for his major work on fishes.

Always ready for a field trip, Ogilby took advantage of an offer of some fishing:

> Tuesday 19th November 1895.
> Up at 5 and worked for an hour at the Elasmobranchs [sharks and rays], and then prepared things for today's trip ... tram to Sandringham, where I met Grant, and we pulled across to and round Towra Point, but seeing no boats trawling went ashore at the half-breed's oyster lease to see the skull of an 11 ft. animal, which had come ashore near his shanty some time ago; it proved to be that of a very old Sea Leopard ... we found three boats lying in the shelter waiting for the tide to fall; they hauled at 3.45, getting a lot of Trevally, Sea Mullet, River Garfish, Flounder and Flathead, while I was fortunate to secure some Diplomystus, Sprattelloides, Gobius, Dicotylichthys, and Tetrodon, [Sandy Sprat, Blue Sprat, Goby, Porcupinefish, Toadfish] as well as heads of a Myliobatis and Carcharias [Eagle Ray and a shark].

He arrived back later than expected and took a train to Sydenham, but 'here not a bus was to be found or a cab so I had to carry the half-hundred-weight [25 kilograms] of fish all the way home as best I could ...' These specimens went in part to the table and in part towards his studies. He turned his attention to any interesting fish that came his way, being satisfied only by describing from fresh or well preserved specimens. This was one of the great strengths of his work: the realisation, not so common at the time, that for a description to be accurate it must be made from the animal itself.

The day after this excursion, Ogilby received a letter asking him 'to write an article on the fish resources of the Colony'. This was paid work, not an offer he could refuse, but yet another distraction from his main work. That night barking dogs kept him awake so he spent a few hours just after midnight in organising his scientific correspondence, read for a while, slept for a while, then 'up again at 5.45, and examined a Stingaree before breakfast, and afterwards cleaned the heads of the Shark and Ray and finished the description of the Flounder. Afterwards I examined and described three examples of Tetrodon hamiltoni [Toadfish, now *Tetractenos hamiltoni*].' After lunch, he went to town on business, but perhaps the lack of sleep made him touchy: '... the arrival of the Governor had put all the myriad fools of Sydney agape and I could not see the persons I wanted ...'

He devoted time each day to the 'Fish Resources' but new distractions appeared:

> Friday 22nd November 1895
> Up at 5 o'c. and did over an hour's work at the 'Resources' ... finally got to the Museum, where after looking over some journals I sat down to a good morning's work at the Index, but had not got fairly settled down before I was summoned to a consultation about some fishes sent up by Bray. They proved to be magnificent specimens of Lobotes surinamensis [Tripletail] and Trachynotus ovatus [Dart]. As Etheridge would not buy them I purchased them for the sake of the descriptions, and having described the latter resold the specimen to the Museum, and brought the other home by the 1.5 train, and spent the afternoon in making a careful generic and specific diagnosis of the specimen, which we eventually had for dinner, finding it excellent ... Worked at Index ...

His sense of humour did not fail him: 'Sunday 24th November 1895. Up at 7 and wrote up this veracious history before breakfast'; but the demands on his time were quite remarkable and one wonders just what it would take to trigger a relapse of his drinking.

Ogilby's publications from 1891 to 1895 (inclusive)

The five years following the loss of his Museum position were not unproductive. *The Edible Fishes and Crustaceans of New South Wales* (1893) was a useful book, virtually a summary of knowledge of commercial fishes of the Sydney region. Ogilby's access to new material may have been curtailed somewhat, particularly following the arrival of Edgar Waite in 1893, but he published descriptions of new species in eight rather brief papers. The element of revision was not marked in these papers but during this time Ogilby was preparing hundreds of pages of the descriptions that he meant to form part of his major work. Most of the specimens for his brief publications were derived from the Australian Museum collection or from his friend, Thomas Whitelegge.

Thomas Whitelegge as a fish collector

The members of the Australian Museum scientific staff of Ogilby's day were much less differentiated, with regard to specialisation within biology, than their counterparts of recent times. For instance, Ogilby and Waite were both able to work successfully on any of the classes of vertebrates; and most of the members of the staff would collect, with some discrimination, almost any group of animals. Several of the Museum zoologists presented fish from time to time, but none collected with more enthusiasm than Thomas Whitelegge (1850–1927). Self-taught, Whitelegge had developed skills in palaeontology, botany and zoology. He migrated to Australia with a young family at the age of 33 and, after a difficult interlude, came to the attention of Tenison-Woods and W.J. Macleay; the latter introduced Whitelegge to Ramsay, thus initiating his association with the Museum. In 1887 he became a permanent member of the staff, in charge of the Department of Lower Invertebrates. He enjoyed field work and collected widely, both plants and animals (McNeill 1929).

Whitelegge was particularly familiar with the intertidal fauna and beach strandings around Sydney; Maroubra Bay, a few kilometres south of Port Jackson, was his favourite collecting ground. His interest in fishes was considerable and in his personal library there was a copy of Ogilby's Bondi Aquarium catalogue, interleaved, with annotations on the fishes of Sydney. The number of new and rare fishes that Whitelegge brought for description was remarkable, and some were named for him by Ogilby and by Waite. Among Whitelegge's many Maroubra finds described by Ogilby were the Surf Sardine, *Tropidostethus* [now *Iso*] *rhothophilus*, and the Threespine Cardinalfish, *Apogonops anomalus*. Whitelegge first discovered *Iso rhothophilus*, washed up at Maroubra and later observed them in large numbers. He took Ogilby to see the little fishes and in his original description Ogilby reports their joint observations:

> They are essentially surf-fishes, coming in with the waves, and being swept up into the gulches and pools on the reefs; they never descend to the bottom, but swim here and there, keeping but a few inches beneath the surface; the pectoral fins

> are always kept fully expanded, at right angles to the body, and motionless, being utilised in fact solely as balancing media; the caudal fin and pedicle [base of the tail] have a distinct downward curvature when the fish is swimming. (Ogilby 1895).

Apogonops anomalus was described by Ogilby (1896a); he acknowledged Whitelegge and the significance of Maroubra Bay:

> It is again my pleasing duty to record yet another new fish from Maroubra Bay, where it was obtained by Mr. Whitelegge early in February. The constant recurrence of new forms of animal life in this small bay, probably the only spot on the Australian coast which has been systematically and scientifically explored, is an additional proof, if one were needed, of how imperfect a knowledge of our littoral fauna we possess.

Steps towards the proposed catalogue of Australian fishes

Ogilby completed his paper on the fish resources of New South Wales and it was published as *Fish Industry* (Ogilby 1896c), a useful summary of the fish inhabiting the waters around Sydney that had a real or potential commercial importance. He could then concentrate on the more important work in progress and the same year Ogilby (1896b) published a *Monograph of the Australian Marsipobranchii* [Lampreys] which was a logical, but very small step, towards his major work. He brought some order to the taxonomy of the Australian members of this group and, in all but one instance, avoided the long-standing confusion engendered by the larval forms of lampreys having very different structure to the adults. This paper is interesting in itself and for a clear statement made by the author regarding the importance of Australian animals being studied in Australia, without too much deference being paid to overseas opinion:

> ... I may as well take this opportunity of entering a strong protest against the practice which is so prevalent among writers on our fishes of copying the descriptions and remarks from the British Museum Catalogue without any attempt being made to test their accuracy, and by so doing perpetuating error, creating confusion, and indefinitely postponing the dawn of that accurate knowledge of our native fauna which every admirer of the native products of our country must ardently desire. (Ogilby 1896b)

Fine sentiments, and also doubtless intended as a snub for Waite whose dependence upon Günther's 'British Museum Catalogue' at this stage was offending Ogilby.

Waite began drawing for Ogilby again in January 1896, 'Ogilby asked me if I would draw him plates of herring for his monograph. Although no price was fixed 15/- for each drawing was mentioned or the lot not to exceed £10.' (WD, January 1896) Then, in May, 'Ogilby told me that the Government had voted £450 to complete the work on Australian fishes for which they will allow but 3 years. they also put down £100 for the illustrations which sum I hope to draw.' But Waite drew only a few more fishes for Ogilby at this time, and was presumably unpaid.

The reference to a three year time limit probably represented Ogilby's own estimate of the time that this enormous work would take, rather than any existing agreement he had with the Government. There was a great deal of acrimony regarding his 'contract'. Towards the end of 1896 Ogilby wrote to the Colonial Secretary with details of his plan to accomplish the new catalogue: when he had finished working on the fishes in the Australian Museum, he would need to visit Melbourne, Adelaide, Brisbane, Hobart, Perth and New Zealand; he would publish section by section in the *Proceedings of the Linnean Society of New South Wales* and, when sufficient had been published to comprise a 'convenient volume', the Government could issue it in book form. He gave a reasonable outline for the composition of his first volume and listed the problems that had delayed the work (no office, no assistant, poor library, insufficient writing materials, meagre funds). Lastly, he included letters of support that he had solicited from the three best known ichthyologists of the day: G.A. Boulenger of the British Museum, David Starr Jordan of the Leland Stanford Junior University and Theodore Gill of the U.S. National Museum (AMA).

Correspondence and influence of Jordan and Boulenger

The remarkable extent of Ogilby's correspondence with other ichthyologists is evident from his letter book for the period September 1897 to February 1898 (SAMA). He copied out his letters in full in this book. Below is a list of some of his interstate and overseas 'fish' correspondents, taken from this book, with the numbers of letters he wrote to them during this six month period:

Molineux, Albert (3)	Agricultural Bureau, Adelaide
Vaillant, Leon (2)	Musée d'Histoire Naturelle, Paris
Langley, S.P. (3)	Secretary, Smithsonian Inst., Washington, D.C.
Spencer, Baldwin (2)	Professor of Zoology, Melbourne University
Kershaw, Jas. (2)	National Museum, Melbourne
de Vis, Charles (2)	Queensland Museum
Morton, Alex (2)	Tasmanian Museum, Hobart
Allcock, A., M.D. (3)	Indian Museum, Calcutta
Sclater, W.L. (1)	South African Museum, Capetown
Eigenmann, Carl (3)	Professor of Zoology, Indiana University
Gill, Theodore, M.D. (3)	U.S. National Museum, Washington
Boulenger, G.A. (3)	Zoology Department, British Museum
Jordan, David S. (3)	President, Leland Stanford Jnr Uni., Cal.
Stirling, E.C., M.D. (2)	Adelaide University.
Woodward, Bernard (2)	Western Australian Museum
Evermann, Barton W. (1)	U.S. Fish Commission, Washington, D.C.
Berg, Dr. Carlos (1)	University, Buenos Ayres, Argentine Republic.

The North American influence in Australian ichthyology had its inception in the last decade of the 19th century when first Ogilby, and later Waite, began corresponding with David Starr Jordan, then President of Stanford University. Jordan was an accomplished and innovative ichthyologist, aware of the importance of establishing a stable nomenclature for fishes (and zoology in general) and pleased to give advice to his correspondents. He knew that Australian ichthyology had suffered from the publications of amateurs and that its taxonomy presented a great handicap to current practitioners. Ogilby put to Jordan a multitude of questions and, tentatively at first, offered criticism of some of Jordan's published work. Jordan accepted the criticism gracefully and replied to the questions with helpful, reasoned opinions. The exchange of ideas was beneficial to the work of both men, but the greatest profit was for Ogilby, who worked in a state of professional isolation that was relieved only by his correspondence. Jordan gave some of his letters to his students to answer and this also worked to Ogilby's advantage; soon he numbered among his correspondents several of the highly trained and strongly motivated ichthyologists who made up Jordan's 'school' of ichthyology.

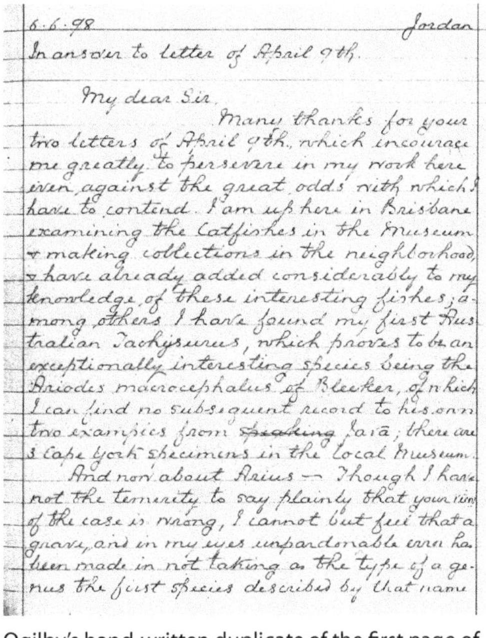
6.6.98 Jordan
In answer to letter of April 9th.

My dear Sir,
Many thanks for your two letters of April 9th, which encourage me greatly to persevere in my work here even against the great odds with which I have to contend. I am up here in Brisbane examining the Catfishes in the Museum & making collections in the neighborhood, & have already added considerably to my knowledge of these interesting fishes; among others I have found my first Australian Tachysurus, which proves to be an exceptionally interesting species being the Ariodes macrocephalus of Bleeker, of which I can find no subsequent record to his own two examples from ~~Speaking~~ Java; there are 3 Cape York specimens in the local Museum.
And now about Arius — Though I have not the temerity to say plainly that your view of the case is wrong, I cannot but feel that a grave, and in my eyes unpardonable error has been made in not taking as the type of a genus the first species described by that name

Ogilby's hand-written duplicate of the first page of a letter to David Starr Jordan, written from Brisbane in 1898. Courtesy of the South Australian Museum Archives.

Some of the long letters written to Jordan by Ogilby are preserved in the form of copies that he kept for his own records (SAMA). These letters were written between June 1898 and July 1899 and presumably came into Waite's possession when Ogilby left for Brisbane. Ogilby undertook a closely reasoned criticism of Jordan's use of generic names. He explains his views clearly and at some length, taking examples from the catfish that he was then studying, and concluding:

> This puts me in the unenviable position of setting up my poor opinion against such authorities as Drs. Gill & Eigenmann & yourself; but much as I dislike the idea I cannot bring myself to make the change which the acceptance of your views would necessitate unless stronger reasons than have as yet been given are brought forward. (SAMA, Ogilby to Jordan, 6.6.98)

Ogilby was arguing against the so-called 'first-reviser principle' of nomenclature which was by no means generally accepted at the time. It was intended to facilitate the resolution of certain equivocal problems in establishing stable scientific names and Jordan made the situation clear in his reply. Ogilby accepted the explanation, thanked Jordan handsomely (SAMA, Ogilby to Jordan, 10.10.98), and applied the principle fairly consistently thereafter.

At this time Jordan complimented Ogilby by naming a new genus of fishes after him, choosing a member of a group with which Ogilby had recently been working. *Ogilbia* was established for a fish from Key West, Florida (Jordan and Evermann 1898). The genus is still accepted although Ogilby thought Jordan's fish so similar to one from Maroubra, for which he had described a new genus himself (the Littoral Cusk, *Monothrix polylepis* Ogilby 1897), that Jordan's name would not stand, 'About Ogilbia; my Monothrix, published some months earlier seems to come perilously near, unless the dentition saves us.' (SAMA, Ogilby to Jordan, 24.11.98) Ogilby returned the compliment a decade later naming a Queensland shark, the Sliteye Shark, *Scoliodon jordani* (Ogilby, 1908b); however, the shark had been described as *Loxodon macrorhinus* by Müller and Henle in 1839.

George Boulenger at the British Museum also responded well to Ogilby's enquiries. In 1895 Boulenger published what was intended to be the first volume of a second edition of the *Catalogue of Fishes in the British Museum*. Albert Günther's first edition of the *Catalogue* (published from 1859 to 1870) had served a generation of ichthyologists, but there had been many advances in the science since its completion. Although only one volume of Boulenger's work was completed, the high standard achieved became a benchmark for ichthyologists. Ogilby was sensitive to this, and it is likely that this publication and the communication with Jordan were the principal reasons for the improvement in his work in the late 1890s.

Ogilby exchanged specimens with British ichthyologists as early as 1890. For instance, he received from the New Hebrides specimens of a little known Flashlightfish (Anomalopidae). He corresponded with Günther, published a detailed and exact description of the fish (Ogilby 1890b), and forwarded a specimen to Boulenger:

> We are sending today per Parcel Post a small box containing a specimen of *Anomalops palpebratus* [now *Photoblepharon palpebratum*], as previously promised to Dr. Günther ...(NHMA, Ogilby to Boulenger, 26.7.1890)

Later, while preparing his review of jawless fishes, Ogilby exchanged specimens of Australian for Chilean lampreys with Boulenger, upon whom he relied for opinions on type specimens held in the British Museum.

Thetis trawling venture

In 1898 the New South Wales government decided, on the recommendation of a Member of the Legislative Assembly, Mr. Frank Farnell, to fund a trawling expedition (see also page 251). Farnell, an enthusiast for developing the New South Wales fisheries, was appointed Director of the enterprise and Her Majesty's Colonial Steamer *Thetis* was fitted out as a trawler. Ogilby was expected to accompany the expedition but he made extravagant demands and proposed conditions that Farnell could not accept. His letter is extraordinary; his conditions included an advance of £25, 'My gazette as 'Government Ichthyologist' before I undertake this duty', 'The sole control of all scientific operations both afloat and ashore to be in my hands absolutely', 'An arrangement to be made with the Government by which the monthly contribution of M.S. to the Australian Museum Board shall be discontinued during the time that I am engaged on these investigations, the salary however being paid as at present' and 'All specimens of natural history collected during the course of these investigations to be my absolute property.' (AMA)

It is hardly surprising that the organisers of the expedition decided to look elsewhere for a biologist. Waite writes:

> To day Etheridge told Whitelegge & myself (assembled) that the Civil Service Board had approached him for the purpose of securing one of the Museum staff as Naturalist to the Expedition. The matter is to be laid before the board on Tuesday & Etheridge wants to know if either Whitelegge or I would go if asked to do so. (WD, 27.1.98)

Waite had not expected to participate in the *Thetis* excursion but he was unfailingly adventurous, rarely missing an opportunity for a new experience, and doubtless saw the professional advantage offered. A correspondence grew around Ogilby's candidature for the trip, including a great number of letters in the Sydney newspapers, but it was Waite who left on the *Thetis* and Waite who had the task of writing up the ichthyological results.

Tribulation and drinking

At this time (September 1898) there was continuing pressure on Ogilby to produce his book on Australian fishes and once again he asked Waite to prepare drawings, saying that if he did not, the book would not be illustrated (WD, 1898). Waite did not always ask permission to carry out paid work outside the Museum and could probably have agreed, but he may have thought that his chance of remuneration was not very great, and he had considered writing a book on Australian Fishes himself. His own little book on Australian snakes (Waite 1898b) was finished at about this time: it had been written and illustrated by him on a contract basis for a Sydney publisher. With Ogilby hoping to have him illustrate his work, Waite must have felt that he could expect a good review of his snake book and he presented it to Ogilby in June 1898 for this purpose. No agreement seems to have been reached between them when Ogilby began drinking to excess once again.

In October of 1898 Waite heard from Rainbow, a zoologist at the Museum, that Ogilby 'has broken out again as a dipsomaniac, after an abstinence of three years'. And two months later 'Rainbow had a letter from Mrs. Corcoran [Ogilby's landlady] asking him to make arrangements for the removal of a cask [of fish] as Ogilby is still imbibing (see Oct 29). He handed the letter to Etheridge, this will in all probability mean that O. will in future be barred entrance to the Museum sanctums.' (WD). At this time there were many obvious reasons for the anxiety that presumably led Ogilby to resume drinking. His private life was in disarray, he was having difficulty in extracting money for the work he was doing, his surroundings were very poor, he had lost the *Thetis* opportunity, saw a younger man making a success of the job he had left, and he had an impossible deadline for his work on Australian fishes.

Preparation and criticism of the *Thetis* account

The *Thetis* fish collections were entered in the Accession Register at the Museum by Waite but Ogilby had a great interest in them. The competitive aspect of their odd relationship is indicated by an entry in Waite's diary. The most spectacular animal taken by the *Thetis* trawls was a Ghost Shark. Waite was delighted with the creature but decided it was similar to, or identical with, a species that had been described some 20 years earlier from Portugal. He had access not to the original description, but to a brief account in Günther's catalogue and hesitated to describe his specimen as a new species. Waite's friends at the Museum, Etheridge (curator and palaeontologist) and Whitelegge (zoologist), 'urge me to describe it as new'. (WD) They told Waite that he should forestall Ogilby who, they were sure, would certainly regard the animal as new and describe it forthwith. They said that he '... might as well name it, as let Ogilby do so.' Waite's humour makes his diaries a pleasure to read; he considered naming the Ghost Shark, and that 'if I decide so to do I will call it Chimaera ogilbyi.' (WD) And so he did.

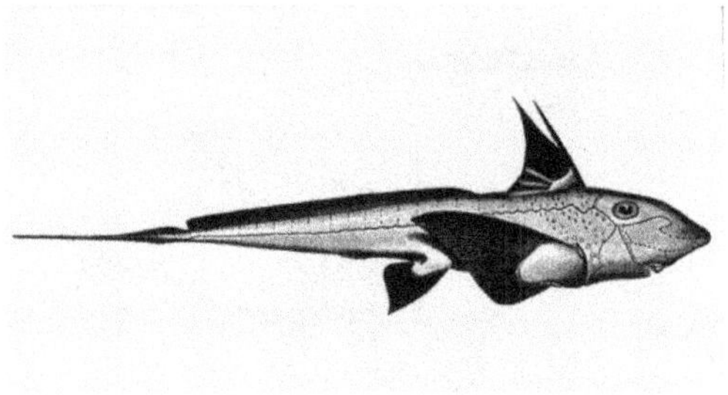

Edgar Waite's drawing of Ogilby's Ghostshark, *Chimaera* [now *Hydrolagus*] *ogilbyi*, trawled by the *Thetis* in 1898 off Port Hacking, New South Wales. From Waite 1898.

Waite had no wish to collaborate with Ogilby on the *Thetis* report which he rightly saw as his most important task thus far at the Museum. He still had his reputation to make as an ichthyologist so he worked on the *Thetis* fishes alone and prepared a preliminary report that was published in 1898. It was referred to on the title page as the 'Scientific Report on the Fishes by Edgar R. Waite, F.L.S.' and was ready for publication in July 1898, just 3 months after the trawling was completed. He had worked rapidly, essentially alone, and with a limited experience in ichthyology. Ogilby resented being excluded and wrote to David Starr Jordan, criticising Waite:

> It is indeed pitiable that such work as Waite's should be allowed to go forth to the world with the stamp of official approval upon it in a country which is – in its own estimation – the most civilised and progressive on the globe; but there is in fact no limit here to official ignorance, incompetence, & intolerance. (SAMA, Ogilby to Jordan, 10.10.98)

This criticism of the *Thetis* report was largely justified: Waite had worked too quickly, his descriptions were inadequate, and he had used an outmoded system of classification derived from Günther's *Study of Fishes*. Ogilby commented to Jordan

> ... Waite refuses to use any but Günther's names, & his official position causes him to be followed by ignorant persons and so perpetuates error.
>
> This gentleman's work on fishes since he came to the Colony has had a distinctly retrograde tendency, and has already annulled much of the educational work which it has been my aim to accomplish during the last 13 years. (SAMA, Ogilby to Jordan, 24.11.98)

Jordan responded to Ogilby's letter by publishing an unfavourable review of the report and Waite wrote:

> June 16. Fri. The American Naturalist, May ... contains for me, a rather nasty criticism of the Thetis Report, mainly dealing with antiquated nomenclature. My obvious course was to quote my own remarks, (Report p 27) and I accordingly sent a rejoinder to the Editor. The note was by D. S. J.(ordan) and was evidently written as a good turn for Ogilby.

Waite's own remarks, to which he referred, drew attention to the fact that this was intended only as a preliminary report. (WD, 19.12.99)

Waite then turned his whole efforts to redeeming himself and this can be seen as the time when he came of age as an ichthyologist. He accepted, and perhaps sought, guidance from Ogilby and referred to him frequently in the final report. Waite indicates his debt to Ogilby in a short anecdote, under the account of *Callanthias platei* '... Mr. J.D. Ogilby placed on my table a reprint of Dr. Steindachner's paper on the Plate Collection, open at page 284. We jointly decided that the species there described was identical with the "Thetis" examples'. He goes on to say that Ogilby later made further comparisons and published a description of the Australian fish as a variety of that from Chile. Waite contributed the general introduction with a frontispiece sketch of the *Thetis* as well as the ichthyology, which included his own drawings of 36 fish. Waite's part of this definitive report was completed in November 1899 and published by the Australian Museum the following month.

The result was a useful and thoroughly professional account of which Ogilby approved. The latter wrote to Jordan again: 'The full report of the *Thetis* ichthyology will be a great improvement & will be copiously & well illustrated; Waite has taken my advice & given up his Güntherian proclivities, & I have therefore assisted him whenever practicable ...' (SAMA, Ogilby to Jordan) Waite was delighted with Jordan's generous review of his work '... the paper is one of the most valuable contributions yet made to the natural history of Australia. The excellent plates are to be specially recommended.' (Jordan 1900). Thereafter, Waite corresponded regularly with Jordan about many aspects of ichthyology. Waite became an effective ichthyologist in his own right and it is clear that Ogilby played a significant part in developing this aspect of Waite's career.

Introduction of Allan McCulloch

In June of 1898 Waite makes brief mention in his diary of an event that was to have an increasing impact on Australian ichthyology over the next 25 years. He had been introduced to a young man who wanted a job in natural history; Waite liked the boy and asked the curator to place him, with the result that 'Etheridge assented to my proposal to take on young McCulloch for a term.' (WD) This was Allan McCulloch, then 14 years old, who eventually made great advances in Australian ichthyology without burdening it with ill-considered publications. McCulloch met Ogilby at about this time and their subsequent friendship and collaboration, which developed fully towards the end of the next decade, was significant to the progress of fish taxonomy in Australia.

Publications and activities from 1896 to 1899 (inclusive)

During this four year period Ogilby became much more active in soliciting specimens. He described fish not only from New South Wales, but from Lord Howe Island, Queensland, South Australia, Western Australia and Tasmania. He also undertook several field trips of his own, both locally and to Brisbane. Early in 1899 he asked for support for a collecting trip to Lord Howe Island but Sinclair, the Museum Secretary, told him to proceed with the fish catalogue rather than visit 'distant localities' (AMA, Sinclair to Ogilby, 23.2.1899).

In Ogilby's published papers of the late 1890s his approach was more organised than in earlier work and there was generally a component of revision. He established a number of new genera, succinctly but adequately defined, with particular species chosen as genotypes. He also proposed some new families at this time, supplied synonymies where appropriate for species descriptions, and constructed keys to aid in distinguishing species within a group. There is left a general impression that he was working towards the goal of eventually integrating this work with his account of Australian fishes.

Ogilby and Waite were working together part of the time in 1899 and Ogilby was considering a second edition of his catalogue of New South Wales fish. He had submitted a substantial part of this, on the sharks and rays, to the Australian Museum Publication Committee. Then the secretary of the Fisheries Commission approached Waite, requesting him to prepare a list of New South Wales fishes. 'As the question is rather an awkward one for both Ogilby & myself, he suggest[ed] joint authorship & I wrote to Brodie to that effect.' (WD, 13.9.99)

Ogilby had been concerned in identifying and describing fishes from Lord Howe Island since shortly after arriving in Australia. He had been sent specimens occasionally since that time and had published *Additions to the Fauna of Lord Howe Island* (1898b) after James Brodie, the visiting magistrate, had brought him some specimens to examine. From these he described five new species and proposed new genera for three of them (three of these species and two of the genera are still regarded as valid). He also wrote to Jordan about these fish, asking for his views of the new genera, particularly of the oddest of the specimens. It was an 8 centimetre fish, 'shining purplish-black the sides of the head and base of the pectoral with a silvery lustre ...' (Ogilby 1898b). It had been picked up dead from the beach and Ogilby named it *Howella brodiei*, after the Island and the Magistrate. He had difficulty in determining its relationships and felt that it deserved its own family, the Howellidae; at present it resides in the family Acropomatidae.

Ogilby's last substantial contribution to ichthyology from Sydney was published by the Linnean Society of New South Wales in 1899. He drew on various sources for the specimens: his examination of type material in the Queensland Museum the previous year; the South Australian catfish collection that Stirling, Director of the South Australian Museum, had loaned; some Port Jackson rarities from Mr. Brodie, Secretary to the New South Wales Fisheries Commissioners; the never-failing Sydney fish market; and a useful collection from Western Australia. The latter was gleaned from the coastal streams south of Perth and from Pelsart Island in the Houtmans Abrolhos by Arthur M. Lea, an entomologist with an interest in fishes who made a number of collections over the years, supplying Ogilby and later Waite.

This publication is worthy of special consideration as it exhibited some of Ogilby's best work. The paper was introduced by a brief abstract and then divided into sections headed by the name of the family of fishes under discussion. Two families, several genera and species of special interest were re-described; a new subfamily proposed and discussed; three new genera established and their type species fixed; five new species were described. The descriptions were detailed, accurate and well organised and, although several of the 'new' taxa have since been placed in synonymy, the descriptions are so clear that later work has been facilitated. The fish named in this paper include the Australian Bandfish, *Cepola australis* from Port Jackson and the Sea Trumpeter, *Therapon* [now *Pelsartia*] *humeralis* collected by Arthur Lea. This paper marked a climax in Ogilby's achievement in ichthyology and there followed a prolonged period of hardship on a scale he had not before experienced. He had later successes, but the quality of this work was not exceeded.

Considering a move: Adelaide, Brisbane

The future for Ogilby looked bleak. Waite was replacing him as the local fish expert and his deadline for the fish book was too close. Over the last few years he had been in contact with naturalists in South Australia and, in early 1898, he considered visiting Adelaide. He was presumably exploring the possibility of moving from Sydney at this time. The Director of the South Australian Museum was Edward Stirling (1848–1919), a well-known physician who had a great interest in natural history and anthropology. Ogilby had asked Stirling if he might come to Adelaide to 'work out' the catfishes and eels in the South Australian Museum. Stirling replied that he had discussed this with A.H.C. Zietz, then the Assistant Director of the Museum, who said that he was already working on the collection and planned to continue. Zietz had a very real interest in the South Australian fishes but had many other demands on his time, working on the fishes in a desultory fashion for many years without substantial results. Ogilby was impatient with Zietz and wrote again to Stirling:

> I have no wish to rob Mr. Zietz of such credit as may be gained by the mere description of new species, but you will forgive me for pointing out that though Mr. Zietz has been connected with your Museum for many years he has as yet done nothing to justify us in the assumption that he intends to take up fishes as a study. Many years ago Mr. Zietz wrote me a somewhat querulous letter complaining that I was describing snakes, though he was making a specialty of that branch of Australian science; since that time, though with plenty of opportunity I have refrained from meddling with them, and the field has been left open to him, but he has done - nothing. Under these circumstances it would surely be unreasonable and subversive of the best interests of science to expect all Australia - all the ichthyologists of the world - to wait longer for his descriptions. (Ogilby to Stirling, 19.4.98, SRSA.)

A rather typical sample of Ogilby's hyperbole but with some justice in it. Ogilby wished at this time to study only the catfish and eels of the South Australian fish collection so he suggested that Zietz should examine them forthwith and describe any new species found before Ogilby began work on them.

In the process of casting about for a new base of operations, but ostensibly just gathering material for his book, Ogilby visited de Vis at the Queensland Museum in April 1898. He wrote to Stirling of his visit, saying that the trip had been a success, and that he had brought back 'large collections among which are many new and rare fishes ...' (SRSA) He presumably wanted to get Stirling's reaction, and perhaps hoped for some kind of offer from Stirling.

Shortly after this, Ogilby asked to have the eels and catfish from Stirling's collection sent to him. He had given up his plan to visit Adelaide but was revising these groups for his book on Australian fishes. At first Stirling offered only duplicate specimens but, when Ogilby explained the scope of his work, sent all of the specimens to him in Sydney. Early the following year Stirling wrote to Ogilby:

> 5 January 1899.
> Dear Sir,
> I am obliged for the 'Fish paper' you were good enough to send me and renew my thanks for your courtesy in dedicating a species to me. I trust you have long since recovered from the influenza which was affecting you when last you wrote. When may we expect the Fish back? (Stirling to Ogilby, SRSA.)

Ogilby (1898a) had described as new an Adelaide River catfish he found in the South Australian collection, naming it *Arius stirlingi* [synonym of *Nemapteryx armiger* (De Vis, 1884)], but Stirling was aware that Ogilby had more problems than influenza; and he wanted his specimens back. Ogilby returned them a few weeks later, together with a donation of extra specimens from New South Wales localities, with a few from Victoria and New Guinea, for the South Australian Museum. These were listed by Zietz in his museum register in October 1899, and thereafter.

Crisis in affairs in Sydney

In September 1900 Waite asserted himself in the matter of specimens. As the Museum ichthyologist he considered, not unreasonably, that he should have any fish collections that came to the Museum. He recorded 'Saw Brodie [of the New South Wales Fisheries Commission] and fishes brought from Lord Howe Is. for the Dept. As these are, as before, to be sent to Ogilby I asked Etheridge to intervene.' (WD, 27.9.1900) Waite may have felt strongly about these specimens since he had spent some time on the island himself in an interlude during the last of his *Thetis* cruises. He had made his own personal contacts among the inhabitants and specimens presented to the Museum might reasonably be regarded as within his province.

Ogilby was capable of the standard of work required for a 'Fishes of Australia' but still deluded himself, and others, never really engaging with the magnitude of the task (over a century later the work has still not been accomplished). His overall plan was noticeably incomplete. He spoke grandly of the first volume, and carried out much of the work for it, but the subsequent volumes were never much more than a blur on his mental horizon. He undertook a number of more or less complete revisions of genera and sometimes families of fishes, but no integrated plan was to be seen in his manuscripts or letters. This situation was finally discovered by Etheridge, perhaps through Waite from whom the Director frequently took advice; Etheridge reported Ogilby's progress as 'Disjointed descriptions of genera and species'. (AMA, February 1899.) Ogilby was resentful and wrote to Etheridge, who noted receiving '... a highly improper, if not impertinent letter.' At this stage Etheridge finally withdrew his support. Ogilby then wrote to the trustees, assembled a large quantity of his manuscript (some 1700 pages) and submitted it for the inspection of the Museum Publication Committee. The Committee's report was relatively favourable 'As a whole the MS. is incomplete and not yet ready for publication, although it shows great pains, knowledge of his subject, and excellent work so far as it goes.' (AMA) They recommended that the scope of the work be reduced to fishes of New South Wales and specified that the first part should be ready for the press by the end of June 1900.

Waite wrote:

> March 21. Ogilby showed me a letter received from the Minister asking him to alter the scope of his work on the 'Fishes of Australia' to the 'Fishes of NS Wales', at which he is much incensed & says he will refuse to comply. The action of the Minister is the result of advice sought by him from the Trustees of the Museum the letter further states that part 1. is to be ready by June 30 & that a continuance of the work will then be considered. Ogilby has already seen Creed and there may be some trouble. (WD, 21.3.1900)

John Creed, physician and enlightened legislator, was a politician of Ogilby's acquaintance, who had several times taken Ogilby's part, and after whom he had recently named an odd little fish, *Creedia clathrisquamus* (Ogilby 1898a); it has proven to be synonymous with *Creedia haswelli* (Ramsay, 1882).

Leaving Sydney

In the latter part of 1900 it appeared that Ogilby might obtain employment as a zoologist at the Queensland Museum. De Vis, Director of the Museum in Brisbane, approached Ogilby, who responded with uncharacteristic brevity and diffidence:

> Many thanks for thinking of me in connection with this appointment, which of course I shall accept if you consider me good enough, not only that I should personally like being under you, but that I know the unknown wealth of your collections & I shall be near my dear sister & her children. I shall await with impatience an official communication. (Ogilby to de Vis, 16.7.1900, QMA)

Ogilby wrote several fretful letters to de Vis over the next few months, betraying a growing disquiet:

> ... I am looking forward with daily increasing anxiety to the action which your parliament will take as regards the estimates, at least so far as such action will affect your 'assistant' ... (Ogilby to de Vis, 20.8.1900, QMA)

When de Vis wrote suggesting that Ogilby should apply formally he had rendered himself unable with drink, but finally sent his application with a letter of excuse and apology:

> I have been very unwell with some sort of low fever for more than two months and have been quite incapacitated from attending to business, hence my apparent neglect to answer yours of August 27. Not having heard from Queensland since then I am in the dark as to whether my application is too late or not, but it can do no harm at any rate to send it in. (Ogilby to de Vis, 12.11.1900, QMA)

In November 1900 Ogilby told Waite that he was definitely going to Brisbane; De Vis had reassured him although he did not receive official notification of the appointment for some time. The Queensland Museum trustees accepted his application in February 1901:

> Sir,
> I have the honour to inform you that His Excellency the Governor etc. has been pleased to approve of your appointment as Assistant in the Natural History Dept. of the Museum, such appointment to take effect from the date of entry on duty, and to advise that your services are desired at the earliest possible date.
> Obediently yours,
> CH de Vis,
> Director.
> (de Vis to Ogilby, 6.2.01, QMA)

De Vis had been without a professional assistant throughout his directorship and this appointment should have marked a sharing of the burden.

For Ogilby, this must have appeared to be the chance for a new start but he was apprehensive and drinking heavily. The handwriting of his letters at this time is generally shaky, their structure and style much more simple than usual, and the content querulous. He wrote (16.1.01, QMA) saying that he was assembling a large collection of fish to bring to Queensland, but that the New South Wales Government owed him money which he needed to pay freight on the goods for Brisbane; he hoped to leave in the first week in February. In February he said that he would arrive in early March (11.2.01, QMA); then:

> ... I was unfortunately taken rather seriously ill before the date of sailing & have been confined to my room since then & am still so weak that I am obliged to get a friend to write this for me... (22.3.01, QMA)

In April he pleaded 'great trouble with the Government' but said he would arrive the following week. (10.4.01, QMA). The friend who wrote his March letter was his landlady, as seen by comparing the handwriting with that in a letter she later wrote to de Vis asking him to 'use your influence with Mr. Ogilby to get him to pay me the sum of £8.0.0 he owes for Board & Residence.' (21.6.01, QMA)

The Australian Museum register reflects Ogilby's preparations for departure to Brisbane. He had decided to take with him the specimens in his private collection most likely to be useful for his continuing work on the fishes of Australia. The remainder he donated, or perhaps sold, to the Australian Museum. These included Australian fishes from various localities, including some type specimens; a few from New Guinea, Norfolk Island and New Caledonia; also small collections from Ireland, England, Scotland, Mediterranean, Persian Gulf, California, Gulf of California, Panama, Gulf of Mexico, West Indies and the Atlantic coast of the United States.

As his worries grew, Ogilby drank excessively. He had ample cause for anxiety: he had no clear plan for accomplishing his ambition for a major work in ichthyology; Brisbane, the Queensland Museum and its curator were to him virtually unknown quantities; and he had no money. He certainly knew that he was in no state to contend with the problems of a new job in a new city, hence the procrastination. He approached his colleagues at the Museum for money and raised a little which he promptly disposed of in drink. When he was due to sail in early March, his landlady came to Etheridge at the Museum to say that she could not get him away to Brisbane. Waite records his dealings with Ogilby during his last few days in Sydney:

> 11. Thurs. ... Ogilby called, he has changed his boat ticket to the 'Eurimbla' sailing on Saturday, asked for £2, being without money & wanting a new hat & boots. As he was in drink I gave him but 2/6.
> 12. Fri. Ogilby came again. I agreed to see him safely away, pay his van & cab hire etc. This he said was no use and left in an injured mood.
> 13. Sat. Found that Sinclair had given Ogilby 10/- and an order for a hat. I went to 189 Albion St & found Ogilby almost helpless. Finished dressing him & packing of his boxes & got him away in a cab, the luggage going in another. I saw him safely in a comfortable bunk & gave to the steward £1 in trust to be handed to Ogilby on arrival. This was given to me by Mrs. W.H. Palmer part proceed of the sale of bookcase. I got Ogilby to sign a p/n [promissory note] for Mrs. P his landlady for £8 in lieu of I.O.U. his signature was illegible & I witnessed it. Had I not taken him in charge he would certainly have been left. He will at any rate have time to straighten up before reaching Brisbane. I was thoroughly exhausted on reaching home & had to walk to Quay. Not a 1d. for tram fare & only paid all by beating down the two Cabbies to something like their proper fares. One demanded 4/- the other 5/- I paid 2/6 & 3/- respectively. The whole lot cost me about 10/-. (WD, April 1901)

Waite was not happy. But he was certainly pleased to have disposed of Ogilby and he settled down to his own routine, obviously relieved to be without the abrasions of their relationship. He also visited Ogilby's landlady and learned that his personal property, including a number of books on fishes, had been deposited with a pawnshop. After some bargaining, Waite bought the books - a considerable collection; he sold some of the duplicates and added the rest to his own library. It was probably at this time that Waite came into possession of some of Ogilby's papers, including sundry notebooks and a diary fragment from 1895–96. Waite made no reference to these in his diary, but they are with his papers at the South Australian Museum.

Charles de Vis, Director of the Queensland Museum

The Queensland Museum, in 1901, had been under the direction of Charles de Vis (see page 147) for nearly 20 years. De Vis had amassed large collections of fishes through purchase and exchange but his knowledge of fish taxonomy was rudimentary; in particular, he did not realise that many of the fishes of northern Australia had very wide distributions through the Indo-Pacific. De Vis worked largely in isolation and, in ichthyology, placed undue reliance on the works of Macleay and Castelnau. He described many 'new' species of fish, but most of his species descriptions were poorly executed, and many of the species had been described long before from other parts of the Indo-Pacific region.

When Ogilby came to work under him, de Vis was over 70. Queensland's economy was depressed, jobs were scarce and the Museum's Director struggled along with minimal staff and a minute budget. De Vis was bitterly disappointed by Ogilby's behaviour and could not, or would not, overlook his misdemeanours; he lasted only one month as Assistant Zoologist. The Brisbane fish collection was mostly in formalin but he obtained spirits, was reported for drunkenness, and was obliged to resign. De Vis wrote:

> Sir,
> I am instructed to inform you that a report charging you with drunkenness on two occasions has been received by the Trustees from the Director.
> Finding that these charges are well founded the Trustees suggest to you the expediency of a resignation of your office. To prevent further action on the part of the Trustees it is necessary that your resignation should be in the hands of the Chairman forthwith.
> Obediently yours,
> C.H. de Vis,
> Director.
> (de Vis to Ogilby, 1.7.01, QMA)

Ogilby responded promptly with the required resignation and an apology (Ogilby to de Vis, 1.7.01, QMA).

Dismissal from the Queensland Museum and desperation, 1901

Waite made occasional reference to his old colleague in his diary:

> Learned from Etheridge and later from Mrs. Palmer, [Ogilby's Sydney landlady] both through letters from De Vis, the [sic] Ogilby was no longer in the employ of the Queensland Museum. His dismissal has come sooner than we even expected. (WD, 8.7.1901)

And towards the end of the year, 'Hall who has been 3 months in Brisbane as Ogilby's successor called on his way back to Melbourne; he "kept" O. during the time in return for his naming fishes.'

The loss of this job signalled a grim period of Ogilby's life. His brother-in-law, who worked in the Audit Office in Brisbane, wrote on his behalf to the museum trustees: he asked for the month's salary to which Ogilby claimed to be entitled, and requested that employment outside the Museum premises be found for him. The claim for salary was allowed but employment in any form was refused (de Vis to R.C. Miller, 3.7.01, QMA). The salary was not forthcoming until the end of the month so, to tide him over, Ogilby offered to sell de Vis his specimens and containers, at the same time asking for the return of his fish manuscripts. De Vis complied and offered £10 which Ogilby accepted (Ogilby to de Vis, 14.7.01, QMA).

In a letter addressed to the Chairman of the Museum Board of Trustees just four days after his resignation, Ogilby explained that exclusion from the Museum meant the cessation of his work. He emphasised its importance and his own position as the 'greatest authority on fishes (my specialty) in the southern hemisphere' and asked for temporary work either within or outside the Museum. He wrote '... when in the future, by the steadiness of my life and my attention to the business entrusted to me, I have regained your respect and my own self-respect, I may then ask you to fully reconsider your verdict and reinstate me in the position which I have at present forfeited.' (5.7.01, QMA) The Chairman was Albert Norton, a conscientious trustee, who apparently had the good of the Museum at heart (Mather 1986). He told Ogilby that he could not recommend to the Board any relaxation in their position (24.7.01, QMA). However, Ogilby maintained a tenuous connection with the Museum, initially by naming fish for his temporary successor, Mr. R. Hall from Melbourne. Early in 1902, after Hall had left, Ogilby again asked the trustees for access to the Museum library (22.1.02, QMA) and this was granted on a strictly limited basis.

This was also the beginning for Ogilby of a protracted process of rehabilitation, a penance that lasted the greater part of ten years. During his short time at the Queensland Museum, Ogilby had begun cataloguing specimens and resumed working on his grand project, the Fishes of Australia. His dismissal deprived him of a place to work and also of the materials essential to his profession. He needed a collection of specimens and access to a good library. De Vis had assembled an excellent range of ichthyology texts at the Museum and Ogilby tried all he knew to regain the privilege of working with them: he wrote to the Chairman of the Museum Board, to de Vis and to Government Departments, his letters varying in tone from pleading and repentant to almost peremptory. He eventually received some concessions but these were hard won and partial. An obvious effect of this restriction was his absence from the published literature for several years (1900 to 1902 inclusive). It was assumed that he would not resume scientific work of any significance and Waite must have reported this to D.S. Jordan, who replied, 'I was much pleased by the excellent character of some of Mr. Ogilby's papers and gave him all the help I could. I am sorry to learn that he is not likely to do much more.' (Jordan to Waite, 17.8.02, SAMA)

Ogilby gave an account of his hardships in several letters written about this time. He wanted sympathy and needed support; his chronicles may have been coloured, but there was little doubt that he was desperate. In December 1901 he wrote to Etheridge from Brisbane:

> In reference to my work on the 'Fishes of NSW' I have to report to you (I do not know to whom else to report) that, in face of very great difficulties, I am

> making such good progress that I fully expect to be in a position to forward to you before the end of the year such an amount of completed ms as will suffice to form a creditable first part, with the promise of other parts to follow speedily ... The main difficulty with me is that, having no room in which to work, I can only now and then get the use of a table, and that only in the day time, while the night – the very time in which I should prefer to work, during the hot weather – is absolutely lost to me ... I am earning now a precarious 10/- a week (for which I give my services from 5 to 11 daily; so that when I have liesure [sic] and opportunities to attend to my book I am already tired out, for the heat affects me greatly.

He requested to be paid at the former rate of 3 pounds per week by the State Government and asked for a ream of foolscap.

> I may state that I do not touch intoxicating beverages now though I have to refuse offers of drink over and over daily. You can if you wish show this letter to Dr. Creed, who could of course assist me greatly if he cares to now and I do not think he would be disappointed if he did so. (Ogilby to Etheridge, QMA)

He wrote that he appeared to have found a new genus of Batrachoididae (Frogfish) and asked for a copy of a paper in the *Proceedings of the US National Museum* for 1886 to help assess this. And again to Etheridge in July 1902:

> 28.vii.02
> My dear Sir,
> I am sending you per same mail some ms of a work on the 'Fishes of Australia and Tasmania' which I shall ask you to kindly submit to Messrs Waite, Cox, and Haswell; these gentlemen with yourself are fully competent to judge the value of the work, small though the portion forwarded is; this is due to the want of paper even more than books of reference, and the want of money to purchase it. If your impression of the work is unfavourable I fear that the vast stores of unwritten knowledge, which I only possess will be lost to science, for things are so bad now in Queensland that even the commonest necessaries of life are now unattainable by me. If favourable I think that the Federal Government should be asked to continue the work from the point where it was dropped by that of NSW. Parts 1 (lancelets, lampreys, sharks, rays, and chimaeras), and 2 (ganoids and physostomes) are nearly complete, and the other parts more or less advanced. A sum of about £200 a year would be required and when this includes all expenses it is surely little enough in comparison with the magnitude of the work. (Ogilby to Etheridge, QMA)

Ogilby received acknowledgement from Sinclair, the Secretary of the Australian Museum's Board of Trustees, but no news of his manuscript. He was aware that the Museum Board was to be approached and wrote back to Etheridge:

> If they can see their way to helping over the present crisis one who whatever his faults, has always proved himself capable in this line, I would thankfully accept the smallest salary on which it is possible to live and work.

He wrote that he would like to return to Sydney to work and would confine himself to the fishes of New South Wales,

> ... and handing in a weekly portion of completed ms for 30/- a week. I need hardly say that the experience of the last twelve months has been such a severe lesson to me that once in work again there can be no fear of me ever again offending as before. Failing the above offer, could you not send me to the Barrier Reef as a collector; this would cost you very little as you have everything necessary in stock; and only the passage money to wherever you decide to send me, money for rations, and an advance of a few pounds (to be deducted from the value of my collections) for a small personal outfit would be required.
>
> Finally I will ask you to pardon me if I speak of my position now with somewhat brutal candor [sic]; briefly:– It is quite impossible for me to get any work which I can do, every place being rushed by people well known here, my clothes are in rags, my boots are held together by pieces of string, and during the pinch of famine prices which now hold sway in Brisbane I am often reduced to a piece of stale bread for a 24 hrs. ration, have not had a decent meal for a month, have been exposed during that time to the greatest cold and the worst fogs of the season, and am now exposed to the rain which seems at last to have set in. If therefore I cannot get employment on the lines suggested I tell no more than the absolute truth when I say that I am face to face with death since I can neither get employment here nor leave here to seek it. (Ogilby to Etheridge, 22.8.02, AMA)

Ogilby continued his efforts to induce the New South Wales Government to resume subsidy of his Fishes of Australia and Waite records:

> Etheridge asked me to 'note' officially, a letter from Ogilby to accompany M.S. as part of his work 'Australian Fishes' suggesting that the MS should be submitted to 'Waite, Cox, & Haswell' & that steps be taken to obtain £200 a year from Federal Government. Etheridge declined to have anything to do with the matter. (WD, 1.8.02)

A little later Ogilby wrote directly to ex-colleagues in Sydney; Waite recorded:

> Letter from Ogilby asking me to support his application for work & saying he is in very low water. Sinclair asked me to join in sending assistance, through some friend. He, Whitelegge, Hedley and self each gave 5/-. (WD, 25.8.02)

Resumption of work despite restricted access to the Museum, 1902

During the next few years, Ogilby's practice of ichthyology was hampered severely by his difficulty in consulting the literature but late in 1902 he worked on catfish with some success; De Vis had descriptions copied out for him, rather than allow him access to the library (Ogilby to de Vis, 19.12.02, QMA). In early 1903, Ogilby tried to get access to the library again and was granted two hours per week which, as he later pointed out to the trustees, was virtually valueless. (Ogilby to de Vis, 21.4.03, QMA). James Tosh (see page 412), the Government Marine Biologist, was relying on Ogilby at this time to identify fishes for his report on the food fishes of Moreton Bay; Ogilby worked on these specimens, when he could get access, in the Museum basement (21.4.03, QMA).

Resilience was assuredly one of Ogilby's most marked characteristics. He made contacts in Brisbane and, by the end of 1902, he was doing occasional fisheries work for the Marine Department and a little work in the Museum. Ogilby wrote again to Sinclair at the Australian Museum towards the end of 1902 to say that he was making some useful friends, and asked Sinclair to return his manuscript and send him a copy of his (Ogilby's) work on the mammals and other papers on vertebrates. 'There are a number of beautiful freshwater fishes here which I will send when possible, as also a nice little collection of spiders and flower beetles which I have for Rainbow.' (QMA) Nine months after he wrote the depressing letter quoted above Ogilby was in better circumstances, feeling cheeky enough to write to Waite, as the latter recorded, 'Letter from Ogilby "it is astonishing what a lot of things I knew but never published that you have found out" (sic)'. The underscored word in parentheses is Waite's only comment on this (WD, August 1903).

In October 1903, on the recommendation of de Vis, Ogilby was being paid 30 shillings a week for his work at the Museum, but this was only assured to the end of the year. De Vis apparently suggested that Ogilby should occupy himself in producing 'a synonymic catalogue of Australian Fishes' and Ogilby, perhaps disingenuously, declared 'the more I consider the scheme the more I am in accord with you as to the advisability of preparing such a work ...' (Ogilby to de Vis, 28.7.04, QMA) This was, after all, the nucleus of the very work that he had been aspiring to for nearly a decade. He asked to be paid at the rate of 25 shillings a week.

In 1904, Ogilby was once again carrying out some of the duties of an ichthyologist in the Museum: registering specimens, researching for his book and writing to donors. His enthusiasm for interesting specimens was unabated; he had a most encouraging manner of responding to donors, offering and soliciting information. For example, his answer to a query accompanying a specimen of an unusual, 2.3 metre long eel:

> Dear Sir,
> The fish which you so kindly forwarded to the Museum on Monday is a very interesting addition to the animals of Australia. Its name is the 'Indian long-tailed eel'. As it will form the subject of a paper in the Royal Society I shall be grateful if you will answer the following questions as soon as you conveniently can.
> Yours truly

> 1. Was it caught in a net or by hook?
> 2. If by the former [sic] did it play hard?
> 3. Was it very fierce when drawn on the bank?
> 4. Was the water fresh, brackish or salt?
> 5. What were your first impressions of it?
>
> Any further particulars will oblige. (6.1.04, QMA)

The donor duly supplied details:

> The fish I sent you was caught by hook. It played hard and was very fierce when landed. It was caught in salt water. I thought it was a kind of Sea Serpent. I was fishing for Jew in the deep water bend, Pine River about 4 miles from the Bay [Moreton Bay] last Sunday morning. About daylight I had a vicious bite which I thought was a Jew, the fish struggled very hard. I managed to get him to the bank and land him. When I went to him he raised the Fins along his back and looked very fierce. I cut the line not caring to touch him and let him lie ... (10.1.04, QMA)

A few years later, Ogilby reproduced a good deal of this information in the *Proceedings of the Royal Society of Queensland*, recording the species as *Rhabdura macrura* (Ogilby 1907a); this fish was probably the Longtail Moray, *Strophidon sathete* (Hamilton-Buchanan, 1822). At this time (1904) Ogilby began to write a series of articles on fishes for a Brisbane newspaper. The articles were interesting and apparently popular: the paper continued to give him space for some time and the remuneration must have been welcome.

Return to the learned journals, 1903

After an absence of three years from the journals, Ogilby published *Studies in the Ichthyology of Queensland* (1903), an account of some genera of Scorpionfish (Scorpaenidae), using the Royal Society of Queensland's journal. He sent a reprint to Waite, who remarked wryly in his diary 'A paper from Ogilby on Queensland Ichth in which 'Mr. Waite' figures largely'. The temptation to assert himself had been too much for Ogilby: he praised some of Waite's work and picked finically at a couple of unimportant points. For instance, he sniped at the logic of Waite's choice of a scientific name. A fish had been named *Gilbertia* and Waite dubbed a larger but related fish *Gilbertella*. Ogilby wrote 'My friend Mr. Waite appears to have become inoculated with some of the prevailing topsyturvydom of Australian nomenclature, since he proposes the diminutive appellation for the larger fish.' Waite had a long memory, referring back to this remark of Ogilby's in a newspaper article that he wrote in Adelaide (WD) many years later. Ogilby's 1903 paper marked a real advance in the study of the Australian Scorpionfish, the taxonomic decisions being well-balanced and generally supported by appropriate data. Ogilby established three new genera, two of which are still accepted, and re-instated *Neosebastes*. In this paper Ogilby described the Blackspot Waspfish, *Liocranium praepositum* from Wide Bay, Queensland.

One of Ogilby's strengths was a healthy scepticism. He checked information himself, taking no other authority's opinion on trust, and where possible going directly

Blackspot Waspfish, *Liocranium praepositum* Ogilby, 1903. Phyllis Clarke drew McCulloch's specimen, trawled in 1910 by the *Endeavour* in the Great Sandy Straight, Queensland. From McCulloch 1916.

to the organism for his descriptions. In this way he uncovered and corrected many of the weaknesses in earlier work. For instance, following his re-description of *Centropogon*, the genus containing the Fortescue, he remarked, 'As will be seen by a comparison of the generic diagnosis given by Macleay (copied for the most part from Günther) and the above definition, the former is far from satisfactory, sins of commission and omission being both plentiful and palpable.' He goes on to specify these errors (Ogilby 1903).

The Amateur Fishermen's Association of Queensland (AFAQ)

In 1905 Ogilby found a position which allowed him to settle down to work steadily. He had lacked a base, a fish collection, steady supporters; he found all of these with the newly formed Amateur Fishermen's Association of Queensland (AFAQ). He attended the meeting at which the association was founded in 1904. The members met in rooms in Brisbane and, within 12 months, Ogilby was appointed honorary curator of their new museum. He occupied this position until 1912 and the members seem to have played no small part in looking after him during that time, providing food and inexpensive lodgings (anon. and undated *Amateur Fishermen's Association of Queensland, Official History, 1904–1997*). In return, he managed their museum, kept a register of specimens, wrote annual reports and enlivened their meetings. He used many of the specimens for his own research.

De Vis retired from the position of Director in 1905 and the running of the Museum was taken over by C.J. Wild, formerly the insect collector and messenger, who had been connected with the Museum since 1889. Wild became Acting Director and, during his tenure, Ogilby continued to do some irregular paid work for the Museum. He would write to the Under Secretary at the Colonial Secretary's Department, asking for payment for work performed; the Under Secretary would ask Wild to certify that work to a certain value had been done; Wild would state that this was so, and finally Ogilby would be paid, usually £5 or £10 at a time. Even up to 1909 Ogilby was still asking for permission to consult the Museum library on a regular basis and was obliged repeatedly to solicit remuneration for work done at the Museum.

His first report as their Honorary Curator was published in the AFAQ's Annual Report for 1905–1906. The tone was rather pompous, but Ogilby was pleased with progress:

> During the year two hundred and ninety-four specimens have been added to your collection. This compares very favourably with the one hundred and sixty-two specimens received during the previous sixteen months of the Association's existence. (Ogilby 1906)

He was congratulating the members and himself; indeed, the presence of an enthusiastic curator, able and willing to identify and preserve specimens, was satisfying for the members. In the report he listed the contributions for the year, mostly fishes and crustaceans, with a sprinkling of other animals. The donors, 57 of them, were all listed by name with a special mention for the top 10 contributors! His own contributions, 34 in all, were included in the register. This published list was a sure and cunning way to stimulate collecting efforts.

As one of the Association's librarians, Ogilby helped to collect a number of books and scientific papers. Some were from Sydney acquaintances, including David Stead, others Ogilby contributed himself. He also lectured the membership on fishes at their monthly meetings, and collared guests whenever he could. His most distinguished catch was David Starr Jordan and the attendance of this famous ichthyologist at the Association was noted in their Annual Report (1906–1907):

> Probably the most notable events of the year were the two visits paid to our Museum by Prof. David Starr Jordan, President of the Leland Stanford Junior University. Dr. Jordan expressed his keen appreciation of the work done by the members of the Association and their friends in forming so creditable a collection. He also expressed himself as pleased with the attention bestowed by your Curator on the classification of the specimens exhibited. (Ogilby 1907c)

Jordan later indicated his respect for Ogilby, referring to him in his autobiography as 'the accomplished curator of fishes in the Queensland Museum' (Jordan 1922). Ogilby's annual report demonstrated the success of his system of publishing donations: in this 12 month period he registered 601 fish from 98 donors.

AFAQ donors and Moreton Bay

The donors in these early days of the AFAQ included some names that became familiar to a wide audience, in particular Edward Banfield (1852–1923) and Thomas Welsby (1858–1941). Both were members of the AFAQ, and both were friendly with Ogilby. Banfield moved from Ararat, Victoria to Townsville, Queensland in 1882, and from Townsville began corresponding with de Vis at the Queensland Museum. His quirky first book, *The Confessions of a Beachcomber* (1908), published a decade after he disengaged from public life and retreated to Dunk Island in North Queensland, gave few indications of his interest in the scientific aspects of natural history; he was an amateur, but very keen, requesting pertinent literature and identifications for his specimens. He corresponded with Ogilby at the Museum, and later at the AFAQ, frequently sending specimens from his retreat on Dunk Island.

Thomas Welsby was a prominent identity and author in the Brisbane of his day, most of his published work being in the form of local history. He worked as a bank clerk and later as a public accountant and maintained an interest in the development of the Moreton Bay district. His first book, *Schnappering*, written in 1905, names the AFAQ in the dedication and paints an interesting picture of life and fishing on Moreton Bay around the turn of the century. For the reader with some leisure, Thomas Welsby can be read with enjoyment; his exposition on 'punt fishing' in the Brisbane River typically includes some fisherman's philosophy regarding the unwanted and troublesome local catfish:

> It has often been argued whether the look of simplicity on the face of a catfish when coming over the side of a punt is more to be admired than the look of contempt on the face of the angler. (Welsby 1967)

He quotes Ogilby extensively in his account of the Australian Lungfish in *Schnappering* and was an active donor of fishes caught in Moreton Bay and which Ogilby lodged in the Queensland Museum and in the AFAQ Museum.

John Colclough was also a member of the Association, a keen collector and fisherman, for whom Ogilby (1908a) named a small Blind Shark, *Brachaelurus colcloughi*. He later had work in the Aru Islands and sent remarkable collections to Ogilby, who praised them in the AFAQ Annual Report and in the scientific literature.

James Tosh (see page 412) contributed fish specimens to the Queensland Museum and to the AFAQ. He collected locally around Brisbane and, while resident at Darnley Island (at the far northern end of the Great Barrier Reef, nearly 200 kilometres north-east of the tip of Cape York), sent large collections to Ogilby (1913c). Tosh himself, as the Government Marine Biologist, published two interesting papers on fishes and acknowledged help from Ogilby 'The nomenclature of the Queensland fishes is, I find, in a somewhat chaotic condition; and I am greatly indebted to Mr. J. Douglas Ogilby, Brisbane, for his kindly assistance in determining and verifying many of the species here mentioned.' (Tosh 1903)

Moreton Bay produced more specimens of interest to Ogilby than any other part of Queensland. The proximity of the Bay to the centre of population was undoubtedly most significant in this, not least because it was the primary sphere of activity for the AFAQ collectors; the bay also offered a wide variety of marine and estuarine habitats, and supported a mixture of tropical and temperate species. Among Ogilby's contributors was James Palmer, chief of the Telegraph Station at Bulwer on Moreton Bay, 'one of my best and most enthusiastic collectors.' After 1907 Palmer's name appeared frequently in Ogilby's publications; his efforts were duly acknowledged and the occasion of his moving from Bulwer was marked by Ogilby in a quaint footnote:

> The Moreton Bay Telegraph Station has just lately been removed from Bulwer to Cowan Cowan, and I seize this opportunity to wish Mr. and Mrs. Palmer

> all happiness and success in their new home, which will, let us hope, prove as fertile a collecting ground as that which they have now left. (Ogilby 1913b)

When Ogilby wrote to Allan McCulloch to let him know that *Coris picta*, a tropical wrasse, had been collected, McCulloch remarked on the diversity of fishes to be found in 'your wonderful Moreton Bay.' (McCulloch to Ogilby, 2.8.19, QMA)

Inland fishes and Thomas Bancroft

The AFAQ was essentially marine based but Ogilby, aware that the fauna of the streams and rivers of Queensland had not been systematically examined, was anxious to obtain freshwater fishes. He made a special plea for the collection of inland fishes but the response was poor, so specimens from Dr Bancroft, a discriminating collector, were particularly welcome (Ogilby 1909a).

Thomas Lane Bancroft (1860–1933) came to Queensland from England with his family in 1864. His father, Joseph, was an enthusiastic naturalist and medical researcher who had a special interest in parasitic diseases. He helped to elucidate the mechanism of transmission of filariasis, work that was carried on effectively by his son, Thomas (Grove 1990). Thomas, who formed an early interest in natural history, attended Edinburgh University, graduating as a surgeon in 1883. He returned to Queensland, researching the pharmacological properties of rainforest plants while working at Innisfail, and made useful contributions to parasitology, some while working as quarantine officer in Brisbane. In 1910 he had a property at Eidsvold on the Burnett River where he became intrigued with the life history of native freshwater fishes.

Bancroft published a few short papers on the larger freshwater fishes of Queensland. The 'Dawson River Barramundi' (*Scleropages leichhardti* Günther, 1864) took his attention and he speculated on the likelihood of it proving to be a mouth-brooder, quoting an acquaintance who had seen 'tiny fish swimming into the mouths of large Barramundi' (Bancroft 1916). He also recorded mouth-brooding in the 'Salmon Catfish, Hexanematichthys australis' (Bancroft 1924). This fish is now identified as *Ariopsis graeffei* (Kner and Steindachner, 1867).

The most interesting of Bancroft's contributions to ichthyology related to the Australian Lungfish, *Neoceratodus forsteri*. Several earlier researchers had attempted to rear the Lungfish from ova; for instance, Thomas Illidge (1894) described the habits of the Lungfish and the naked-eye development of its embryos, but rarely had them survive longer than a few weeks or months. Richard Semon (1859–1918), a German doctor and zoologist, came to Australia in 1891 to make zoological collections, with the hope of elucidating the embryology of the monotremes and the Australian Lungfish. He spent two years in Queensland, described in his book *In the Australian Bush and on the Coast of the Coral Sea* (1899). Semon brought back preserved embryos and histological material and published a remarkably complete series, showing the early development of the Lungfish (Semon 1893).

Bancroft published a number of papers on Lungfish between 1912 and 1933, in the most notable of which he described his success in rearing them, concluding:

> The cardinal points in rearing larval Ceratodus are: firstly to keep them in a very confined space supplied with abundance of food ... and secondly, to keep the depth of water as near half an inch as possible, and to have a diminutive sand-bank on which the fish can rest out of water. (Bancroft 1928)

Preliminary professional contact with Allan McCulloch, 1906

In the 1906–1907 AFAQ Annual Report Ogilby acknowledged the services of Allan McCulloch of Sydney in identifying the crustaceans in the AFAQ collection. They had met before Ogilby left Sydney and were eventually to collaborate in a most productive manner. Waite had taken the position of Director of the relatively small Canterbury Museum in Christchurch, New Zealand and this left the ichthyological field in Sydney open. McCulloch had worked as Waite's assistant for eight years and had developed remarkable skills as an ichthyologist. While Waite was still in Sydney, McCulloch was obliged to subordinate his efforts to those of Waite and he had become more than a little resentful. Initially, he had concentrated on crustaceans but, when Waite left, lost no time in applying himself wholeheartedly to ichthyology.

Developing resources for the practice of ichthyology

By 1907 Ogilby had considerable resources to draw on for his studies of Queensland fishes. He had his 'own' museum with a small army of collectors, he had built up a small library at the AFAQ, and he was once again allowed to use, albeit somewhat irregularly, the Queensland Museum collections and library. Most important of all, he had a potential collaborator in McCulloch; this implied access to the Sydney collections. Ogilby now began to publish again in earnest, utilising the *Proceedings of the Royal Society of Queensland*. He provided exhibits for their meetings and read papers on Queensland fish that were duly published. The first of these recorded newly registered species in the AFAQ collection and Ogilby gave the members full credit for their activity. He also deposited the type specimens in the AFAQ Museum, which gave it a larger importance, lost only when the fish in question were removed to the Queensland Museum collection many years later.

Personal names in scientific nomenclature

From 1905 to 1910 Ogilby was dependent on Wild for access to the Queensland Museum and Wild tended to be restrictive. Ogilby's efforts to placate him included dedicating a species of fish, a *Pseudochromis* 'Named for Mr. Charles James Wild, Acting Curator of the Queensland Museum, to whom I am indebted for the opportunity of describing the fishes contained in this paper.' (Ogilby 1908c). The common practice of complimenting colleagues and sponsors by naming animals for them was used freely by Ogilby and all of the more active members of the AFAQ had one or more fishes named for them: *welsbyi*, *banfieldi*, *bancrofti*, *colcloughi*, *palmeri*, *jeffi*, *raymondi* and *jamesoni* are just a few examples of specific epithets derived from friends. He appeared to take this process rather seriously but there are vicissitudes inherent in celebrating a name in this way, as seen in the case of Mr. Jameson.

In 1884 William Macleay named a small fish that had been sent to him from a tributary of the Brisbane River. He was particularly interested in '[t]his curious little fish which I name after its discoverer, Mr. Jamieson of Ipswich ...' and called it *Atherinosoma jamesoni* (Macleay 1884e). Twenty-four years later Ogilby wrote 'Mr. J.T. Jameson, of Woody Point, having kindly obtained for me some of the fishes on which Macleay founded his *Atherinosoma jamesoni* I am reluctantly obliged to announce their identity with *Pseudomugil signifer*, Kner.' (Ogilby 1908b). Rudolf Kner had examined fishes brought back to Vienna by the *Novara*, which visited Sydney in 1858 (see pages 74–5); he named *Pseudomugil signifer* from this collection (Kner 1866) and Ogilby, who had access to a copy of the zoological results of the *Novara* voyage, recognised the fish from Kner's description and illustration and so submerged Macleay's species in the synonymy of *Pseudomugil signifer*. But Ogilby thought that Jameson deserved better: on receiving a new fish from Moreton Bay, it was '[n]amed for Mr. Jonathan Thompson Jameson, an enthusiastic collector, who has brought me many interesting zoological specimens,' adding in a footnote, 'I am the more pleased at this opportunity of naming so handsome a species after my friend Mr. Jameson, because it was through my instrumentality that the supposed species *Atherinosoma jamesoni* was reduced to a synonym of *Pseudomugil signifer*.' (Ogilby 1908a) Ogilby's name for Jameson's Seaperch, *Hypoplectrodes jamesoni* appears to be secure.

The lists of new acquisitions for the AFAQ museum were drawn up by Ogilby and published in the association's annual reports. Those specimens intended to be the designated types of new species were indicated by the word 'type'. The descriptions were made prior to the publication of the lists, but often these descriptions were not published for a year or more and in the interval Ogilby sometimes changed his opinion of the status of the fish. For instance, Jameson's Seaperch (*Hypoplectrodes jamesoni*) was first listed in the AFAQ Annual Report for 1905–1906, but the description of the fish was not published until 1908. *Epinephelus jonesi* of the AFAQ Report of 1906–1907 was eventually published as *Epinephelus raymondi* (Ogilby 1908b). Perhaps Mr. Audrey Raymond Jones, one of Ogilby's AFAQ friends, thought his given name more appropriate. At any event this little grouper, the Chinaman Rockcod of Queensland, now bears the scientific name of *Epinephelus rivulatus* (Valenciennes, 1830). It is widely distributed in the Indo-Pacific region and a small specimen was brought back to Paris from the Indian Ocean island of Réunion by the surgeon-naturalists Quoy and Gaimard. It has received new names both before and since Ogilby's publication (Heemstra and Randall 1993).

Manuscripts in the Queensland Museum

The Queensland Museum Archives holds many documentary relics of Ogilby's life in Brisbane and Sydney, including ichthyological manuscripts. These papers, thousands of pages, deal with a wide variety of fish groups and include formal descriptions of genera and species, keys for identification, lists of references, and family synopses. Some essay-like material appears to have been adapted for his newspaper articles. Parts of this work represent various stages in the preparation of scientific papers, some of which he eventually published; other parts are mere fragments, but most were doubtless intended to contribute to his ultimate work on Australian fishes. Some of these manuscript pages date back to Ogilby's time in Sydney when he was submitting papers to Etheridge for assessment of his progress towards the Fishes of Australia.

Among his preparations, Ogilby had produced indices to the works of de Vis, Johnston, Ramsay, Macleay, Castelnau, Bleeker, Stead, Waite and other authors; and separate indices to ichthyological items in the main Australian journals. It appeared that he had virtually completed work on the sharks and rays, and on mullet and catfish; and was well advanced with many of the tropical families, particularly Mullidae (Goatfish), Siganidae (Rabbitfish), Apogonidae (Cardinalfish), Pomacentridae (Damselfish) and Blenniidae (Blennies). Much of this work he was able to use for his *Edible Fishes of Queensland.*

Early papers for the Royal Society of Queensland, 1907–1908

The manuscripts include drafts of papers which Ogilby prepared for the Royal Society of Queensland. He read papers for them in 1907 and 1908, listing many fishes that had not previously been recorded from Queensland. The value Ogilby placed on the work he was doing, on the AFAQ's contributions, and on the future importance of fisheries is shown in his rather sententious introduction to a paper *On New Genera and Species of Fishes* (Ogilby 1908a):

> The addition of these 36 species to the fauna of the State is due to the enterprise and acumen of the members of the Amateur Fishermen's Association of Queensland ... The fact that such an addition has been made to our fauna during the short space of twelve months through the medium of a comparatively small and struggling Association, calls attention to ... the magnitude of the work which still remains to be done in elucidating the problems connected with the marine zoology of the State; and it has to be remembered that with this elucidation is intimately connected an industry, which ... though now small and neglected, cannot fail in time to be one of the most reliable as well as the most valuable assets of the State. Secondarily it proves the utility of this and kindred Associations, so long as they keep ever in view that the primal reason of their existence is to encourage the preservation of our fishes, and to foster the development of our fisheries; these two objects can only be successfully accomplished by increasingly protecting the nurseries of our food-fishes from the onslaughts of ignorance and greed.

Ogilby often attempted to synthesise his material in papers written at this time. For instance, to a publication describing a new species and two new genera of Anglerfish (Antennariidae), he appended a key to the Australian genera that he assigned to this family (Ogilby 1907b). He treated the Moray Eels similarly, giving a key to the genera and a list of the species 'at present known to inhabit the seas of Australasia.' In the same paper, he recorded *Amphipnous cuchia*, the first known Swamp Eel from Australia, now identified as *Monopterus albus* (Zuiew, 1793); the specimen, collected in Queensland, had been sent to him in a collection of eels from the South Australian Museum and he discussed the group (Ogilby 1907a). In the following year he published a *Revision of the Batrachoididae*

(Frogfish) *of Queensland*, including keys to the known genera (and proposing two new ones) and to the Australian species (Ogilby 1908d).

Publications from 1900 to 1908, and a visit by David Jordan

In the first five years of the new century Ogilby published only a single paper on fish. However, he had some work at the Queensland Museum, being directed by de Vis to study crocodiles, tortoises and turtles in the collection: he published the results in 1904 and 1905. Whenever possible Ogilby worked on fish and, after de Vis retired, continued to visit the Museum under Wild's eye. He found a great number of interesting specimens in the old collections that had been gathered by Kendall Broadbent, de Vis' chief collector, a decade or more earlier in North Queensland and elsewhere. These included a small Dragonet, *Calllionymus* [now *Repomucenus*] *limiceps* collected in Moreton Bay. He read two other papers (Ogilby 1908a and 1908b), prepared from the AFAQ collections, to the Royal Society of Queensland. In the latter papers Ogilby listed large numbers of fish newly recorded from Queensland, of which he described 18 as new species. The descriptions were detailed and accurate but the value of his taxonomic conclusions was reduced in part due to limited access to the literature. All Australian zoologists up to that time, and for many years thereafter, were hampered by varying degrees of geographical and professional isolation. Ogilby's access was also restricted by the consequences of his alcoholism. In these papers he may also have been hastening to make up for lost time. Be that as it may, barely a third of his 'new' species from these three 1908 publications are still recognised. Among the fishes described were the beautiful Harlequin Smiler, *Merogymnus* [now *Opistognathus*] *eximius* from Moreton Bay, presented to Ogilby by an AFAQ member, and the large Estuary Stingray, *Dasyatis fluviorum*.

David Starr Jordan visited Queensland in 1907, when Ogilby's revival was well under way, and later observed:

> At the Queensland Museum I met Ogilby, its curator of fishes, an enthusiastic ichthyologist engaged in studying a rich local fauna then very little known. To him I brought the flying fish I had caught on the wing near Walpole Island, and when we found it to be new, called it *Cypselurus ogilbyi*. (Jordan 1922).

Jordan was impressed by Ogilby's enthusiasm, and the flying fish (*Cypsilurus ogilbyi* Jordan and Snyder, 1908) was one of several nomenclatural tributes he paid to his colleague. Indeed, Ogilby's energy at this time was remarkable and well shown by the variety of his publications between 1907 and 1909: 5 papers for the Royal Society of Queensland, 3 for the Annals of the Queensland Museum, 1 for the Royal Society of New South Wales, Curatorial Reports for the AFAQ, a report for the Marine Department, and an article for the *Queensland Naturalist*.

Recovery as a scientist, 1908

Ogilby used McCulloch's expertise with Crustacea shortly after establishing himself as the curator/librarian of the AFAQ, sending specimens for identification. He worked at the Queensland Museum part of the time and, although he still had to apply for remuneration, his scientific rehabilitation was virtually complete; and, by most of his scientific colleagues, quite unexpected. At this stage, Ogilby was still carrying out almost all of his scientific work unaided, but this was to alter within a few years.

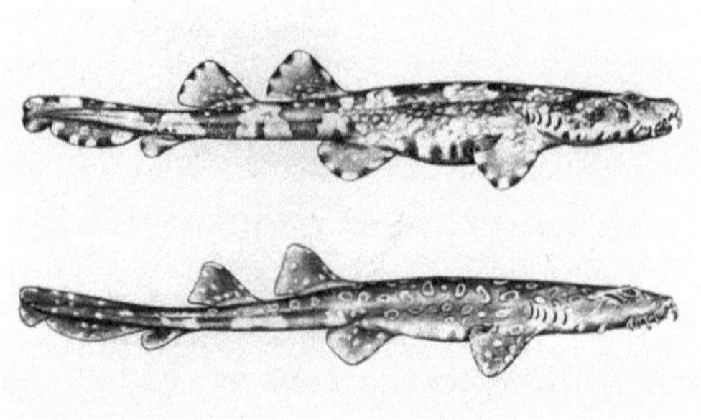

Drawings of Carpet Sharks (Wobbegongs) from the first published collaboration of Ogilby and McCulloch. McCulloch drew the sharks: Banded Wobbegong, *Orectolobus ornatus* (De Vis, 1883) (above) and Spotted Wobbegong, *Orectolobus maculatus* (Bonnaterre, 1788) (below). From Ogilby and McCulloch 1908.

The association between Ogilby and McCulloch appears to have begun with their joint revision of the Wobbegongs (Ogilby and McCulloch 1908) in which Ogilby carried out the greater share of the work. The specimens available were divided equally between the Australian Museum and the Queensland collections (Queensland Museum and AFAQ Museum). With evident satisfaction, Ogilby wrote 'The senior author of this paper has enjoyed control of a large series [of Wobbegongs] gathered by the Amateur Fishermen's Association of Queensland.' Ogilby had worked on those in the Australian Museum on a number of occasions, most recently in 1896 and he had the AFAQ specimens before him. He also had a considerable background in taxonomic work on sharks, some published (for example, *Catalogue of the Fishes in the collection of the Australian Museum. part I. Recent Palaeichthyan Fishes*, 1888). The hallmarks of McCulloch's contributions are seen in the consistent organisation of the synonymies and keys and in the accurate illustrations.

James Stevens, the Inspector of Fisheries

The Queensland Inspector of Fisheries, Mr. James Stevens, had been using Ogilby, since shortly after the latter's arrival in Brisbane, to identify fishes and for general advice on fisheries. In 1903 Stevens sent specimens of mullet to Ogilby, who duly prepared a detailed report on them, written from the 'Royal Standard Hotel, Wickham Terrace' from which address Ogilby was obliged to work at the time (Ogilby to Stevens, 6.6.03, QMA). In January 1908 Stevens retained Ogilby's services in determining the cause of damage to oysters in the Wide Bay district, some 200 kilometres north of Brisbane. Ogilby found that the fish responsible was a large Dart, now known as the Giant Oystercracker. He described the fish in his report for the Fisheries Department (Ogilby 1909b), naming it *Trachinotus anak*. The relationship with Stevens was satisfactory to both parties. They had fishing excursions together and Ogilby received some paid work, contributing his expertise on fisheries matters. Stevens presented collections of fish specimens to the AFAQ Museum and Ogilby, in turn, made exhibits available for the 'Marine Department'. One such was used at Rockhampton, which the Inspector acknowledged in fine form:

> Marine Department,
> Brisbane, 28th June, 1910.

> Sir,
> I have the honour to write thanking you for your kindness and courtesy in connection with loan of specimens of your valuable collection of Queensland fishes in glass jars, which were received and placed on view in a conspicuous position at the Rockhampton Agricultural Society's Annual Show, held on 10th to 13th June. The exhibit aroused considerable interest among the visiting public, to say nothing of its educational value. Again thanking your Association through you,
> I have the honour to be, Sir,
> Your obedient servant,
> JAS. H. STEVENS,
> Inspector of Fisheries.

This acknowledgement was reproduced at the end of Ogilby's 'Honorary Curator's Report' for 1909–1910 (Ogilby 1910b). The following year Ogilby prepared, at Stevens' instigation, another report for the Marine Department of Queensland: this was a basic systematic *List of the Edible Fishes of Moreton Bay and its Affluents* comprising 183 species (Ogilby 1911b).

Changes at the Queensland Museum: Ogilby comes in from the cold, 1910

There was great advantage for Ogilby when Ronald Hamlyn-Harris, a highly trained zoologist, became Director of the Queensland Museum in 1910. He and Ogilby became friends and Hamlyn-Harris appears to have felt some responsibility for the wayward ichthyologist's welfare, seeing to his clothes and accommodation at times. The greatest advantage to Ogilby was access to the Museum facilities: Ogilby had regular employment at the Museum from 1912 and this privilege was maintained when Heber Longman took over the directorship in 1917. Each of these directors recognised Ogilby's value as a zoologist and seem to have liked him as a person as well. They were also keen fishermen and collectors and Ogilby registered their donations to the Museum collection.

Tom Marshall, cadet at the Museum, 1910

Tom Marshall (1896–1976) was appointed to the Queensland Museum as a cadet in 1910. He prepared and painted casts, and learned a great deal of ichthyology from Ogilby. In 1916 Ogilby published an essay, *The Commercial Fishes and Fisheries of Queensland*, and Marshall, in his 'editor's note' to a revised and expanded edition (1954), acknowledged his debt to Ogilby:

> Although long since out of print, the original Essay on 'The Commercial Fish and Fisheries of Queensland,' by the late James Douglas Ogilby, has not been forgotten, as evidenced by the many enquiries still being made for its possession. It is this fact which has prompted me to revise, illustrate, and reprint the Essay and I do so with warm appreciation of its author, to whom I had the honour of being his humble ichthyological student whilst on the staff of the Queensland Museum, Brisbane.

The Queensland Museum staff in 1912. Standing are attendants, stenographer, librarian, entomologist and taxidermist. Seated from left to right are Heber Longman, Ronald Hamlyn-Harris and Douglas Ogilby. The young Tom Marshall is at their feet. Courtesy Queensland Museum.

Marshall did not develop a profound interest in taxonomy (see page 413) but he contributed to Australian ichthyology in Museum curation, as ichthyologist to the Fisheries Branch of the Department of Harbours and Marine in Brisbane, and in authorship of popular works on Queensland fishes.

The *Endeavour* exploring Queensland waters, 1910

Douglas Ogilby was an enthusiastic field worker, undertook small collecting trips whenever possible, and attempted to participate in major expeditions. Some of these ambitions were frustrated by his own unguarded behaviour; an example is the *Thetis* trawling affair (see page 192). Another occurred in 1899 when Ogilby heard that Professor Alexander Agassiz was to lead an expedition in Queensland waters and he wrote to Jordan 'Is there any possibility of my being able to accompany him in his researches off our shores; my knowledge of the Australian fish fauna should be of some value to him, while the knowledge gained in such a trip would be of inestimable advantage to me.' (Ogilby to Jordan, 23.7.99, SAMA) He went on to raise the difficulties of his financial position, a topic which he could surely have left to the discretion of the Americans. He was not included in the Agassiz party. Nevertheless, the Queensland Museum Archives possess his journal for an expedition that was probably nearer to his ambition than any other.

From Sydney, a serious investigation of the species composition and distribution of potential Australian trawl fisheries was being undertaken. The vessel chosen was the

Endeavour and the fish catch came to Allan McCulloch at the Australian Museum for evaluation and description. He made his findings known in a series of publications by the Department of Trade and Customs: the first of these was entitled *Zoological Results of the fishing experiments carried out by the F.I.S. "Endeavour" 1909-10 under H.C. Dannevig, Commonwealth Director of Fisheries*. The bulk of McCulloch's work on these cruises appeared in this series between 1911 and 1916, with a final part in 1926 after the author's death. The *Endeavour* worked mostly off New South Wales, Victoria, Tasmania and South Australia, but there were cruises into Queensland waters in 1910. From this vessel experimental trawling investigations were carried out as far north as Bowen, and Ogilby was aboard.

Ogilby's *Endeavour* journal (QMA) provides useful insight into the way in which this investigation was performed, and his onboard observations are full of interest. Reading the journal, there is a strong feeling that he was in his element, combining field observations with on-the-spot scientific enquiry. He fished with lines when possible, sorted the trawl catch, made detailed descriptions of the fishes and chortled over the many species that were unfamiliar to him.

At Hervey Bay on Sunday, 10 July 1910, Ogilby, Dannevig and some of their companions landed from the *Endeavour* to collect specimens from a coral reef and found some sizeable fish in a large enclosed pool. They failed to catch them with a landing net, and 'we who had hitherto stood shivering on the brink were "shamed through all our nature" by our intrepid Inspector [Dannevig] who without an instant's hesitation plunged in media balneas[14] and hunted the wily prey from the sheltering caves in the coral rock ...' A week later, he recorded 'a good 3 hours writing in the evening, though my hands were so sore from the two days hauling up the lines that I could hardly hold the pencil.' Dannevig was disappointed in the commercial value of the trawl catch and Ogilby remarked:

> The catch was rich from a scientific point of view but lamentably poor from an economic, only 16 of 776 fishes being in all honesty marketable. Of the notabilities caught an Aploactis (possibly new), a Paracentropogon ... were the most worthy; the latter signified his dislike of being lionized as a sp. nov. [new species] by jabbing me to the bone in the ball of my right thumb, a piece of attention which I did not appreciate ...

The next day he was back fishing: they steamed northward to Perforated Rock, began fishing 'and the fun was soon fast and furious.'

Ogilby's *Endeavour* results and their reception

On board, Ogilby had worked hastily and his efforts included drawing up a detailed synopsis of the hauls of fishes and a large number of manuscript descriptions. He rushed his results into print as the fishes were taken back to the Australian Museum for McCulloch's attention. The last of the *Endeavour* trawl hauls was made in early September 1910, he read his paper to the Royal Society of Queensland towards the end of November, and it was due to be published before the end of 1910 in their Proceedings. He distributed his

14 *in media balneas* = into the bath

author's reprints as soon as they came to hand but the authorities acted to stop publication. A special Council Meeting of the Royal Society of Queensland was held in March 1911 'to consider a request from the Commonwealth Customs, Melbourne, that we should refrain from publishing accounts of the *Endeavour* trawling expedition' (QMA). Ogilby was invited to explain his position but the Council decided to withhold the paper. At their next monthly meeting (April 1911) they considered an 'Urgent Telegram from Comptroller of Customs, Melbourne, protesting against our publication of "Endeavour" matter' and a letter from Charles Hedley (for the Curator of the Australian Museum) also requesting that the paper be suppressed. Essentially, the Commonwealth Department of Trade and Customs had financed the *Endeavour* expedition and they expected the specimens to be forwarded to the Australian Museum where Allan McCulloch, their ichthyologist, was to work on them. Ogilby had been a guest on the Queensland cruise. The situation was clear enough and the Queensland Royal Society notified Ogilby that the publication had been officially suppressed.

McCulloch treated Ogilby very gently in his own *Endeavour* report for 1914:

> An item requiring special reference here is the status of a paper by Mr. J. Douglas Ogilby, entitled 'Endeavour Series, No. 1 – On some new Fishes from the Queensland Coast.' This was read before the Royal Society of Queensland in November 1910, and was to have been published in Vol. XXIII. of the 'Proceedings' of that Society. It was set up in type, the author's reprints were issued and distributed[15] before it was withdrawn from publication for reasons stated on page 85 of the same volume. It is fully quoted in the 'Zoological Record' for 1911 as though it had appeared in the 'Proceedings,' but, of course all the references there given are unfortunately incorrect. The only copies of the paper extant therefore are a few of the author's reprints which were distributed by him before its withdrawal, so it becomes a matter of opinion as to whether the paper had been published or not.
>
> I have accordingly treated it from both points of view, quoting it and using the new names as though they had appeared according to the rules of nomenclature, while redescribing and figuring the actual specimens, so far as is possible, on which the names are based.
>
> Mr Ogilby was on board the 'Endeavour' during the greater part of the time she was in Queensland waters, and his paper deals with some of the more striking fishes that attracted his attention. Some of his descriptions were prepared on board the ship, and the type specimens were forwarded to Sydney without labels or other identification marks, and without being again examined by him. The preparation of a second description of such specimens as can be identified as his types has enabled me to rectify errors in the original diagnoses which were doubtless the result of their hurried preparation.

15 Edgar Waite's copies are in his old reprint collection at the South Australian Museum.

If there was much disagreement between McCulloch and Ogilby over this issue, it does not seem to be preserved in their extant correspondence. They were soon working together in a friendly spirit.

Although confined to the description of new species and lacking the reasoning which Ogilby usually advanced in support of such descriptions, his *Endeavour* paper (Ogilby 1910a) was of an acceptable standard. More than two thirds of the 29 species he described as new are still thought to be valid. This may be due partly to the meagre sampling of the trawl fauna of southern Queensland up to that time; and certainly in part to Ogilby's extensive experience. Among the many new species he described were four gurnards (*Lepidotrigla* spp.), a Roguefish, *Paracentropogon vespa* (*vespa* = wasp: this was the fish that spiked his thumb), a small Skate, *Dipturus polyommata,* and the Rosy Deepsea Boarfish, *Antigonia rubicunda.*

It is curious that Ogilby apparently made no attempt to collaborate with McCulloch in this work. Rushing into print as he did gave an appearance of high-handedness and deceit but this was not consistent with the relationship he was developing with McCulloch. Perhaps it is possible to understand his conduct. Ogilby was a keen and careful observer, familiar with the available literature and still possessed of an excellent memory. He had developed a wider experience of the Eastern Australian fish fauna than any of his predecessors or contemporaries. Thus he was more able to appreciate the significance of the fish specimens that came into his hands than anyone else at the time, and he knew it. He was hampered by poor material resources in that his access to the literature was limited, his working conditions were inadequate and, thanks to his past misdemeanours, his relationship with the Queensland Museum was still shaky. His colleagues regarded him as personally unreliable and this attitude was naturally and inevitably, but not justifiably, extended to his work. Had he appreciated the balance of these factors, Ogilby might have been a little more cautious; but caution was not among his virtues. He demonstrated the lack of insight in dealing with people that is characteristic of sufferers from alcoholism. His ambitions, given his past record, could fairly be described as grandiose.

Ample material for study 1910–1912

The number of specimens available to Ogilby around this time was prodigious. He wrote *On New or Insufficiently Described Fishes* (Ogilby 1911a), concerning 34 species, primarily AFAQ specimens collected locally; some that his friend, John Colclough, had sent from the Aru Islands and from Croker Island, Northern Territory; and a few from the older collections of the Queensland Museum. The following year Ogilby (1912a) published *On Some Queensland Fishes,* dealing with about 40 species: there is a significant contrast with the paper just mentioned (Ogilby 1911a) as the research elements in the 1912 paper are much more prominent and Allan McCulloch's influence is manifest. In particular, McCulloch supplied three elegant sketches of Hardyheads (Atherinidae) to illustrate a portion of the paper; Ogilby also quoted McCulloch's opinions and acknowledged his assistance with several matters. Two stingrays from Moreton Bay were among the new descriptions in this period: the Purple Eagle Ray, *Myliobatis hamlyni* Ogilby, 1911, named for Ronald Hamlyn-Harris, and the Australian Cownose Ray, *Rhinoptera neglecta* Ogilby, 1912.

Some difficulty in relations with the AFAQ

In his last report to the members, Ogilby (1912b) notes that '... the authorities of the Queensland Museum approached the Association with a view to obtaining the "type" specimens.' Ogilby presumably instigated this approach and certainly supported it but the members did not want to part with their specimens. In the same report Ogilby, drawing attention to the reduction in the number of specimens donated by the members, accused them of 'slacking.' At about this time the AFAQ members discovered that their honorary curator was sleeping at their museum premises and he was told that his services were no longer required. However, he maintained close contact with the association until his final illness: he still identified their fish, donated some books to their library and had enduring friendships with many of the members. In 1914 the association found that their museum was in a poor state; they asked the Director of the Queensland Museum for Ogilby's services in making identifications and labelling the specimens, and the following year made him a Life Member of the Association (AFAQ Official History, undated).

Opportunities and modified objectives in ichthyological work

Ogilby eventually seemed to come to terms with the prospect of limiting his major work primarily to the fishes of Queensland. However, he continued to maintain a broader context, referring frequently to species related to those with which he was dealing, though they might inhabit other parts of Australian waters, or those of other countries. His *Edible Fishes of Queensland* appeared in 15 parts between 1913 and 1920, throughout which period he and McCulloch sustained a productive collaboration. Ogilby's experience and familiarity with the fishes and the literature were at McCulloch's service and Ogilby could examine type specimens in the Queensland Museum. Although all parts of the *Edible Fishes of Queensland* were published under Ogilby's name alone, McCulloch contributed advice, literature searches, illustrations and specimens to most of them. The species descriptions of William Macleay provided as rich a source of confusion as did those of de Vis, and most of the relevant type specimens were to be found at the Macleay Museum, located in the University of Sydney. McCulloch frequently consulted the collections in that Museum for Ogilby. The illustrations which complemented the *Edible Fishes* were of a high quality and most were drawn by McCulloch himself, by his assistant Dene Fry, or by Phyllis Clarke (see page 378), who worked under contract and with McCulloch's guidance as a scientific illustrator at the Australian Museum.

Ogilby was under no illusion as to his own drawing skills, remarking to Regan at the British Museum:

> I enclose a sort of outline figure, but as my knowledge of drawing is on a par with my knowledge of Sanscrit, you will have to make the best of it.

He must have had second thoughts as on the back of this letter he referred Regan to a published figure of what he thought was a similar fish, listing the differences from his specimen, 'Instead of attempting a figure ...'.(Ogilby to Regan, 22.4.14, NHMA)

Characteristic style and unabated enthusiasm

Ogilby's writing style was consistent, be it in letters, diaries, popular literature or scientific publications. It was flowery, with an undeviating tendency for over-embellishment, and repeated interpolations of a personal nature. Many of his scientific publications include matter of a kind which has come to be regarded as extraneous. Enthusiasm for his subject often surfaces in his papers, as in the initial part of the *Memoirs of the Queensland Museum*, where he presents the first record of the Butterfly Mackerel (*Gasterochisma melampus* Richardson, 1845):

> During the month of August, 1911, while passing along Queen Street, my attention was attracted to the dismembered portions of a large fish exposed for sale in the window of one of our leading fishmongers. Inquiries of the proprietor elicited the fact that the fish was caught in the Bay along with five similar examples; also that they were quite unknown both to fishermen and dealers. Unfortunately all six had been cut up and sold before I became aware of their presence, and I was only able with difficulty to secure the head of this specimen for the Queensland Museum. (Ogilby 1912a)

Ogilby had speculated that a fish now known in Queensland as the Beach Salmon, *Leptobrama muelleri* looked as though it was pelagic and surface-living but in 1913 he reported, surely with a grin,

> The capture ... of a specimen well up the Aird River, British New Guinea, apparently demolishes the pelagic theory, while supporting that of the surface-swimming, since the example in question jumped into the boat of its own accord. (Ogilby 1913a)

In *Edible Fishes of Queensland*, Ogilby took advantage of an indulgent editor to present an enormous amount of information. In the first two parts, which were published at the same time (Ogilby 1913a), he wrote exhaustively on the Bullseyes (Pempheridae) and the River Blackfish (*Gadopsis marmoratus*). The format in these accounts was used throughout the *Edible Fishes*: besides historical reviews, species lists, synonymies, keys, descriptions and distributions, he speculated on the biogeography of the families and gave all the information he had gathered on the biology of the fishes involved, including lengthy quotes from various local 'authorities'.

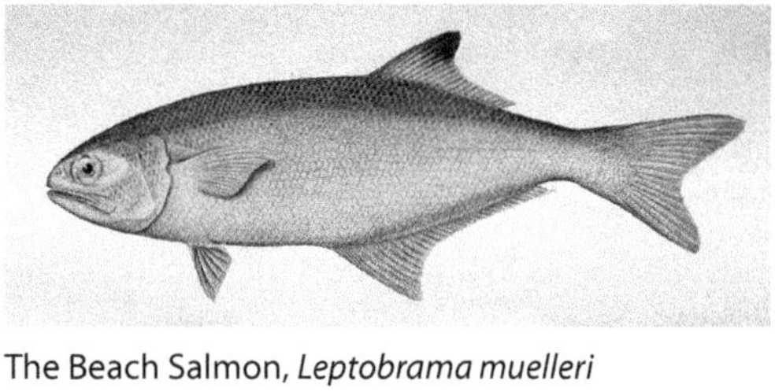

The Beach Salmon, *Leptobrama muelleri* Steindachner, 1878 of northern Australia, drawn by Allan McCulloch for the first part of Ogilby's *Edible Fishes of Queensland*. From Ogilby 1913.

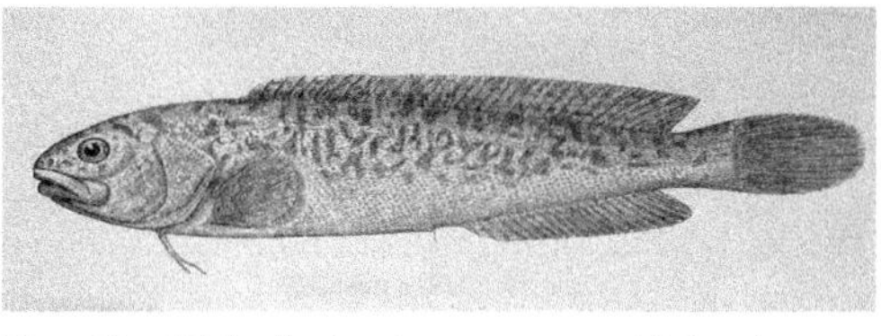

River Blackfish, *Gadopsis marmoratus* Richardson, 1848 drawn for the second part of Ogilby's *Edible Fishes of Queensland* by Dene Fry, McCulloch's promising new assistant. From Ogilby 1913.

Correspondence between Ogilby and McCulloch, 1912–1920, content and format

The correspondence between Ogilby and McCulloch over this period is represented in the Queensland Museum Archives by at least 44 letters, all but one sent by McCulloch. The single letter of Ogilby's was returned by McCulloch with extensive annotation by way of reply. The correspondence is also incomplete in that no letters are preserved prior to 1912, there are obviously many missing letters, and some are lacking pages or parts of pages. Perusal of this correspondence offers insight into the relationship between these two ichthyologists, with some hints as to how it altered with time. Analysis also yields a profusion of data regarding their work, particularly details of the subjects dealt with and the methods used. Information on collections, relations with other ichthyologists and institutions, personal problems, and a variety of other incidentals emerge as well.

The format of most of the letters is answers, followed by questions: Ogilby's letters must generally have included a list of queries so McCulloch began by more or less systematically responding to these. Once he had disposed of Ogilby's questions, McCulloch tended to ask some of his own, often interspersed with comments on his work in progress. He sometimes found Ogilby too demanding:

> Your assumption that I can do the work of three amuses me quite a lot. Even if I had no other work, I don't think I could get through all the things you want me to do. (14.9.14, QMA)

At other times he was frustrated when Ogilby ignored questions, concluding a letter with:

> P.S. There are <u>two</u> (2) questions in the above. I draw your attention to them as you are apt to overlook them. (2.8.14, QMA)

McCulloch's motivation

The motivation for Allan McCulloch's close association with Ogilby, in spite of the latter's manifest weaknesses, is many-sided and becomes clearer as this series of letters is considered. Ogilby in 1910 had studied the Australian fish fauna with extraordinary intensity for 25 years. He had become familiar with the early literature and had recognised the importance of the contributions being made by modern ichthyologists to their science and to zoological nomenclature. Part of the bond between Ogilby and McCulloch was their shared views as to the methods by which fish taxonomy should be carried out. They were the only zoologists studying fish in Australia in this way. McCulloch knew that Ogilby could appreciate his work, would sympathise with and share the problems inherent in it, and that he would help when possible. In June 1914 McCulloch was working with a group of Tropical Snapper (Lutjanidae) and thanked Ogilby for putting him on the right track in a difficult identification:

> Your note just received. Very many thanks for the identification. I have compared the large specimen with a small one from Madras identified by Day as <u>L. johni</u> and find no structural differences, though without your suggestion I should not have thought of associating the two.

The key to the identification lay in consulting an old specimen in the Australian Museum: a specimen that Ogilby had himself registered 29 years earlier! A few months later (22.11.14, QMA) McCulloch put before Ogilby a problem in nomenclature, introducing the facts with 'Now for a matter on which I want your carefully considered opinion.' He proceeded to state the history of the names attached to the Australian species of Jack Mackerel (*Trachurus*) and concluded by asking for Ogilby's advice. In a later letter (10.12.14, QMA) he acknowledged Ogilby's response:

> Many thanks for note on T. novae-zelandiae, & the trouble you have taken over it. Since you, Waite, and I are agreed, the 'Cowanyoung' will henceforth be called T. nz, Hutton.

In April 1915 McCulloch had completed a report on some of the fishes taken by the Commonwealth trawler, *Endeavour*. He wrote:

> By the same post I send you part 3 of my Endeavour Report. There is something about it which leaves me dissatisfied, but I do not exactly know what it is. Therefore all criticism, especially adverse, will be much appreciated.

Ogilby also brought to the association the advantage of his situation in Brisbane. He had access to fresh specimens of many Queensland species, receiving contributions regularly from the fishermen of Moreton Bay, fisheries officials, and collectors in North Queensland and elsewhere. He had a large number of specimens in the museum of the AFAQ and, after 1910, had ready access to the Queensland Museum collection; this included the type specimens of Charles de Vis, the accurate identification of whose many 'new species' was an essential part of establishing a sound basis for Australian ichthyology. McCulloch often asked for help. In September 1914 he wrote:

> Do all you can to sort out de Vis' types of Carangids. They are evidently easy things to deal with if one only has specimens enough, and between us we ought to clear the atmosphere somewhat.

In August 1918, McCulloch was jubilant as he and Ogilby had both been working on yet another rather difficult group, the Jewfish (Sciaenidae): 'Considering everything, we are doing quite a heap of clearing among the Jewfishidae' (McCulloch liked neologisms) and he asked Ogilby to 'clean up' two of the species described by de Vis, and to let him know their actual identity. The following year he asked Ogilby to help with his work on another family, the Girellidae: 'Sincerely hope you will be able to fix De Vis species

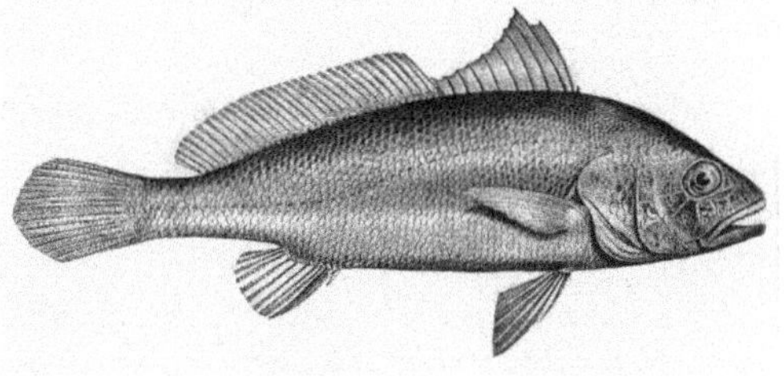

The Smallscale Jewfish, *Sciaena* [now *Nibea*] *leptolepis* Ogilby, 1918 was sent to Ogilby from Croker Island, Northern Territory by his friend John Colclough. Ogilby sent the type specimen to McCulloch in Sydney who had Miss Phyllis Clarke draw it. Ogilby was delighted by the quality of her drawings. From Ogilby 1918.

up for me.' A little later he suggested that since Ogilby was working with the Lutjanidae, 'why not hunt up nigricauda & regia and notata var sublineata of De Vis, and get these straight.' These were all obscure at the time and proved to be synonyms of known species of *Lutjanus* (Eschmeyer 1998).

Their initial correspondence was businesslike, essentially an exchange of information. However, by late 1914 McCulloch sounded a personal note by complaining about his work load and its effect on him. For instance, 'I have two lots of Endeavour material to register now and work everywhere. Cannot carry on at the high pressure I have been working under during the last few weeks…' (14.9.14, QMA). Within another 12 months, McCulloch appeared to see Ogilby as a confidant. In 1915 he wrote at some length about his hopes of enlisting in the army, concluding 'Hope you are not bored by all these details, but it's a relief to tell ones friends of ones disappointment.' Over the next few years, the tone of the letters altered subtly, and the personal element became more prominent. When Ogilby became seriously ill, McCulloch wrote no more than the truth, 'I was fairly bowled over when Longman told me of the fearsome happenings, since the fun of working at fishes without somebody to argue over them with is relatively small.' (8.12.19, QMA, and see page 236). McCulloch described his hopes and disappointments, his plans and his frustrations. He complained bitterly about colleagues and administrators, expecting and receiving sympathy from the older man. McCulloch appeared to benefit from this outlet for his feelings.

Ogilby's motivation

The advantages to Ogilby in corresponding with McCulloch are plain. He could obtain a great deal of information on the specimens under McCulloch's care in the Australian Museum collection, which included some important components derived from Day, Castelnau and Bleeker. In his hasty retreat from Sydney, Ogilby had shed his books and the Queensland Museum library naturally lacked some of the texts which were held at the Australian Museum (including some of his own early papers). McCulloch had access to these and frequently checked references for Ogilby, sometimes even copying out lengthy original descriptions for his friend.

On many occasions Ogilby asked McCulloch to consult the Macleay Museum collections on his behalf. This process seemed simple enough as McCulloch frequently visited the University, where the collections were located, but he was reluctant to arrange loans:

> Am afraid I cannot check the identification of the sharks in the Macleay Mus. for you. I have no satisfaction out of my determination[s] unless they are made with both library and specimens at hand, and the proceedure [sic] necessary to borrow from that museum is both complicated, and uncertain in its results. Again as often as not, an examination is being held in the hall and one is not allowed to enter. I cannot think how pleased poor old Macleay would be if he only knew the uses his museum is put to. (19.4.12, QMA)

Seven years later, when Ogilby requested information on the ambassids (Glassy Perchlets) in the Macleay Museum, McCulloch looked at them but found that he would have to take them to the Australian Museum for examination 'with a good binocular' (microscope). The procedure for obtaining loans from the Macleay Museum necessitated an application to the management. McCulloch made it clear that he would help but was unwilling to approach his Curator, Etheridge. The staff members at the Australian Museum were not permitted to make loan applications directly; their Curator had to apply on their behalf:

> ... you will have to request this loan from the Committee of Management for the purpose. If you do this, I will undertake to prepare outline figures and descriptions for you. Why should I not make the request myself, says you? Well, our old man is so touchy these days that I might get my face pushed if I asked him to write officially, so I must leave it in your hands. (3.11.19, QMA)

Prior to his association with McCulloch, Ogilby's papers had either outline drawings or, more often, none at all. However, he understood the importance of effective illustration and was delighted with the drawings that McCulloch prepared for him, by his own pen or through Phyllis Clarke. Miss Clarke prepared scientific illustrations by contract and McCulloch supervised her drawings for his own and Ogilby's papers:

> ... I set Miss Clarke to work upon a beautiful specimen [of *Alectis ciliaris*] in our collection about 10 ½ ins long [26 cm]. The figure is finished, and she is now doing the A. gallus (largest specimen) [now *Alectis indica*]. (1.9.14, QMA)

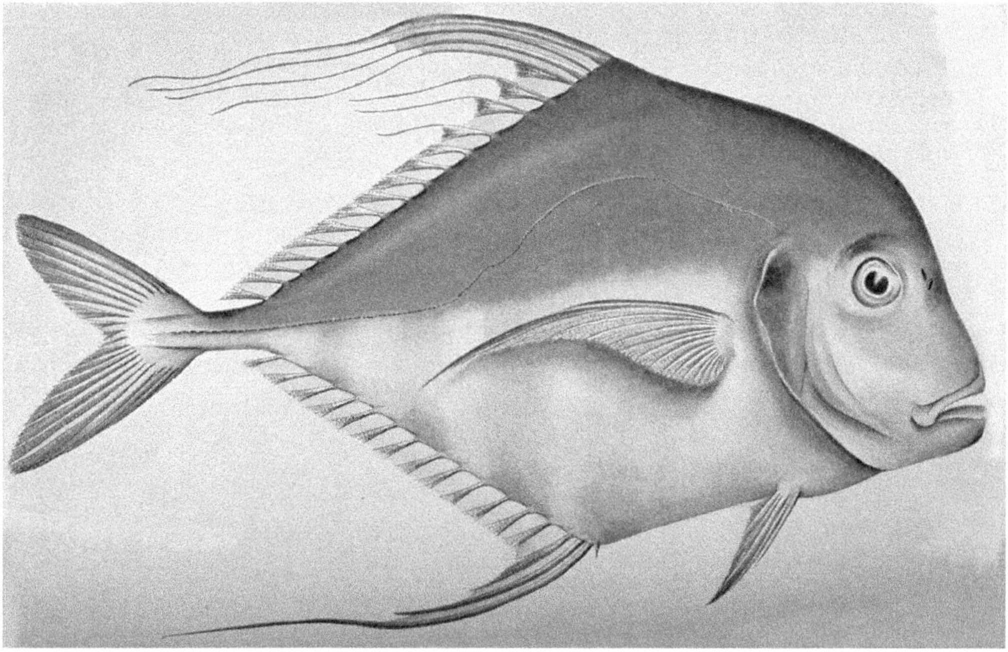

Juvenile Diamond Trevally, *Alectis indica* (Rüppell, 1830) drawn by Phyllis Clarke for Ogilby's *Edible Fishes of Queensland*. From Ogilby 1915.

Another benefit to Ogilby lay in McCulloch's access to, and familiarity with, the enormous amount of material that was coming to the Museum from the *Endeavour* cruises. Many of the groups of fishes that Ogilby chose to work on were represented by fresh material in the *Endeavour* collections, and McCulloch frequently used this. For instance, the reference to 'largest specimen' above implies that McCulloch had picked out a good example of the species for Miss Clarke to represent. McCulloch made the situation clear in a hasty note to Ogilby (9.9.14, QMA):

> Having dug out all the Endeavour Carangids I find usually 50 each of the several species you sent down to be figured. They usually include a larger & sometimes better specimen than yours, and I think it would be better to prepare the drawings from them. Therefore, have I your permission, & of course Dr. Harris' also, to draw the best specimen available? In nearly every case I can send you the actual specimen figured as part of the 'Endeavour' distribution. Please answer this question by return post, as Miss Clarke is getting on fast with the others.

Hamlyn-Harris, as the Queensland Museum curator, edited the *Memoirs of the Queensland Museum* wherein Ogilby was publishing; the *Endeavour* distribution was an arrangement whereby the state museums received specimens from the trawling.

An indication of the value set on McCulloch's expertise by his older colleague is found in a notebook kept by Ogilby which includes an index to fish references found in McCulloch's letters. Ogilby frequently asked for McCulloch's opinion; for instance, Ogilby tried to apply the new rules of nomenclature to an old problem - which of the two current names for the main genus of the Ambassidae (*Chanda* or *Ambassis*) ought to be used:

> Dear McCulloch,
> Herewith I send my defence of Chanda v. Ambassis as sketched out for my revision of the Australian Centropomidae, & I beg of you to point out remorselessly any flaws or weak points therein. I do not bind myself to take your advice, but I promise the most careful consideration of it, be it what it may. (7.6.19, QMA)

McCulloch responded by extensive annotations on Ogilby's letter, which he returned by way of answer. He had excised one of Ogilby's wordy digressions and a conjectural comment, inserted some relevant references and generally behaved like a good editor.

McCulloch's role as the dominant partner in the correspondence

McCulloch often performed a balancing act between impatience and indulgence with Ogilby, as seen in this correspondence. Although their relationship was mutually beneficial, McCulloch frequently assumed a dominant role. At times he more or less insisted on Ogilby following his 'advice', as in the case of the Waspfish, *Apistops caloundra* (4.12.14, QMA). Ogilby (1911a) established the genus *Apistops* but McCulloch could not see sufficient grounds for separating it from the related *Apistus*. He explained his reasoning and concluded 'I should therefore be very glad if you could look into it again, as it would be more satisfactory to have a note from yourself admitting its invalidity (if you do, as it seems from your letters), than to knock it out entirely off my own bat.' However, McCulloch (1915a) eventually supported the separation of these genera. McCulloch sometimes found Ogilby obtuse but remonstrated with forbearance. He introduced a correction to Ogilby's remark on the distribution of a Goby, *Asterropteryx*: 'How successfully you miss my point at times! Or is it that I express myself so badly?' McCulloch also made it clear when he wanted Ogilby to leave a topic alone, in fact to reserve it for himself:

> Of course that note on the young <u>Glaucosoma</u> is worthy of record, but maybe you had better leave it to me, since the Endeavour lot includes a whole series which I hope, some day, to write up and figure. But don't think I want to cut you out, and the Lord only knows when I shall be able to touch it. (probable date 10.10.19, QMA)

Perhaps McCulloch's attitude was due partly to Ogilby having put himself in the wrong so completely over the 1910 *Endeavour* paper.

For every comment of McCulloch's that may be seen as negative in this correspondence, there are many positive ones. He was conscious of the support he was getting and showed this repeatedly. For instance, on receiving Ogilby's latest parts of the *Edible Fishes of Queensland* (1918), he responded:

> Having just card-indexed your splendid paper, I'm feeling very bucked at the great advance it has made in our knowledge of the groups dealt with. I shall have the devil of a shuffle to bring my cards into line with all the conclusions arrived at, while there are not a few alterations and additions to go into my N.S. Wales list (which by the way still awaits publication, though completed over a year ago). Many thanks for your numerous references to myself, and its most satisfying to note how well we are agreed in the various matters dealt with. (probable date 11.8.18, QMA)

He followed this with constructive criticism, but was clearly pleased with the paper. McCulloch broke the rules for Ogilby from time to time, as in the case of a Glassy Perchlet, *Ambassis jacksoniensis* that Ogilby was studying. McCulloch abstracted a specimen from a registered batch:

> And just to show how I love you, I am sending you one of No. I.5116 [Australian Museum specimen accession number] to play with. This is but a trifle compared with the results it will bring about, but according to the rules, regulations, and ideals of this institution it can't be done, so for the love of Mike and others of his ilk, you must say no word about it. (2.8.19, QMA)

Contact with overseas ichthyologists: Charles Regan and David Jordan

Both Ogilby and McCulloch had frequent contact with overseas ichthyologists, in particular Charles Tate Regan of the British Museum and David Starr Jordan of Stanford University, California. Regan (see page 169) had been appointed assistant to George Boulenger at the British Museum in 1901 and his initial task was the curation of fishes at that institution. He soon began a series of anatomical and osteological studies that contributed to the understanding of the relationships within and between groups of fishes. He undertook some taxonomic work on Australian fishes and was viewed as something of a competitor by the Australasian ichthyologists of the day (Ogilby and McCulloch; and Waite who was in New Zealand from 1906 to 1913). However, they all respected Regan's judgment and experience and corresponded with him.

Between them, McCulloch and Ogilby puzzled out the relationships of many rare and unusual fish, but on one occasion they were both at a loss, and this proved to be a minor embarrassment. Ogilby (1915c) published *On some new or little-known Australian Fishes*, including the first description of a beautiful fish from Thursday Island. McCulloch had the specimen sent to Sydney where he examined and drew it himself. Ogilby and McCulloch exchanged letters about the likely taxonomic position of the fish but could not agree. Finally Ogilby decided that the fish probably belonged to the family of Bullseyes (Pempheridae) and that it merited a new genus. He complimented his British correspondent, C. T. Regan: 'I have much pleasure in naming this fine genus after Mr. C. Tate Regan, B.A., of the British Museum of Natural History, whose taxonomic work has been invaluable to students of this class' [i.e. perciform fishes]. So the fish became *Reganichthys magnificus* but Regan, with McCulloch's excellent drawing and Ogilby's detailed description, recognised the fish as a member of the Pearl Perch family (Glaucosomatidae). Ogilby gracefully and promptly acknowledged his error (Ogilby 1916d); but it is worthy of note that the Pempheridae and Glaucosomatidae are now regarded as closely related (Nelson 2006).

David Jordan, who was less threatening than Regan since he had no apparent ambition to work with Australian fish, was a source of advice and encouragement, and conducted a voluminous correspondence with Ogilby and his colleagues. He was of particular service in discussing the niceties of the rules of scientific nomenclature, some then in the process of being established; Jordan helped formulate some of these rules and so led Ogilby and McCulloch to an early appreciation of them. In one friendly and informative letter, Jordan answered a series of Ogilby's queries on taxonomy and said 'The Zoological Commission, to my regret, adopted the compromise rule of the first reviser, instead of the first species.' Ogilby accepted the ruling, as explained by Jordan, who also told him 'I have in press a catalogue

The Threadfin Pearl Perch, *Reganichthys magnificus* from Thursday Island. C.T. Regan of the British Museum assigned the fish to the genus *Glaucosoma*: it is now *Glaucosoma magnificum*. From Ogilby 1915.

of the generic names from Linneaus to Cuvier with types of each and a full discussion. When the world becomes sane again, I shall ask for all possible criticisms and for a final judgment.' (Jordan to Ogilby, 17.4.17, QMA)

McCulloch and Ogilby collaborating: Grunters and Gobies, 1916

The joint revision of a difficult group, the Grunters (or Therapon Perches) (Ogilby and McCulloch 1916), demonstrates again the importance of their collaboration. This can be estimated by considering the magnitude and nature of the problems presented by this group of fishes, the approach that the authors took towards resolving the difficulties, and the result of their work.

There are about 50 species of Grunter (Terapontidae) found in coastal, estuarine and freshwater habitats around the Indian Ocean and the Western Pacific. They are particularly common in the inland waters of northern Australia and New Guinea. The collections to which Ogilby and McCulloch had access contained a little less than 20 species of Grunter but their identification, using the existing literature, was impossible.

Many of the early zoologists had described and named new species, often in publications that were difficult of access, and frequently had given inadequate descriptions. Their type specimens were often inaccessible or lost, adding to the opportunities for confusion. In several publications between 1872 and 1878 Count Castelnau had given descriptions, brief and of varying accuracy, of 6 supposedly new species. William Macleay, in compiling his *Descriptive Catalogue of Australian Fishes* (1881), collected 9 species from the pages of Günther's Catalogue, 1 from another of Günther's recent papers, 5 from Castelnau, and described 3 new ones himself: thus providing a list of 18 species. Over the next 3 years, Macleay described 2 more species, Edward Ramsay described 1, and Charles de Vis described 6. De Vis' descriptions vied with those of Castelnau for brevity and ambiguity, and his type specimens were often misplaced. None of these authors made any serious attempt to derive synonymies, having only a poor understanding of the work of their predecessors. They often described the same species under different names, did not trouble to consult type specimens, and had little appreciation of the range of variation present within each species.

The Coal Grunter, *Hephaestus carbo* of northern Australian freshwater coastal drainages. In their revision McCulloch and Ogilby named this fish *Therapon carbo*. It had been taken by Dr Thomas Bancroft, who wrote 'I found this black fish at Gregory Plains, south-west of Burketown, where it is common ... It is a fine game fish, growing to a weight of four pounds and over, and is of excellent quality for the table.' From Ogilby and McCulloch 1916.

To this series of problems, Ogilby and McCulloch brought a recognisably modern approach. They had an appreciation of the literature and could see the weaknesses of the work of their predecessors. They were prepared to sort out the confusion by making new descriptions from fresh specimens, and from the type and other specimens in their respective museums. Ogilby located many of de Vis' types in the Queensland Museum;

McCulloch tackled the specimens of Macleay and Castelnau in the Australian and Macleay Museums. A good series of fresh specimens from Queensland was provided by Ogilby's contacts; once again his most useful collector in inland waters was Dr Thomas Bancroft.

The *Revision of the Australian Therapons with notes on some Papuan Species* (Ogilby and McCulloch 1916) presented detailed descriptions, dichotomous keys for identification, full synonomies, distribution data, and often extensive supplementary notes. The final paper was enhanced by 10 of McCulloch's accurate and attractive illustrations. The authors demonstrated that the majority of the species described by Castelnau, Macleay and de Vis were invalid, finally recognising 16 species; a measure of their success is that all but 3 appear under the same specific names today.

Their joint publications were not always so successful. The last was an assault on the complexities of the gobies and gudgeons (McCulloch and Ogilby 1919): the authors resolved many problems (and contributed a few) but recognised the limited nature of the success that they had attained with this difficult and prolific group (see page 342).

Ogilby's diary for 1917: field trip and ichthyology

A fragment of Ogilby's diary for 1917 has been preserved in the Queensland Museum Archives. It covers just 40 days, including a field trip to the fresh waters inland from Noosa. The entries open with Ogilby hard at work on the Gudgeons and Gobies, describing those specimens to which he had access and continuing a long term project which he saw as vital to the preparation of any major work on Australian fishes: the preparation of a series of indices to the papers of other workers. He was taking their papers journal by journal, and at this stage he had just finished what he referred to as his 'Index Ichthyologicus' of the periodical of the Linnean Society of New South Wales, and was working on Castelnau's *Researches on the Fishes of Australia* (1875).

Noosa gave some scope to Ogilby, now 64 and still a keen fisherman, revelling in the opportunity for a trip with his mates. He was with J.H. Stevens, the Director of Fisheries from Brisbane, among others. They cast, trolled and netted, albeit without any very remarkable success, then Ogilby writes:

> The next excitement was caused by an unexpected commotion under the north bank, whereupon our doughty gunner let fly both barrels into the thick of it, just for luck. On investigation we discovered 4 large jewfishes (Tandanus tandanus) [Freshwater Catfish] in a comatose state and promptly annexed them ...

He recorded his observations of birds, the trials of camping in the rain, catches (and is bitten on the thumb by) a small snake. He likened the sensation to that of being 'stung by a peculiarly virile wasp.'

Back in Brisbane Ogilby received a new collection of gudgeons from his reliable friend, Thomas Bancroft of Eidsvold, and resumed work on 'the gobioid paper'. Amid details of his progress with the gobies, he chronicled some war news and the prospects of the Federal Elections and State Referenda. His correspondence remained extensive and shows his persistent, if diffuse, enthusiasm:

> Received letters from (1) Rev. Ed. J. Nash, who writes most kindly re Synaptura selheimi; [the uncommon Freshwater Sole (see page 137)] having, however, left the Peninsula [Cape York] he is unable to make the necessary inquiries, but has forwarded my letter to his successor, the Rev. W.M. Wilkinson & also suggests that I write to the Telegraph Operators at Musgrave & elsewhere; (2) Robt S. Archer, who promises me a series of Eleotris crescens, [the Sleepy Cod, *Oxyeleotris lineolata* (Steindachner, 1867)] known there as 'black-fish', as soon as the warm weather comes, & gives me the following interesting information concerning it – 'This fish is never seen in the winter; in summer it lies close to the surface in the shade of a log or under a boat, & is easily caught with a landing-net; it is very sluggish & I have never known it to take a bait. it is the only fish worth eating in the Mere, having white, firm, flaky flesh, & not at all muddy; all the other fishes found in the Mere are almost uneatable because of their muddy flavor'. Pity one can't get more correspondents like Mr. Archer; & (3) Dr. T. Bancroft on the recent destruction of Ceratodus. [Australian Lungfish] Answered 1 & 2 & passed on 3 to Dr. Harris [Queensland Museum Director] with a recommendation to stir up the Government on the subject.

The following day (9 May 1917) he received from McCulloch an extensive progress report on the gobies, Ogilby commenting 'He tells me he is getting on well with the figures of the gobies & has, the Lord be praised, worked out the Australian synonymy of Paragobiodon echinocephalus (lovely name for a little divil that scarcely grows more than an inch long) ...' Another letter from McCulloch 12 days later included a correction to an identification of Ogilby's; his comment on this was good-humoured 'but the most astonishing information of all is that Trach. bankiensis is really a scleropeid after all, being probably Amphiprionichthys unipinna; the rotten little spoofer, taking me in bald headed with its mimicry of a Gobiodon. Selah.'[16] De Vis (1884e) had described the fish as *Trachycephalus bankiensis* and had placed it in the correct group; Ogilby had sent the little fish to McCulloch as a goby! It is known as the Coral Croucher, *Caracanthus unipinna* (Gray, 1831).

More on elections; a trip 'down the Bay'; indexing gobies in Günther's works and others; too cold in his room (57° Fahrenheit); a great deal of correspondence, including David Starr Jordan; careful transcription of a key to a genus of gobies (*Valenciennea*) sent by McCulloch for his comment; registration of Museum specimens; much more on his gobies; and the last entry:

> Saturday, June 2. 56°F.
> Worked at home all day. Beatrice's [his sister] second boy is killed in France. Would to Heaven that all the Hohenzollerns & Hapsburghs, & the rest of the dirty gang were safe in hell. (QMA)

16 *Selah* is a Hebrew expression, used by Ogilby occasionally, rather as a philosophical 'such is life'.

Edible Fishes of Queensland and later work, 1913 to 1920

Three major factors allowed Ogilby to resume significant scientific work, as demonstrated by the number and scope of his publications, particularly from 1910 to 1919. His own efforts at the AFAQ, sympathetic treatment by the Queensland Museum directors from 1910, and the association with McCulloch were the key elements. Most of his work appeared in the *Memoirs of the Queensland Museum*, where he published the various parts of his *Edible Fishes of Queensland.* This series appeared from 1913 to 1920 and comprised partial revisions of the Queensland representatives of many groups of fishes. He tackled 15 groups, as indicated by their part numbers; for instance, 'Part V. – Heterosomata (No. 1)' (Ogilby 1916c). The order Heterosomata comprises the flatfish: flounder, sole and their relations. This paper gave a general introduction with a key to the several families of the order but dealt with only a portion of one of these families in detail, the subfamily Soleinae. He left the other subfamily (Brachirinae): 'As Mr. McCulloch is now making an exhaustive study of the members of this subfamily, I shall await the result of his researches before dealing with the Queensland members.' So 'Heterosomata (No. 1.)' was presumably to have been followed by other numbers in due course. The same applied to the family Carangidae: 'Part III. – Carangidae (No. 1.)' (Ogilby 1915a) dealt with 10 of the 45 species he planned to describe. Ogilby generally indicated by 'No. 1' that there was more to follow, but only the Carangidae saw a 'No. 2.' and that dealt with just two more species.

While working on the *Edible Fishes* papers, Ogilby prepared a number of other publications, some of which were of considerable significance. For example, the studies of Grunters (1916) and Gobies (1919) with McCulloch, and a review of Queensland Angelfishes (1915b), with a supplement (Ogilby 1916e). Ogilby kept up with new records of fish for Queensland in *Ichthyological Notes*, published in the *Queensland Museum Memoirs* (1913, 1915, 1916 and 1918). New species he generally described, in more modest numbers than previously, in occasional papers. He was asked by J.H. Stevens, the Chief Inspector of Fisheries, to write a popular summary of the main commercial species of fish found in Queensland waters. This appeared as a 60 page essay, *The Commercial Fishes and Fisheries of Queensland*, published by the Treasury in January 1916 (Ogilby 1916a).

On completing the essay, Ogilby applied to the treasurer for remuneration of '£100 to £105' (18.5.16, DPIQ). Eventually a gratuity of £50 was passed for payment, '[t]he money to be placed to the credit of a joint a/c in the names of Dr R. Hamlyn-Harris [Director of the Queensland Museum] & Mr Stevens.' (22.6.16, DPIQ) In August 1916 the Under Secretary for the Treasury answered Ogilby's requests for payment, advising him that the 'Applications for advances from the Account should be made by you to Mr. Stevens.' (7.8.16, DPIQ). Apparently Ogilby, for all his scientific output, still required looking after.

Final assessment of Ogilby's ambitions

Ogilby's ambition to produce a major work on Australian fishes had first been made clear by his correspondence and diary entries through the 1890s. For instance, in a letter to the Director of the South Australian Museum, Edward Stirling, Ogilby referred to '... the work on the "Fishes of Australasia" on which I am engaged on behalf of the Government of New

South Wales, which is at present supplying all the funds for what will, when completed be a great, valuable and costly national work.' (19.4.98, SRSA) Around the time of his move to Brisbane, in spite of his disastrous alcoholic illness, he still invoked the plan frequently in his letters although his rhetoric rang hollow. There is, throughout the later correspondence between Ogilby and McCulloch, evidence of their intention ultimately to publish a 'Fishes of Australia'. To this end, they tackled the taxonomic problems posed by early work, described new species, and revised one group of fishes after another. However, Ogilby's opportunities were clearly based in Queensland and his publications were mostly confined to the fishes of that state.

Ogilby (1916b) published the first part of a check-list of the fishes of Queensland. This was a modification of part of the check-list that was to have been the framework for his great work on Australian fishes; but it was, essentially, a list of the sharks and rays of Queensland.

Did he still hope to write the definitive work? The numbering of sections of his *Edible Fishes of Queensland* gave the impression that he intended to bring them together in book form eventually. Such a book would probably have been his more immediate personal aim at this time, the larger work on the whole fish fauna of Australia being left to McCulloch, or perhaps imagined as the ultimate joint publication. His sister Edith wrote from Dublin in 1918, worried about the activities of Sinn Fein and the stringency of the peace terms. She asked, 'When will the book be ready? I must have it.' But the same letter suggests that Ogilby may have already had a minor stroke, as she also says, 'Dear Boy, I am sorry for what you tell me about yourself, you must be very careful and I hope will be spared for some years yet.' (undated letter, QMA)

Last work, 1919–1920

Ogilby's correspondence for 1919 gives an impression of his activities. He appears to have written each month to McCulloch, judging from the replies, and these letters give an ample view of the work upon which Ogilby was engaged: last minute investigation for their joint paper on Gobies; revision of the family of Glassy Perchlets (Ambassidae); construction of a key to the difficult family of Tropical Snapper (Lutjanidae); reference to the Queensland Museum collection to satisfy McCulloch's several queries. He was also carrying out a large number of smaller projects as specimens came to hand from his collectors. There was correspondence with Inspectors of Fisheries in Sydney and in Perth; with David Stead, who was the General Manager of the State Trawling enterprise in Sydney; with David Starr Jordan; and with his many collectors and informants in Queensland.

In November 1919 Ogilby had a serious stroke and was hospitalised in Brisbane. McCulloch was very upset and wrote offering support:

> It can afford you neither pleasure nor profit, my dear Mr. Ogilby, to get up to such tricks as you have been trying lately, and I sincerely hope you'll cut hospitals & all related thereto from your programme for a while. I was fairly bowled over when Longman told me of the fearsome happenings, since the fun of working at fishes without somebody to argue over them with is relatively small.

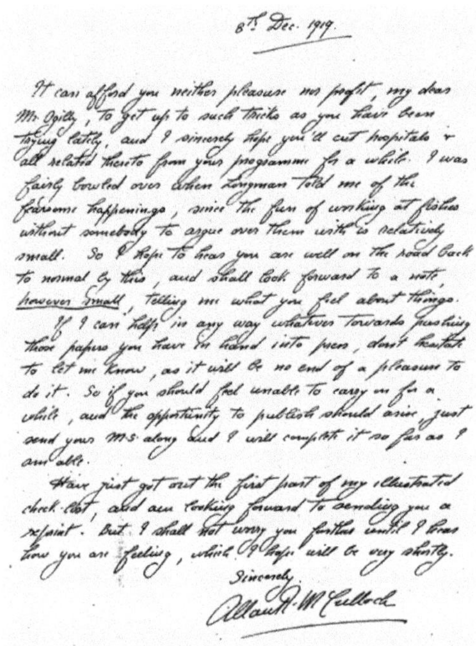
8th Dec. 1919.

It can afford you neither pleasure nor profit, my dear Mr. Ogilby, to get up to such tricks as you have been trying lately, and I sincerely hope you'll cut hospitals & all related thereto from your programme for a while. I was fairly bowled over when Longman told me of the fearsome happenings, since the fun of working at fishes without somebody to argue over them with is relatively small. So I hope to hear you are well on the road back to normal by this, and shall look forward to a note, however small, telling me what you feel about things.

If I can help in any way whatever towards pushing those papers you have in hand into press, don't hesitate to let me know, as it will be no end of a pleasure to do it. So if you should feel unable to carry on for a while, and the opportunity to publish should arise, just send your M.S. along and I will complete it so far as I am able.

Have just got out the first part of my illustrated check-list, and am looking forward to sending you a reprint. But I shall not worry you further until I hear how you are feeling, which I hope will be very shortly.

Sincerely,

Allan R. McCulloch

McCulloch to Ogilby when McCulloch heard that Ogilby was in hospital after suffering a stroke. The 'illustrated check-list' McCulloch refers to was his major work on the fishes of New South Wales. Courtesy of the Queensland Museum Archives.

So I hope to hear you are well on the road back to normal by this, and shall look forward to a note, however small, telling me what you feel about things.

If I can help in any way whatever towards pushing those papers you have in hand into press, don't hesitate to let me know, as it will be no end of a pleasure to do it. So if you should feel unable to carry on for a while, and the opportunity to publish should arise, just send your M.S. along and I will complete it so far as I am able.

Have just got out the first part of my illustrated check-list, and am looking forward to sending you a reprint. But I shall not worry you further until I hear how you are feeling, which I hope will be very shortly.

Sincerely,

Allan R McCulloch.
(8.12.19, QMA)

McCulloch had his reply, complete with corrections for the checklist which he had sent to Ogilby for his consideration, by the first week in January. (8.1.20, QMA). A few weeks later McCulloch was answering some new questions from Ogilby, but these later letters from McCulloch employ a more solicitous tone than previously, 'Hope you are much better, and able to move along at your old pace.' (8.2.20, QMA) Then Ogilby turned again to the Eels, a group upon which he had worked on a number of occasions in the past, and McCulloch tried to guide him; for instance:

> As for Muraenesox myrus. [Eschmeyer (1998) indicates that this is a synonym of *Muraenesox ferox* (Costa, 1850), not known from Australia (Hoese *et al.* 2006)] If your only record of this species rests upon your uncertain identification of the specimen secured by you in the '90s., and supposed to be preserved as a skeleton, I sincerely hope you will not publish it. I cannot find the specimen, and therefore beg you to leave it alone unless you have additional material to go upon. (26.2.20, QMA)

Ogilby must have written several times in fairly rapid succession, trying to get his paper on the eels in order for publication, but McCulloch had his own problems:

> You will put me down as a low sort of cow for having left your several letters unanswered, but I have been so unwell lately that things have gone all awry. Further, owing to the annual gassing [fumigation] of the library during the holidays it has not been possible to work upstairs, and so your Conger questions have had to take a back seat. But I shall look into them at the first opportunity, and will send you the answers before long. (5.4.20, QMA)

The final letter is undated, and the first page is missing, but it seems to have been written shortly after that quoted above. McCulloch gave Ogilby's questions careful treatment, supplying some descriptions, helpful criticism and a key to the Conger Eels.

In his last major paper Ogilby (1920) reviewed some fishes then assigned to the families Serranidae, Lutjanidae and Opistognathidae, including a lengthy treatment of the Estuary Perch, *Percalates* [now *Macquaria*] *colonorum*. This species had many synonyms and Ogilby attempted to sort them out, with reasonable success. He recorded some perceptive speculations on natural variation in freshwater fishes:

> None of the authors, who have so overloaded this species with useless synonyms, seem to have taken into account that owing to diversity of environment; climatic conditions, due not only to their geographical range but also to the varied altitude at which they may reside; the character of their native streams, whether sluggish and discolored with a muddy or weedy bottom, or clear and sparkling as it flows over a bed of rock and gravel and sand; the nature of the geographical strata over which the stream flows; the great diversity of the food supply; and many other fortuitous circumstances, much greater latitude of variation should be allowed to fluviatile fishes than to their more equably situated marine relatives. And the variation is apt to be greatly accentuated when the particular fish under review leads, as in this case, both a fresh-water and a brack-water existence.

Final illness

Ogilby had suffered a series of strokes, continued working for a while, but a severe episode finally stopped his work on fishes in 1920. He spent several years incapacitated in an institution and died in August 1925 in the Diamantina Hospital. He was buried in the Toowong Cemetery in Brisbane. His friends in the Amateur Fishermen's Association of Queensland named their club house on Bribie Island after him, and the 'J. Douglas Ogilby Cottage' it remains to this day.

McCulloch's tribute

Allan McCulloch and his apprentice in ichthyology, Gilbert Whitley (see page 398), completed the preparation of a checklist of Queensland fishes early in 1925. McCulloch wrote a draft introduction, including a tribute to Ogilby, which he handed to Whitley for the printer:

> The study of Queensland Fishes was elevated to a much higher plane when Mr J.D. Ogilby, at one time Ichthyologist of the Australian Museum, established himself in Brisbane. Under the auspices of the Amateur Fisherman's Association of Queensland, he gathered together a fine collection of local fishes and wrote several important papers upon them. He was later appointed as Zoologist to the Queensland Museum, to where the types of his many discoveries were transferred later. Though Ogilby added considerably to the already long list of synonyms in the Queensland list, he did more towards disentangling the errors of his predecessors than any other investigator. The senior author of the present contribution here gratefully records his indebtedness to Mr Ogilby for unfailing assistance and lasting friendship, which began when the first-named was a lad of thirteen years and flourished until his elderly friend was overtaken by misfortune. A profound appreciation of those over-inspiring products of the vine, which have been so happily lauded by the poet Omah, gradually undermined and finally ruined Mr Ogilby's otherwise splendid record, and finally brought him, to that state in which he now lies, oblivious of his friends, unknowing, and waiting but for the Reaper who is so slow to relieve him of his burden. (ML)

McCulloch was then suffering from severe depression which tended to impair his judgment. Whitley showed the good sense to excise his mentor's florid, final sentence before sending the introduction to the printer. Only a few months after this, McCulloch's illness became much worse and he killed himself.

An appreciation

Douglas Ogilby's contributions to Australian ichthyology were immense. Between 1885 and 1920 he published approximately 100 papers on Australian fishes, including in this work the descriptions of over 300 species which he regarded as new. He prepared his descriptions directly from the fish, never being satisfied to copy the work of other ichthyologists, but checking every detail and dissecting whenever practicable. He frequently included synonymies and keys to groups of fishes, in an effort to synthesise as he described; the review element in his publications became more evident in his later years, partly but not wholly due to his association with Allan McCulloch. Ogilby sought and published biological information on the fishes he studied whenever possible and emphasised the importance of gathering this sort of information in his publications and correspondence.

His achievement was limited by the chaotic state of fish taxonomy in Australia at the time he began his studies, by the lack of support for scientific institutions, and by the effects of his alcoholic illness. The latter resulted in restricted access to the libraries and collections upon which the accuracy of his work depended. A great work on the fishes of Australia was his ambition, but this could only have been achieved by a less encumbered personality. The success of David Jordan's *Fishes of North and Middle America* (1896–1900) was comparable to the work towards which Ogilby strove. The American project was brought to completion by Jordan's own powerful and intelligent influence through which he was able to engage

J. Douglas Ogilby in about 1919, virtually at the end of his long, erratic but successful career as an ichthyologist; still employed at the Queensland Museum, respected for his science and with a wide circle of friends. Courtesy of the Amateur Fishermen's Association of Queensland.

the support of colleagues and governments. Ogilby's attempts to organise on this scale were doomed by his own lack of insight into his personal problems.

Although as McCulloch mentioned 'Ogilby added considerably to the already long list of synonyms in the Queensland list', his descriptions were of a high standard which generally has allowed subsequent investigators to identify his species. He exercised considerable discrimination in correcting the work of his predecessors in Australia and worked at a professional level unattainable by them. He put Australian ichthyology in touch with the advances that were being made in North America and Britain, realising the importance of keeping pace with modern systems of nomenclature and classification. Modern ichthyology in Australia owes a significant, and only partially recognised, debt to J. Douglas Ogilby.

Waite

E.R. Waite was a museum manager and ichthyologist of significance in Australia and New Zealand. His work is readily divided into his time in Sydney (1893 to 1905), in Christchurch (1906 to 1913) and Adelaide (1914 to 1927).

Diaries and early life

Edgar Ravenswood Waite (1866–1928) left an extensive record of his life and work in the form of a large number of diaries, maintained methodically from 1885 for the rest of his life. His wife and son survived him, keeping intact his diaries and notebooks. His grandson still has charge of the diaries which present a remarkably complete account of his grandfather's life and incidentally of the progress of ichthyology in Australia between 1893 and 1927.

Born in Leeds in 1866, Edgar Waite had one brother, Wilfrid. Their father was the foreign correspondent in a Leeds bank and had an enthusiasm for painting. His sons' education was unremarkable but Edgar had some instruction in biology at the Victoria University in Manchester and then began work as a clerk. He had many interests and, although he was successful with his clerical work, he was attracted to natural history and pursued its study in every way that was open to him.

Holiday in Whitby

The first book in the diary series, which Waite himself labelled 'No. 1', is an account of a family holiday at Whitby in 1874 when he was eight. The little book preserves an encounter with marine fish and perhaps an indication of a methodical nature:

> Then we went out fishing. I caught nothing but Willie caught a young cod-fish & a repulsive thing with a very big head & some spikes along its back. We then came in & had tea, then went out to fish a while again without success. it began to rain so we came in & had supper. I went to bed. It rained very heavily the greater part of the night.
>
> Wed ... met with Mr Holroyd his son & a friend – who were just going out fishing so Willie & I jumped in & we went about 2 miles out & then cast anchor & threw in our lines. Willie caught the first two both these being whiting Mr Holroyd then caught 1 gurnard & his son 1 plaice & I caught 1 gurnard & 1 plaice Mr Holroyd's friend caught nothing so that we had,
>
> 2 whiting 2 gurnard 2 plaice making a total of 6. It began to thunder so we drew up the anchor & put up the sail when we got about 1 mile in it began to rain in torrents so that we were pretty wet when we got in ...

This notebook ends with the return to Leeds, '... & got safely home thus ended our holiday in WHITBY.' (WD, August 1874)

Early interest in natural history: Leeds Naturalists' Club

A notebook for 1883–84 and the diary sequence that ensues give indications of Waite's activities in natural history, particularly ornithology. Over the next few years it is clear that he developed skills that would contribute to his potential as a naturalist. He worked in the Accounts Office at the Leeds Town Hall but at home practised taxidermy, drawing and photography, transcribed notes on the classification of birds and made frequent field trips in the country around Leeds. In 1885 Waite joined the Leeds Naturalists' Club and participated in field trips and lectures. His interests within natural history throughout the 1880s were very wide, reflected in his many references to the insects and vertebrates that came to his notice. The club included 'microscopists' and Waite soon had his own instrument. A highlight of a brief holiday he took to London was '3 hours visit at the South Kensington Natural History Museum'. He was very active in the Naturalists' Club, exhibiting specimens and speaking, most often on ornithology; in November 1887 he ran a session, '... I having charge of the evening did the spouting. I did not use any notes & satisfied myself much better than I had ever done previously in that line ...' Edgar's father, John Waite, wrote to the British Museum in April 1888, exploring the prospects for his son. The belated reply came in December 1888, suggesting that Edgar could take an exam for 'a vacant assistantship of the second class in the Department of Zoology.' (4.12.88, NHMA) However, in May 1888 he had left the Leeds Corporation to become the sub-curator of the Leeds Museum.

Leeds Museum and a little ichthyology

While employed at the Museum, Waite was immersed in natural history. He collected animals for himself and for the Museum, attended classes in biology, and prepared lectures and papers on ornithology. He also began learning German, a study that he pursued intermittently for many years. His interest in ornithology grew: he shot birds and skinned them, collected their eggs, observed their habits closely, drew and painted them, prepared skeletons, and gave lectures on his work. He prospected for specimens further afield and began to assemble a collection of British mammals at the Museum. By this time Waite saw his future in terms of Museum curatorship and, on a holiday to Ireland with his father in 1889, visited the Museum of Science and Art in Dublin, introducing himself to the Curator.

Fish were not completely neglected but the opportunities for their study were few around Leeds. Waite made some occasional notes on freshwater fish and had a field excursion to York with the Leeds Naturalists' Club in August 1889; invertebrates were taken by a 'dredge and tow-net' from a small boat, while other members of the club spent their time 'working the rock pools.' Waite's ignorance of marine fish at this time is shown by his having to consult Yarrell's British Fishes (as Ogilby had done some 20 years earlier in Ireland) to identify a small shark, which proved to be 'the Common Tope (Galeus canis)'; and he 'made out a small fish to be the Shanny (Blennius pholis).' [17] (WD, 16.9.89) He made a water colour drawing of the shark, perhaps his first attempt at representing a fish.

17 Now *Galeorhinus galeus* (Linnaeus, 1758) and *Lipophrys pholis* (Linnaeus, 1758).

Waite frequently attended lectures on scientific topics and at a meeting of the Geological Association in Leeds on 16 January 1890, he read a manuscript on fossil fish in the author's absence:

> ... it was far too technical for the audience. Afterwards they asked me to give them something about the anatomy of fish but I said the subject was too big to be even mentioned in a few minutes...

A bad moment for Waite? (WD, 16.1. 90)

Making a career in natural history

Having negligible financial resources, Waite was ever conscious of the need to economise. He had been taking classes in biology but in 1889 wrote 'The Biol: Class at the College commenced yesterday but I did not go, the xpence being greater than I can at present bear.' (WD) He was paid for his lectures, generally a guinea a time, and made his own lantern slides to illustrate them. He earned extra money drawing illustrations and maps, selling some of his collections and operating the lantern for other lecturers. Whenever possible he purchased books, mostly second hand volumes on natural history from 'the Market', and he tried his hand at bookbinding. He found Cuvier's *Animal Kingdom*, Wallace on zoogeography and many others. He contrived election to the Linnean Society: '6 pounds entrance fee & 3 pounds first year's subscription.' His study of choice was still ornithology and he prepared a series of bird skeletons. He also undertook a large project, preparing a paper on the vertebrate fauna of his district, the Western Ainsty (Waite 1891). This involved some freshwater fish, but for these he relied on the work of others, treating this section rather superficially.

By 1890, Waite was looking for a curatorial position outside Leeds; he had been given 'a vague hope of an appointment as Curator of the Museum which would be organised in connection with the College in a few years time.' (WD, 28.5.91) In August 1891 he took a carefully planned, three week tour through Germany, Austria, Bohemia, Bavaria and Holland. Waite visited the usual tourist sites and introduced himself at the Museums and Zoological Gardens, being shown over them, and writing critical notes on the organisation of the displays for his own instruction. The Natural History Museum in Vienna, in particular, met with his approval. Shortly after his return, Waite was elected as joint editor of *The Naturalist*, the journal of the Yorkshire Naturalists' Union and, unexpectedly, the curator of the Leeds Museum resigned and Waite was appointed to the position in October 1891.

Waite's calendar had become very full. Aside from his curatorial work, he was an active member of various naturalists' associations, including the Leeds Naturalists' Club (Council Member), the Yorkshire Naturalists' Union (Assistant to General Secretary), the Geological Association (Council Member) and the Conchological Society (Librarian). He pursued field work with vigour and began, on his own initiative, a collection of British freshwater fish for the Leeds Museum. Over a number of years 'the Misses Green' had figured in his diary; Edgar Waite and Rose Green were married in April 1892. He had considered applying for curatorial posts at Scarborough and Edinburgh, but finally noticed an advertisement in the journal *Nature* for the position of Assistant in Zoology at the Australian Museum. 'Showed this to Rose who agreed with me that I ought to go in for it.' (WD) He organised an impressive array of 'testimonials' and sent his application to the trustees of the Australian Museum. The duties associated with the position were 'to determine, arrange & look after the Specimens of Fishes and Reptiles.' He had worked on reptiles for several years, but his knowledge of fishes was rudimentary. In common with many naturalists of the time, he felt confident that he could work with any group of animals.

Move to Sydney

In November 1892 Waite learned that his application had been successful. He ordered Günther's *Introduction to the Study of Fishes* which arrived three weeks before sailing in February 1893, in time to be read on the boat. Günther (see page 86) had in 1870 completed his major work, the *Catalogue of Fishes in the British Museum*, and had made little attempt to bring his *Introduction to the Study of Fishes* (1880) up to date in spite of significant advances that had been made by other ichthyologists. The fish literature available in 1893 was considerable but the most recent ideas were to be found only in specific journals, and Waite was largely unacquainted with these.

The Waites arrived in Sydney in April 1893 and Edgar reported at the Museum, expectant, full of enthusiasm and keen to begin work. E.P. Ramsay was still the Curator but had been unwell for some time and initially provided no instructions for Waite. However, he was introduced to the trustees, who included William Haswell, then Professor of Biology at Sydney University. Haswell invited Waite to look over the University and the Macleay Museum, where Waite met George Masters, Macleay's old Curator, who was still in charge of the Macleay collections. Another of the Australian Museum trustees whom Waite met at this time was Dr. James Cox, an old colleague of Macleay's with an abiding interest in fishes. J.D. Ogilby still attended the Museum on an irregular basis and received at least some of the collections that were sent there, although since Ogilby's dismissal Thomas Whitelegge (see page 188) had been the de facto curator of reptiles and fishes. Waite was soon briefed about his predecessor's problems but was placed in an odd position; although nominally in charge of fishes, he was virtually in competition with an experienced ichthyologist who maintained a loose attachment to the Museum and was actively practising ichthyology. Waite, faced with the prolific and diverse fish fauna of Port Jackson and the surrounding waters, was in no position to assert himself but applied his efforts mostly to the study of reptiles and mammals.

Confronting ichthyology in Sydney

Although Ogilby's presence inhibited Waite with regard to ichthyology, there arose many opportunities for its study. There were irregular donations from the public, apart from the larger collections which Ogilby still received; these donations of occasional fishes prompted Waite to consult the available literature, such as it was. Whitelegge studied marine invertebrates and also had a long-standing interest in fishes. He brought interesting specimens to the Museum from his collecting ground at Maroubra with remarkable regularity and Waite notes 'Whitelegge lent me his MSS 'Fishes' for the purpose of copying'. Both

Whitelegge and Charles Hedley, the Museum's conchologist, shared their shore excursions with Waite: they informed and encouraged him in marine biology. The inspection of anglers' catches gave Waite a little experience of the local fishes, whose common and scientific names he began to learn. A typical diary entry made just two weeks after landing in Sydney described his first tram outing to Maroubra with Whitelegge. He noted the unfamiliar plants, birds and lizards, then:

> We next had a walk round and looked at the people fishing. The species caught were Green Eel (Muraena afra), Rainbow Fish (Coris lineolata) of beautiful colour, and Sweep (Scorpis aequipinnis) ... Afterwards I visited the Rock-pools with a hand net, but the tide was too high for much work to be done and we got but little. Quite a number of Toadoes were swimming in a little creek in which I was wading, they were easy to catch in the net but I took only one away (Tetrodon hamiltoni.). There were numbers of Sea Slugs and anemones in the pools, caught a small blenny (Lepidoblennius geminatus) and a young Black Fish (Girella tricuspidata).[18] (WD: 23.4.1893)

All very new to the naturalist from Leeds.

Drawing the animals was an effective means of becoming familiar with them and Waite drew some of the lizards and fish that he saw, including a water colour sketch of the toadfish from the creek at Maroubra and a shark from the museum collection (see colour plate 13, upper).

During the following week three fishes of interest to Waite were brought in to the Museum and he drew them all: an adult Luderick that he identified as *Girella simplex* [now *G. tricuspidata*]; a Flute-mouth, *Fistularia serrata* [now *F. petimba*]; and a Black Cod, *Epinephelus daemeli*. He also began to visit the fish markets in the early mornings, where he often encountered Ogilby looking for specimens for his descriptions. With Ogilby's occasional help, Waite made lists of the fishes he saw, with the scientific names; the early morning litany for 6 May, 1893 reads 'Schnapper, Flathead, Leather Jacket, Garfish, Nannygai, Yellow-tail, Sea Mullet, Sole, Rock Cod, Mackerel, Bream', each with the scientific name entered opposite; on the 10th he added 'Jew Fish, Tailor, Sand Mullet, Blackfish, Tarwhine, White Trevally, Sergeant Baker, Green Eel.' These were the commonest of fish, but new to Waite, and his ignorance would have been quite plain to Ogilby. However, the latter hoped to have Waite draw fishes for his own work, and they came to a tentative agreement for this purpose. Before long, Waite found that little reliance could be placed on Ogilby and gradually realised that he must eventually replace him altogether at the Museum (which was the purpose of his appointment).

Work with terrestrial vertebrate groups

At first, although he was gradually familiarising himself with the fish fauna, Waite found himself more at home with terrestrial vertebrates. He made a number of trips into the country around Sydney, usually invited by new acquaintances for weekend excursions to

18 Fish identifications from this passage: Green Eel, *Gymnothorax prasinus* (Richardson, 1848); Maori Wrasse, *Ophthalmolepis lineolatus* (Valenciennes, 1839); Silver Sweep, *Scorpis lineolata* Kner, 1865; Toado, *Tetractenos hamiltoni* (Gray & Richardson, 1843); Eastern Jumping Joey, *Lepidoblennius haplodactylus* Steindachner, 1867.

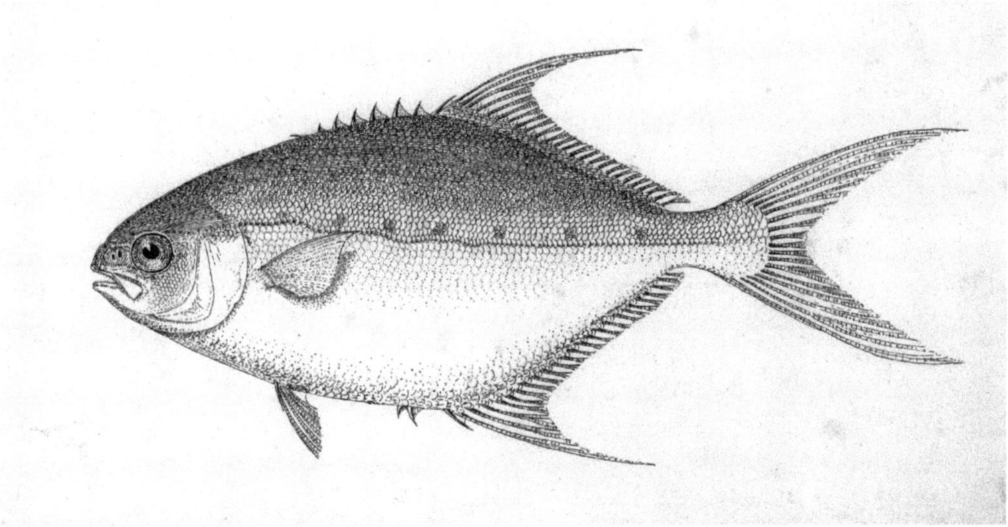

Waite's drawing of the Common Dart, *Trachinotus botla* (Shaw, 1803), the first of a series of fish Waite drew for Ogilby. In his diary for 25 May 1894 Waite wrote 'Ogilby bought a Dart (*Trachynotus russellii*) from the market. I took it home & drew it in the evening, it taking me 3 hours.' From AA356 Waite Collection, courtesy of the South Australian Museum Archives.

their land selections, and he revelled in hunting and collecting. In this way he became familiar with many of the native mammals, reptiles and birds and he has left detailed accounts of his experiences. At the Museum in the mid 1890s the near disastrous economic problems of New South Wales took a toll. Waite was indeed fortunate to have been appointed at all: there was no money for new positions for some years after his arrival and several staff members were dismissed. In this way, mammals and osteology were added to Waite's responsibilities; he had charge of the accession registers for these disciplines as well as those of reptiles and fishes.

At the Australian Museum, Waite's time was occupied in overseeing and adding to the collections under his care, while keeping the registers up to date. His chief research project emerged from tardy instructions given to him by Ramsay two months after his arrival in Sydney:

> Sir,
> You will be good enough to take up a minute examination of the Australian Snakes including their habits & life histories more particularly of the specimens in the Museum. To describe such new species as you find in the collection; & to have a well labelled set exposed for view in one of the show cases in the Museum. I would point out to you that it is intended to bring out a work on the Australian Snakes & with this object in view it would be as well for you to note all literature & original descriptions & figures so far recorded of existing species. These instructions must not interfere with your duties in seeing to the welfare of the Collections of Fishes, Reptiles &c under your care generally.

I am,
Yours truly
Ed. P Ramsay.
Curator

This led Waite to the study of a poorly understood group that was well represented in the Museum collection, the Blind Snakes (Typhlopidae). The first of several of his papers on these snakes was published within a few months (Waite 1893). Studying this group also led him to establish contact with zoologists at the other Australian natural history museums.

Adequate illustration was often lacking in the Australian Museum publications at this time so Waite's artistic skills were much in demand. Within a short time he found himself drawing not only the vertebrates under his care but also birds, insects, spiders, mollusc shells, and fossils for his colleagues. In 1897–98 Waite referred frequently to drawing specific fishes at Ogilby's request, but none of these drawings was published in Ogilby's papers, although some appeared in later publications by Waite himself. Presumably the financial rewards that Ogilby proposed for this work were not forthcoming. Although the value of Waite's illustrations was appreciated by his colleagues, he by no means saw himself primarily as an illustrator.

During his first five years in Sydney Waite wrote relatively little on fish, being more concerned with reptiles and, in the latter part of this period, with rodents. In 1894 an expedition had been undertaken as a joint project of the state governments of South Australia, Victoria and New South Wales: an examination of the natural history of Central Australia, with experts drawn from the Museums and Universities of each of the participant states. This was the Horn Expedition, named after the philanthropist who provided the initiative and the bulk of the funds. The results in terms of anthropological data, and the stimulus this provided for original research into the structure of Australian aboriginal societies, were without precedent. Professor Baldwin Spencer of the University of Melbourne led the expedition, participated in the collecting and research, and finally saw the reports on the expedition through the press. Various animal groups were collected, including some fishes that were sent to Zietz at the South Australian Museum (see page 291), and native mice that Spencer sent to Waite, who duly reported on them (Waite 1896).

Field trips in Port Jackson

Ramsay was still organising occasional trawling excursions in Port Jackson in the mid 1890s and Waite was delighted to be invited to participate. A tug was hired and the gear stowed aboard. They trawled in several parts:

> We first made for Rose Bay & put down the trawl & slowed the Engines. In half an hour it was drawn up heavy with Ascidians. Flounders, 'Fiddlers' and 'Stingarees' were all the fish taken – Next time new ground was trawled but the net was so heavy that it seemed impossible to raise it, & we lost quite 2 hours. it was simply laden with multitudes of Ascidians, each as large as a Cocoanut & masses of Seaweed. When landed the fish were more numerous

> but chiefly Rays. A Jew-Fish, Fortescue, Bream and others ... Thence up the Paramatta river & round Cocatoo [sic] Island, secured Cat-Fish & more rays. Landed at 5.30. I went to Museum & pickled the captures. (WD, 22.3.94)

He lists the species of fish, 18 in all.

The field trip from which Waite derived the most excitement and satisfaction, and ever after recalled with pleasure, was a dive in Sydney Harbour in 1894. He and his friend Charles Hedley were taken out to Shark Reef, which was then a hazard to shipping and being reduced by blasting. They donned the massive, clumsy dress of hard-hat divers and were initiated into the noisy workings of pumps and hoses, and then into the beauties of the harbour underwater. His diary account is well over 4000 words, half for the mechanism and preparation, and half for the underwater experience, 'the most marvellous of anything I have before undergone.' In theory, he was looking for the egg-cases of Port Jackson Sharks, which had been reported by divers here as attached to algae. The impression from reading his account is that the experience was so overwhelming that he forgot to look for them.

Influence and connections at the Australian Museum

Although Ramsay was the Curator until 1895, during Waite's tenure the Museum's palaeontologist, Robert Etheridge, was most active in administrative affairs within the Museum. It was not initially a job that he relished and he found Waite's experience and advice helpful. The influence that Waite exerted in this quarter attracted some unpleasant comment, but his approach to problems was generally that of common sense, with an occasional hint of the devious. The rapport with Etheridge was very useful, and when the time came for re-organising the displays of reptiles and fishes, Waite quoted Etheridge in his diary with satisfaction, 'You will have your own way in the arrangement entirely & any assistance I can give you, you shall have.' The Australian Museum was a good place to develop contacts: at one time or another Waite met all of the men, and they were exclusively men, who were influential in Australian natural science. These included some who were to have an effect on Waite's own career, most particularly Edward Stirling, Director of the museum in Adelaide.

First publications and related activity in ichthyology

Waite's initial contribution to Australian ichthyology was a paper based primarily on stranded specimens that Whitelegge had brought from Maroubra (Waite 1894). The fish were supposedly new records for New South Wales, but they were not all well described, the identifications incompletely substantiated, Günther's discredited concepts of eel larvae were quoted and the introduction to the paper was rambling and inconsequential. However, a Driftfish, *Psenes* [now *Cubiceps*] *whiteleggii* was described as new and was adequately characterised. Waite made a useful attempt to define the differences between some unusual pipefish, with effective illustration. The following year he reported the egg-cases of an uncommon species of Port Jackson Shark, *Heterodontus galeatus*, once again with useful illustrations (Waite 1895).

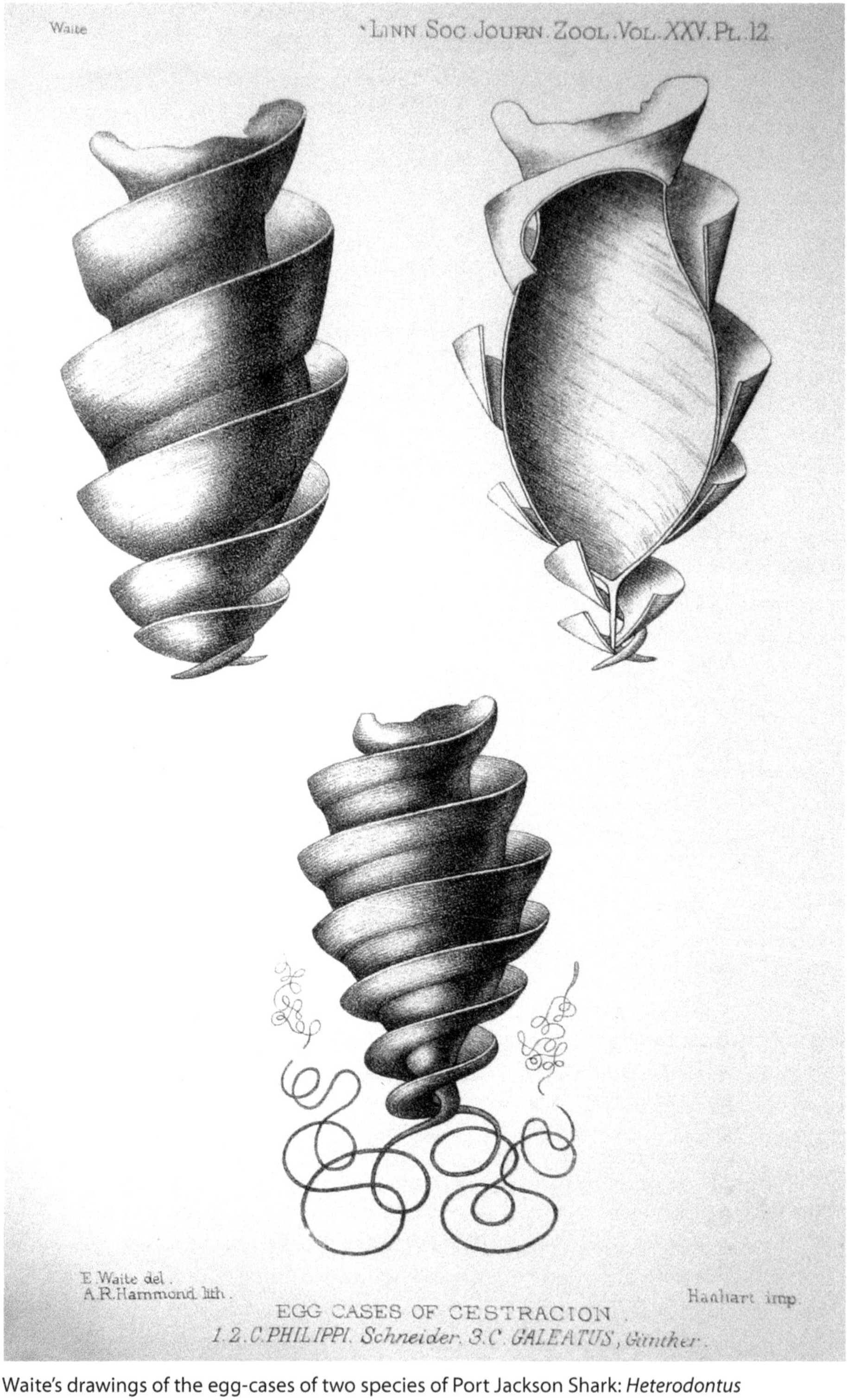

Waite's drawings of the egg-cases of two species of Port Jackson Shark: *Heterodontus portusjacksoni* (above) and *Heterodontus galeatus* (below). From Waite 1895.

Waite made frequent field trips to the nearby coast, noting the fish and trying his hand at catching them. He continued to visit the markets in the early morning, 'I bought an alarm 4/- in order that I might get up for the Fish Market. Etheridge told me I might buy fish for mounting purposes.' He was drawing fish consistently, but it was not clear that Ogilby would utilise this work. He and Waite were on reasonable terms and Ogilby gave Waite reprints of many of his papers. Waite also began corresponding with overseas zoologists: Steindachner in Vienna; Boulenger and Günther at the British Museum.

Settled in Sydney

In May 1895 applications were invited for the position of Curator of the Western Australian Museum. The salary was better and there must have been some temptation to be in charge, but Waite wrote 'not sufficient inducement for me to leave beautiful Sydney.' Not long after this the Waites purchased a block at Mosman and built a cottage. Their son, Claude, was born in April 1896 and they appeared to be well established in Sydney. Rose had many friends and Edgar was fully occupied with his work, with excursions, friends, photography, music and with a variety of birds, reptiles and mammals that were kept as pets and for natural history observations.

His salary from the Museum was reasonable (£250 per annum) but Waite was looking for opportunities to supplement his income. An entrepreneurial publisher in Sydney, Thomas Shine, commissioned illustrations of Australian native animals and Waite began drawing for him at a guinea an illustration. In this way, he drew large numbers of Australian reptiles and mammals; later, the publisher suggested a small book on Australian snakes and for this Waite not only drew the illustrations but also wrote the text. It was a useful, inexpensive production although Waite was a little disappointed in its final appearance when it was published (1898b). He also undertook a series of popular articles, for which he was paid, for the *Town and Country Journal* on Australian reptiles, mammals and sharks. This was moonlighting, although he kept Etheridge informed and most of the work was performed at home. Never was there any suggestion that the Museum did not get good value from Waite; rather, he did not distinguish his private time from his time at work once a project was under way.

Funafuti

Charles Hedley, conchologist at the Australian Museum, joined a party that was carrying out experimental boring of coral substrates in the Ellice Islands in 1896. They worked at Funafuti and among the specimens Hedley brought back was a small collection of fishes. Waite (1897) reported on the mammals, reptiles and fishes from this expedition but the collection of fishes was unsatisfactory in several ways: many of the fish were juveniles and so difficult to identify, their preservation was imperfect and the Ellice Islanders could not be prevailed upon to bring non-edible fish. However, it was a significant collection, comprising over 50 species, from a locality that had not previously been sampled and Waite appreciated that he should identify the specimens as accurately as possible, and at least record distribution extensions. The paper was of some interest but the identifications were not well substantiated and his descriptions were generally insufficient. Illustrations would have eased the situation, but the Museum budget at this time restricted the number of plates that could be published; some fish sketches were crowded onto a page with portions of a rat. Generally when he was doubtful about the identification of a specimen Waite speculated, but without providing a description. He was hampered by lack of experience, by the limited nature of the literature available to him, and by his lack of familiarity with that which was available. Ogilby could have guided him but perhaps Waite did not wish to appear dependent upon his colleague.

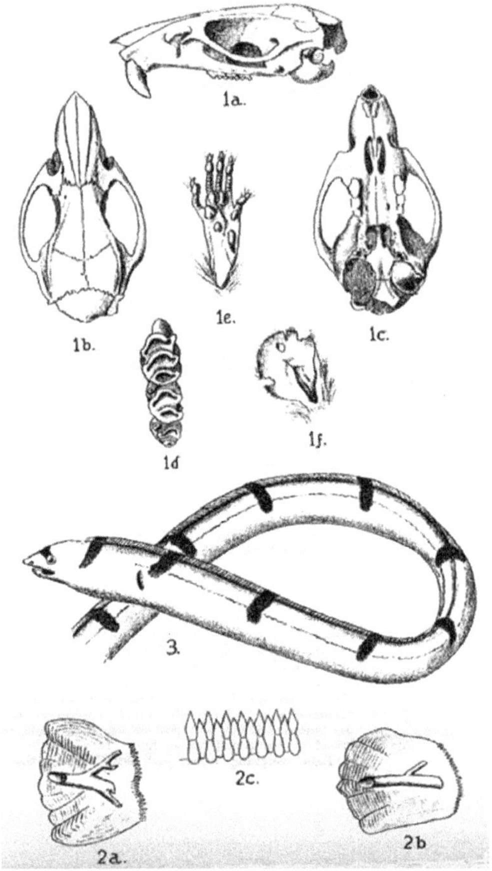

Waite's report on the mammals, reptiles and fishes of Funafuti included these cramped illustrations: the skull, teeth, feet and ears of a rat are crowded with an eel and the scales of a 'Red Mullet'. Waite identified the eel as *Ophichthys* [now *Myrichthys*] *colubrinus* but it is probably the Saddled Snake Eel, *Leiuranus semicinctus* (Lay and Bennett, 1839). From Waite 1897.

Aquaria

Having his own house prompted Waite to make more provision for keeping animals. He was already maintaining an aviary, and now constructed a 'bush house' and began to take an interest in aquaria. Initially, he collected small fish from the streams around Sydney. A typical excursion took place on 30 December 1897:

> Took Randwick tram at 9.30 to Kensington where I met Whitelegge, Ogilby, Gale & 3 others joined us & we went down a stream fishing – both with rod and net. Galaxias were very plentiful but Eleotris & Carp were scarcer & much more wary & few were obtained, I bought back some Galaxias, alive, for the tank & also some Gyrator Beetles. (WD, 30.12.97)

A few days later 'Whitelegge bought me some young Carp caught by one of his boys to whom I sent 1/-.' He made copious notes on all of his captive animals, including the 'Native Trout' (*Galaxias*), and his observations were acute and often captivating; he dropped a large worm into the aquarium, '... whereupon my only 2 Galaxias at once seized it one at each end. they did not pull like animals but side by side shot around the tank now diving to the bottom, now jumping clean out of the water, they continued their antics until the worm was divided & each retained a portion.' (WD) Later Waite established marine aquaria, collecting animals from the adjacent harbour.

Although he continued to keep Australian fishes in his aquaria, Waite became fascinated by the behaviour of the nest-building fishes of Asia. He began to keep 'Fighting Fish', *Betta pugnax*, the first batch brought from Singapore by a friend in 1904; and later 'Paradise Fish', *Macropodus opercularis* in 1905. Both species bred in his tanks. The aquaria allowed him scope for his ingenuity, 'I wrote to H. McLachlan Sec. for Railways & asked for order to Managers of Electric depots to allow me to take any damaged Accumulator Cells as when patched-up the[y] make very good Aquariums.' (WD) He experimented with home-made aerators and thermostats, made a study of aquarium water chemistry, and documented original observations on the reproductive behaviour of the fishes. Waite (1904d and 1905a) published these observations.

Albert Gale, a friend of Waite's and fellow aquarium enthusiast, published accounts of some Australian freshwater fish, including behavioural observations of the Purple-striped Gudgeon, *Mogurnda adspersa*, a popular aquarium species (Gale 1914). He also helped to popularise aquarium keeping and the breeding of aquarium fishes, writing a brief guide (Gale 1915). Gale sent a copy to Waite, who commented in his diary 'Received Gale's book on Aquaria. Written in short irritating sentences.'

Thetis cruises

Early in 1898 there arose an opportunity from which Waite profited through a series of more or less fortunate circumstances. Frank Farnell, M.P. instigated an investigation of the 'Offing Fisheries' of New South Wales. He was made superintendent of the venture, which employed an experienced trawling master from the North Sea and Douglas Ogilby was his first choice to accompany the vessel, the *Thetis*, on experimental trawling cruises; but Ogilby spoiled his own chances by ill-judged demands. Next, the Civil Service Board asked the Australian Museum Curator, Etheridge, to supply a naturalist. Hedley was the obvious choice, having had considerable experience with Museum expeditions, but he was honeymooning in Tasmania; Etheridge asked Whitelegge and Waite if either would go, and Whitelegge declined. Waite discussed the situation with his wife; Rose was nervous about being left alone but they decided that the opportunity should not be lost. Once the decision had been made, Waite became enthusiastic, but found that the Civil Service Board had given '... no information as to what was required, duration of expedition & probable cost ... so Etheridge wrote an adverse minute for the Board [of the Australian Museum] which sat today & he took the opportunity of emphasing [sic] the fact that I could ill be spared, vetoing my chances as I thought.' (WD) However, the information was forthcoming and Waite was told to prepare. The process was then obscured by Ogilby's efforts to be reinstated. His chief supporter was the politician, Dr Creed, who waged a considerable war on his behalf through the press and the administration. However, on Friday 18th February, 1898: 'Punctually at 4 pm the Thetis moved away from Circular Quay' (WD, 18.2.98) with Farnell, Waite and reporters from several Sydney newspapers aboard on the first of a series of cruises which lasted nearly two months in all.

The *Thetis* was not designed for trawling and Waite observed that he had little space in which to work. He had a variety of gear, including fine mesh nets to pull through the surface waters and a small dredge with which he hoped to supplement the trawled specimens. Rough weather was often encountered and the heavy seas and rocky sea bottom, previously uncharted, caused serious damage to the gear. The first day was an indication of the difficulties that beset the excursions. Waite was obliged to come up on deck in the middle of the night to help secure his gear when the *Thetis* began rolling in the heavy seas. Next morning, working north from Sydney:

Waite's drawing of Her Majesty's Colonial Steamer *Thetis*, used for experimental trawling in New South Wales in 1898. From Waite 1899.

> All on deck at 5 am with a promise from the Captain of 'dirty' weather. Left Broken Bay an hour later & put down the trawl … The sea was lumpy & I ran out a tow net astern on a shark line, the rough sea broke the line & set the net free. Next got out the log line & sent out my second iron ring, this time getting a good haul of Copepods & other small organisms. A run was next made for the 30-40 fathoms, shortly after the trawl was lowered the vessel began bumping & soon came an unmistakable signal, the net was fast. The engines were stopped and after some manoeuvring a clearance took place very suddenly & the winch was got into play & rapidly hauled the two steel warps aboard. At length the otter-boards appeared & it was a relief to find that some of the net, at any rate was attached, one pocket was full of fish, the other empty, indicating damage. One of the otter chains had snapped & the net was badly damaged.

The net was brought ashore for repairs and the next day they were trawling again with encouraging results. Waite noted:

> I was kept very busy with specimens brought up by the trawl. Still going north we worked off Tuggerah Lakes. In 3 hauls we got consecutively 745, 530 & 750 fish, and worked night and day & whenever the trawl was lifted I was called for attendance & while sorting out the immense mass of material, the reporters & others could sleep if they wished.

After six days, the *Thetis* returned to Sydney, then a day's work at the Museum for Waite, and back to the boat for the second cruise. This was the pattern; trawling was carried out at 59 stations between the Manning River and Jervis Bay, and offshore reefs were fished with lines. Off Port Stephens, Waite tried dredging, but the catch was meagre and the dredge was 'twisted & broken beyond recognition.' (WD)

Lord Howe Island, first visit

The last cruise of the series was to Lord Howe Island where the weather proved too boisterous for trawling. Waite, Farnell, Etheridge and some others were landed on the island but the weather deteriorated and they were stranded for 11 days. This was an unexpected introduction to Lord Howe for Waite who regretted the lack of adequate collecting and preserving gear during the stay. However, the party carried out some collecting on the reef and used their seine net in the lagoon, making a collection of fishes. In the preliminary report on the *Thetis* venture, Waite listed these Lord Howe Island fishes, providing some associated notes. He planned to return, better prepared.

Thetis reports

Farnell, who was concerned to demonstrate the potential for commercial trawling around Sydney, prepared a report on the economic implications of the trawling results; to accompany this, Waite (1898a) wrote a *Report on the Fishes*. He had about ten weeks for this, but during that time he was also occupied with many other matters: the snake book for Shine; at least 11 articles on Australian mammals for *Town and Country Journal*; continuing work on the Central Australian native mice; modification of his photographic equipment. His report ignored recent developments in fish classification, relying on Günther's system, and the descriptions and references were incomplete. Waite did not do justice to the material and his work was severely criticised by Ogilby and, more significantly, by David Starr Jordan, the leading North American ichthyologist of the day.

Waite defended his report merely by stating that it was a preliminary account, and then set about revising it. He sought advice from Ogilby and consulted all of the relevant literature that he could find. This marked his realisation of the resources and effort that were needed to produce a valuable publication on Australian fishes. Illustration was one of Waite's strengths, so he set out to draw many more of the *Thetis* fishes; the initial report illustrated 12 species, the final report showed nearly 40. The descriptions were detailed and informed by the literature, and the classification was no longer bounded by Günther's work.

The *Scientific Results of the Trawling Expedition of H.M.C.S. 'Thetis'* included papers by Hedley and Whitelegge on various invertebrate groups; Waite wrote the general introduction (1899e) and the report on fishes (1899f). In the introduction he gave a history of the 'marine investigations of the Colony', patched together skillfully from published reports of expeditions that had visited Sydney, from newspaper accounts and state government records. He essayed some taxonomic exercises in proposing a new generic name, *Pterygotrigla*, for a Gurnard whose scientific name was preoccupied, having been used for a genus of beetles; and in establishing a new genus,

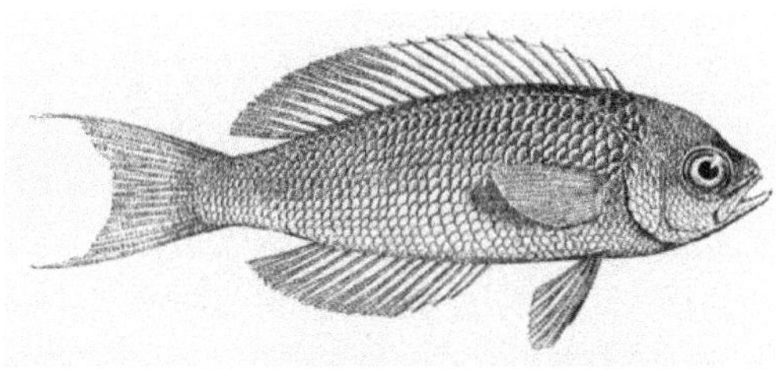

'One of the few Thetis fishes that bothered me, I determined to bottom today. I have up to now regarded it as a Labroid but found it to be one of the Pomacentridae & a new genus & species for which I propose the name Thetidosia typica, Thetis fish.' (WD) But he had not quite reached 'bottom', as acknowledged in a later marginal note: he found the fish to be a species of *Callanthias*, thanks to Ogilby. The species is now *Callanthias australis*, as it was re-described and named by Ogilby in 1899. From Waite 1898.

Paratrachichthys, for the Sandpaper Fish, *Trachichthys macleayi* (Johnston, 1881) which he noticed had a distinctive anterior placement of the vent. Nine new species were proposed, most of which are still regarded as valid. Several rare fish were re-described and, in all, 108 species were recorded. The final publication was well received by Ogilby, Jordan and others, and this represented Waite's acceptance as an ichthyologist.

An assistant for Waite: Allan McCulloch

While Waite was completing his preliminary *Thetis* report, a young man was introduced to him, a friend's friend's nephew, 'with the idea of having the latter at the Museum as he is said to have 'a bent' in Natural History.' (WD) This was Allan McCulloch whom Waite had Etheridge appoint as a cadet. McCulloch worked for Waite within the Museum and proved to be exceptionally talented. Waite was practising ichthyology at a professional level by this time and there was much that McCulloch could learn from him, including scientific illustration, photographic techniques, collection management, and general zoology. Waite enjoyed instructing and McCulloch, an apt student, was soon contributing to the work. McCulloch was also included in the Waite family's activities, often visiting the Mosman house to use Waite's home photographic laboratory and participating in outings and field trips.

After McCulloch had been his assistant for 16 months, Waite wrote to the Museum Board and enlisted Etheridge's support to try to have him paid. His first remuneration was £5 for indexing the *Thetis* report and other small payments followed, but his status remained that of an unpaid cadet.

The literature of ichthyology

The study of fishes around 1900 was at a much less advanced stage in Australia than in Europe or North America. However, the relevant literature was extensive and familiarity with these older publications was essential if real advances were to be made. Ogilby had grasped this situation several years before Waite arrived in Sydney and felt that Waite's relatively naive early publications were undoing some of the good work that he, Ogilby, had performed.

For the second *Thetis* report Waite had been obliged to consult the old publications in detail. Always a bibliophile, he began to assemble a major library of fish literature. There were several methods at his disposal and he utilised them all. He purchased such works as were available in the Sydney second hand shops; it was surprising how helpful these were at times. In this way, he gradually assembled a full series of the *Proceedings of the Linnean Society of New South Wales*, duplicates of which he used for exchanges. When Dr George Bennett's library was sold, Waite appeared and made off with some useful items. He traded works with his contemporaries at the Australian Museum, obtaining useful books from those who had an interest in fishes, particularly Whitelegge and Hedley. He wrote to de Vis asking for his old papers and was delighted when Edward Ramsay, the old Curator, invited him to choose copies of some of the works from his own extensive collection. The Museum library contained many duplicates and Waite offered some of the collections that he had brought from England in exchange; by this means, he obtained some important volumes.

For instance, he offered an insect collection to the trustees who, taking Etheridge's advice, agreed to trade:

> At yesterday's Board meeting my exchange, insects for books was passed at value £10. ... Selected following books from Museum for my exchange.
> Voy. Novara. Vertebrates [Kner's *Voyage of Austrian Frigate, 'Novara'*, 1864-65]
> Balfour's Embryology. 2 vols.
> Supp. Macleay's Fishes. [supplement to Macleay's *Catalogue of Australian Fishes*, 1884]
> Hutton's Fishes of N.Z. [1872]
> Woods' Fishes of N.S.W. [Reverend Tenison-Woods, 1882]
> 24 plates Bleeker's Atlas. [plates for Pieter Bleeker's Atlas of Ichthyology]
> Voy. D'Astrolabe. [French voyage of 1834] (WD, 1899)

By any standards he had done well. Since his trip to Germany ten years earlier, Waite had been in contact with Friedlander's of Berlin, a reputable second hand bookshop, and he now applied to them for old fish books. The response amazed him:

> Received letter from Friedlanders offering exchange value 300 mks for duplicate books etc. They sent me
> Bonnaterre... 3/- [*Dictionnaire encyclopédie méthodique*, 1782-1832]
> Gouan........ 6/- [*Historia Piscium*, 1770]
> Klein.......... 9/- [*Historia Naturalis Piscium*, 1740-49]
> Broussonet.. 4/- [*Ichthyologia sistems piscium descript. et icones, decas I*, 1782]
> and 5 Australian papers.......4/-
> ... The prices seem to be ridiculously low for such old works. They will send Bloch & Schneider [1801] shortly. (WD, 1902)

He dealt many times with Friedlander's who often discovered rare works for him at short notice, including Johann Forster's *Descriptiones Animalium*, 1844. (WD, 1903)

When Ogilby left Sydney in disorder and confusion in 1901, he had pawned many of his personal effects, including his books (and some books belonging to other people as well). Waite was contacted by the dealer who wanted advice on the goods. He did not resist the temptation to buy back the books. The diary entries give the impression that he was originally doing this for Ogilby's ultimate benefit, but the books and papers stayed in Waite's possession. The collection must have been considerable as Waite worked several nights making a catalogue of the books and papers, and discovered books belonging to other colleagues, which he duly returned. It was probably at this time that Waite came into possession of some of Ogilby's private papers, including the 1895–96 diary (see page 184) which contained material of a very personal nature, and which he also retained.

Waite kept abreast of the progress of ichthyology in his day in the usual way: he established exchanges of 'separates' (separate reprints of papers made available to authors, for their personal distribution, by the publishers of journals) and other publications. The Americans were almost embarrassingly generous and he received some lavish and valuable productions from David Jordan, his associates and the U.S. Fish Commission; from Samuel Garman; from Bashford Dean; and from Henry Fowler. A typical package arrived in 1902:

> Received from Jordan
> 1. Fishes of Formosa.
> 2. Trigger Fishes of Japan.
> 3. Blennioid Fishes [of Japan]
> 4. Gobiesocidae [of Japan] (WD)

Boulenger, at the British Museum, saluted Waite's enthusiasm for ichthyology with 'a mass of reprints'. (WD) He corresponded with ichthyologists in France, Holland, Germany, Austria, Italy, Denmark, South Africa, India, Japan, Chile, Brazil, Uruguay and New Zealand. Waite was a lively correspondent, frequently offering constructive criticism of the publications he received, and responding with his own publications whenever possible. Substantial remnants of Waite's efforts may be found in the South Australian Museum, where much of his library has been preserved.

A natural extension of this trade in books and papers was a trade in specimens. Individual troublesome fishes were dispatched to the interested specialists for advice, others were sought for comparative purposes; from Jordan, in particular, whole collections from various regions were sent as exchanges for Australian fish. Jordan recommended to Waite a seaman, Loudin, of the *Sierra* and through him messages and collections were exchanged, thus avoiding freight charges.

Increasing output in ichthyology (from 1899)

The excellent reception of the final *Thetis* report encouraged Waite to focus his efforts increasingly on the study of fishes. He soon enlisted Etheridge's assistance in ensuring that he had access to all of the fish collections that were sent to the Museum; this may have contributed to Ogilby's decision finally to move from Sydney. Waite's confidence increased, as seen by his handling of the occasional fish specimens of special interest. Within months of the *Thetis* publication, Waite had published several short articles: he noted the first New South Wales occurrence of an Oarfish, a specimen over four metres in length (Waite 1899b) found inside Port Jackson; a Leopard Shark, *Stegostoma tigrinum* from the Hawkesbury River (Waite 1899a); a fish from Funafuti, rumours of which had puzzled Hedley: the 'Palu', (Waite 1899d); and a magnificent Opah which had been taken in New Zealand waters (Waite 1899c).[19]

Aside from new records of this kind, Waite began some numbered series of publications which he continued as appropriate specimens came to hand over the remainder of his time at the Australian Museum. These were *Additions to the Fish-Fauna of Lord Howe Island*, *Notes on Fishes from Western Australia*, *Studies in Australian Sharks* and *New Records or Recurrences of Rare Fishes from Eastern Australia*. Although there was little continuity in these publications, they did loosely define Waite's current fish studies, and they represented real advances in the understanding of the Australian fish fauna.

---

19 Oarfish, *Regalecus glesne* Ascanius, 1772; The Leopard Shark is now called Zebra Shark, *Stegostoma fasciatum* (Hermann, 1783); Palu is the Oilfish, *Ruvettus pretiosus* Cocco,1833; Opah, *Lampris guttatus* (Brunnich, 1788).

Western Australian ichthyology and Bernard Woodward

The fishes of Western Australia had been very little studied up to 1900. The early voyagers had collected some species, particularly in Shark Bay. John Richardson had reported fish sent to him from King George Sound and the north-west. Count Castelnau had given a list of dubious value in 1873 and Arthur Lea, then the government entomologist, had sent specimens to Ogilby that he had described in 1899. Ogilby had also initiated correspondence with the Curator of the Western Australian Museum, Bernard Woodward; but Ogilby, who was drinking heavily and expecting to move to Brisbane, was relinquishing his interests in New South Wales and Waite was taking them over. The Lord Howe Island collections now came to Waite, and Woodward sent him a small collection from the Swan River. Waite (1900a) used the new collection to verify a few of Castelnau's old records, and extended the list. Oddly enough, he overlooked Ogilby's paper (1899), in which the latter recorded eight Western Australian species, but Waite otherwise made good use of the references available to him. The collection featured a Knifejaw, *Hoplegnathus* [now *Oplegnathus*] *woodwardi*, the first of its kind described from Australian waters. Waite's work on this fish was thorough, but he was more than ever aware of the lack of important references, expressing his frustration:

> In common with other Australian workers, I have at times referred to the difficulty experienced by zoological writers living at prohibitive distances from European literary centres, and the hopelessness, in many cases, of bringing an undertaking to a satisfactory conclusion. Such disabilities are caused by a lack of necessary literature, and many are the instances in which a train of research has to be abandoned owing to the impossibility of consulting some particular paper.

This complaint was provoked by his having no way of obtaining certain articles, relevant to the Knifejaw, published in journals of Austria, Holland and Denmark.

Bernard Woodward (1846–1916) had been in Western Australia since 1889, holding the positions of Government Analyst and Curator of the Geological Museum. The latter institution was reconstituted as the 'Western Australian Museum and Art Gallery' in 1897 and, there being no applications deemed suitable from outside the state, Woodward was appointed as Director. He was committed to the arts and to natural history, and was an active advocate for the Museum and for the establishment of reserves to protect the native plants and animals (ADB). His frustration at official indifference was evident in his postscript to a letter addressed to Waite:

> My staff is still limited to one clerk notwithstanding the growth of the Collection but to the Minister of Education (vide paper) a Library is a luxury, a Museum a toy & they are of no educational value. He has never been over the Museum. (Woodward to Waite, 17.6.02, SAMA)

In the Western Australian Year-Book for 1900–1901 he presented a list of just over 100 species of fish recorded from the state, generally with a more or less broad locality when known; he used the scientific names and supplied the common name when it was available.

In 1902 Woodward dispatched a second small collection of fish upon which Waite reported, preparing beautiful illustrations of those that had not previously been figured (Waite 1902b). Yet another collection came from the west in 1905, supplemented by some trawled specimens from the Chief Inspector of Fisheries, Mr. C.F. Gale. These had been taken in some recent trials, the first trawling undertaken in Western Australia. Waite acknowledged the specimens but was critical of the trials, 'The services of a professional Zoologist, in an undertaking of this kind, should be recognised as a necessity in Australia equally with Europe and America.' The trawling had been performed by a small sailing vessel, which had worked from Cape Naturaliste to Shark Bay, and off the Houtmans Abrolhos. In this paper (Waite 1905b) nine species were described as new and, remarkably, eight of these are still regarded as valid. Allan McCulloch was contributing significantly to Waite's publications at this time, preparing some of the descriptions and locating references. His artistic talent can be judged by the excellent quality of the illustrations in this paper. It included a 'new' fish, named for him by Waite, 'I associate with the species the name of Mr. A.R. McCulloch to whom I owe the figure of this and the other species illustrating this paper.' Waite called the fish, a small Leatherjacket of peculiar form, *Chaetoderma maccullochi* [synonym of *Chaetodermis penicilligera* (Cuvier, 1816)]. Other species in this paper were dedicated to Woodward (Museum Curator) and C.F. Gale (Chief Inspector of Fisheries in Western Australia). Among those McCulloch drew for Waite was a leatherjacket from Shark Bay: *Pseudomonacanthus* [now *Meuschenia*] *galii*, named for Gale.

Waite named the Western Pomfred, *Bramichthys* [now *Schuettea*] *woodwardi* from a single specimen that was among the fishes sent by Bernard Woodward. McCulloch's drawings illustrated all of the new fish in this paper. From Waite 1905b.

McCulloch's increasing contribution

The scientific illustration which Waite regarded as essential for his publications was very demanding on his eyesight. He had been experiencing increasing difficulty with close work and, by early 1900, found that he had to curtail it. Up to that time, he had illustrated papers for many of the Museum staff, including Whitelegge's Crustacea and Sponges for the Funafuti report. In February 1900, Waite writes 'Whitelegge asked me to make drawings of the *Thetis* Crustacea, but I declined owing to the strain on the eyes.' Within 12 months, he was rationing the drawing for his own papers, 'Commenced drawing in line of Chaetodon tricinctus but my eyes will not stand the strain.' (WD, February 1901) In the same month, 'I am now giving Allan lessons in drawing.' The results pleased them both.

Shortly after McCulloch began drawing fishes, Waite tried the Board again, this time suggesting that McCulloch be 'assistant to the Curator', but it was not until January 1902 that he was finally employed as a 'Mechanical Assistant' at £50 per year (with a retrospective payment of £25 for the previous six months). Shortly before this success, McCulloch had won a prize (21 shillings) for his drawing and had asked Waite's advice on spending it; Waite wrote 'I selected books for him "Günther's Study of Fishes" 15/- and Natural History Vol. of Concise Knowledge Library.' (WD)

Need for a New South Wales fish catalogue

New South Wales lacked a usable catalogue of fishes. Macleay's catalogue (1881), with its supplement (1884), was the basis for Ogilby's 1886 catalogue which listed 498 species of fish from New South Wales. These works provided certain names and references but could not be used to identify the fishes caught in the state. Government backing for Ogilby had not produced a useful result and in June 1900 Waite was asked by Doctor Cox (see page 134), on behalf of the Fisheries Commission, to write and illustrate a book on the fishes of New South Wales. He agreed, but no funds were forthcoming; he was approached again in November, the Fisheries Commission asking for a fish list for the state. He conferred with Ogilby and Etheridge, and it appeared that a joint production might be possible, but once again the funds were not committed. Waite was approached again informally in 1900; he was participating in an inspection of the trout hatchery, then located in the National Park south of Sydney:

> Brodie [Chief Inspector of Fisheries] told me that by some mistake a sum had been put on our estimates for a book on fishes. He intends to ask Cox [the Museum Trustee with most interest in fishes] to apply for this to remain & ask me to prepare a semi-popular work. (WD)

Waite heard no more for some time but began the work,

> At night commenced to digest P.L.S. NSW[20] with view to preparing Catalogue of Fishes of N.S.W. (WD, 11-3-00)

20 Waite here referred to the many fish papers that had been published in the *Proceedings of the Linnean Society of New South Wales*, using 'digest' in the precise sense: 'reduce into a systematic form, usually with condensation.' (OED)

He was preparing to tread the same path that Ogilby had attempted, and started his literature survey in much the same way; Ogilby had prepared indices to the fish papers of the same journal in 1895, and had enlarged the project, making an index for each of the journals and authors whom he considered significant to Australian ichthyology up to that time. Waite continued to work on the fish catalogue in his spare time and in August 1902 Sinclair, the Secretary at the Museum:

> ... told me that at yesterday's Board Meeting of the Fisheries Commission it was decided to put £250 on estimates for fish-book. The last twice such a movement was made the Minister threw the scheme over & history repeats itself. (WD)

He was right: financial support for the catalogue was never guaranteed, and Waite eventually let the work lapse.

Waite did prepare a considerable refinement of the basic list but was initially frustrated by difficulty in having even this published; it finally appeared in the *Memoirs of the New South Wales Naturalists' Club* as a *Synopsis of the Fishes of New South Wales* (1904e). His list had a reference to the original description of each fish, and a supplementary reference indicating where an illustration could be found. 526 native species were listed, including the new records that had been established since 1886 and excluding records that seemed doubtful. The classification adopted was that used by George Boulenger in the *Zoological Record* for 1902. Waite wrote '... the space limit placed upon the publication precluded the possibility of giving any synonymy.' This was a little disingenuous as he had certainly worked out detailed synonymies for no more than a small proportion of the species listed. He prepared some manuscript forming the basis of a larger work on New South Wales fish but it was never published; it is held in the South Australian Museum Archives.

Waite's sharks

Their reputation and their size have always ensured that sharks and rays attract public and press attention, and many publications have been devoted to them. Waite wrote a series of popular articles for the *Town and Country Journal* in New South Wales on sharks; he also wrote occasional scientific papers, in particular, those entitled *Studies in Australian Sharks* between 1901 and 1903 (with an afterthought in 1922). Prior to this he had dealt with the Port Jackson shark egg-cases and, in the first of the 'Studies', continued his interest in reproductive processes in sharks: he noted viviparity in the so-called Blind Shark, *Hemiscyllium modestum* [now *Brachaelurus waddi*], and illustrated the cartilaginous knobs on the spines of foetal Dogfish, *Squalus megalops* which protect the female from damage during parturition (Waite 1901). This theme was extended in his next paper (Waite 1902a) in which the foetus, yolk sac and uterine membranes of the Gummy Shark, *Mustelus antarcticus* were described and illustrated. William Haswell, then Professor of Biology at Sydney University, had studied sharks intermittently over 20 years and Waite obtained his advice before publishing these papers. The third paper in the series (Waite 1903) included an attempt to assign different forms of egg case of oviparous catsharks to their species.

Bloch described the Spotfin Lionfish, *Scorpaena* [now *Pterois*] *antennata* from a specimen sent from Ambon, Indonesia. From Bloch 1787. Courtesy Museum Board of South Australia.

Bloch and Schneider described the Moonfish, *Zeus* [now *Mene*] *maculata* from an Indian specimen. From Bloch and Schneider 1801. Courtesy Museum Board of South Australia.

The Peacock Rockcod, *Cephalopholis argus,* another common northern Australian fish, was described from a Japanese specimen. From Bloch and Schneider 1801. Courtesy Museum Board of South Australia.

The Southern Roughy, *Trachichthys australis* was described and illustrated by George Shaw from a Port Jackson specimen. It is confined to southern Australia. SLSA: Special Collection 574.9 S534, 1789–1813 – Vivarium naturae … The Naturalists' Miscellany, Volume 10, Plate 738, image courtesy of the State Library of South Australia.

The Spiny-tailed Leatherjacket, *Acanthaluteres brownii* (Richardson, 1846) was caught at Princess Royal Harbour, King George Sound, Western Australia in December, 1801, and illustrated by Ferdinand Bauer. From the Zoological Drawings of Ferdinand Bauer, plate 30, V3934/R, Courtesy of the Natural History Museum.

The Shark Mackerel, *Thynnus* [now *Grammatorcynus*] *bicarinatus* (top), described from a Shark Bay specimen by Quoy and Gaimard. These illustrations also show the Tailor, *Pomatomus saltatrix* (Linnaeus, 1766) (centre) and the Rainbow Runner, *Elagatis bipinnulata* (Quoy and Gaimard, 1825) (bottom); both are well known Australian fish, but are here illustrated from specimens taken elsewhere in the Indo-Pacific. SLSA: Rare Book Room 508.3d++-Zoologie/par J.R.Quoy et J.P. Gaimard (folio atlas), 1824, Plate 61 Poissons, image courtesy of the State Library of South Australia.

Plate 19 of the Voyage of the Astrolabe shows (from top to bottom): Birdnose Wrasse, *Gomphosus caeruleus* Lacépède, 1801 collected at Mauritius; Longray Weed Whiting, *Malacanthus* [now *Siphonognathus*] *radiatus*, described by Quoy and Gaimard from a drawing of an Adelaide specimen; a wrasse, *Julis gibbifrons* (synonym of *Coris aygula* Lacépède, 1801) from Mauritius; a Mimic Blenny, *Aspidontus taeniatus*, described from a Guam specimen but common on northern Australian coral reefs. SLSA: Rare Book Room 508.3d++-Zoologie/Mm. Quoy et Gaimard. (Atlas), 1833, Plate 19 Poissons, image courtesy of the State Library of South Australia.

The Soldier Fish, *Apistus* [now *Gymnapistes*] *marmoratus* was described by Cuvier in 1829 from a Tasmanian specimen. From Griffith and Smith 1834. Courtesy Museum Board of South Australia.

Richardson described the Barber Perch, *Serranus* [now *Caesioperca*] *rasor* from Lemprière's collections and illustrated this fish in a later paper (Richardson, 1841b) where he also showed a juvenile Jackass Morwong, *Nemadactylus macropterus*. He described the latter as a new species, *Nemadactylus concinnus*; it was 8 cm long and still retained some larval features. From Richardson 1841. Courtesy Museum Board of South Australia.

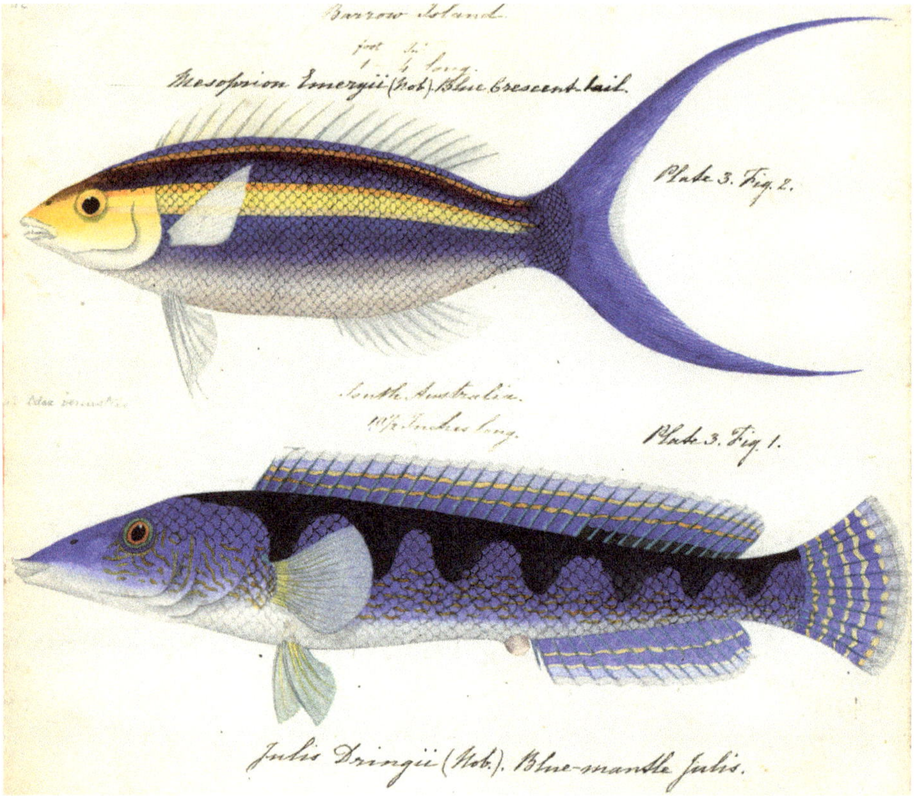

Emery's drawings 42 and 43 from Barrow Island illustrate the magnificent Purple Threadfin Bream, *Mesoprion* [now *Pentapodus*] *emeryii* and from Safety Bay, south of the Swan River, the Blue Weed-whiting, *Julis Dringii* from South Australia. Richardson described both as new in his *Icones Piscium* but the latter is a synonym of *Haletta semifasciata* (Valenciennes, 1840). The earlier writer, whose annotations are dim but discernible, suggested the name *Odax venustus* for this fish. SLSA: Rare Books Room 597.0994 E53d+++ - Sketches of Australian fishes, etc./by Commander J.B. Emery, Nos. 42 and 43, image courtesy of the State Library of South Australia.

Bleeker described the Steephead Parrotfish, *Scarus* [now *Chlorurus*] *microrhinos* in 1854 from Jakarta specimens and illustrated it in his *Atlas Ichthyologique* (1862). This species is common around northern Australia, occurring from Sydney to Rottnest Island. From Bleeker 1862.

Emery's drawings 10, 11 and 12 from the Abrolhos Islands showing a large Silver Toadfish, *Lagocephalus sceleratus* (Gmelin, 1789), the Sergeant Baker, *Aulopus* [now *Latropiscis*] *purpurissatus* and an unidentified labrid. Richardson's annotations indicate his choice of *Aulopus* for the *Icones*. SLSA: Rare Books Room 597.0994 E53d+++ - Sketches of Australian fishes, etc./by Commander J.B. Emery, Nos. 10, 11 and 12, image courtesy of the State Library of South Australia.

Emery's drawings from Turtle Island, 'North West Coast': the top drawing may represent a juvenile Indonesian Snapper, *Lutjanus bitaeniatus*; centre, a juvenile Brownbarred Rockcod, *Cephalopholis boenak* (Bloch, 1790); below, a Speckled Catshark, *Hemiscyllium trispeculare*. The earlier writer, who numbered the drawings, sometimes suggests names, in this instance *Squalus ocellatus*. Richardson described this shark in the *Icones*. SLSA: Rare Books Room 597.0994 E53d+++ - Sketches of Australian fishes, etc./by Commander J.B. Emery, Nos. 39, 40 and 41, image courtesy of the State Library of South Australia.

Günther described the Harlequin Tuskfish, *Xiphochilus* [now *Choerodon*] *fasciatus* from small, dried specimens from Cape York, Queensland. From Günther 1867. Courtesy Museum Board of South Australia.

George French Angas drew a juvenile White Ear, *Parma microlepis* Günther, 1862, a very common fish in Sydney Harbour. From Bennett 1859. Courtesy Museum Board of South Australia.

Colour Plate 10

The Common Seadragon, *Phyllopteryx taeniolatus* drawn by G.H. Ford for Günther. From Günther 1865. Courtesy Museum Board of South Australia.

Pimelodid catfish from Castelnau's 1855 report of his journey along the Amazon: Günther ignored Castelnau's illustration (the lower figure) of the Redtail Catfish, *Phractocephalus hemiliopterus* (Bloch, 1801). The upper image, labelled *Arius*? *longibarbis*, has not been identified with any known species to date. From Castelnau 1855. Courtesy Museum Board of South Australia.

Female Sergeant Baker, *Aulopus* [now *Latropiscis*] *purpurissatus* (Richardson, 1843) from Hobson's Bay, Victoria. From McCoy 1881.

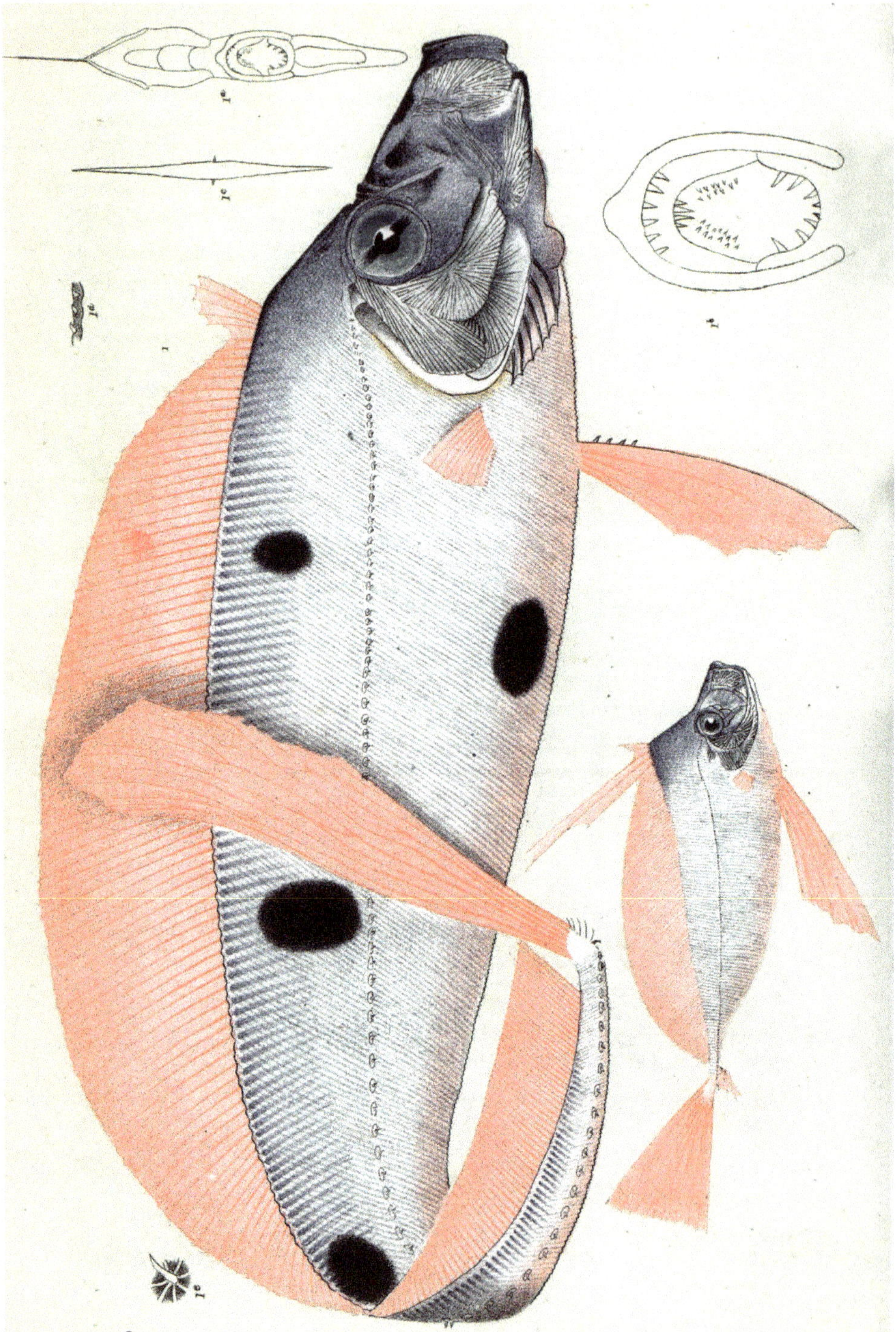

Juvenile Southern Ribbonfish, *Trachipterus taenia* [synonym of *Trachipterus jacksonensis* (Ramsay, 1881)], showing specimens 35 cm and 9 cm in length from Portland, Victoria. From McCoy 1886.

Waite noted in his diary for 22 August 1893 'At the Museum I made water colour drawing of a Shark (*Parascyllium collare* R & O), the only specimen known here besides the type.' The drawing is annotated by Waite: 'Parascyllium collare. ♂. R & O. (from fresh specimen Registered No. I.3153). Edgar R. Waite del. 21. Aug 1893. 1/3. off North Head Sydney 50 fathoms.' From AA356 Waite Collection, courtesy of the South Australian Museum Archives.

The Rainbow Fish, *Heteroscarus acroptilus* (Richardson, 1846) of southern Australia. Waite drew and painted it several times in Sydney and in Adelaide and used it as the frontispiece of his fish catalogue. From AA356 Waite Collection, courtesy of the South Australian Museum Archives.

Colour Plate 14

Whitley (1929b) published a paper with many elements that comprised a memorial to Allan McCulloch. It was illustrated in part by colour plates of McCulloch's beautiful paintings. One of these plates shows the Malayan Eel Blenny, *Congrogadus subducens* (Richardson, 1843); a juvenile Senator Wrasse, *Pictilabrus laticlavius* (Richardson, 1839); the Halfbanded Seaperch which Whitley here named *Ellerkeldia* [now *Hypoplectrodes*] *maccullochi*; and the Bandfish, *Cepola australis* Ogilby, 1899. The Eel Blenny was collected at Darwin, the other fish near Sydney; the Senator Wrasse was painted in 1921, the others in 1902. Whitley also used the occasion to propose *Maccullochella* as a necessary generic replacement name for the Murray Cod. From Whitley 1929. Courtesy Museum Board of South Australia.

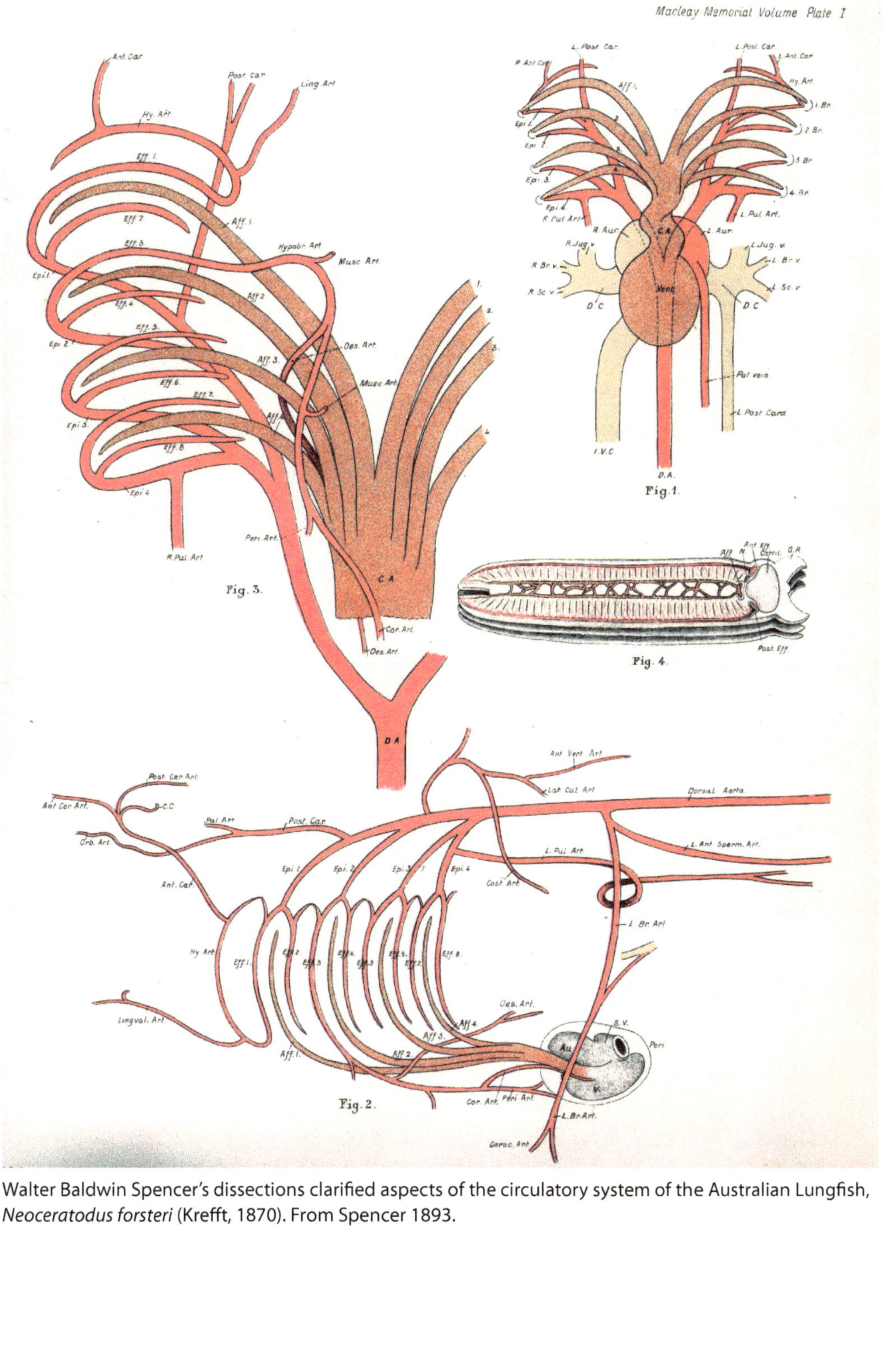

Walter Baldwin Spencer's dissections clarified aspects of the circulatory system of the Australian Lungfish, *Neoceratodus forsteri* (Krefft, 1870). From Spencer 1893.

The Freshwater Catfish, *Tandanus tandanus* (Mitchell, 1838) was among the fine illustrations prepared by Roughley for his *Fishes of Australia and their Technology*. From Roughley 1916.

The Bluefish, *Girella cyanea* Macleay, 1881 illustrated in Roughley's *Fishes of Australia and their Technology* (1916). From Roughley 1916.

Waite took a particular interest in the foetal stages of sharks, illustrating the Spiky Dogfish, *Squalus megalops* (Macleay, 1881) (left) and the Gummy Shark, *Mustelus antarcticus* Günther, 1870 (right). From Waite 1902.

Lord Howe Island re-visited

Lord Howe Island, 630 kilometres off the coast of New South Wales, has been of special interest to zoologists of the Australian Museum since their first visit in 1887. Its marine environment includes well developed coral reefs at a uniquely high latitude and the associated fauna has distinct tropical and temperate components. Waite's first stay on the island was incidental to the *Thetis* trawl venture and he planned a second visit in 1902. He applied for extended annual leave, intending to combine a family holiday with an ambitious collecting expedition, and requested permission to take Allan McCulloch as his assistant (WD, July 1902). As usual, there was a battle with the administration: Waite and McCulloch should not be absent simultaneously. Waite wrote to the trustees, recruited Etheridge to his cause, and McCulloch was eventually granted leave for the trip.

Waite's preparations were extraordinarily thorough; he was in his element organising an expedition. He obtained permission from the Fisheries Department for the use of poisons and explosives for the purposes of fish collection at Lord Howe Island. He wrote to David Jordan for advice and Jordan, jubilant at the time over the enormous collections he had made in Samoa, replied:

> ... this is the first time a coral reef ever received such an overhauling ... we used dynamite continuously both in Apia and Pago Pago ... The use of poisons was also vitally important. We have tried many of these, but the best and cheapest is a solution of commercial 'chloride of lime'. (Jordan to Waite, August 1902, SAMA)

Whitelegge, McCulloch and Waite experimented with lime in the rock pools at Maroubra and had encouraging results. He also received instructions in the use of the explosives from the Works Department (but did not practise around Sydney).

The party, consisting of Edgar and Rose, with their young son Claude, and McCulloch, embarked on December 1st 1902 for the island. McCulloch was then 17 and had been Waite's assistant for four years. They planned to stay one month but rough weather prevented the boat from collecting them and the visit was extended to two months. Waite and McCulloch were active throughout, sketching and photographing, marching through the bush, walking the beaches, wading on the reef and collecting fish whenever possible.

They inspected the beaches for stranded fish, line fished from the shore and from small boats, set fish traps and nets, poisoned and dynamited the rock pools. They walked to the south end of the island: 'Our quest was for freshwater fish in Soldier's Creek but we found none.' The lime was very effective but the dynamite, although spectacular, was disappointing: 'We put dynamite into a large rock pool on 3 occasions but it proved a failure 2 or 3 fishes only being turned up.' The island residents, who for many years had contributed fishes to the Museum, were keenly involved, presenting Waite with jars of fishes, advising and helping with the collecting. Some of Waite's diary entries for a few days after 30 December, 1902 show the theme of the trip, barely interrupted by the New Year's celebrations:

> ... In afternoon Allan & I to Middle beach, low tide limed pools & got a number of small fishes. After tea with the girls took dinghy out got Mullet etc but a small shark scared them away. I put out a line for the shark but Susie falling sick with the chopping motion we returned about 10.30 pm.
>
> 31. Wed. Again went fishing this time outside the reef got Epinephelus fasciatus, a beautiful fish salmon color with pink bands & the tip of the dorsal black [Blacktip Rockcod, *Epinephelus fasciatus* (Forsskål, 1775)]. Saw 2 turtles, one close to the boat. The other fish taken were common to us. In the afternoon Allan & I to Middle beach limed pools & got fish on the lines. At night we had a Cheveau, song & dance. at 12.0 all went to the lagoon beach & lit up a large fire of dead palm wood & leaves. Other fires were burned along the beach by the islanders and on the stroke of midnight guns were fired all over the island. We got to bed at 2.30 and thus we closed the year 1902.
>
> 1903.
> January.
> 1. Thurs. After attending to fish etc we went to Middle beach got fish with lime and line. Noted crabs living in Placostylus shells.
> 2. Fri. After our swim saw several Scopelus on the beach & after breakfast I walked with Allan to the S. end getting more Scopelus and three Machaerope, while he went further to Kings beach I returned & worked N to the end but got nothing of note ... At night went fishing in the lagoon but obtained Apogons only.[21]
> 3. Sat. Searched the beach got more Scopelus including 2 larger examples with luminous organs above the tail also 6 more Machaerope. In the afternoon with Mr. Potts & Susie we went to the 'Clean Place' fishing, got Cirrhitichthys splendens [see plate on page 179] among others. At night went to the Comet hole in the lagoon but did not get anything. George caught two small sharks one of which we preserved.

The specimens were carefully preserved, labelled and packed; the fish collection was enormous and made the subject of a paper (Waite 1904a).

21 Oceanic fish are frequently stranded at Lord Howe Island. Waite's *Scopelus* were Lanternfish (Myctophidae) and *Machaerope* was the Black Snake Mackerel, *Nealotus tripes* Johnson, 1865. The 'Apogons' were small Cardinalfish (Apogonidae), a nuisance when fishing as they took relatively large baits.

Lord Howe Island fish results

Adding formally to the fish fauna of Lord Howe Island provided Waite with the subject for four papers, published over as many years; the last and largest included an account of the 1902–03 visit, from which ten species were described as new and a further 17 species recorded from the island for the first time (Waite 1904a). Waite appeared as the sole author of this last paper although McCulloch contributed much of the work, including most of the illustrations. Waite acknowledged McCulloch's help, mentioned here as contention later arose from Waite's lack of acknowledgement of some of McCulloch's work.

During the visit to Lord Howe Island one of the residents presented Waite with a specimen that was not readily identifiable. It was a brown fish with yellow fins, over half a metre in length, with large eyes and impressive fangs. He could not ascertain, from the external features, to which family the fish belonged. A vertebral count was required, but the specimen was unique and Waite did not want to damage it:

> Working at fish taken by Stevens at L.H.Is. being unable to decide whether Scombridae or Carangidae. took it to Dr. Sandes [demonstrator in anatomy at the Sydney University] & had Radiographs taken of it in order to see vertebrae. (WD)

And a few days later:

> Received X-ray negatives of fish & found it to be of family Gempylidae ... Found the fish to be of a new genus which I propose to call Xenogramma carinatum ... Wrote description of genus & species of the fish & Allan made me an outline drawing and reduced it by pantograph. (WD)

Waite could be excused for missing the original account of this fish, as it had been described initially from the Cape of Good Hope in 1843 (as *Cybium flavo-brunneum*) by Andrew Smith in *Illustrations of the Zoology of South Africa*, a work to which Waite did not have access. Günther's Catalogue included the fish by name in the second volume (1860) but his description was taken from the original and only specimen, which had been stuffed, and some of the creature's most distinctive characteristics are omitted; for instance, the undulating lateral line, separate finlets behind the main vertical fins and prominent keels at

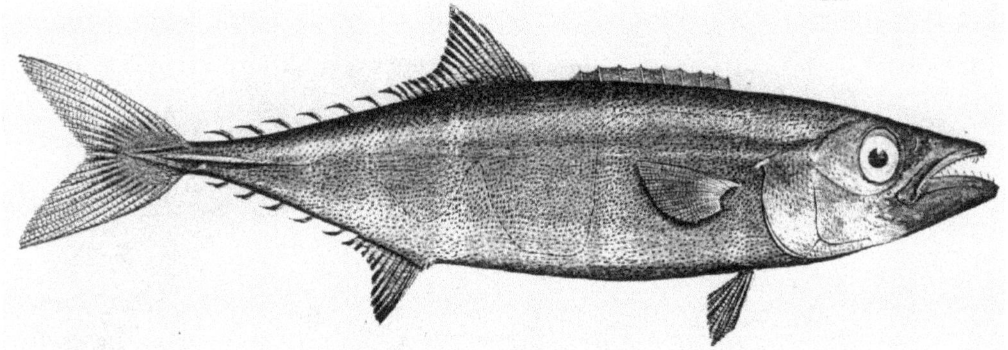

'Xenogramma': the Escolar, *Lepidocybium flavobrunneum* (Smith, 1849) from Lord Howe Island. From Waite 1904a.

the base of the tail. Theodore Gill (1862b) had published a succinct account of the group of fishes to which '*Xenogramma*' belonged but, although Waite had corresponded with Gill, he lacked many of that zoologist's early papers, probably including the one that may have set him on the path to identify his fish. However, the description and drawing in Waite's publication were excellent, so that the next comer could identify the fish with ease.

Waite's paper included original observations of value. Among the most prolific of the edible fishes of Lord Howe Island is the Bluefish, *Girella cyanea* to which Waite paid special attention:

> The Blue-fish is the staple food fish of the island and is generally caught with hand lines off the rocks; a shoal of perhaps fifty individuals or more may be seen in a deep gulch, the colour of those in the deeper water appearing of the most intense blue, while those near the surface or just drawn from the water are the colour of the sky, but the beautiful tint fades before the fish is dead. The young abound in all the rock pools but all are dusky in colour, the characteristic blue not appearing until the fish is five or six inches in length. The Blue-fish is a good species for smoking and curing, and recent consignments realised twenty-eight shillings net per hundredweight in the Sydney market. (Waite 1904a)

He described dense schooling behaviour in young Eel Catfish, *Plotosus lineatus*:

> At the north end of the lagoon are small masses of seaweed three or four feet in diameter, their fronds oscillate in the wash of the waves and at low tide are barely covered. Occasionally when wading through the water one such apparent mass of seaweed will be noticed to become active and pass slowly over the sand. After the first investigation of such phenomenon this moving mass is known to consist entirely of the striped catfish. Though the water telescope reveals the striking markings of the species, the unaided eye perceives the yellow stripes only as disconnected patches rendered irregular by the ever-present ripple on the water. The resemblance to the seaweed becomes thus so complete that I imagine great protection is obtained from the attacks of Gannets and other birds to whom the fish would fall an easy prey in such shallow water. (Waite 1904a)

Waite and McCulloch made many friends at Lord Howe Island with whom they communicated regularly and the fish records for the island increased, the residents sending Waite regular kegs of specimens. Waite produced a list of the fishes as found to 1904, 180 species being represented. This was a model paper of its kind, citing the original description, reference to an illustration where possible, and indicating the authority for the record of the species.

Rare fishes of eastern Australia

The fish fauna of New South Wales in the early 1900s still held numerous novelties for its investigators: Waite continued to report them more or less as they were brought to the Museum. Several of his papers entitled *New Records or Recurrences of Rare Fishes of Eastern Australia* formed an irregular series in which these fish were described. The origins of the specimens were various; for instance, several from catches sent to the Sydney fish market from the Richmond River district on the north coast of New South Wales; some contributed by the Amateur Fishermen's Association of New South Wales, for whom Waite sometimes gave lectures illustrated by lantern slides. Waite also had numerous specimens from his colleagues at the Museum, particularly Whitelegge and McCulloch.

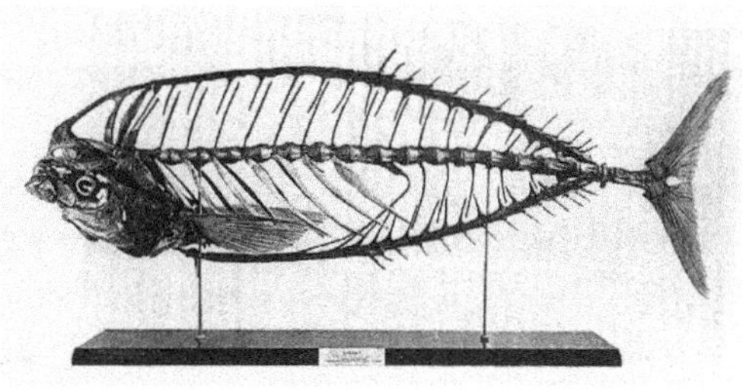

Waite supplied the first report of the Louvar (*Luvarus imperialis* Rafinesque, 1810) in Australian waters when one was cast up at Bermagui, south of Sydney. The Australian Museum 'articulator' prepared, assembled and mounted the skeleton. From Waite 1902.

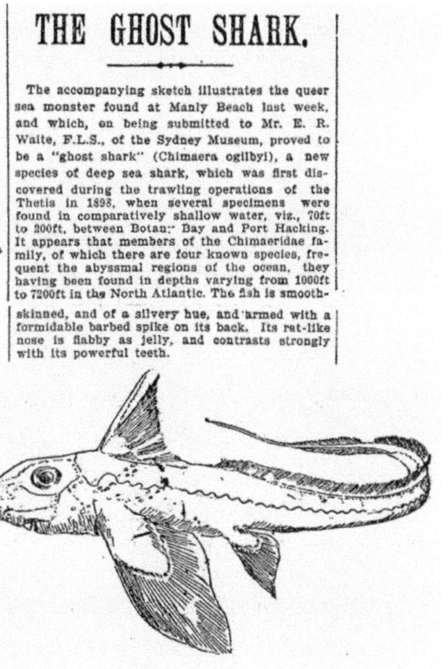

THE GHOST SHARK.

The accompanying sketch illustrates the queer sea monster found at Manly Beach last week, and which, on being submitted to Mr. E. R. Waite, F.L.S., of the Sydney Museum, proved to be a "ghost shark" (Chimaera ogilbyi), a new species of deep sea shark, which was first discovered during the trawling operations of the Thetis in 1898, when several specimens were found in comparatively shallow water, viz., 70ft to 200ft, between Botany Bay and Port Hacking. It appears that members of the Chimaeridae family, of which there are four known species, frequent the abyssmal regions of the ocean, they having been found in depths varying from 1000ft to 7200ft in the North Atlantic. The fish is smooth-skinned, and of a silvery hue, and armed with a formidable barbed spike on its back. Its rat-like nose is flabby as jelly, and contrasts strongly with its powerful teeth.

A Ghost Shark, *Hydrolagus ogilbyi* taken at Manly. Waite supplied some details and a sketch for the *Evening News* and other Sydney papers. News clipping from Waite's diary, courtesy of Duncan Waite.

In 1902, south of Sydney, there were some interesting strandings of large pelagic animals, including a three metre Basking Shark, *Cetorhinus maximus* and an enormous, silvery-pink fish. The local fisheries inspector drew a rough sketch of the fish from which Waite was able to identify it provisionally as a Louvar, a species which had not until then been seen in Australian waters. It was two metres in length and was thought to weigh well in excess of 100 kilograms. The skeleton was preserved and Waite included a description of it as well as recording the occurrence (Waite 1902c and 1902d). Waite's account benefited from his overseas contacts, Edwin Starks of Stanford University having recently sent him *Synonymy of the Fish Skeleton* (Starks 1901). Starks was one of several American ichthyologists with whom Waite was then exchanging papers.

The *Thetis* catches of 1898 had included a chimaera which Waite described as a new species, *Chimaera* [now *Hydrolagus*] *ogilbyi*. Although the species is generally an inhabitant of the waters of the continental shelf, several were caught on lines at Manly in 1902. They were discarded by the fishermen but one that was washed back on the beach was brought to the Museum. Waite used the damaged specimen to prepare a skeleton and supplied a sketch and some details for the newspapers. He was not in a position to use the press very much at this stage, due to his subordinate position in the Museum, although later in his career he placed great emphasis on his relationship with journalists.

Problems and gudgeons

In the late 1890s Waite had experienced health problems and his diary showed that he was worried. He carefully recorded his symptoms and the advice that he received from the doctors consulted (WD). The state of medical knowledge at that time was insufficient to give him assured answers to his questions, and he received poor advice. For instance, he complained of spells of dizziness and rapid heartbeat: he was diagnosed as having heart disease and a regime of medicine (digitalis and strychnine were its most impressive constituents) and rest was prescribed. The physical signs of heart disease were almost certainly absent. After taking the medicine, he began trembling, and reasonably attributed this to the medicine. He was told to limit his physical activity, but soon returned to brick-laying and carting soil without adverse effects, his instincts being of a higher order than his medical advice. On other occasions, he described severe headache followed by transient visual disturbances, consistent with a form of migraine but not recognised as such.

Glasses were prescribed when he complained of trouble with close work, but prisms were prescribed and these appear to have aggravated his difficulties. McCulloch's skills were developing apace. He still illustrated under Waite's supervision but his work was at least as good as that of his mentor and, by 1903, McCulloch was doing more of the art work for Waite's publications than Waite himself.

McCulloch had begun his own research and expected to have it published. He and Waite had been collecting gudgeons (Eleotridae), small fishes common in the streams around Sydney, and Waite had been observing some in his aquaria. Although Ogilby had studied the group years before, it became obvious that his work was incomplete. To Waite's discomfiture, McCulloch's enthusiasm led him to attempt independent publication of his work on these fishes. The story emerges through extracts from Waite's diary between March and September 1904:

> [March] 16. Wed. Allan is now making drawing of Eleotrids. he has an idea of publishing drawings in his own name and asked and obtained leave from Etheridge. I could scarcely object though it is encroaching on my domain.
>
> [March] 22. Tues. I am now working at the Eleotrids which despite Ogilby's papers need revision.
>
> [April] 5. Tues. Etheridge not approving of Allan on my subject (fishes) is suggesting to the publication committee that the work of no one under 21 shall be published.
>
> [June] 1. Wed. Have also in view paper on our Eleotrids ...
>
> [Aug] 18. Thurs. A circular sent round for papers for 'Records'. [Etheridge often solicited papers at short notice, when money was available for an issue of the *Records of the Australian Museum*] I handed in title of paper on Eleotrids of which plates are ready & many notes for writing-up.

> [Sept] 5 Commenced paper on Eleotrids.
>
> [Sept] 15. Thurs. Concluded Eleotrid paper.

Thus Waite asserted himself through Etheridge and eventually published the Gudgeon paper in his own name (Waite 1904c). McCulloch's work was recognised in this publication only through acknowledgement of specimens donated and illustrations drawn. Much the same pattern was displayed in Waite's description and publication of two new species of fish that McCulloch had discovered in rock pools near Sydney. They were included in a paper published hurriedly in 1906, a tying up of loose ends in which Waite dealt with his last projects before leaving Sydney. Once again he utilised McCulloch's drawings. One of the new species was the Blackback Snake Blenny, *Ophiclinus gracilis*, found and illustrated by McCulloch. McCulloch himself had begun to concentrate on marine crustacea, while still assisting in the ichthyological research.

Why was Waite so restrictive? His pre-eminence in ichthyology in New South Wales was recognised generally since the departure of Ogilby and he intended to remain the sole authority. He was protecting his position in his chosen field, but was perhaps creating problems for himself that would be manifest in future years.

Invitations

David Starr Jordan, with his associate Barton Evermann, produced one of the most ambitious ichthyology books ever written, *The Fishes of North and Middle America* (1896–1900). On more than one occasion, Jordan approached Waite, suggesting collaboration. In 1904, after collecting in Samoa, he wrote to Waite from Stanford:

> I am having a desperate time with my fishes of Samoa. I can only give a little time each day, and my eyes will not stand even the work I would like to give, and so I have to fall back on others. I wish you could take your vacation and spend the Winter here. (Jordan to Waite, 9.8.04, SAMA)

This letter continues with pages of discussion of fish and ichthyology texts, and he mentions that 'I have a pretty pipe fish which I am calling Corythoichthys waitei.' Jordan was well known for having most of the details of his research carried out by his junior associates and he clearly saw Waite in this capacity. Waite declined, presumably not being overwhelmed by the 'pretty pipefish' (which proved to be a synonym of the Messmate Pipefish, *Corythoichthys intestinalis* (Ramsay, 1881)).

Early in 1905 Jordan tried again, suggesting that Waite complete some work, which he said had been begun by himself and Evermann, on a check-list of the fishes of Europe. 'I know that you are skilful in this kind of work & that you seem to enjoy it.' (Jordan to Waite, 28.11.05, SAMA) He said that the work would be published as 'Jordan, Evermann & Waite.' In January 1906 Waite received another proposal of this sort and recorded Jordan's letter in his diary: 'Letter from Jordan asking me to join himself & Evermann in preparation of a list of the fishes of Europe, on Modern lines.' (WD, 2.1.06) Waite was moving in another direction and declined the offer. Jordan responded, a little disappointed:

> I am sorry that you cannot join with us in getting out the catalogue of the fishes of Europe. There would have been lots of fun in bringing them under the control of the International Code. (Jordan to Waite, 7.2.06, SAMA)

Jordan never undertook the project. Waite's participation may have been a condition of the work proceeding. However, these propositions were unequivocal signs of Waite's success and his acceptance as an ichthyologist.

Life in Sydney; moving to New Zealand

In 1905, the Waites had been residents of Sydney for 12 years. Neither liked grand social occasions but they had many friends and activities that afforded them great pleasure. Rose was a theatre-goer and there were always new productions to see. Edgar had become a proficient flautist and participated in private and public performances regularly. Their Mosman house was in a beautiful situation, not far from the water, and he had modified the grounds by building large aquaria, aviaries and enclosures for his animals. The family had annual holidays to various parts of the state and weekend trips around Sydney with their friends. Waite had earned an excellent reputation as a zoologist and seemed very well established in his Museum position.

In 1905 Waite applied for a curatorial position at the Canterbury Museum, a small natural history museum in Christchurch, New Zealand. Why? Waite probably felt that his career was stalled. He had occupied the same position at the Australian Museum since leaving England 12 years earlier. Several members of the Museum staff had recently taken much better paid government jobs but his own chances were limited as he had no tertiary qualification. He had to look for promotion within the fields of natural history and museum management. At the Australian Museum, Etheridge, 58 years old in 1905, had the Curator's position (which he held for a further 15 years, dying in harness). Charles Hedley, a likeable and respected zoologist of wide experience, had been associated with the Museum for many years and it may have seemed likely to Waite that Hedley would be promoted first. Waite himself had recently secured his first increase in pay, but only by threatening to leave; not a very safe manoeuvre, and not to be repeated. So he discussed the situation with Rose, applied for the New Zealand position, was appointed, and they left for Christchurch.

Christchurch: the Canterbury Museum

When Edgar Waite arrived in Christchurch in April 1906, his reputation as the leading Australian ichthyologist seemed secure. He had a better grounding in the science than any previous zoologist who had worked in New Zealand, but the study of fishes, although high on his list of priorities, was not necessarily his principal concern. He certainly had not left Sydney solely to pursue the study of fishes in a new environment. He wanted to take charge of a Museum and thereby enhance his reputation as a museum manager. He threw himself into Museum work: the displays and the collections were prime considerations and he ensured that his activities were noticed through a good relationship with the press and with the community.

That Waite saw his New Zealand position as a stepping-stone is evident: he had acquired two Sydney properties and kept both, maintaining a fluctuating income by renting through agents, and showed no sign of wishing to own his own home in Christchurch. This was something of a trial to him, as he had become accustomed to modifying his home by his own hands, revelling in the construction of ponds, 'bush houses' and workshops, installing electrical apparatus and furnishings. Many of the Waites' possessions remained in storage at the Canterbury Museum throughout their seven years in New Zealand. Rose made the best of the situation, making some friends and attending such entertainments as offered, but she became decidedly homesick for Sydney.

Museum zoologists, in their writings, often leave the impression that museum administrators and trustees have been created to make life difficult for them. Waite, however, found in Christchurch that funds were usually made available promptly, at least for his more modest projects, and he was given a free hand in the day to day running of the Museum. Innovation, if not actively encouraged, was at least tolerated. His first impressions of the Canterbury Museum were favourable, although he was disappointed when he realised that there were only three other employees. He found '... a large establishment, the exhibits being numerous & in many cases good, but the cases and arrangement are antiquated.' (WD) He was introduced to the job by Professor Charles Chilton (1860–1929), an expert in Crustacea, who then occupied the Chair of Biology and Palaeontology at Canterbury College. The Museum was associated with the College, the College Board having a Museum Committee to see to the running of that institution, but requests for special leave or funds were referred to the Board.

The New Zealand fish fauna and the earliest work thereon

The marine fish fauna of New Zealand has strong affinities with that of southern Australia but it is less diverse. The South Island has a distinct Antarctic component and the north of the North Island includes among its fishes some that have been derived recently from warmer waters. The inland waters of New Zealand support an even more depauperate fish fauna than those of Australia. There is a significant endemic element in both marine and freshwater environments.

The discovery of the New Zealand fish fauna followed much the same course as that of Australia. The naturalists of early voyages collected more or less at random and brought back collections to Europe. Some of the fishes common to Australia and New Zealand had been found by Parkinson and Solander, or a few years later by the Forsters. John Richardson brought together the work that had been done, identified and described new finds, and presented a *Report on the present state of ichthyology of New Zealand* to the British Association for the Advancement of Science in 1843. In the same year he also published, with John Gray of the British Museum, a checklist with descriptions of new species collected by Dr Ernest Dieffenbach, a traveller naturalist, in New Zealand.

Gilbert Whitley (1968) expressed the opinion, in a historical note included in his *Check-List of the Fishes Recorded from the New Zealand Region*, that:

> [t]he history of New Zealand's ichthyology differs from that of Australia in that there was not the competition amongst foreigners to describe and name the species in the early years.

It is well observed that the synonymies of New Zealand endemic fishes are often less tortuous than those of their Australian counterparts but there has been no lack of problems for New Zealand ichthyologists to solve. They have been largely spared the ministrations of Castelnau, de Vis and Macleay, but their own people have caused more than a little confusion.

New Zealand ichthyology, 1870 to 1905

From 1870 to 1905, seven New Zealand zoologists studied fish and published their findings: Julius van Haast, Frederick Hutton, James Hector, F.E. Clarke, Augustus Hamilton, W.B. Benham and T.J. Parker. Between them, they described and named about 100 New Zealand species as new, of which nearly half are now accepted as valid. Hutton described almost half of the new species, and Hector about a quarter of them, by far the greater part of their work being performed in the 1870s. F.E. Clarke was the next most active in describing new species, but his work was of a lesser standard judging from the small number of his species now recognised.

The other zoologists mentioned described only a few species each. T.J. Parker, Professor of Biology at the University of Otago, collaborated with William Haswell of Sydney University in producing a well-known textbook of zoology that made full use of Australian and New Zealand animals. Parker recorded some new occurrences of fish in New Zealand in the 1880s, and incidentally described the seminal vesicles of the Elephant Fish, *Callorhinchus milii* (1893). Augustus Hamilton, Director of the Dominion Museum, Wellington, described a new Blind Electric Ray, *Astrape* [now *Typhlonarke*] *aysoni*. William Benham, Professor of Biology at the University of Otago and writer of authoritative works on general zoology, recorded some rarities and described a Streamerfish, *Regalecus* [now *Agrostichthys*] *parkeri* (Benham, 1904).

A minor contribution from a non-zoologist may be mentioned: R.A.A. Sherrin (1886) prepared, under instructions from the Commissioner of Trade and Customs, a *Handbook of the Fishes of New Zealand*. As he stated in the introduction, this was a compilation on fish & fishing. Nearly half of its 300 pages were occupied by information on seals, whales and introduced fishes (mostly Salmonidae) and he warned the reader: 'The nomenclature of ichthyologists conveys no meaning to the bulk of mankind. What information is here gathered is put in a simple form, for simple people to easily find.'

Julius von Haast (1822–1887) migrated from Germany where he had undertaken some geological studies. He came to New Zealand as an adviser to a shipping company on German immigration but became geologist for the Canterbury province, and then Curator of Christchurch's Canterbury Museum. The Museum had been founded in 1861, but occupied only two rooms in council buildings. A new Museum was built, largely due to Haast's energy, and opened in 1870 (Sheets-Pyenson 1988). Haast described eight or nine fishes as new, but his claim to ichthyological significance rests mainly on his description of the strange freshwater Torrentfish, *Cheimarrichthys fosteri* (Haast, 1874).

Frederick Wollaston Hutton (1836–1905) was from Lincolnshire. His education included attendance at a naval academy and studies of applied science in King's College,

London. He served in the British Army in the Crimean War and in India, where he was at the relief of Lucknow, leaving the army with the rank of Captain in 1865. He had studied geology while in the army and, on coming to New Zealand with his wife and children in 1866, took part in the Geological Survey. Hutton lectured in Geology and Zoology at the University of Otago and became Professor of Biology at Canterbury College in 1880. In 1887, on the death of Julius von Haast, he was appointed Curator of the Canterbury Museum, holding the position until his death in 1905 (DNZB). He was Waite's immediate predecessor.

Hector (1872) prepared a *Catalogue of the Colonial Museum* and included a fish list, the nomenclature of which was drawn from Richardson (1843). Hutton published *Fishes of New Zealand. Catalogue, with Diagnoses of the Species* (1872). The classification was derived directly from Günther's works (the *Zoological Record* and the British Museum catalogue), which Hutton referred to as 'the fountain head of systematic Ichthyology.' Hutton included a key to the families of fishes and brief descriptions which were intended to be sufficient to separate the species. 140 species were listed, and 49 of these were depicted by line drawings. The text was published with separate notes on the edible fishes supplied by James Hector (1834–1907), the founder of the New Zealand Institute (later the Royal Society of New Zealand) and Director of the 'Colonial Museum' in Wellington.

For the preparation of the catalogue, Hutton had visited the British Museum to inspect Solander's manuscript notes and Georg Forster's drawings. He was an active zoologist and his publications included several on fish. In 1873 he published a supplement to the *Fishes of New Zealand.* Papers listing new records followed at intervals so that Hutton was able to produce a revised list of 226 species in 1889. His last synopsis of New Zealand fishes was published in 1904 shortly before he died. This list provided the foundation for Waite's work on the New Zealand fish fauna.

Waite's early activities in New Zealand

Among Waite's first concerns was participation in the New Zealand International Exhibition, due to open less than six months after he arrived in Christchurch. He set about importing exotic fish and amphibia, writing to friends in Sydney, America and Japan for the animals. Shortly after his arrival the members of the Museum Committee called on him and he was appointed to the Natural History Committee of the Exhibition. He found that £500 had been voted for a Natural History Court, to which other New Zealand cities would contribute, and he set about organising displays. Among a multitude of tasks he personally arranged the aquaria and aviaries, and supervised the acquisition of 'vegetable sheep' (intriguing and bulky alpine plants which grow in the form of large, woolly cushions). Three days of his concise diary entries show the new curator in action:

> 10. Sat. Got up early, walked up a small creek & obtained aquatic plants. Found nest of Lark with 3 eggs. Left with C. [Cockayne, a new friend] at 8.45 and cycled to Porter's Pass, passing through grand river beds, showing well-marked levels. Had a stiff climb up Mt. Torlesse to 4000 ft where we encountered Vegetable Sheep on the shingle. Selected specimens for removal &

> considered methods of transport. We did not reach the summer snow line but met patches in the hollows. The plants on the shingle are most interesting & I got lizards and jumping spiders. Having lunch a Weka [ground dwelling bird, *Gallirallus australis*] came close to us several times & picked up crumbs just as at L.H. Is. Came down shingle slips but found the back of my shoes cut by the process. Reached Springfield 8.0.
>
> 11. Sun. Up at 4.30 again. Very chilly. We went to foot of mountain in company with 5 men in a cart, repeated yesterday's proceedings, & the men removed the 'Sheep' with pick & crowbar & then had a tremendously hard carry down the steep rocky sides. Saw Keas. C took photos of operations & we were thankful when the Sheep were aboard the cart in which we drove the 9 miles to Springfield.
>
> 12. Mon. Collected water plants before breakfast & left by 7.30 train. Eucalyptus globulus [Australian blue gum] lines the railway for miles. Reaching Ch.ch. [Christchurch] at 10.0 I took Sheep to Exhibition. Saw Hulsen, arranged to get marine life on Thursday. Rose had got Blue Eyes [*Pseudomugil signifer*] from Cherry [Sydney friend] by Saturday's boat.

His interests were as wide as ever. Waite was a consummate organiser with great energy and an eye for the spectacular. The Exhibition ran for over four months.

During the Exhibition, Waite learned that the Governor, Lord Plunket, was going on a tour of the southern islands. At short notice, he contrived an invitation and his Museum Committee voted leave of absence and £15 towards the expenses. He obtained spirit, formalin, chloride of lime, fishing nets and large wooden collection tanks and joined the *Tutanekai* for the two week trip to Auckland, Enderby, Campbell and Chatham Islands.

Waite was independent enough in his new position to deal directly with the press. He wrote an account of the southern islands tour, for which he was paid, and from this time regularly sent articles and photographs for publication. He occasionally gave interviews but was very critical of the reporters' interpretations and sometimes provided the script of an 'interview' ready for publication. His contact with the press had a number of favourable aspects. Waite saw public education as an important part of his work at the museum in Christchurch, and later in Adelaide; his efforts in museum work could be kept in the public eye and remuneration for the articles was always welcome.

Waite's introduction to New Zealand ichthyology

Within a few weeks of his arrival, Waite settled down to revise Hutton's New Zealand fish list, the latest version of which had been published only four years earlier. He gave his reasons for this in the introduction to the work which he produced:

> Every worker in systematic zoology appreciates the value, almost the necessity, of having at hand, a list which supplies exact references to the original descriptions of the constituents of the fauna he is studying.

> The catalogue of fishes in the '*Index Faunae Novae Zealandiae*' prepared by the late Captain F.W. Hutton, F.R.S., forms the foundation of this list ... The 'Index' does not supply references to the original descriptions ... (Waite 1907)

For this reason Waite located the original references and, with only slight changes to Hutton's list, published *A Basic List of the Fishes of New Zealand* (1907). Waite incidentally modernised Hutton's list in another way: he used as a framework the classifications of the British Museum ichthyologists: Regan's for the sharks and rays, and Boulenger's for the bony fishes, with some adjustments to accord with the ideas of Jordan's American school of ichthyology.

The Canterbury Museum had no journal when Waite arrived. He had appreciated the importance of the *Records of the Australian Museum*, produced by the Museum primarily for the publication of research carried out by the staff. He wrote in his diary, 'Am now working at my list of N. Zealand Fishes. Hamilton [Curator of the Dominion Museum, Wellington] wants it for his "Bulletin" but why should I not publish it for benefit of Canterbury Museum, the first of a running publication?' (WD, 23.2.07) He wrote to the chairman of his Museum Committee, who asked him to prepare a report for the Committee. Waite 'Obtained estimate fr. Whitcombe & Tombs for printing my list of N.Z. Fishes, viz £9.15.0 [for] 37 pages and cover.' (WD, 6.3.07) He then put his case to a meeting of the Museum Committee who decided 'to recommend my publication proposal to the [Museum] Board.' Number one of Volume one of the *Records of the Canterbury Museum* was duly published on 25th April, 1907 and consisted of the Basic Fish List. Waite had, of course, pledged himself and his Board to an ongoing commitment and he provided the material for publication in the journal for the duration of his residence in New Zealand.

At the Canterbury Museum, Waite had a chance to learn the technique of making plaster casts to represent the fishes that he wished to display. The time-honoured methods of exhibiting fishes involved either stuffed skins or whole animals in spirit. The taxidermy exhibits were commonly imperfect due to discoloration and distortion of the shape as the skin dried and shrank, and to fraying and breaking of fin rays. Spirit specimens lost their colour and were subject to distortion when viewed through the glass vessels. The casts had none of these disadvantages and, if prepared with care and painted accurately, depicted the species in a most satisfactory manner. Waite soon mastered the techniques of casting, and frequently painted his own specimens, gradually building up a large display of fish casts. The first cast that he coloured was that of a small Catshark, *Cephaloscyllium isabellum*. That the casts were essentially artefacts tended to upset only taxidermists; Waite had at least one newsprint battle with members of this profession.

The process of becoming familiar with the New Zealand fishes proceeded at the level of field work as well. Waite carried out a number of informal collecting trips in his own time, visiting the littoral rock pools and the streams, wading after his quarry, fishing with lines and handnet. The trip to the southern islands provided more exotic fare, and soon another and still more ambitious project loomed.

The *Nora Niven*

For Waite, trawling was the fish collection method of choice. He had agreeable memories of the *Thetis* spoils and, when the opportunity to accompany an experimental trawling venture arose in New Zealand in 1907, he was swift to act. He contacted the organiser, L.F. Ayson[22], Chief Inspector of Fisheries for the Dominion, who agreed that the presence of a zoologist would be appropriate. After some manoeuvring on Waite's part, the Chairman of the Museum Board received a letter from the Marine Department suggesting that Waite accompany the trawling expedition. Waite was not appointed by the government, as he had been hoping, but became a 'guest' on board. This weakened his position, but initially Ayson was present, and he supported Waite and facilitated his work. When Ayson was obliged to leave through illness, he was succeeded by Thomas Anderton, who also backed Waite.

On boarding, Waite was disconcerted to find that his berth was far from comfortable:

> No bath room not even lavatory accommodation on this craft ... To think that I have left a comfortable home & warm bed for 3 months on a wretched boat & a bunk so hard that ¼ hr on either side stops the circulation & produces 'pins & needles' & then I try the other. (WD)

He did not enjoy the 'lurch, twist & corkscrew motion' and new difficulties emerged:

> Ayson tells me that the Minister has prohibited any information being given to the press. All my collecting gear has been put into the freezing room & I don't see where I am to work. (WD)

He was soon busy recording the trawl catches, preserving specimens, dredging when possible and photographing the proceedings. The weather was poor, with heavy seas; they went ashore at Ulva Isle, to send letters home from the most southern post office in the world, and were 'caught in a snow storm as we pulled back to the "Nora".' While ashore, Waite collected from the rock pools, poisoning when practicable, and took geological specimens for Speight, a Christchurch geologist who later replaced Waite as Curator of the Canterbury Museum.

The trawl take was astounding, yielding frequent large catches and a great variety of fish species that were unfamiliar to Waite. In the Bay of Plenty, he recorded:

> great hauls of schnapper [*Chrysophrys auratus*] which floated the cod [end of the net]. Alward jumped overboard into the net & was supported by the fishes, sinking only to about his waist.

When the vessel was at anchor, the crew fished with lines, often with very good results, and Waite invariably participated. Near Chatham Island, he was taken out in a small boat by some 'islanders':

22 Lake Falconer Ayson (1855–1927) was an enthusiast for the acclimatisation of northern hemisphere species in New Zealand. He was instrumental in establishing the Porto Bello Marine Station at Otago in 1904, primarily as a research facility to promote introductions. He is credited with the successful introduction from Canada of the Chinook Salmon, *Oncorhynchus tshawytscha* to New Zealand waters. Ayson encouraged Waite's activities as he supported a scientific approach to fisheries.

> I never had such rapid fishing in my life. Before the sinker reached the bottom it was jag, jag, jag & three Blue Cod [*Parapercis colias*] on the hooks almost every time, we soon had the boat half-full & after 1 ½ hours pulled back to the 'Nora'. (WD, 2.8.07)

The trawler *Nora Niven* was chartered by the New Zealand Government from the Napier Fish-Supply Company. She was a 56 ton vessel of 30 metres, and supplied with a 'complete ice-making plant, capable of turning out about 2 tons of ice every twenty-four hours' and 'an acetylene gas lighting plant.' From Waite 1909.

Off Whangaroa, Waite was delighted to catch the heaviest fish for the day, a Hapuka (*Polyprion oxygeneios*) of '62 lbs'; [28 kilograms] but he frequently used small hooks for the sake of collecting.

Bad weather was often encountered; a south-east gale on the night of 9 July off Akaroa impressed Waite:

> I scarcely slept, was once thrown out of my bunk & was so often on the verge of it that I had to be constantly grabbing the sides which were not high enough. At 4 O'clock I got up & put a table fender across, then I got to sleep. Mountainous seas running, when day broke I found nearly all my cases on deck adrift. One of the large tanks overturned & the fishes scattered about in the wash on the deck. I don't know what I may have lost through the scuppers. (WD)

On one occasion, the high seas offered some compensation: 'Some Myctophids [Lanternfish] were washed aboard during the night in the heavy seas.' (WD, 13.8.07)

The expedition occupied nearly three months, trawling at 96 stations from Stewart Island in the south, along the east coast, into the Bay of Plenty and at Chatham Island. The maximum depth of the hauls was about 200 metres, with most of the stations being worked between 30 and 90 metres. As the net was brought out of the water, Waite was generally standing by with a fine mesh hand net to collect any small fishes that dropped through the coarse trawl net. The invertebrates were sent to various specialists and Waite settled down to study the fish.

Problems with New Zealand fish taxonomy

As he began to work on the trawl fishes some of the disadvantages of Waite's situation emerged. He was working essentially alone again, no assistant being available to help with the references and drawings. He had amassed a considerable personal ichthyology library but missed some of the books, and the foreign journals, held at the Australian Museum. The absence of a reference collection of fishes was also a serious handicap.

Waite found that New Zealand fish taxonomy was complicated. A single species had often been described more than once by Hutton, his contemporaries and his predecessors from local specimens; or new descriptions had been published by New Zealand workers for trans-Tasman species that had already been named in Australia; or sometimes a species similar to one found in Australia was thought to be identical, but later proved to be different. To sort out the last two problems, it was necessary to compare specimens from both sides of the Tasman.

Trans-Tasman cooperation

Cooperation with McCulloch was essential. By the time Waite was fully engaged with study of the *Nora Niven* fishes, McCulloch had his own problems: he was handling the mammal, reptile, fish and crustacean collections at the Australian Museum and the first large collections were coming to him from the trawling of the fisheries research vessel *Endeavour* in Australian waters. Waite's re-location had meant rapid advancement for McCulloch who, as a consequence, was bound to see himself as a fully fledged zoologist. One result of this was his understandable reluctance to resume a posture of submission and dependence. He had, in a sense, been abandoned and meant to assert himself. There was no suggestion of confrontation but his responses to Waite's frequent requests for help were sometimes delayed or incomplete (see below).

To any form of management by remote control, as it were, McCulloch was definitely resistant. Waite's policy of, where possible, providing accurate illustrations of new or previously unfigured species prompted him to make a suggestion:

> Letter from Allan, with list of fishes he is figuring. As 4 are duplicating mine I suggested that I drop Centriscops and Seriolella for Cyttus & Hippocampus[23] proofs from the blocks of which I sent him. (WD, 16.3.11)

Waite intended to avoid duplication but trans-Tasman cooperation did not extend this far: 'Letter from Allan, he declines to suppress his drawings duplicated by mine.' (WD, 28.3.11) (See also page 362.)

Within the limitations implied above, the relationship between Waite and McCulloch was one of cooperation, the results of which improved the work of each. McCulloch suddenly had access to useful New Zealand collections; Waite could obtain specimens for comparison or frequently have the comparison done for him, and he still had reasonable access to the Australian Museum library through McCulloch. Working on the rays that had been taken by the *Nora Niven*, Waite found that he needed to distinguish one from a species held in the Australian Museum and 'wrote to Allan asking for outline drawing of Astrape japonica [an Electric Ray].' (WD, 18.10.07) Working on two species of trawled 'Stargazer', Waite sent specimens to McCulloch for comparison with Australian Museum fish, 'Letter from Allan, he says that my Gnathagnus is innotabilis, but that Kathetostoma is not laeve.' (WD, 20.10.10) In the report, Waite acknowledged this, 'I am obliged to Mr. A.R. McCulloch for comparing specimens with Australian examples.' (Waite 1911a) He went on to quote directly from the information that McCulloch provided.

23 Fish referred to were: *Centriscops humerosus* = Bellowsfish; *Seriolella brama* = Warehou; *Cyttus novaezealandiae* = Silver Dory; *Hippocampus abdominalis* = Seahorse.

Interrupted ichthyology

The Chief Inspector of Fisheries, Ayson, asked Waite to prepare a scientific report on the trawled fish but Waite referred the matter to his Board, saying that he had participated in the excursion not for the government but for the Museum. Ayson said that he regretted that the government had not paid Waite as then he would have had to produce the report when asked. Waite began to examine the fish within a few weeks of returning to Christchurch. He commenced with the most 'primitive' form taken, the hagfish, and moved on to the sharks and rays. He took photographs of these, made casts and was proceeding with the work when there came the first of a long series of interruptions that delayed his *Nora Niven* report for nearly four years.

A two week trip was proposed to the islands south of New Zealand and Waite was invited. Once again leave was arranged and he embarked on the *Himenoa* at Invercargill in November 1907. The trip took Waite to the Snares, and south to Enderby and Campbell Islands. At Disappointment Island, they found survivors of a shipwreck who had been stranded for five months and brought them back to New Zealand. Waite was fascinated by their story, retrieved relics of their ordeal for the Museum and helped to raise money for the rescued seamen. He collected in rock pools, fished and operated a dredge. The trip yielded only a small fish collection but it did give him further experience of the Sub-Antarctic Islands. Waite also had a lesson in fishing from an expert. He and the other members of the party had been unsuccessful in trying to take specimens with lines at the Auckland Islands, and they decided that the seals had driven the fish away:

> ... one day a member of our Maori crew brought me three examples which he had caught. He took me to a tumbled mass of sunken rocks, and by means of 2 ft. [60 cm] of line attached to a short stick dropped the bait into a crevice and drew out a Notothenia. Adopting this plan, I obtained as many fishes as I required. (Waite 1909a)

In December 1907 Ernest Shackleton and his party were in Christchurch en route to Antarctica in the *Nimrod*. Waite found that some of his Sydney acquaintances were with Shackleton, including Professor Edgeworth David; they visited him at the Museum where he showed them his work and the collections. This connection led to Shackleton sending Waite the fish collection from his expedition, providing his first experience in Antarctic ichthyology (Waite 1911b).

Waite undertook short field trips as a routine. One of these excursions, undertaken when perhaps he should have been working on the trawl fish, produced some interesting observations. He noted the way in which different species of freshwater fish preferred different habitats, a rudimentary but important ecological observation:

> We did some river wading and put in a drag net, got Galaxias in the streams & Retropinna [New Zealand Smelt, *Retropinna retropinna* (Richardson, 1848)] in the river & where it ran over rapids, Cheimarrichthys, it is surprising how these little fishes can hold their way among the smooth boulders swept by rushing water.

Paranotothenia microlepidota, a representative of the subantarctic fish fauna that Waite caught among rocks at the Auckland Islands. From Waite 1909.

Waite's drawing of the New Zealand Torrent Fish, *Cheimarrichthys fosteri* Haast, 1874. Waite collected and drew the fish in 1908 and published notes on its habitat. From AA356 Waite Collection, courtesy of the South Australian Museum Archives.

Cheimarrichthys, the Torrentfish of New Zealand, was worthy of special attention and Waite wrote a newspaper article on it and prepared an accurate illustration (Waite 1910b).

The way in which the trawling report was to be funded was obscure, and Waite procrastinated. The stranding of a 27-metre Blue Whale at Okarito, a remote part of the coast, provided another diversion. The whale was offered to the Museum and Waite found himself in the invidious situation of being in charge of fund-raising to pay for the retrieval of the whale's skeleton. He inspected the whale and supervised the project.

Then came another form of distraction. Charles Hedley, Waite's old friend from the Australian Museum wrote to him in August 1908, asking for a reference for the newly-created position of Assistant Curator at the Museum. Waite was embarrassed. He wanted the job for himself. He acted deviously, having Rose write to the Curator in Sydney for information, and giving the impression that she did this in her husband's absence. He applied, and began to believe that he would get the appointment, although he was acutely aware that Allan McCulloch, in the normal course of their correspondence, had made no mention of the Sydney position. Waite's machinations were interrupted by an excursion that had been long planned and was well funded by the Canterbury College: a tour through northern New Zealand in order to obtain Maori artefacts. As the Maori people had become increasingly separated from their traditional way of life, their crafts were practised less, and the objects of these crafts became uncommon and sought after. Waite took an assistant and travelled widely, accumulating a massive hoard of old carvings and other craft items, which he sent back to the Museum.

There had been an odd encounter with fish in the course of travelling around the North Island:

> ... we left this day, at 8 am for Opotiki. Passed through great diversity of country, mountain passes, mud flats and sea beach. We had six miles of beach and saw two frost-fishes coming ashore. They seemed to be writhing helplessly and were carried further inshore by each succeeding wave. We could not wait, but the driver intended to pick them up on his return. In one place the sea sand was very loose, so we walked to ease the horses. Striding ahead, I disturbed a flock of gulls at the carcase of a frostfish. (WD)

These one to two metre long, silvery Frost Fish, *Lepidopus caudatus* were collected from the beaches as they are excellent food; in Waite's day they were taken almost exclusively in strandings, as he describes.

Waite also chose this time to become fascinated by 'motors' and bought a primitive machine for £14. Dismantling, reassembling and modifying this motor cycle may have been therapeutic as he waited for the museum trustees in Sydney to announce their decision. After more than six months suspense Hedley was appointed and Waite was devastated. The news came just as he was about to leave on a trip to the Southern Sounds of New Zealand, and the diary entry reveals near desperation:

> By the afternoon's post I received advice that my application for the Sydney post had failed. This is a crushing blow & I don't know how I shall get on with Rose who had built entirely on again going to Sydney. Hedley is appointed. Took Rose & Claude to Wests pictures, passed a sleepless & feverish night. I shall not tell Rose of failure until I return. The prospect of spending the rest of our lives in ChCh will break her up.

He was wrong. Waite left on the tour, the account of which in his diary is for once mechanical and lifeless. Rose discovered the news in her husband's absence and, on his return, restored his confidence, telling him not to give up hope of returning to Sydney. The Sounds trip had lasted only two weeks but Waite brought back another interesting selection of fish.

An accumulation of ichthyology

Six months after the cruise of the *Nora Niven* Waite had completed the introductory part of the *Nora Niven* report and, by 1909, the work on the sharks and rays; this was published in the *Records of the Canterbury Museum* (Waite 1909b and 1909c). In early 1909, Waite was faced with an accumulation of work. He had to see the smaller *Nora Niven* papers through the press and still needed to write up several collections: the major work on the bony fishes from the trawling; the vertebrates of his Sub-Antarctic expedition; the fish from the Southern Sounds. He then received a collection from an expedition to the Kermadec Islands and Shackleton, on his way back from Antarctica in the *Nimrod*, assured him that he would soon have their fishes to describe. Late in 1909 Waite was completing a term as president of the Canterbury Philosophical Institute and wrote 'Presided at C.P.I. meeting & read! a paper on the Kermadec Is. fishes not yet written.' (WD)

The literature on Antarctic fishes was sparse but essential for his paper on Shackleton's fishes. Reports on French and Swedish expeditions were proving difficult to get, 'Stuck for want of literature, one of Shackleton's fishes is probably Artedidraco a genus described by Lönnberg [Lönnberg 1905] in Swedish Antarctic Exp. report.' When the Swedish paper finally arrived, he decided that his fish was indeed *Artedidraco* and described it as a new species, naming it after Shackleton (Waite 1911b).[24]

As president of the Canterbury Philosophical Institute in 1909, Waite was required to deliver a presidential address. Initially he chose a biogeographical topic, provisionally entitled 'Comparison of fishes of Southern lands'. While he was still in Sydney, Ogilby had given Waite a reprint of Theodore Gill's *Comparison of Antipodal Faunas* (1893), in which he presented a tabulated comparison of the fish fauna of New Zealand and that of the British Isles, and drew interesting conclusions. Waite's intention to compare the southern faunas was potentially significant to biogeography. He had amassed a great deal of original data on Australian, New Zealand, Antarctic and Sub-Antarctic fishes, but needed information on South America and South Africa. He wrote to Regan, and to Lahille at Buenos Aires but found that the list for South America was too imperfect for his purpose. Then Waite decided on an easier option for his talk: he chose fish nesting. His own experience with exotic aquarium fish was the basis of this and he wrote to Theodore Gill, who had published some semi-popular papers on this topic. Gill sent reprints containing information that Waite incorporated in his address.

The Kermadec fishes were proving interesting but difficult and Waite was again hampered by the lack of some literature and specimens for comparison. He turned to McCulloch for assistance, 'Sent some Kermadec Is. fishes to Allan & asked him to look over my Norfolk Is.

24 There are about 25 species of 'Plunderfishes' now recorded from Antarctic waters but when Waite described *Artedidraco shackletoni* only Lönnberg's two species were known.

list.' At first sight innocuous, this was a request for a great deal of work. McCulloch replied that he was going on holiday and later, when Waite followed up the request, that he was too busy to make the comparisons wanted. Waite proceeded to write this paper (Waite 1910a), and others reporting new fishes from New Zealand.

Final report on the trawled fishes

The larger part of the *Nora Niven* report was finally published in 1911, with special attention to the flounders and their relations as requested by the Chief Inspector of Fisheries, Ayson. Waite had difficulty with the close work for the illustrations that he regarded as integral to the project. He had been allowed an assistant by the Board in 1908 but the pay was presumably poor and, one after another, the new 'boys' found work elsewhere. However, Frank Pollard was with him for two years; in April 1909, 'Am now giving Frank drawing lessons. I started him on Cyttus, he is doing well.' But Frank resigned in 1911 to enter his father's cabinet-making business. Waite credited Frank's work in the second part of the trawling report, 'The illustrations which accompany the report will be distinctly useful, and in the preparation of these I desire to acknowledge much help rendered by my assistant, Mr. Frank A. Pollard.' There were 40 excellent illustrations in this report. Another assistant, 'Harry', in 1912 prepared outlines for many of the drawings for short publications.

The Silver Dory, *Cyttus novaezealandiae* (Arthur, 1885) was among the trawl fishes of the *Nora Niven*. The fish was drawn in outline by Waite's assistant and the detail completed by Waite. From Waite 1911.

Waite preferred drawings for publication but occasionally used photographs. The sharks and rays of the *Nora Niven* report were illustrated by photographs; and he used a photograph of a Bellows Fish, an animal so compressed that it is almost two-dimensional. The remaining fish illustrations in the New Zealand trawling report were drawings.

The Banded Bellowsfish, *Centriscops humerosus* (Richardson, 1846). Waite thought his large specimens distinct and named them as a subspecies *Centriscops humerosus obliquus*; later it was found that there was variation in shape and coloration with growth. From Waite 1911.

In this part of the trawling report, dealing with the bony fishes, Waite recorded 15 species from New Zealand for the first time, ten of which he described as new. Several of the latter have proved to be synonymous with species already described. Waite had too many projects running simultaneously to do full justice to the *Nora Niven* fishes but his publication was a valuable contribution to New Zealand ichthyology (Waite 1911a).

The Longsnout Pipefish, *Syngnathus* [now *Leptonotus*] *norae* (top). Waite described this pipefish from specimens he took from the stomachs of trawled Hapuku and Elephant Fish. He noted the colours as green with brown cross bands and wrote 'I have associated with this pretty species the name of Miss Nora Niven, after whom the trawler was named and from the owner of which I received many kindnesses while in Napier.' The plate also shows Waite's drawings of *Leptonotus elevatus* (which he identified as *L. blainvilleanus*) (centre) and a Lanternfish, perhaps *Myctophum phengodes* (which he identified as *Myctophum humboldti*) (bottom). From Waite 1911.

New Zealand fish book

Over the next three years, Waite continued to record new species and published *Additions and Corrections to the Basic List of the Fishes of New Zealand* (Waite 1912b). He was asked repeatedly, by members of the Marine Department to prepare a book on New Zealand fishes, George Thomson of Dunedin being the most persistent: 'G.M. Thomson called with his new wife, he is keen on an illustrated catalogue of fishes, but does not see how it is to eventuate.' (WD, 12.7.10) And, 'Ayson called & discussed perennial question of a book on N.Z. Economic fishes.' (WD, 16.8.10) The funding was never guaranteed in spite of the best efforts of the fisheries people, whose Minister impeded several of Waite's projects. As well as this financial difficulty, Waite imparted always some sense of being a sojourner in New Zealand, and he may have been reluctant to commit himself to so large a project.

Intimations of opportunities and a visit to Sydney

In March 1911 a friend showed Waite an advertisement for the position of Chief Inspector of Fisheries for New South Wales. He lacked the practical experience for this but considered that nothing would be lost in applying. A few months later he heard that Edward Stirling, Director of the South Australian Museum, was considering retirement and this presented another possibility. In November 1911 Waite had a working holiday in Sydney, being commissioned to arrange exchanges and purchases at the Australian Museum. He renewed old friendships and an expedition was mentioned: Edgeworth David, an eminent geologist who had been to Antarctica with Shackleton, asked Waite to lunch and said that he would like to propose that Waite accompany Douglas Mawson's sub-Antarctic expedition.

McCulloch and Waite compared notes and had some outings together; McCulloch displayed his *Endeavour* work and they examined the Kermadec Island fishes. Waite described an incident that had occurred on his way to Sydney. He had been given a very odd looking fish, and thought he recognised it as 'Areosoma' from an illustration he had seen in *Oceanic Ichthyology* (Goode and Bean 1896). Waite and McCulloch looked at the fish:

Waite's drawing of a juvenile Oxeye Oreodory, *Oreosoma atlanticum* Cuvier, 1829. While visiting the Newtown Museum in Wellington Waite 'noticed in one of the exhibition cases a small fish which seemed familiar.' It was 8 cm long and had been washed up alive on a nearby beach. He recognised it, recalling an illustration in Goode and Bean's *Oceanic Ichthyology*. He found that Boulenger of the British Museum believed it to be the young of 'a fish allied to *Cyttus*.' Waite was sceptical but Boulenger was correct. From AA356 Waite Collection, courtesy of the South Australian Museum Archives.

> Spent morning at the Museum. Examined the Newtown Museum Fish which I found to be as suspected Oreosoma though I spelled it wrongly. Allan thought it wonderful that I should be able to identify such a queer creature from the remembrance of Goode and Bean's figure.

McCulloch, it appears, could still please his former mentor when he wished to.

Illness followed by work

1911 ended badly for Waite. Towards the end of November he had become very ill and 'enteric fever' (typhoid) was diagnosed. He wrote nothing in his diary for six weeks, during most of which time he was confined to bed. Early in the new year, while he was still shaky, the family took a holiday at Akaroa. They had a quiet fortnight, fishing and relaxing, but Waite seemed uncomfortable without a project on hand, so he decided to try to recover an old trypot (whaling relic):

> ... we went by launch to Island Bay with a party from the Hotel & after much trouble got one of the Trypots down to the water, but could not get it into the dinghy, it weighs about 6 or 7 cwt. [about 300 kilograms] ... Went again next day & erected shear-legs: by means of block & tackle we got the trypot into the dinghy and towed it behind the launch in the open sea. As it shipped water we put Haynes in the stern to bail the boat, and I was not sorry when we landed it on the wharf at Akaroa.

He seemed to have recovered.

Back at the Museum, Waite worked on the *Oreosoma*, which he included in a paper (Waite 1912a). The usual host of projects was under way, including several fish papers that had been stalled by his illness. The recurrent theme of a fish book surfaced again, 'G.M. Thomson called, he said that the Fish Catalogue was a necessity and would have to be taken in hand.' (WD) Waite retrieved another stranded whale for the Museum, examining it and making a cast.

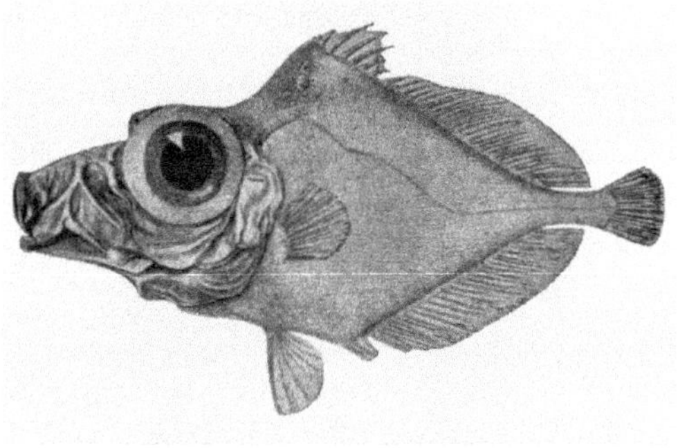

The adult of *Oreosoma atlanticum*, described as *Cyttosoma boops* in 1903 from South African specimens, is trawled in the continental slope waters of New Zealand, southern Australia and South Africa. From Waite 1921.

Macquarie Island

Then came a cabled invitation:

> Aurora leaving Sydney subantarctic oceanographic Cruise about May fifth, lasting six weeks. Mawson most anxious you accompany ship as biologist in charge for cruise, sincerely hope you will consent - David. [Edgeworth David] (WD, 13.4.12)

Waite consulted his Board and accepted. At the beginning of May he was back in Sydney where he learned that the cruise was to call at Macquarie Island, and that he would be able to collect while ashore. A fortnight was spent seeing his friends, assembling gear and working with Allan McCulloch at the Museum.

The *Aurora* left on 17 May, 1912. Waite made observations of the seabirds, collecting some. The seas were impressive; on 3 June he wrote:

> This is the roughest day we have had ... The enormous waves seemed as though they would overwhelm the boat but just as they reached us the vessel rose & was now on the summit & the next moment 40 feet lower, in the trough while the twist we received moved everything movable, including the ladder to the main deck & the grating for the wheel-man. The indicators showed that our inclination at times exceeded 45°. I attempted to photo the boat at this angle but could not control my own movements sufficiently to enable me to do so.

Five days after this he landed on Macquarie Island.

Waite was met by the son of the Director of the Dominion Museum, Harold Hamilton, who had been collecting zoological specimens. He had only a few species of fish for Waite, Macquarie Island offering few days in the year when conditions were suitable for shore fishing or collecting. Waite spent most of his five days on the island observing the seabirds and elephant seals. He and Haynes, the Museum taxidermist, shot a sea leopard for the Museum collection. They had brought lime for the rock pools and used it with some success, 'Limed 2 pools in the afternoon and got a lot of Notothenia and Bovichtus.'[25] (WD, 12.6.12) Hamilton had a fish trap that Waite used but 'Got only a couple of Notothenias and a crab.'

Back on board the *Aurora* Waite was disappointed to find that the weather was too rough for dredging or trawling. The following day he landed again, spending four more days ashore, this time taking an elephant seal four metres in length and a variety of birds for their collection. Back aboard, the one and only attempt was made to trawl. They needed water shallow enough for the trawl to be towed along the bottom:

> Sounded in 450 fath. [830 metres] & put down the trawl after much loss of time: as we rapidly drifted from land and as the depth increased somewhat suddenly, 1000 fath. [1850 metres] of wire were let out, but it still kept nearly vertical.

The trawl was let down in mid-morning and retrieved late in the afternoon without a catch; it was assumed that it had not touched the seabed. The expedition returned via the Auckland Islands, Waite paying his third visit.

As they approached the landfall, '[w]hen south of Adam's Island, a small fish unknown to me was washed aboard.' A seaman handed Waite the little fish still alive; it was just over five centrimetres in length, 'intense blue with colourless fins, the frontal organs whitish and translucent.' (WD) Eleven days yielded a considerable collection, but few fish. Waite made a brief second trip to Macquarie Island in the *Tutanekai* in August 1913, collecting a 'drum of fishes' from Hamilton and once again calling at the Auckland Islands on the return trip.

25 The 'Bovichtus' proved to be a species of Spiny Plunderfish, genus *Harpagifer* (see page 296).

Last endeavours in New Zealand ichthyology

Shortly after returning from his first trip to Macquarie Island, Waite participated in more discussion of a fish book: 'Thomson & Hamilton told me that the fish catalogue would be hung-up for 12 months, but the latter thought that if instead of asking for a lump sum (£600) a small amount (£150) a year was requested it might go through.' (WD) He continued preparation of short publications, adding new records to the New Zealand fauna as they were found. Zoologists from the other museums in New Zealand sent fish for his identification and, when warranted, he drew them, published descriptions and frequently made casts. He also started indexing the papers of his predecessors, beginning with articles in the *Transactions of the New Zealand Institute* in which Hutton, Hector, Haast and Clarke had published most of their ichthyological papers. He then moved on to an analysis of New Zealand fish records from Richardson's *Voyage of the Erebus and Terror*. This work was certainly done with a fish catalogue in view, but motivation was feeble while the finances were unsure. And there were so many distractions.

Waite continued his series of *Notes on New Zealand Fishes* with two more publications under that name. The first of these (Waite 1913a) included New Zealand records of the Butterfly Mackerel, *Gasterochisma melampus* with a discussion of the remarkable changes

Waite's original drawing of the small blue fish washed aboard the *Aurora* on the return from Macquarie Island. He named it *Aurion effulgens*, the generic epithet referring to the ship, the specific epithet taken from Latin 'to shine out' as he interpreted the pale region on the snout as a luminous organ. The fish is now regarded as the pelagic juvenile of a Thornfish, *Bovichtus psychrolutes*, the adults of which inhabit shallow rocky areas around the Subantarctic Islands of New Zealand. Waite's annotations '23 VII, 13 A.A.E. S. Adams Island' indicate that the fish was collected on 23 July 1913 in the course of the Australian Antarctic Expedition south of Adams Island (about half way between Macquarie island and New Zealand). From AA356 Waite Collection, courtesy of the South Australian Museum Archives.

it undergoes with growth, and illustration of an adult fish. In the second (Waite 1914), among some new records and supposedly new species, was the description and drawing of an embryo of 'Lord Plunket's Shark', *Centrophorus* [now *Proscymnodon*] *plunketi*, so named (Waite 1910b) in compliment to the New Zealand governor who had invited Waite to accompany him on a cruise of the southern islands of New Zealand in 1907. He also revised some New Zealand blennies (Waite 1913b) of the families Clinidae and Tripterygiidae, redescribing, illustrating and giving synonymies as far as he could work them out from the available specimens and literature.

Antarctic fish

In May 1913 a new task loomed, 'Letter from David & Haswell asking me to undertake description of all of Mawson's fishes.' (WD) The Australian Antarctic Expedition, led by Douglas Mawson, took place between 1911 and 1914. The zoological collections amassed were considerable and included fishes from the shores of Antarctica, Macquarie Island and the Sub-Antarctic Islands of New Zealand. There was also a competitive element. The members of the British *Terra Nova* Expedition had stayed at Christchurch on their way to Antarctica in October 1910. Waite had '[a]ttended reception at Council Chambers to Capt. Scott & members of Terra Nova Exp.' and a month later 'Wilson, Garrard [Scott's companions in the South Pole expedition] and Lillie of Terra Nova came to bid goodbye.' (WD) Then came news that Amundsen had been first to the South Pole, and in February 1913, 'Early in the morning it was rumoured that the S. Pole party had perished, an "extra" edition of the evening papers at noon confirmed the report ...' (WD, 11.2.13). The *Terra Nova* had made large collections and the fish were taken to C.T. Regan at the British Museum: a report could be expected soon. Waite began writing to ichthyologists who had reported Antarctic fishes, requesting papers in exchange for his own. Regan sent Waite his *Scotia* report (Regan 1913b) of the Scottish National Antarctic Expedition (1902–04). Waite wrote to Pappenheim for his account (Pappenheim 1912) of the German South Polar Expedition (1901–03) and to Dollo for the *Belgica* report (Dollo 1904) of the Belgian Antarctic Expedition (1897–99).

Douglas Mawson (1882–1958), geologist and Antarctic explorer, invited Waite to participate in a trip to Macquarie island and later sent him the fishes from his Antarctic expedition. From Hale 1956.

Adelaide opportunity

In June 1913, Waite received news from Arthur Lea in Adelaide, informing him that the directorship of the South Australian Museum would soon be vacant. Next came a letter from Etheridge, in Sydney, advising Waite to contact the South Australian Museum Director. He did so and Edward Stirling advised him to come to Adelaide to negotiate for the position. He visited Sydney two months later, taking the Melbourne Express to Adelaide where he stayed a week, establishing his interest in the director's position. Stirling was helpful and frank and 'painted the black side of the directorship', which Waite decided was mostly 'red tape'; he was sure he could manage. He also met Archibald Watson, the eccentric and high-spirited Professor of Anatomy, lunching with him in the Botanic Gardens after which 'we adjourned to the dissecting room.' They continued their chat over the assembled corpses, Waite impressed by Watson's demonstration of the effects of 'tight-lacing' on the female liver. Back in Sydney, Waite prepared and posted an application and, on his return to New Zealand, the waiting began. He had contradictory messages from Adelaide and the situation was complicated by the death of Augustus Hamilton, Director of Wellington's Dominion Museum, which made another relatively desirable job available. Two months after his return to Christchurch Waite received an unsettling cable from Stirling:

> Museum Committee unanimously & strongly recommend your appointment, but since find strong efforts being made appoint quite impossible local man, so impossible guarantee action of Board which meets tomorrow. Friday afternoon. Your friends working hard your behalf & hope for success. Alternative horrible to contemplate!! (WD)

Two days later came the news of his appointment. Then to packing and completion of the outstanding fish work: the last paper on New Zealand fishes and colouring of the latest batch of fish casts. He took a one week excursion to the Tasman Glacier by way of bidding farewell to New Zealand. The family sailed for Sydney on 13 March, 1914 and arrived in Adelaide a fortnight later.

Waite's contribution to New Zealand ichthyology

In writing to Ogilby, McCulloch belittled Waite's work in New Zealand (and elsewhere); he judged him purely as a fish taxonomist and meted out harsh criticism. Waite's accomplishment in New Zealand was circumscribed, but he left New Zealand ichthyology better than he found it. He built up the Canterbury Museum collections, resolved a number of taxonomic problems, offered good descriptions of many poorly known species and accurately described new species without burdening the literature with an excessive number of synonyms. His illustrations of previously unfigured or poorly represented species made a welcome contribution. He established a useful pattern for the 'basic fish list', which was a boon to the taxonomists who followed him.

It was unfortunate that Waite did not attempt revisionary work on more than a few New Zealand fish families as this kind of study was badly needed. However, he did revise the flatfish in the *Nora Niven* report. The flatfish work took several months of intermittent study and the result was well worthwhile. Previously, fisheries workers had been unable to identify their catches. Waite re-described and illustrated the species of which he had specimens and constructed a key that allowed easy separation of the common species (Waite 1911a).

Several decades after Waite's departure an illustrated work on New Zealand fishes was published (Graham 1953), but only a small proportion of the fauna was described; approximately half of its illustrations were taken from Waite's drawings. As in Australia, a complete illustrated catalogue is still wanting.

South Australian ichthyology

The shores of the state of South Australia make up the central part of a marine biogeographic province, generally described as warm temperate, which extends approximately from Wilson's Promontory in Victoria to Geraldton in Western Australia. The South Australian gulfs comprise bodies of protected, warmer, more saline waters within this province. South Australian waters were not explored biologically before settlement so only a small proportion of South Australian fish species was first collected from South Australia itself. Port Phillip and Western Port Bays in Victoria were sampled intensively and in south-western Australia early collections were made at King George Sound. These large bays, populated and sampled early, provide the 'type localities' for many southern Australian fishes. Others were described from Port Jackson and a few, particularly cool temperate forms, were discovered first in Tasmania. The range of some species includes New Zealand and some of these were collected by Cook's naturalists and their successors.

Waite's departure from New Zealand attracted friendly notice in the local paper and he was delighted by their cartoon which he kept in his diary. News clipping from Waite diary, courtesy of Duncan Waite.

Pelotretis flavilatus was trawled at almost all stations on the *Nora Niven* cruise. Waite remarked 'Though so common and sold in all the fish shops of the Dominion as "Lemon Sole," it does not appear to have been previously named.' He described the fish in the 1911 report of the trawling expedition. From Waite 1911.

The South Australian Museum

The South Australian Institute, comprising a public museum and library, was opened to the public in 1862 by which time a variety of mineral, anthropological, botanical and zoological specimens had been assembled. The collections increased rapidly under the long curatorship of G.F. Waterhouse (see page 112) but the Museum gradually became disorganised through lack of space. This problem was mitigated by the completion of a new north wing in 1893. From 1884 to 1913 Edward Stirling (1848–1919) was Director of the Museum and, in spite of a remarkable range of interests in medical and scientific affairs in Adelaide, he played a vital part in its day to day running (Hale 1956).

Waterhouse and Haacke (see page 155), the museum's first two curators, sent most of their interesting fish overseas or to Count Castelnau in Melbourne.

Zietz, Verco and Molineux

Amandus Heinrich Christian Zietz (1839–1921), appointed preparator by Wilhelm Haacke (Director from 1882 to 1884), became the Assistant Director of the South Australian Museum in 1888, working under Edward Stirling who was then the Honorary Director. Zietz was born in Schleswig-Holstein, Denmark and had been a preparator and later a curator in the Zoological Institute Museum at Kiel. Birds and fishes were his main interests so ichthyology had its first South Australian proponent who was in a position to study the fauna. Zietz was obliged to work on such a wide range of zoological and palaeontological material that he could spend only a small proportion of his time on fishes, although he was intermittently active in marine collecting. He took specimens in the South Australian gulfs and from offshore islands, usually in the intertidal zone. Commander Walcot of the ship *Protector* (which has ended its days as a breakwater at Heron Island in Queensland) supported the museum: he supplied some large sharks and conveyed Zietz to collecting sites.

Frederick George Waterhouse. From Hale 1956.

Edward Charles Stirling. From Hale 1956.

Joseph Verco (1851–1933), an Adelaide physician and honorary curator of molluscs at the South Australian Museum, conducted excursions over many years, dredging from a ketch in Gulf St Vincent and elsewhere along the South Australian coast (Verco 1935). As an enthusiastic conchologist, Verco's main objective was the Mollusca, but he had success in collecting other marine groups as well. Among other small fish, he brought to light many of the remarkable Pipefish species that inhabit the waters of South Australia. He was the most active of the amateur naturalists who provided Zietz with specimens and Zietz travelled at Verco's invitation on collecting trips on several occasions. Seven of the fish recorded by Zietz bear the notation 'dredged (Dr. J.C. Verco)'.

Zietz carried out ichthyological researches in South Australia over a period of 30 years but the work was intermittent and unfocussed. Rather late in his tenure he began work on a catalogue of the fishes in the collection; the catalogue, *A Synopsis of the Fishes of South Australia*, was in three parts, published by the Royal Society of South Australia (1908a, 1908b and 1909). This work was unfinished but improved on Castelnau's 1872 list that attributed only 27 species to South Australia. Both of the lamprey species were included in Zietz's list, together with nearly 30 sharks and rays and about 170 bony fishes. He followed Boulenger's classification but the project came to a halt less than half way through the perciform fishes. The anglerfishes, flatfishes, leatherjackets and their kin were omitted.

Zietz collected the Gulf Catshark, *Asymbolus vincenti* from Investigator Strait in the vicinity of Kangaroo Island during 1887–88 (Glover and Olsen 1985) and eventually described it (Zietz 1908c). Zietz's *Synopsis* also included some freshwater fish from the Horn Expedition of 1894, sent to him by Professor Baldwin Spencer (see page 416). Among the specimens brought back from the Finke drainage system were seven of the nine species of fish now known to live there; several of these were new to science and descriptions were published by Zietz (1896); some notes on the habits of these fishes were provided by Spencer.

Amandus Heinrich Christian Zietz. From Hale 1956.

Joseph Verco. From Verco 1935.

In the rare book room at the South Australian Museum is a copy of Macleay's *Descriptive Catalogue of Australian Fishes*, interleaved by Zietz, with illustrations and sketches of many of the fishes. The volume is heavily annotated, demonstrating that Zietz took an interest in Australian fishes as a whole; he may have been compiling his own 'Fishes of Australia'. The book had been purchased by the Museum from Zietz on his retirement.

An impression of the activities at the South Australian Museum during Zietz's tenure is given in an unusual publication by Albert Molineux,[26] a keen Adelaide naturalist. He contributed items on natural history to the *South Australian Register* and the *Observer* under the name 'Amateur Naturalist' and published a collection of these articles (Molineux 1887). He mentioned in the introduction that he had noticed errors and 'trusts that ... the wish to be useful must be his excuse for being often wrong.' He included an index from which it is simple to locate the articles on fishes and these are interesting. For instance, there are observations on the fishes to be found in the River Torrens, indicating a richer fauna than has been seen there for many years. He refers to an Elephant Fish, *Callorhinchus milii* washed ashore at Brighton: 'A German war vessel happened to be here at the time, and

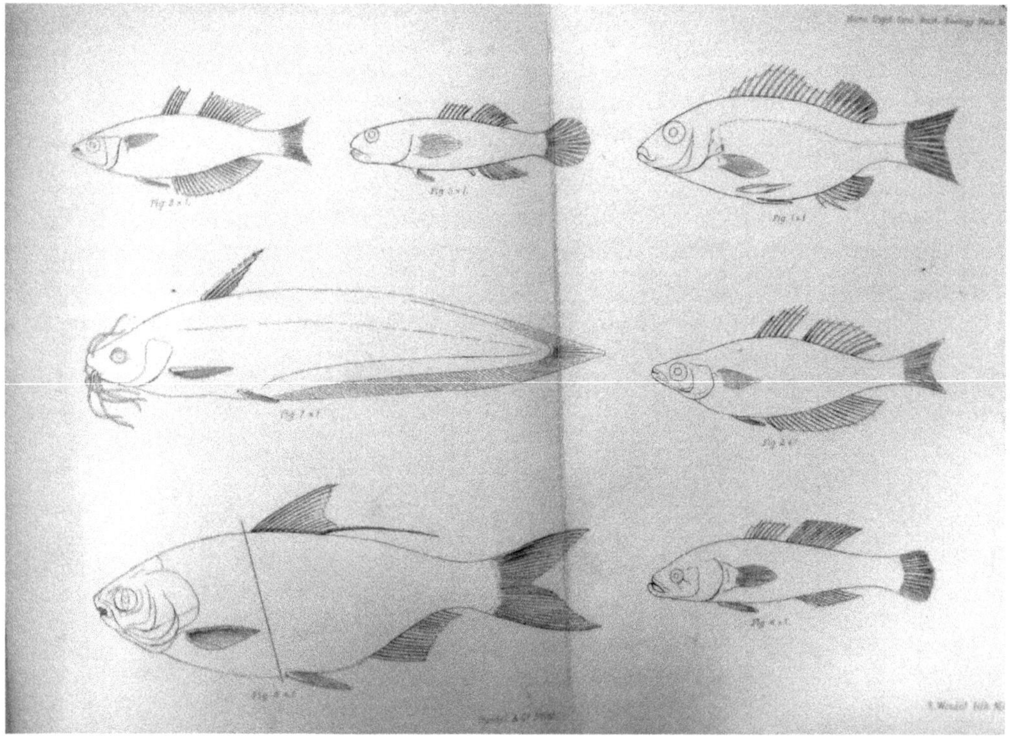

Zietz sketched fish from the Horn Expedition, describing as new species those in figures 2 to 7. Fig. 1 *Therapon* [now *Amniataba*] *percoides* Günther, 1864; fig. 2 *Nematocentris tatei* and fig. 3 *Nematocentris winneckei* [both now *Melanotaenia splendida tatei*]; fig. 4 *Eleotris* [now *Mogurnda*] *larapintae*; fig. 5 *Gobius* [now *Chlamydogobius*] *eremius*; fig. 6 *Chatoessus horni* [*Nematalosa erebi* (Günther, 1868)] fig. 7 *Plotosus* [now *Porochilus*] *argenteus*. From Zietz 1896. Courtesy Museum Board of South Australia.

26 Albert Molineux (1832–1909), a South Australian farmer known as a strong advocate for experimental agriculture, constructed his own trawl nets (ADB); the South Australian Museum accession registers indicate the many fish specimens that he donated.

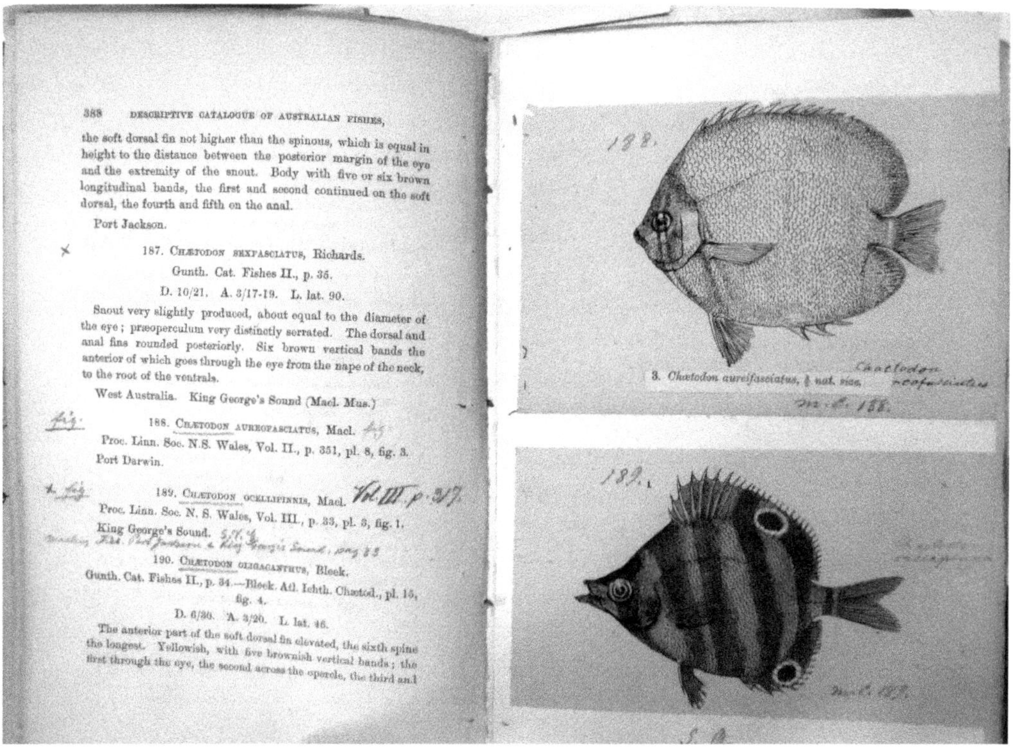
388 DESCRIPTIVE CATALOGUE OF AUSTRALIAN FISHES,

the soft dorsal fin not higher than the spinous, which is equal in height to the distance between the posterior margin of the eye and the extremity of the snout. Body with five or six brown longitudinal bands, the first and second continued on the soft dorsal, the fourth and fifth on the anal.

Port Jackson.

187. CHÆTODON SEXFASCIATUS, Richards.

Gunth. Cat. Fishes II., p. 35.

D. 10/21. A. 3/17-19. L. lat. 90.

Snout very slightly produced, about equal to the diameter of the eye; præoperculum very distinctly serrated. The dorsal and anal fins rounded posteriorly. Six brown vertical bands the anterior of which goes through the eye from the nape of the neck, to the root of the ventrals.

West Australia. King George's Sound (Macl. Mus.)

188. CHÆTODON AUREOFASCIATUS, Macl.

Proc. Linn. Soc. N.S. Wales, Vol. II., p. 351, pl. 8, fig. 3.

Port Darwin.

189. CHÆTODON OCELLIPINNIS, Macl.

Proc. Linn. Soc. N. S. Wales, Vol. III., p. 33, pl. 3, fig. 1.

King George's Sound.

190. CHÆTODON OLIGACANTHUS, Bleek.

Gunth. Cat. Fishes II., p. 34.—Bleek. Atl. Ichth. Chætod., pl. 15, fig. 4.

D. 6/30. A. 3/20. L. lat. 46.

The anterior part of the soft dorsal fin elevated, the sixth spine the longest. Yellowish, with five brownish vertical bands; the first through the eye, the second across the opercle, the third and

3. Chætodon aureifasciatus, nat. size.

Pages from Macleay's *Descriptive Catalogue of Australian Fishes* annotated by Zietz who also inserted his own drawings of the fish or illustrations from publications, often by Macleay. Macleay (1878) described *Chaetodon aureofasciatus* and *Chaetodon ocellipinnis*. Zietz inserted illustrations from this paper into his copy of Macleay's *Catalogue*. *C. ocellipinnis* proved to be a juvenile of the Moonlighter, *Tilodon sexfasciatus* (Richardson, 1842). Courtesy Museum Board of South Australian.

one of the passengers, who was ashore, having never seen or heard of such a curious fish, was induced to purchase the rarity for four sovereigns.' Molineux also listed some of the specimens that had been sent to Castelnau in Melbourne and noted the Pouch Lamprey, *Geotria australis* from the Torrens, offering the information that 'The lampreys get their living like lawyers and leeches by sucking the blood of their victims.' He prepared a similar collection of articles in 1889.

From the time Zietz retired in 1909, at the age of 70, to Waite's appointment in 1914, there were no significant additions to the South Australian fish collections. Stirling continued as Director of the museum until Waite arrived.

Waite's early work in South Australia

Waite's priorities in Adelaide did not at first include ichthyology. After finding temporary accommodation he began house-hunting. This differed from the early days in Christchurch as he and Rose were prepared to settle in Adelaide. As Waite assumed his work at the Museum, he began meeting that part of the community with an interest in natural history and familiarising himself with Adelaide. He was a little disappointed to find that the new East Wing, only recently constructed for the Museum, had been partly occupied by the Art Gallery and that the arrangement of display cases was rather inflexible. However, the museum in Adelaide had several advantages over that in Christchurch. The greatest of these was that, although each institution maintained only a few paid staff, the South Australian Museum was served by a number of remarkably dedicated and active Honorary Curators. The fish collection from Mawson's Australian Antarctic Expedition arrived three months after Waite came to Adelaide and this was his first ichthyological work in South Australia.

The fishes of Antarctica soon had Waite intrigued, 'Commenced sorting Mawson's fishes, some strange forms among them.' As he spent more time comparing the fishes with the descriptions just published in a preliminary report on the recent *Terra Nova* collection (Regan 1914), Waite was a little disappointed, 'Most of the novelties have been snapped-up by Regan ...'

The work was interrupted by a spectacular occurrence: an enormous shark had been captured at Fowler's Bay on South Australia's west coast and brought to Port Adelaide. Waite went to the Port to see the animal: it was a Basking Shark, *Cetorhinus maximus* nearly eight metres in length and Waite made arrangements for it to be delivered to the Museum. He decided to make a cast of it, a mammoth task, which occupied him and most of his staff for four days. They worked over a weekend, his diary entry for the Saturday reading:

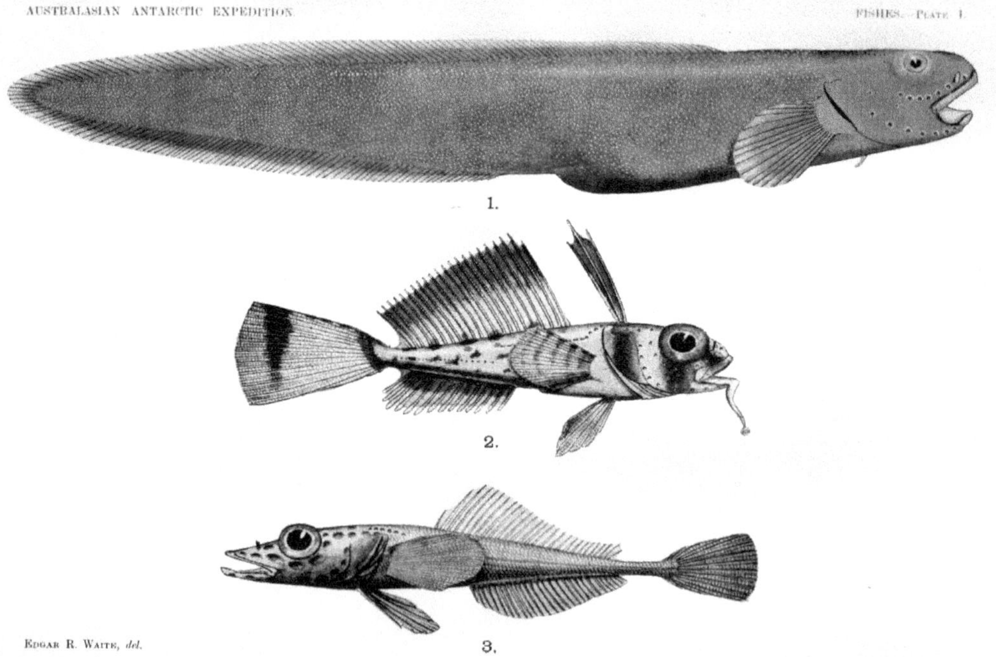

Waite drew some of the 'strange forms' that he found among Mawson's collection from the Shackleton Ice Shelf. He identified *Austrolycichthys brachycephalus* [now *Pachycara brachycephalum*] (Pappenheim, 1912) (top) and *Dolloidraco longedorsalis* Roule, 1913 (centre); and described *Bathydraco* [now *Akarotaxis*] *nudiceps*, a new species (bottom). From Waite 1916.

Basking Shark, *Cetorhinus maximus* cast in 1914 with Waite to the right of the dorsal fin and Hale on the far right. From Hale 1956.

> Spent the whole day at the Shark with the following staff, Limb, the two Rau's, Conroy and Hale. Got the body sewn up and filled out & arranged in position, the head partly done and the fins cast separately. Arranged to work again tomorrow & for disposal of the carcase on Monday.

On Sunday they worked 'from 8 am to 7 pm' and on Monday removed the mould to find that the procedure had been successful. The next day Waite heard that 'Britain declared War against Germany, a European upheaval - Attended a patriotic meeting Flags, songs & cheers.' A week later the newspaper informed him that work on the shark would have to stop as all resources were to be committed to the war effort. Waite commented, 'What a serious thing war is!!' (WD) He proceeded to cast the shark from the mould and coloured the cast preparatory to exhibiting it.

The *Simplon*: trawling again (1914)

In September 1914, Waite was informed of an imminent trawling trip. The Commonwealth experimental trawler *Endeavour* had made good catches in the Great Australian Bight and the South Australian Government hired a vessel, the *Simplon*, to explore further. Waite assembled his tow nets and photographic gear, including 'a dark lantern for plate changing'. He found the *Simplon* crowded and dirty and arranged his sleeping quarters in a small boat on the deck, but '[a]s the sea became rough I removed my "bedding" to a seat in the cabin, the Capt deeming the boat unsafe.' (WD) Waite found plenty to interest him but the results were disappointing from a commercial viewpoint. They steamed 1600 kilometres on the round trip between Adelaide and the trawling grounds. As Waite remarked in his report, the cost of coal and wages could only have been offset by phenomenal catches. However, he recorded nearly 50 species of sharks, rays and fish, including several new species. The biggest catches were of Swallowtail, *Centroberyx lineatus*, a beautiful fish but not one that found a ready market.

Mawson's fishes

Back in Adelaide Waite resumed studying Mawson's fishes. He wrote to McCulloch and to the Linnean Society of New South Wales in Sydney for the loan of literature; and to L. Roule in France for the report of the second French Antarctic expedition (Roule 1913). Meanwhile, Regan sent Waite his definitive report on the *Terra Nova* fishes (Regan 1914). Waite found that the small blue fish which had been washed on board the *Aurora* in stormy weather on the return from Macquarie Island proved to be of special interest. He correctly deduced that it belonged to the family of Thornfishes (Bovichtidae) but regarded it as a new species, establishing a new genus for it. He named this fish *Aurion effulgens*, but it is now regarded as the juvenile stage of another Thornfish, *Bovichtus psychrolutes*. The Antarctic fish study was progressing as one project among many. Waite had purchased a house and was building a workshop; he was reporting on a collection of reptiles from a recent expedition to Central Australia; writing long articles for the newspapers, for which he was paid; modifying his latest motorbike; participating in local field trips; designing a system to suspend the massive Basking Shark cast.

Two months after returning from the *Simplon* trawling cruise Waite had completed the identifications and descriptions of the Antarctic fishes. Then he turned to those from Macquarie Island, which were to be included in the same report: 'Now commenced examination of the Macquarie Is. fishes. Harpagifer & Zanclorhynchus[27] are the plums.' Months later he wrote 'the end is now in sight!!' The interesting forms had given him more than a few headaches but the paper was impressive. As various scientists completed their portions, Mawson made moves to publish. As so often, for a large work, there was difficulty but Mawson organised a deputation that included Waite and they put their case before the Premier of South Australia, whose Government eventually published the report (Waite 1916a).

The drawings were still to be done, but he settled down to them after a trip to Sydney. His diary entries show that the illustrations were each done over the course of one to three days. While in Sydney in the 1890s, Waite used sometimes to note (in his diary) that a drawing did not satisfy him, and he would then start again; by the time he was in Adelaide his skill had become more certain. He commonly drew in 'wash' (water colour painting) and prepared some drawings that he coloured, although cost generally precluded the publication of coloured illustrations. Photography, at which Waite was very experienced, usually displeased him as a means of representing fishes due to foreshortening of the image extremities.

Collaboration with McCulloch: initial work (1915)

Waite went to Sydney early in 1915. He was charged with Museum business for Stirling in Melbourne on the way, and in Sydney he was to arrange exchanges. He packed a selection of the *Simplon* fishes and meant to use McCulloch's expertise in dealing with them, and in helping to prepare a catalogue of the fishes of South Australia. He clearly saw McCulloch still as an assistant; a respected assistant, but junior for all that. They laboured together for a week at the Australian Museum, the work being resolved into three components, each of

27 Waite identified the Spiny Plunderfish, *Harpagifer bispinis* (Forster, 1801) but this species was later described as *H. georgianus* Nybelin, 1947 (Miller 1993); the Spiny Horsefish, *Zanclorhynchus spinifer* had been described from Kerguelen by Günther in 1880.

which was to result in a publication. The *Simplon* fishes comprised 40 species, five of which were described as new, and included a particularly good series of Boxfish.

The Boxfish provided the basis of a separate study, for which specimens of the group from both the Australian and South Australian Museums were examined (McCulloch and Waite 1915). The southern Australian Boxfish belong to a subfamily, the Aracaninae, of exotic form and colour. They have a rigid dermal skeleton of hexagonal bony plates, armour that is lacking only around the mouth, vent, gills, eyes, nostrils and the bases of the fins. Species of this group are found from Hawaii to South Africa but it is best represented in the waters of southern Australia where seven species are found. The *Simplon* had taken five of these species, and some were found among the catches of the *Endeavour* and the *Thetis*. In their paper, McCulloch and Waite gave accurate descriptions and a succinct generic key. McCulloch also presented 10 drawings, in the belief that he was representing 8 species. Boxfish are a deceptive lot, changing form with growth and exhibiting extreme sexual dimorphism and dichromatism. The authors of the paper had regarded the males and females in several instances as distinct species; in reality, McCulloch's drawings showed only five distinct species.

McCulloch and Waite gave accurate descriptions of 'Boxfish' but did not appreciate the variation with growth and between the sexes. They identified the Spiny Boxfish, *Capropygia unistriata* (Kaup, 1855) (left) but thought the male and female of the Humpback Boxfish, *Anoplocapros lenticularis* (Richardson, 1841) (right) were different species. From McCulloch and Waite 1915.

McCulloch and Waite identified the female Shaw's Cowfish, *Aracana aurita* (Shaw, 1798) (left) but followed Richardson in regarding the male (right) as a separate species, *Aracana spilogaster* (Richardson, 1840). From McCulloch and Waite 1915.

There was initially to be a single paper on the *Simplon* fishes but McCulloch, who continued work on them in Sydney, wrote to Waite, 'Letter from Allan, he asks that our joint paper may be by McCulloch & Waite, he has done the bulk of the work.' (WD) Not unreasonably, McCulloch wanted to appear as the senior author. Waite delayed answering for one month and then said that he would ask the Publications Committee to hand the project over to McCulloch. McCulloch's response was to suggest splitting the paper, Waite having sole authorship of the *Simplon* fishes (to which he had made a greater contribution) and McCulloch authorship of the Boxfish. Waite then suggested a relatively equitable arrangement whereby they would have joint authorship of both papers, but Waite being senior author of the *Simplon* and McCulloch senior author of the Boxfish. McCulloch agreed and in this way the papers were finally published by the Royal Society of South Australia (McCulloch and Waite 1915; and Waite and McCulloch 1915). There was some strain in this bartering of credit but both men were apparently satisfied by the final decision.

South Australian Fish Catalogue: work begun (1915)

The collaboration between Waite and McCulloch in Sydney in 1915 also involved the first steps towards a catalogue of the fishes of South Australia. Waite wrote, '26. Tues. Public Holiday. Spent the day at Museum making Catalogue of S.A. Fishes.' and again, on the morning of the day he caught the train back to Adelaide, '29. Fri. Spent morning at the Museum on Catalogue of Fishes S.A.' (WD) Waite then appears to have handed the work over to McCulloch to pursue while he returned to drawing Mawson's fishes in Adelaide.

Fish catalogues, at least in Australia, always appear to have presented special problems. Ogilby's efforts demonstrated some of the possible reasons for failure: personal disorder (drinking); an over-ambitious, inclusive approach (he tried to incorporate every detail imaginable); lack of a plan (no clear design or schedule for his work was ever evident). Such projects also have inherent problems: they are likely to result in a work too large for the usual scientific periodicals to publish, and this reluctance may be compounded if illustrations are included. The need for special publication usually means that the government will be asked to provide finance. If the workers are staff in a Museum, they may need to have on-going permission to work on the project. As a result Australian fish literature has more than its share of unpublished manuscripts and 'first parts'.

There were several reasons why McCulloch found himself doing the lion's share of the work for the South Australian catalogue. He had established a valuable card index at the Australian Museum containing many of the references vital to the synonymies, the museum fish collection had most of the relevant species and his access to the literature was superior to Waite's in spite of the latter's best efforts. Waite had great respect for McCulloch's abilities as an ichthyologist, consulting him about difficult problems and frequently deferring to his judgment; but he still regarded McCulloch as a colleague to whom he could delegate work.

Two months after Waite returned to Adelaide from his Sydney trip, McCulloch dispatched a first section of his work on the S.A. fish catalogue. Waite had resumed consistent work on the Antarctic fishes and seemed to make no further contribution to the catalogue himself at this stage.

A new assistant: Herbert Mathew Hale

Shortly after his arrival in Adelaide, Waite was approached by a young man, Herbert Hale, wanting a job as a cadet at the Museum. Hale gave an account of this:

> At this time the writer [Hale] had gathered a small private marine zoological collection and had hoped to find a place on the staff of the Museum, in which much of his spare time was spent; there had seemed little chance of this and he was being trained in accountancy ... Immediately after his appointment Waite asked for a cadet to function as his 'personal assistant,' to use his own words. Application was made by the writer but Waite was hesitant, having visualised a younger lad for the position and moreover one who was not on the threshold of a career completely dissociated, apparently, from Museum activities. (Hale 1956).

Herbert Hale (1895–1963) became Waite's assistant in 1914. He worked with Waite on fishes but made crustaceans his specialty, rarely publishing on fishes after Waite's death. Courtesy Museum Board of South Australia.

Waite accepted Hale's application and they worked together until Waite's death. Hale was involved in most of Waite's projects. He was soon occupied in making an index to Waite's published papers and another of his reprint collection, which by this time comprised many hundreds of his colleagues' publications. Like McCulloch before him, Hale had the benefit of Waite's friendship and instruction. He was tutored in drawing, photography and aquarium keeping, learnt museum procedures and taxonomy, and many other skills. He also participated in the Waite family outings and in field trips. Hale pursued the study of Crustacea, a more comfortable occupation than competing with Waite.

Reprints at the South Australian Museum

The ichthyological part of Waite's reprint collection survives at the South Australian Museum in a damaged and depleted state. However, it gives an interesting view of his contacts with ichthyologists around the world. Most of the papers were sent in exchange or as gifts during the lifetime of their authors, although a few of the older works were obtained through purchase. The papers were derived from approximately 50 correspondents, from 15 countries. Jordan's students and associates, to the number of 20, comprised the largest group of correspondents. From Jordan alone, Waite received over 90 papers. Nearly 120 of C.T. Regan's papers are included and Franz Steindachner contributed at least 60 reprints. Waite first wrote to

Steindachner in 1898 and the reply came with reprints of Steindachner's important articles, and a suggestion that a collection of Red Sea fish would follow. (Steindachner to Waite, 12.11.98, IDSAM) Many ichthyologists sent their books as well, and found Waite a very grateful recipient, 'I received from Garman his magnificent work on Selachians [Garman 1913].' (WD, 4.10.14). Among others, he wrote to Max Weber of Amsterdam suggesting an exchange of reprints, which duly occurred; then Waite received 'from Max Weber his "Fishes of the Siboga Exped." [Weber 1913] an imposing and valuable work. additional to papers received on the 8th. Supplemented my packet to him with Cat. Fish S.A.'(WD).

Aquaria in Adelaide

Waite had shown a special interest in maintaining aquaria since he first kept small native fish from the streams around Sydney. He tried marine aquaria and bred nest-building exotic freshwater fish. Shigeho Tanaka, an associate of David Jordan and Lecturer in Zoology at the Tokyo Imperial University, had sent Waite enormous numbers of 'Japanese goldfish' for the New Zealand Exhibition.

In Adelaide, Waite attempted to interest the Zoological Society in aquaria, but with only limited success. At the Museum, he '[f]itted up a 'sample' Aquarium in one of the Museum windows, for inspection of the President of Zool. Soc.' (WD, 11.1.15). The Museum Board approved of the aquarium and a series of 22 were purchased and installed in the Museum. Waite collected locally both marine and freshwater fishes for these aquaria and for others that he was building at home. He noticed an advertisement for an old tramcar, bought it for four pounds and arranged for it to be delivered to his back yard: 'It would do well for an Aquarium House: Venetian shutters etc.' (WD, 2.4.16). He set to work and the new aquarium house flourished, together with aquaria that he maintained at the museum.

Early in 1918, Waite began to look for support for an aquarium society and in March noted:

> Inauguration of Aquarium Society, as convenor, I took the chair & was afterwards elected Chairman, with Hale as Sec. We had an attendance of 12.

In his presidential address, Waite drew attention to the importance of keeping native species of fish to make observations on their life cycles and behaviour. Under Waite's guidance the Aquarium Society of South Australia conducted excursions, regularly visiting the Murray River at Murray Bridge and other collecting sites. Waite encouraged the members to keep native fish, as he was aware that their life histories were almost entirely unknown. These efforts bore occasional fruit, as in the case of Clarence Blewett who published chatty but useful observations on the life histories of native Gudgeons (*Mogurnda adspersa* and *Hypseleotris klunzingeri*) and Rainbow Fish, *Melanotaenia fluviatilis* from the River Murray (Blewett 1929).

Waite became very enthusiastic about the possibility of starting a public aquarium close to the shore. He envisaged the project as half public aquarium and half research centre but support was difficult to enlist. He kept the project in mind for years and in 1922 began canvassing support from the community, referring to the advantages of a large public aquarium in public lectures, and discussing ways and means with Councils. Waite wrote to the directors of aquaria in Honolulu and San Francisco for support and information. Waite had Hale making lantern slides of aquaria and biological stations around the world: the lecture was finally illustrated with 64 slides.

Once the public aquarium issue had been discussed for a few months, Waite began to write for a wider audience, submitting articles for the newspapers and preparing a pamphlet intended for wide circulation. He had support from the Naturalists' and Aquarium Societies and a 'Glenelg [beachside suburb] Aquarium Committee' was formed. Unfortunately, the project was not brought to a satisfactory conclusion in his lifetime although an aquarium was established in the middle of the Glenelg jetty in 1929.

Casts for exhibition

As fish were brought to the Museum Waite had casts made for the galleries, often painting them himself. He had an arrangement with Daw's fish market: they brought him fishes of special interest. The casts were an effective means of interesting the public and the work was noticed by his Museum Board, 'At its last meeting, the Board recorded its appreciation of the way in which casts are being prepared for the Museum.' (WD, 23.3.15) He could rarely resist a large fish, plaster replicas of a Wobbegong and a large Stingray, the latter '5 ft. [150 cm] across the disc', being prepared (and still exhibited). (WD, January 1926).

Relationship with New Zealand zoologists

Waite's connection with New Zealand zoologists and fisheries personnel was maintained and they frequently sent him descriptions or photographs of unusual fish for identification, and sometimes sent specimens. The new curator at Canterbury was Speight, a geologist; he relied on Waite for help with the fishes and seemed to be well served. Harold Hamilton, from whom Waite had received fishes on Macquarie Island and who later became curator of the Newtown Museum at Wellington, wrote a paper on *Trachipterus*, a rare deepsea fish, and sent the manuscript to Waite for comment and criticism; Waite obliged, and his help was acknowledged: '[I] ... extend my thanks to Mr. E.R. Waite, of the Adelaide Museum, for his sound advice on the arrangement of the subject-matter.' (Hamilton 1916). In April 1917 Waite received photographs of two fish from Allan Thomson in New Zealand, 'Wrote to Allan Thomson his fishes are Germo [Albacore] & Leptoscopus macropygus [Sandfish].' They no longer asked him to prepare a catalogue but used him for advice since no one with a comparable knowledge of fishes resided in New Zealand.

Field trips

Field naturalist, ichthyologist, herpetologist and museum manager are the titles most indicative of Edgar Waite's work. When he arrived in Adelaide a reporter asked 'Have you a preference for work in the open rather than in the laboratory?' and he replied 'I love my work whatever it be.' This appears to have been the simple truth. He had chosen an ill-paid profession but revelled in all aspects of the work, with the exception of the squabbles that arose around administrative problems. In the field, his enthusiasm was unbounded. He seized opportunities for excursions whenever possible and the Adelaide position fortunately provided him with some fine journeys. His two longest trips, Cooper's Creek and New Ireland, are largely incidental to ichthyology, but worthy of brief accounts.

Cooper's Creek (1916)

Waite led a small party to Cooper's Creek in northern South Australia, assisted by Captain S.A. White, an ornithologist. They used camels, which Waite seemed rather to enjoy: 'My camel, which is a big one, counted several times again to-day takes exactly 36 steps of each foot per minute, that is 72 swings, in the saddle.' (WD) The journey took seven weeks in September/October, 1916 and the collections were sent back to the Museum. The expedition was generally orientated towards ornithology and provided many chances to observe and collect birds. However, they collected omnivorously: anthropological material, insects, reptiles and a few mammals. The sole ichthyological opportunity arose at Cooper's Creek where Waite tried fishing: 'Caught a Therapon, & then another and a tortoise on the same line with two hooks.' The fish appeared to be 'new'.

Back at the Museum, Waite looked through the preserved specimens and located another odd 'Therapon Perch' (Grunter) in the 'old collection'. This specimen had also been taken from Cooper's Creek, but no other details were available. He thought it was another undescribed species but the relevant collections were in Sydney so, once again, it was McCulloch who had the task of describing the two species. The descriptions were published with the *Results of the South Australian Museum Expedition to Strzelecki and Cooper Creeks* (McCulloch and Waite 1917). The fishes, both restricted to central Australia, were Welch's Grunter, *Bidyanus welchi* and the Barcoo Grunter, *Scortum barcoo.*

New Ireland expedition (1918)

When the prospect of a trip to New Guinea arose, Waite could hardly believe his good fortune:

> Stirling has been very keen on getting Ethnological material from (late) German Territory & he proposed an Expedition to New Guinea, New Britain, and Solomons. This was agreed to, the Director [Waite] & A.C. Davis late administrator of the natives in New Ireland to form the party. If the Board approves I am likely to realise one of my most fervent dreams; to visit the land of the Bird of Paradise & other groups I had not dared to even dream of. (WD, 1.5.18)

Waite spent ten weeks collecting ethnological and zoological specimens: the trip was very successful although most of the time was spent in New Ireland. He tried collecting small fishes from the onshore reefs and made interesting observations of the use of 'rope', an organic fish poison extracted from a plant. Most of his small collection of fishes, very much a minor by-product of the enterprise, was lost when the flask containing them was smashed in stormy weather in Torres Strait on the return journey. In New Ireland Waite was infected with the malarial parasite and, from that time on, he was subject to episodes of fever.

Brief excursions

Short field trips with members of the aquarium society and occasional dredging ventures with the South Australian Field Naturalists' Club sometimes produced useful specimens, particularly for the aquaria. The inspection and retrieval of whale skeletons were occasions for trips, which usually involved wider collecting as well. In this way, Waite saw Streaky Bay (27 metre Blue Whale, September 1918), Kingston (5 metre Strap-toothed Whale, February 1919) and Port Wakefield (7 metre Blue Whale, September 1925). His friends liked to include him on field excursions; he was good humoured, uncomplaining in adversity and always active. On the trip to recover the whale skeleton from Kingston in 1919 he was accompanied by Archibald Watson, whose noxious sense of humour was generally a source of pleasure to Waite. He suffered an episode of malaria while visiting Mount Gambier, writing:

> 6. Thurs. By motor to wonderful lakes only 1¼ miles from town, after breakfast. We had arranged for one to Pt. McDonnell in afternoon but while visiting the Museum!!! [he thought it disorganised] an attack of malaria came on & I had to lay up. Watson picked up an Archdeacon & took him instead. He said he would bring him along if I felt like dying. (WD, 6.2.19)

Weekend field trips with friends frequently took Waite to visit the creeks and shore, and he collected fish whenever possible. These excursions extended from the south-east of the state, to sites in both gulfs, and the west coast. There were also occasional short excursions by boat, and longer ones to Kangaroo Island and the smaller islands off the west coast. Waite enjoyed the sea, not least because he was never subject to sea-sickness, or so he claimed. Hale (1956) recalled an occasion at variance with this:

> Rough weather and Waite—as well as others—suffering as a consequence. Aboard was Archibald Watson, Professor of Anatomy at the University of Adelaide and perhaps the most extraordinary personage who has ever held a Chair at the University. At the height of our discomfort Watson appeared from below with a large tin of sardines and a tumbler of whiskey - neat. He ate the fish, drank the olive oil in which they were preserved, followed it with the spirits and remarked 'If you had a meal like Christians you wouldn't feel ill' —and *he* wasn't. (Hale 1956)

In 1918 land reclamation was being undertaken near Murray Bridge. Fish in the enclosed area were affected and Waite formed one of a party from Adelaide who inspected the damage:

> The water in one place is confined to a wide deep drain which is being pumped out. The fishes, mostly Carp, are in millions all gasping at the surface. We walked up bank of the drain for ½ mile and their [sic] was no diminution of the fishes the whole way ... all heading the (falling) current they gradually congregate at the pump intake & are taken out at the rate of 5 tons a day. Murray Cod, Catfish, <u>Pseudaphritis</u>, [the Congolli, *Pseudaphritis urvillii*] <u>Eleotris</u>, <u>Galaxias</u>, etc. all dead the Carp [presumably the Common Carp, *Cyprinus carpio*] excepted. (WD, 23.3.18)

This scene was enacted before the ecology of the Murray was degraded by overextraction of water and the present system of dams, locks and barrages.

South Australian Fish Catalogue: work resumed (1917 to 1920)

Waite turned his attention to the South Australian fish list again early in 1917 and in May of that year sent it back to McCulloch prior to visiting Sydney in July. The list travelled to and fro a couple of times, with annotations from McCulloch. At this time Ogilby and McCulloch were working on gobies and, at his request, Waite extracted these fish from the Museum collection and sent them to Sydney. The fish list was progressing well but Waite was then informed that 'the Minister will defer publication of the S.A. list of fishes for the present'. This was the first of a series of delays that McCulloch interpreted as vacillation on Waite's part.

'Am taking a lot of spiders for Pulleine [an Adelaide spider expert] to Sydney.' (WD) Waite generally had a multitude of tasks when he travelled interstate and his Sydney trip of July 1917 was no exception. However, he still wanted to publish a South Australian fish catalogue and wrote: 'Spent day at Museum working over S.A. fish list with Allan.' (WD, 11.7.17) This trip included a rapid visit to Brisbane, where Waite was establishing a collaboration with Heber Longman at the Queensland Museum for the purpose of preparing a book on snakes. Waite had six days in Brisbane and his time was packed with meetings and minor excursions; these included an afternoon with Tom Marshall, Ogilby's young protégé, who took Waite to Ithaca Creek for the purpose of netting small fish. He managed to bring most of these back to Sydney, and eventually to Adelaide, for the aquaria.

Once in Sydney again, Waite spent time at the Museum with McCulloch and:

> made definite arrangements with Allan for two papers as follows
> (a) Illustrated Cat. Fish of S.A. by W. & McC.
> (b) Studies in S.A. fishes, McC and W. (WD).

Their first collaboration, which resulted in the *Simplon* fish papers had been a success, although McCulloch complained to Ogilby that Waite had done little of the work. The prospect of further joint papers on the fishes of South Australia apparently did not daunt McCulloch. On Waite's return to Adelaide, he approached the President of the Royal Society of South Australia, suggesting that they publish the fish catalogue. The initial response was favourable and Waite made another advance:

> Meeting of Museum Committee. I obtained authority to expend £10 for drawings of fishes by Miss P. Clarke of Sydney to illustrate papers by Allan and self conjointly. (WD, 1.8.17)

He wrote to McCulloch, telling him that money was available for him to have some illustrations prepared, and that:

> I should apply for more with a view to publishing illustrations of all unfigured S.A. fishes and incidentally working through all our species thus producing a series of joint papers. (WD)

The scheme was growing, almost 300 illustrations being contemplated, and the revision of many groups of fishes, but it was not clear just how much of the work Waite was prepared to do himself.

Shortly after Christmas 1918, the Waites visited Sydney again and Edgar, as usual, went directly to the Museum with the manuscript of the fish catalogue. He and McCulloch worked on it in the first week of 1919, but then Waite became ill:

> While working at the Museum with Allan, an attack of fever came on & I had to make haste to the house and go to bed. (WD, 8.1.19)

By this time, Waite was becoming accustomed to episodes of his recurrent malarial fever, and expected them to pass over the course of a few days, although they left him feeling weak and shaky. He returned to work at the Museum after a week:

> 15. Wed. To Museum in morning. Arranged details for Cat. of Fishes of S.A. by McC. & W. and saw Etheridge re use of original drawings. (WD, 15.1.19)

Financial difficulties associated with the war appeared to have eliminated the chance of substantial expenditure on a catalogue. McCulloch, who had become very frustrated by the intermittent bursts of energy from Waite, allowed the work to lapse.

In October 1919 Waite found that the South Australian Fisheries Department had allowed £100 towards the work and the Museum would find the balance if necessary. He wrote to McCulloch, telling him that the work could proceed. By this time McCulloch was well advanced with preparing his own illustrated catalogue of the Fishes of New South Wales and in researching the fishes trawled by the *Endeavour*; however, he responded, assuring Waite that he would press on with his part of the work. Six weeks later McCulloch wrote again:

> Letter from Allan; he withdraws from joint authorship of Cat. of Fishes, wanting to get on with the 'Endeavour' fishes. (WD, 24.12.19)

McCulloch sent the manuscript to Waite and, not surprisingly, work on the project slowed even further, although Waite made occasional notes of his progress. In May 1920, he asked the Fisheries Department if the £100 might be held over until the next year.

Eyes on Sydney again (1920–1921)

Robert Etheridge, the Director of the Australian Museum died in January 1920. Once again Waite considered his chances, and once again Charles Hedley was a candidate as well. In canvassing support, Waite found that several influential zoologists whom he approached for references intended to support Hedley. However, he assembled an impressive set of testimonials and, when the position was advertised, he applied. He did not seem particularly anxious this time, perhaps because Hedley had been appointed Acting Director and he could be expected to be confirmed in the job, perhaps because Waite was happy enough in Adelaide. Be that as it may, after a hiatus of more than 12 months (perhaps too long for suspense), Waite heard that Dr Charles Anderson had been installed as Director. Waite's comment was: 'The Australian Museum produces some surprises, here is the latest ...' In his diary he included the newscutting reporting the appointment.

Records of the South Australian Museum

Most zoologists in Adelaide before 1918 published their work in the *Transactions of the Royal Society of South Australia* but Waite decided to found a Museum journal, along the lines of the *Records of the Australian Museum*. He had established the *Canterbury Museum Records*, which had been a limited success, and believed that the South Australian Museum needed a publication of this kind. He approached his Museum Board, who proved cooperative:

> Board granted me the £25 requested for preparing drawings of fishes and approved my suggestion to issue a Museum 'Bulletin.' I am to draw up necessary details as to size and general 'get up.' My present idea is for No. 1 to contain my review of Aust. Typhlopidae, [he had returned to the study of Blind Snakes] notes on fishes by McC. & W. and possibly a Catalogue of Australian Lizards now being prepared by Zietz. [Robert Zietz, the son of A.H.C. Zietz, who had retired in 1909]

The first volume of the new journal was published in four numbers between 1918 and 1921. There were 18 articles by 12 contributors, making over 400 pages in all. Waite was the editor and a major contributor. His work at the time is reflected in the variety of his contributions:

Waite	1918	Review of the Australian Blind Snakes
Waite	1918	Description of a New Blind Snake ...
McCulloch & Waite	1918	Some New and Little-known Fishes ...
McCulloch & Waite	1918	Description of Two New Australian Gobies
Lea	1918	On Australian Coleoptera, Part I
Stirling & Waite	1919	Description of Toas, ... [28]
Waite	1919	Two Australasian Blue Whales ...
Turner	1919	Description of Three New Species of Thynnidae
Waite & Longman	1920	Descriptions of Little-known Australian Snakes
Zietz	1920	Catalogue of Australian Lizards
Rainbow	1920	Arachnida from Lord Howe and Norfolk Islands
Lea	1920	On Australian Coleoptera. Part II
Waite	1920	Notes on Radiographs of Two Mice
Waite & Hale	1921	Review of the Lophobranchiate Fishes ...
Cilento	1921	Observations on ... Artificially Distorted Skulls
Berry	1921	A Review of the Cephalopod Genera ...
Ferguson	1921	a List of the Tabanidae ...
Lea	1921	On Australian Coleoptera, Part II

To volume 2 Waite contributed seven of the 21 articles; of these seven, there were four on fishes, and the first was the long awaited *Catalogue of the Fishes of South Australia*.

28 Stirling died in 1919 and the paper on toas, aboriginal artefacts, was written by Waite from Stirling's notes at the request of Stirling's family.

Collaboration with McCulloch (1916 to 1918)

McCulloch and Waite published two papers in the *Transactions of the Royal Society of South Australia* in 1916. One of these recorded corrections and additions to the fishes reported from Lord Howe Island (McCulloch and Waite 1916), including the description and illustration of several new species. The collections upon which this paper was based came mostly from the residents of the Island although Arthur Lea, on a recent visit, had also made a small collection. Waite's contribution to this publication was slight but he appeared as the sole author (Waite 1916b) of the second paper, *A List of the Fishes of Norfolk Island and Indication of their Range to Lord Howe Island, Kermadec Island, Australia, and New Zealand*).

Arthur Lea (1868–1932) was an entomologist with a long-standing interest in fishes. He visited Lord Howe and Norfolk Islands in 1915, studying entomology, and made collections of fishes for the South Australian Museum at each island. Waite received the collections in March 1916 and noted 'One of Lea's fishes is probably new, a long-anal Apogon (Amia) perhaps Archamia.' It was essential to compare the species collected with those in the Australian Museum, so Waite contacted McCulloch. This entailed a great deal of work for McCulloch: he made his own lists of the Australian Museum specimens from Norfolk and the Kermadec Islands, compared the specimens with those sent by Waite, and described at least one of the new species, providing an illustration as well. Waite (1916b) made full acknowledgement of all aspects of the help that McCulloch gave but did not include him as co-author.

In December 1917 Waite organised a one week Museum expedition to Kangaroo Island during which fish collection was a major objective. The party fished from the shore and a

McCulloch contributed a drawing of a new species to Waite's paper on Norfolk Island fishes, the Smallscale Kelpfish, *Chironemus microlepis*. The drawing of the type specimen bears McCulloch's signature and the legend 'Waite's Norfolk Is. Paper. Plate C. *Chironemus microlepis* sp. nov. I 5411 Type. Nat size. 1.10.16'. From AA356 Waite Collection, courtesy of the South Australian Museum Archives.

boat, used nets, 'limed' pools and waded the shallows at night, spearing fish. They collected a large number of species, from many of which casts were prepared on the spot. Waite had brought Robert Limb (preparator) and Gustave Barnes (artist) from the Museum, and they worked from the fresh specimens: 'What with own and fishermen's effort we are keeping Limb and Barnes occupied.' The collections from shallow rock pools proved to be of special interest, yielding species that had not previously been described, and some which had not been recorded since their original description.

McCulloch and Waite (1918a) joined forces again to produce a valuable paper on South Australian fish. It formed a significant portion of the first issue of the *Records of the South Australian Museum*. Although McCulloch was the senior author, Waite provided all of the specimens: he had ransacked the 'old collection' of the Museum and added his rock pool fishes from the Kangaroo Island excursion of the previous year. The fish were described by McCulloch, and illustrated by him and Phyllis Clarke. Twelve new species were described, most of which are still recognised today, and several new genera established. Some of the species described by Günther, Castelnau, Steindachner and Macleay were re-described and illustrated. This paper was an important step towards the catalogue of South Australian fishes. Among the new South Australian fishes were the Blindfish, *Dermatopsis multiradiatus* from Kangaroo Island; and the Whitespotted Anglerfish, *Histiophryne* [now *Phyllophryne*] *scortea* and Handfish, *Sympterichthys verrucosus* from Gulf St Vincent. A second co-authored paper was published at the same time, in which two species of northern Australian gobies were described from specimens in the South Australian Museum collection (McCulloch and Waite 1918b).

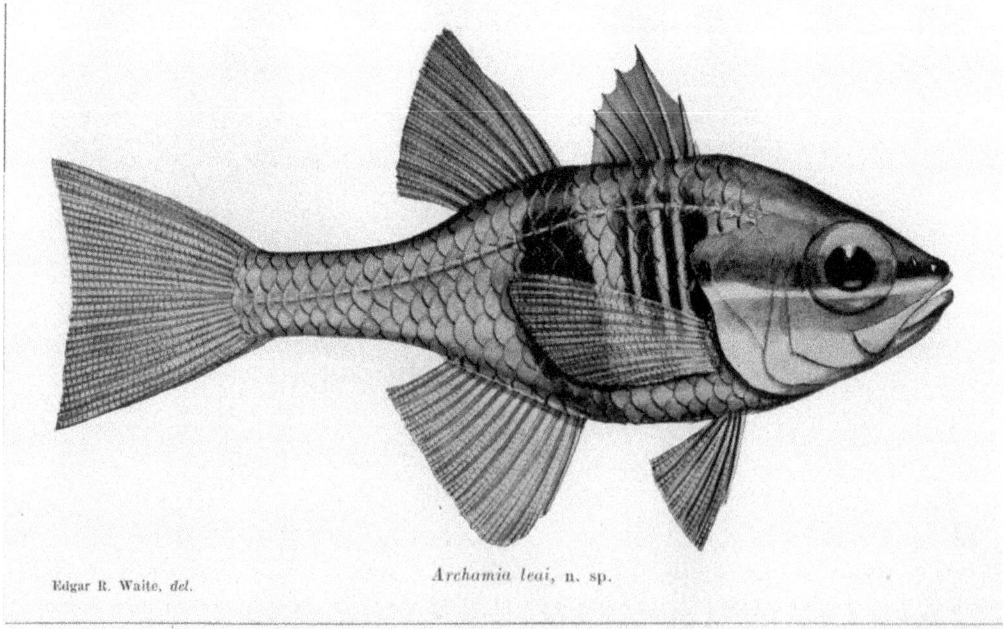

Lea's Cardinalfish, *Archamia leai* sent to Waite from Norfolk Island by Lea. From Waite 1916.

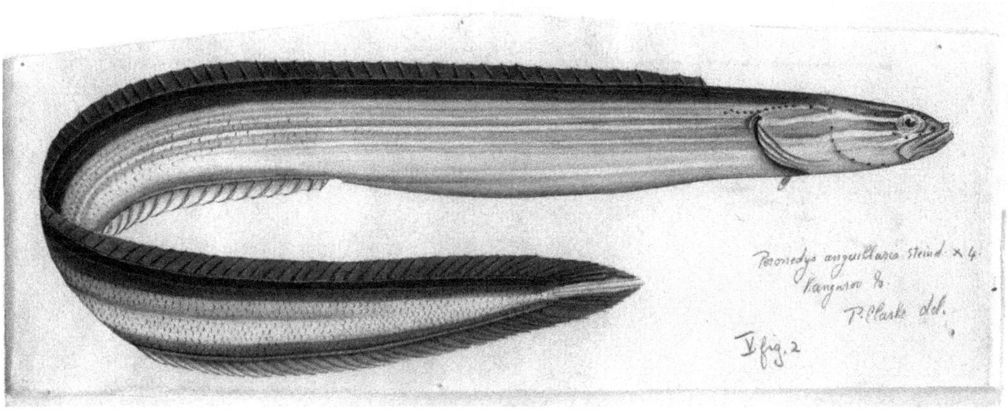

The Eel Blenny, *Peronedys anguillaris* Steindachner, 1884 from Kangaroo Island drawn by Phyllis Clarke. From AA356 Waite Collection, courtesy of the South Australian Museum Archives.

Western Jumping Joey, *Lepidoblennius marmoratus* (Macleay, 1878) from Kangaroo Island drawn by Phyllis Clarke. From AA356 Waite Collection, courtesy of the South Australian Museum Archives.

Collaboration with Herbert Hale

Waite and Hale worked together on many projects, including the Aquarium Society and whale skeleton retrievals. Hale contributed to many of Waite's papers, not least to the *Catalogue of the Fishes of South Australia*. Their only joint publication was a review of the fish family Syngnathidae (Pipefish and Seahorses). The paper was based on a re-evaluation of the fishes of this family held in the collection at the South Australian Museum (Waite and Hale 1921).

The Syngnathidae is a widespread group of small fishes with many anatomical peculiarities, including a cuirass of bony rings around the body and a brood pouch on the undersurface of the males. Southern Australia is a centre of diversity for these fishes, more than 30 species having been found there. The Pipefish were an excellent choice for revision for several reasons: they are common in South Australia; the Museum collection contained specimens of many species; there was useful work to be done in the form of species descriptions and re-descriptions. Although a significant proportion of the work was done by Hale, Waite's influence and experience are detectable throughout.

Zietz had listed 14 species from South Australia but his list included several mis-identifications which Waite and Hale corrected. Joseph Verco's old collections at the Museum yielded species missing from Zietz's list and these were included in the new paper. The authors acknowledged the source:

> In preparing this paper we have examined all the cognate material preserved in the South Australian Museum, a large proportion of which was dredged at various times in St. Vincent Gulf, by Sir Joseph Verco, President of the Royal Society of South Australia. (Waite and Hale 1921)

Waite and Hale recorded 18 species, 5 of which were described as new; all of the latter are still recognised as 'good' species. Most of the illustrations for the paper were in the form of photographs, taken by Hale, who also prepared some of the species descriptions. When the paper was nearly complete the Museum '[r]eceived a fresh Phyll. Foliatus [*Phyllopteryx taeniolatus*] male, with eggs, like little rubies, we photo'd it for the paper on the Lophobranchiates.' (WD, 11.12.20) Before publishing, Waite sent this paper to McCulloch for his comment but did not record McCulloch's response.

This paper included original descriptions of some remarkable forms from the South Australian gulfs: the Deepbody Pipefish, *Leptonotus* [now *Kaupus*] *costatus*, the Knifesnout Pipefish, *Histiogamphelus* [now *Hypselognathus*] *rostratus* and the Pygmy Pipehorse, *Acentronura* [now *Idiotropiscis*] *australe*.

South Australian Fish Catalogue: work completed (1920 to 1923)

Through 1920 Waite kept the catalogue in mind, as shown by regular diary entries: 'Working steadily at the Fish Cat. as opportunity offers.' (WD, 11.5.20) He was still hindered by lack of references, consistently writing to other ichthyologists, particularly McCulloch, for assistance, 'Wrote to Melbourne for loan of Muller & Henle [1842 text on sharks and rays]' (WD, 22.7.20) and 'Wrote to Allan and asked for loan of papers by Gilchrist [South Africa], and Weber [Indonesia].' (WD, 23.7.20) He was checking references and choosing illustrations to copy. His elation is clear and he is nearly gloating, 'Received "Muller & Henle" on loan from Melbourne & sent to block makers. We shall now have 240 illustrations of fishes.' The work by Müller and Henle was of special value: Waite was able to obtain useful illustrations of several sharks and rays therefrom. The illustrations for the catalogue were derived from previous publications, from photographs of specimens, from photographs of casts that had been made and coloured at the South Australian Museum, and from the drawings of McCulloch, Phyllis Clarke and Waite himself.

In August 1920, Waite was injured:

> Was knocked down by a horse and delivery cart at noon, from my bicycle, rendered unconscious. Attended to by [Doctor] Cilento, who took me home in a taxi. Slight concussion, shock & 2 ribs broken. (W.D.)

He was given a certificate for three weeks sick leave but soon settled down to work from home, writing the Annual Report for the Museum Board, editing papers for the records and preparing a drawing to be used as a coloured frontispiece for the catalogue. Waite chose the beautiful Rainbow Fish, *Heteroscarus acroptilus*, found only in the shallow waters of southern Australia and common around reefs in Gulf St Vincent and around Kangaroo Island. He had first drawn the fish in 1898, from a Sydney specimen, and now prepared a drawing to show the remarkable form and coloration of the male of the species (see colour plate 13, lower). Shortly after returning to work he began indexing the catalogue and writing short notes on the fishes.

The catalogue was finally published in the *Records of the South Australian Museum* on 23rd April, 1921 and Waite sent a copy to Allan McCulloch on the same day. Almost at once, Waite began to consider a version of the catalogue to be published as one of a series of handbooks on the fauna and flora of South Australia:

> At a Committee of the Fauna & Flora handbooks we constituted ourselves the editorial Committee. Prof's Cleland and Wood-Jones, Dr. Pulleine & self. (WD, July 1921)

The 'Handbooks Committee' met regularly in Waite's room at the Museum to discuss the production of handbooks on South Australian fauna and flora. The handbooks were government funded and issued by the South Australian Branch of the British Science Guild. In December 1921, Waite began to prepare a handbook on the fishes. The catalogue illustrations could be used, but much more text was required.

The first of the Science Guild's handbooks was issued in 1922, a Flora of South Australia by J.M. Black. After the Handbooks Committee had been in action for some time, they decided that the body should be 'properly constituted as follows:'

Chairman	Wood-Jones
Secretary	Cleland
Treasurer	Pulleine
Editor	self (WD, 11.8.22).

Waite's own contribution was progressing and in August 1922 he wrote, 'Am now about half-way through the fishes for the handbook.' By November, he had finished the bulk of the work and was furnishing some embellishments, 'Have commenced to supply meanings of [scientific] names of fishes.' He had some other adjustments to make, 'Owing to difficulty of imposing large block and short text I have to write some padding for the handbook.' He filled in the gaps admirably. After several more meetings with the Government Printer and the Premier's representatives, the Handbooks Committee had welcome news:

> Business of Handbooks finally settled, the Premier will print up to 300 pages annually in as many parts as may be offered.

Waite's fish handbook was to appear in 1923, directly after the first part of Frederick Wood-Jones' work, *The Mammals of South Australia* (1923).

McCulloch's reaction to the handbook

The Fishes of South Australia was finally published in May 1923 and, as usual, Waite sent a copy directly to Allan McCulloch. He also sent copies to his family, local newspapers, the international journal *Nature*, and to other ichthyologists interstate, in the United States and Japan. He was very proud of the work and the handbook had a good reception nearly everywhere. It was not well received by McCulloch. In the 1921 Catalogue, Waite had acknowledged use of 'the beautiful drawings made, in illustration of Mr. McCulloch's papers, either by himself or by Miss Phyllis Clarke ...' There was no other mention of McCulloch's work in the preparation of the text, but he apparently made no complaint. His reaction to the Handbook was very different: he wrote an emotional letter, stating that he believed that Waite had set out to deny him credit for his work; he also had his Curator write to the Handbooks Committee with a strongly-worded complaint. Waite was amazed and wrote in his diary: 'A remarkable letter from McCulloch in which he says I have 'studiously ignored him in the Handbook of fishes.' (WD, 5.6.23) The Handbooks Committee met and, with Waite still as their Editor, drafted a careful reply to the Australian Museum's Curator and had a slip printed to include with the book giving credit for the illustrations.

Waite's personal response to McCulloch was subdued but unrepentant. He had not set out to offend McCulloch and did not appear to think that an apology was appropriate. Waite himself was used to giving a great deal to his colleagues, often without published acknowledgement of his efforts. The relationship between Waite and McCulloch, in some ways, had never matured. For instance, 'Dear Mr. Waite' was McCulloch's unfailing form of address in their correspondence and he received 'Dear Allan' in reply. The fact remains that Waite had been guilty of a real lapse of courtesy and the added page, which listed sources for the illustrations in alphabetical order, did little to mitigate the offence.

The professional relationship between McCulloch and Waite was altered in that there was no further collaboration in scientific publications although they continued to cooperate in routine matters of ichthyology. Their personal relationship was uneasy for at least a short time, although friendly overtures were soon in evidence. In 1924 McCulloch was to come to a meeting in Adelaide, and Waite wrote to him in June offering him accommodation at his home. McCulloch's attendance was ultimately precluded by his duties at the Museum. On McCulloch's 40th birthday he was on Lord Howe Island, recovering from an illness; he wrote a friendly letter to Waite who made a prompt and cordial reply. McCulloch's death a few months later was a shock to Waite. *Nature* published a brief notice of McCulloch's death which distressed Waite as it had several inaccuracies. He was indignant on his friend's behalf and wrote to the editor, correcting the errors; the correction was duly published. McCulloch died in 1925 and, in publishing a supplement to the South Australian catalogue, Waite included a brief tribute to McCulloch's skill:

> As an early mentor, in Ichthyology, of the late Allan McCulloch, I take this opportunity of briefly expressing my grief at his early demise and of adding my testimony to the excellency of his work on Australian Fishes with both pen and brush. (Waite 1927)

Even here, perhaps, there is a hint of the proprietorial and still no real acknowledgement of McCulloch's contribution to the South Australian Handbook.

Field trips with Wood-Jones; handbooks and conservation

During Waite's incumbency at the Museum, some far-sighted and responsible biologists resided in Adelaide. Professor John Cleland, born in Adelaide, was a pathologist with a special interest in botany, mycology and ornithology. Doctor Frederick Wood-Jones (1879–1954) came to Adelaide in 1920 as Professor of Anatomy; as an accomplished zoologist he carried out a useful study of the native mammals which culminated in the publication of *The Mammals of South Australia*. Professor T.G. Bentley Osborn arrived in Australia in 1912 to take the position of Professor of Botany. These men witnessed the pace and extent of degradation of the natural environment and appreciated the importance of studying the fauna and flora, and of establishing reserves in the interests of conservation. They saw Waite as essentially like-minded and he cooperated with them in approaching the government of the day, helping to form appropriate institutions and protected areas. There was a contradiction in that all were collectors, most of whom cheerfully set out, gun in hand, on expeditions to collect animals. Waite's diaries recount many scenes of slaughter, happily carried out in the spirit of the collector/sportsman.

The lack of the ready means to identify the native plants and animals of South Australia was seen as critical and Cleland took the initiative in obtaining support from other biologists to remedy this. Through the South Australian branch of the British Science Guild a deputation visited the Premier of the day to obtain finance for a scheme to provide handbooks on the fauna and flora. It was proposed that the authors of the handbooks offer their expertise without charge, and that the books be printed by the state Government Printer and made available to the public at low cost (Cotton 1964).

Waite was associated with the Fauna and Flora Protection Committee from 1921 to 1927, and the State Advisory Committee on Faunal Protection, initially as the Honorary Secretary. He addressed at least one public meeting on the subject of faunal protection and, in 1922, was invited to accompany the Fauna Board on inspections of the new reserve at Flinders Chase. He was elected to the Flinders Chase Board in 1924, representing the Royal Society of South Australia, and took part in tours. When Waite was in attendance on such tours, as was frequently the case, the party combined the excursion with fish collecting. They all participated, using fine mesh seines, lines and poison, sometimes making significant collections, and occasionally Waite was able to record species new to the state.

Association with the biologists who appreciated conservation brought opportunities to collect in remote localities. Several of these trips had some significance for ichthyology. A week on Pearson Island in January 1923, with Wood-Jones, Cleland and others, produced specimens for the Museum. Waite's collection included two Silver Spot, *Threpterius maculosus* caught by handline from the rocks. This was the first time the species had been recognised since described by John Richardson (1850) and Waite recorded and illustrated the find (Waite 1923a). Richardson's specimen was a skin, sent with a sketch, from King George Sound in Western Australia. Waite (1921) had used the old figure of Richardson's in his *Illustrated Catalogue* but replaced it with his new drawing for the *Handbook* (1923b), remarking on the failings of Richardson's representation; however, the latter had the virtue of showing the tufts of cirri on the dorsal interspinous membranes, overlooked by Waite.

Waite caught the Silver Spot, *Threpterius maculosus* Richardson, 1850 at Pearson Island. In his 1921 catalogue he used Richardson's illustration but drew from his fresh specimens for his 1923 handbook. From AA356 Waite Collection, courtesy of the South Australian Museum Archives.

In January 1924 Wood-Jones' party spent a week on Flinders Island off the west coast of South Australia, Waite again being included. Summer 1925 saw Wood-Jones and his colleagues thwarted by heavy seas in their intention to pass a week on Price Island; they landed instead on the north end of the Coffin Bay Peninsula, near Point Sir Isaac. Here they collected fishes by lines but Waite's chloride of lime was unsatisfactory for poisoning; he complained, 'the lime is an absolute fraud.' On both of these trips Waite expended much time and effort capturing fish, but found no species to publish.

Later work on South Australian fishes (1922–1924)

Waite was, as usual, occupied by a variety of projects aside from ichthyology. Foremost among these was the preparation of a handbook on South Australian reptiles and amphibians, but he also wrote some anthropological papers based upon Edward Stirling's notes and collections and still found time to retrieve a couple of stranded whales.

The *Catalogue* was his main fish project at this time but he described the Whitenose Pigfish, *Congiopus* [now *Perryena*] *leucometopon* (Waite 1922a) and made useful drawings of a number of species that had previously been but poorly represented (Waite 1924).

Visit to the United States (1926)

The ichthyologists of North America did not lose sight of Waite. He still exchanged reprints with Jordan and others, and Jordan sent his major work on the *Genera of Fishes* as each of the four parts was published between 1917 and 1920. Waite participated by contributing obscure generic names to Jordan, and consulted a few works at Jordan's request to which the latter had no access. He was also sent another impressive opus, the *Bibliography of Fishes* by Bashford Dean (1923), with a request that he review it.

In August 1921 Professor William K. Gregory of the American Museum of Natural History visited South Australia and spent a morning at the Museum with Waite. Waite was staggered at Gregory's parting suggestion that he visit the American Museum of Natural History:

> Gregory left saying he wanted me to spend 12 months in N. York 2 years hence to superintend the fitting of an Australian Court. (WD, 24.8.21)

Gregory kept in touch by correspondence and asked Waite to plan his visit to the United States.

In 1926 Edgar and Rose travelled to England, where Rose stayed while her husband visited several European countries in company with Archibald Watson, his surgeon friend from Adelaide. At the end of this tour Waite spent some time in England and, after seeing Rose off on her way back to Australia, he sailed for New York late in August. Gregory introduced him to many of the ichthyologists of the east coast, with whom he had corresponded for decades, including Bashford Dean, Samuel Garman, and Eugene Gudger. Gregory cabled Wood-Jones at the South Australian Museum, 'May we retain Waite 3 weeks. pay salary and expenses.' So Waite spent a busy three weeks working on displays of Australian animals in the American Museum of Natural History and then travelled by train to San Francisco. By 11 October he was becoming a little anxious about funds, 'My present wealth this a.m. is $26.10.' (WD) Fortunately he received his $500 pay from New York the following day. In San Francisco, Waite was shown over the Steinhart Aquarium by ichthyologist Alvin Seale, who drove him to Palo Alto where he met David Jordan and some of his people: Starks, Snyder and Tanaka. This visit was essentially for courtesy, as Waite left on his way back to Australia the next day.

Last work

Waite's last publication on fishes appeared in 1928: he had been requested to provide an updated list of the fishes of South Australia for the Pan-Pacific Research Institution. The list he submitted consisted of the list from the Handbook with additions from the supplement (Waite 1927) published the previous year. His last work was a handbook, *The Reptiles and Amphibians of South Australia*[29] and this was largely completed when he left for a meeting of the Australian Association for the Advancement of Science in Hobart in January 1928. He was ill before the time came to depart but he assumed that he was suffering from another bout of malaria; he left as planned, going to Melbourne by train, and sailing to Hobart on 11 January. In Hobart, the illness became much worse and, by the time medical advice was sought, he was beyond help. He died on the 19 January, 1928 (Hale 1928).

Waite's ichthyology

Waite contributed to many branches of natural science, including anthropology, but his main work was with vertebrates, particularly reptiles and fish. A good balance between field and laboratory work is uncommon but Edgar Waite achieved this. He was an enthusiastic and innovative collector who took a delight in personally organising excursions large and small. He collected many specimens that were new to science and provided many new records for the regions in which he worked. Like other good field workers, he also made original observations of ecological and behavioural significance.

He was effective in public education through his many newspaper articles, public lectures, museum guidebooks and displays, and through his handbooks of reptiles and

29 After Waite's death, Hale collated Waite's manuscript, notes and illustrations; he had *The Reptiles and Amphibians of South Australia* published in 1929.

fishes. He established two scientific journals; the *Records of the South Australian Museum*, in particular, was a valuable publication. He was an active participant in the learned societies of his day and contributed a great deal to them.

Edgar Waite near the end of his life: Herbert Hale inserted this image of his old friend and mentor as the frontispiece of *The Reptiles and Amphibians of South Australia*. From Waite 1929.

Waite did not delight in taxonomic studies but worked effectively on many groups of fishes. This work was accurate, within the limits of the references available; unlike some of his predecessors, he did not clutter the literature with hosts of synonyms. In New South Wales, after 1898, his work was of high quality. He provided New Zealand with the framework for a catalogue of fishes and some useful studies of individual species and groups. He made a useful contribution to the ichthyology of Antarctic and Sub-Antarctic regions. Waite was the main force behind the South Australian fish catalogue; McCulloch made a large contribution that was not properly acknowledged at the time, but the project was essentially Waite's. The foundations for an understanding of the South Australian fish fauna were laid. Waite rightly saw accurate illustration as an integral part of species definition and, in his 24 years in Australia and New Zealand, published his own drawings of approximately 150 species of fishes.

Edgar Waite sought guidance from David Starr Jordan and his 'school' of ichthyologists, and from C.T. Regan at the British Museum; he kept up with the achievements of the ichthyology of his day. The significance of his own contribution to Australian ichthyology is evident from his many useful publications and from the part he played in informing and encouraging Allan McCulloch.

McCulloch

Early years

Born in Sydney in June 1885, Allan Riverstone McCulloch's schooling was abbreviated by an early apprenticeship in the field of his choice. His father was a barrister but Allan had a fascination for natural history. Presumably there was a choice to be made: entrance in due course to Sydney University, where he may have been able to take a course in Biology, or a more direct approach. The Australian Museum, following the practice of its British counterpart, had an irregular scheme of volunteer positions that could lead to virtual apprenticeships, and this was the career path chosen by Allan McCulloch.

Meeting Edgar Waite: 1898

At 13, Allan was brought to meet Edgar Waite, then a zoologist at the Australian Museum. McCulloch's father, Herbert Riverstone McCulloch and his mother, Ella Maude Backhouse, were Australian born. Backhouse was a well-known Leeds name and Waite frequently showed a particular readiness to help visitors or acquaintances who had a Yorkshire connection. McCulloch's entry to Waite's life is marked by an incidental diary note for June 1898:

> 5. Sun. In morning took dogs out for a stroll & brought home moss etc for Bush house. In afternoon Riches brought Mr. Clarence Backhouse & his nephew with the idea of having the latter at the Museum as he is said to have 'a bent' in Natural History. (WD, 5.6.98)

Waite was sufficiently impressed to use his influence with the Curator, Robert Etheridge, to obtain a volunteer position for McCulloch; a week later, 'Mr. Backhouse brought his nephew Allan Mc.Culloch who is now to work as my cadet.' (WD)

Waite had a great deal to offer a new recruit to natural history: unbounded enthusiasm and skills of extraordinary significance that he was willing and able to impart. Essentially a practical zoologist who delighted in field work, Waite was largely self-taught but widely read. He was developing an appreciation of the importance of precise taxonomy in his chosen field and his work with the trawled fishes was leading him to an understanding of the complexities of the literature of Australian ichthyology. Waite employed a disciplined approach to scientific illustration and taught McCulloch accordingly.

Australian Museum mentors

There were advantages in an introduction to zoology through immersion in it, as provided by the museum. Aside from Waite, two other zoologists at the Australian Museum were experienced in marine biology: Charles Hedley and Thomas Whitelegge. They were agreeable companions and, like Waite, had an enthusiasm for field work.

Charles Hedley, conchologist at the Australian Museum. McCulloch accompanied him on field trips to Torres Strait and the Great Barrier Reef. From Walkom 1925.

Thomas Whitelegge specialised in marine invertebrates at the Australian Museum. He and Waite took the young McCulloch on field trips around Sydney. From McNeill 1929.

One of the more significant early influences by which McCulloch profited was that of Douglas Ogilby. In the last year of his own life McCulloch, as senior author of a paper on Queensland fishes, 'gratefully records his indebtedness to Mr. Ogilby for unfailing assistance and lasting friendship, which began when he [McCulloch] was a lad of 13 years and flourished through the succeeding years.' (McCulloch and Whitley 1925b) When McCulloch joined the Museum, Ogilby was nearing the end of his connection with that institution, having damaged his health and his reputation by excessive drinking. However, he was a genial man who said himself that he always got on well with the young. Unlike Waite, Ogilby would have been undemanding, probably enjoyed McCulloch's company, and the new recruit would have appreciated Ogilby's experience and knowledge of fishes. McCulloch soon showed a preference for the study of fishes over other specialties in zoology.

Why ichthyology?

It is hardly surprising that marine biology was McCulloch's choice. Sydney Harbour had long been recognised as home to a greater variety of marine organisms than any other site in temperate Australia. The new recruit received active encouragement from the several active biologists working with fishes and marine invertebrates. Ogilby was loud in voicing his belief that a book of Australian fishes was long overdue and Waite's preoccupation with the taxonomic problems presented by the fishes of the *Thetis* trawls gave McCulloch valuable experience.

The fishes of Port Jackson had been studied intensively and a great deal published about them. However, species new to the Harbour, or even new to science, were still discovered with some regularity. Confusion over the nomenclature of even the most common species provided a wide and obvious field of study. The creatures themselves were readily observable and endlessly fascinating in their variety of form and behaviour.

Edgar Waite's cadet: relationship, pay, drawing and field trips: 1898–1902

Within a few months of starting as Waite's assistant McCulloch, then living with his family at Potts Point, was visiting the Waites at Mosman, and sometimes participating in their outings. Some entries from Waite's diaries give an impression of the relationship:

> 2. Fri. Asked Allan to tea for Sunday. (December 1898)
>
> 4. Sun. Allan spent the afternoon with us ... (December 1898)
>
> 31. Good Friday. holiday. In the afternoon took Rose & Claude also Allan & his sister Doris for a picnic to Balmoral. Had a swim with the boys ... (March 1899)
>
> 24. Wed. Queen's Birthday. Public holiday. My week on late duty. At noon I rode to Pott's Point (having Bike in town) & called to see Allan who is ill. He was in bed & had been delirious. (May)
>
> 4. In the afternoon Allan came & after tea we rigged up the lantern and got 4 nice enlargements 15 x 12 of negative of Claude. (June)

Waite's own son, Claude, was only three and Allan appeared to occupy the place of an elder son. Waite enjoyed his company and took pleasure in demonstrating and sharing his knowledge of photography:

> 7. Wed. Attended Palace Theatre Amateur performance, 'The Schoolmistress' in which Allan and his sister had characters. (June 1899)
>
> 11. Sun. ... Rose, Claude and Allan went by tram I rode to the Zoo ... (June)
>
> 30. ... Had tea with Allan & attended meeting of the Linnean Soc. ... (August)
>
> 15. Sun. In the afternoon Allan & Harry came & we had a lesson in photography. I gave the former a small cabinet, a Belemnite & other small things. (October)

Eventually Waite decided that Allan should be paid, 'At to-day's Board meeting Etheridge was at my request to apply for some remuneration to Allan who has been 16 months with me.' (WD) The Curator did as he was asked but 'Etheridge told me he had had a row with the trustees re Allan and that the matter was held over.' (WD, 8.11.99) For Christmas, the Museum Trustees gave Allan '£5 as a recognition.' (WD)

The informal side of the relationship flourished:

> 26. Fri. Anniversary day. A Public holiday. Allan & his sister Doris came by 2.30 boat and we picnicked at Balmoral, they returned to 'Headingley' [Waite's house] with us to tea ... (WD, January 1900)

> 25. Sun. Went to town by 2.0 boat with Bike. Met Allan & his friend Ester Humphrey beyond Rose Bay where they changed from tram to 'bus. I pedalled to South Head rejoined them & together we visited Barraclough's Ostrich Farm. (WD, March 1900)

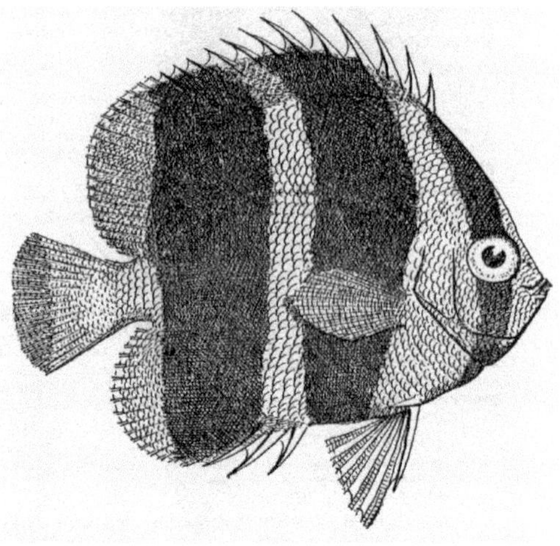

Waite's drawing of the Threeband Butterflyfish, *Chaetodon tricinctus* of Lord Howe Island. From Waite 1901.

In April there was a small victory, '27. Fri. Obtained work of indexing the Thetis Report for Allan £5 for completion.' (WD) A month later Waite and McCulloch had a few days at Cowan Creek, where '[w]e camped in a small cottage provided & cooked our own provisions.' They collected birds and reptiles, fished in the creek and examined an aboriginal kitchen midden: 'we turned to the thick deposit of shell & excavated some feet but failed to obtain either bones or implements.' (WD, 31.5.00) Waite noted the markings of young Snapper, *Chrysophrys auratus*, the juveniles of which are known as Red Bream:

> On Cowan Creek. Had a day's fishing & obtained half-a-dozen small Red Bream. At this age, they have distinct vertical bars, a feature I was unaware of. (WD, 1.6.00)

Early in 1901, Waite wrote 'I am now giving Allan lessons in drawing.' (WD, 13.2.01). Difficulty with close work had been a problem for some time; Waite had sought advice from an optician and an ophthalmologist, and had begun to use spectacles, but the problem was not resolved, 'Commenced drawing in line of Chaetodon tricinctus but my eyes will not stand the strain.' (WD, 14.2.01) And a few days later, 'Finished line drawing of Chaetodon, the work is not very careful as I dare not try my eyes too much.' (WD, 18.2.01)

This difficulty may have prompted Waite to try again to have McCulloch paid: 'Museum Board meeting will consider Allan's case.' (WD, 4.6.01). The result was encouraging, 'At yesterday's Board Meeting it was decided to put £50 on the estimates for Allan as assistant to the Curator.' (WD, 5.6.01). A few months later, 'Etheridge told me that the estimates were passed & that sum to engage Allan was assured.' (WD, 27-9-01) Finally:

> Allan received letter officially appointing him 'Mechanical Assistant' @ £50, he was paid £25 being wage from 1st. July last. (WD, 9.1.02)

Field trips around Sydney were a significant part of McCulloch's life, sometimes with Waite or other members of the Museum staff, sometimes alone. Aquarium animals were frequently the object of these excursions:

> Took day off with Allan ... took Botany tram and walked to La Perouse. Our object was Galaxias & Eleotris. We got a good supply the species being Mesites attenuatus [Common Galaxias, now *Galaxias maculatus*] & Krefftius australis [Striped Gudgeon, *Gobiomorphus australis*]. (WD, 2.11.01)

> lunched with Cherry, Allan joined us & all went to Gale's thence to Cook's River. Got Pseudomugil signifer [Pacific Blue Eye] (the object of our quest), a few Mugil [Mullet], Galaxias, etc. all small. (WD, 11.1.02)
>
> Mr. White went with me to Balmoral where we were joined by Hedley. Hammered out sea Anemones with rock to which they were attached, & placed them in Aquarium with 2 Hippocampus [seahorses] sent by Allan from Elizabeth Bay. (WD, 31.3.02)
>
> With Whitelegge, Hedley, Allan to Watson's Bay got some beasts for the Aquarium. (WD, 26.4.02)
>
> Met Allan at 9.0 boat, went to Balmoral a[t] low tide, got a Tripterygium [Threefin], young Mullet & Blackfish & a bit of the beautiful Alcyonarian, Telesto smithii [soft coral] with pink stem and large white polyps. (WD, 4.5.02).

> Bank holiday, with Whitelegge & Allan to Long Bay. Got some small fishes, Caranx, Atypichthys, Lepidoblennius [Trevally, Mado, Jumping Joey] etc for Aquarium. (WD, 4.8.02)

The Museum accession registers give some indication of McCulloch's pursuit of fishes around Sydney. The fruits of his efforts are recorded in Waite's hand, beginning with some specimens from Balmoral Beach in June 1901.

Waite continued instructing McCulloch in photography and drawing. The visits to the Waite household continued, and McCulloch must have absorbed a great deal of useful knowledge. Waite was a practical person, and delighted in working with wood and metal, constructing his own aquaria, thermostats, pumps and other contrivances. McCulloch participated in some of these exercises and Waite mentions 'Allan came in the evening and I made him a water telescope.' (WD, 29.6.02)

Lord Howe Island: McCulloch's initial visit: 1902–1903

Since his brief trip to Lord Howe Island in the *Thetis* in 1898, Waite had maintained contact with some of the residents from whom he occasionally received specimens. He was sure that he could accomplish a great deal if he could organise a trip and decided to combine collecting with a family holiday, taking his wife and son, and also Allan McCulloch as his assistant. He began to arrange this in July 1902:

> Received letter from Mrs. Nicholls [of Lord Howe Island], she will accommodate the four for £2 per week. Wrote Trustees applying for extended Annual leave & permission to take Allan. I also suggested financial aid. (WD, 24.7.02)

The museum trustees granted Waite the leave, but no allowance for his expenses and no assistant: 'I am refused Allan.' Once again, he enlisted the Curator, 'Interviewed Etheridge re Lord Howe Is and Allan, and brought him round to promised [sic] to support my application ...' (WD, 18.8.02) Permission to take McCulloch was eventually granted and five pounds allowed for expenses.

The preparations began in earnest. Waite sought and received advice from David Starr Jordan regarding collecting on reefs and set about obtaining explosives and poison. Allan was involved in these preparations:

> With Whitelegge & Allan to Maroubra to try effect of chloride of lime as fish-poison. We selected a large rock pool containing perhaps 3000 gallons [about 14 kilolitres]. 1 lb [0.45 kilograms] of lime turned up the fishes, Lepidoblennius [Eastern Jumping Joey, *Lepidoblennius haplodactylus*] was the first to move, they made for shallow water & jumped out on to the weeds. The main occupants were Black-fish these also made for shallows & curled themselves in small holes or hid in crevices. We caught many in the nets & returned to pure sea-water; the tide rising big waves swept into the pool & revivified it but many fish were beyond its influence. The experiment was a great success & will be of great value on the Island reefs. (WD, 20.9.02)

Photography was an important part of the work and Waite 'finished rehabilitating "Rover" camera which I intend for Allan's use on the Island.' (WD, 12.10.02) McCulloch practised with the camera and took the exposed plates to Waite to be developed. Finally Waite '[o]btained tickets for Allan & ourselves £15.' (WD, 28.11.02)

The trip was intended to be of one month's duration, but lasted two months as bad weather prevented the return boat from landing in December. The influence on Allan McCulloch was immense. Judging from the account in Waite's diary, McCulloch matched him for energy and enthusiasm and they had a magnificent working holiday. McCulloch met the residents for the first time and made enduring friendships. Again, extracts from Waite's diary paint the picture:

> With Mrs. Potts, Allan & I sailed the Dinghy to the North beach. We visited two caves in search of bats but found none, the roots of banyan trees had penetrated the soil above the cave & reentered the ground beneath. (15.12.02)

> Allan & I sailed to North Beach, limed pools got small fish Parapercis [Grubfish], Acanthoclinus [Basslet, *Belonepterygion fasciolatum*], Tripterygion [Threefin] etc. Took photos Echinoderms etc in water & developed at night. (16.12.02).

> Walked beach to south rocks caught Salmon [*Arripis trutta*] & Black Rock-cod [*Epinephelus daemeli*]. In the afternoon strolled through the gardens. After tea with Mrs. Potts & Susie, Allan & I went fishing in the 'Comet' hole got Apogons only. (20.12.02)

Boxing day ... Allan went up the Mt. [Mount Eliza] got 3 bosun birds off the nest & succeeded in bringing down one egg. The next move was luncheon spread on a carpet of palm leaves and before leaving we had picnic tea. A late day was finally closed by a dance in the seed shed my principal efforts being in the Musical items. (26.12.02)

Allan & I went to Middle beach, made sketch of S[outh] end of Island therefrom with Mutton Bird Island, the Pyramid & the mountains. Stroll on the beach at night. (28.12.02)

After dinner walked to Ned's beach and made a sketch of the Admiralty Islets therefrom. Returning by the lagoon beach picked up some tiny red crabs which were being washed up in some numbers. In the evening fishing in the lagoon, Dawson's Point. Salmon, Mullet & a few rock fish, next a Shark & fishing was off. An Octopus on being hauled up to the boat shot up jet after jet of water to the height of three feet, sending spouts into the boat as it was drawn to the gun'ale. (8.1.03)

Spent the day with Mr. & Mrs. Retinock, and Susie also going. To Stephens Look-out had fine fishing off the rocks but got drenched from top to toe by several waves. One such carried all our fish off the high rocks into a pool, we failed to recover three large Blue fish [*Girella cyanea*]. We caught 20 in all including Blue fish, Salmon and a Shark. (9.1.03)

Pulled across to Rabbit Island & took photo. of the mountains. Between the island and the reef I saw three turtles, they were feeding over some rocks remaining at the surface a short time only, while there they kept the head under water but raised it for 2 seconds or so 5 or 6 times, then flopping the head down they dived to (apparently) feed upon the vegetation on the rocks. We were within 120 yards of them & somewhat above them. Having good opportunities for seeing them I believe them to be Hawksbills though the 'Greenback' only is recorded. After dinner we walked to Stephen's Look-out and fished catching Blue fish. (13.1.03)

The Marblefish, *Aplodactylus etheridgii* (Ogilby, 1889) of Lord Howe Island, collected by Waite and McCulloch from a rockpool in 1902–03 and drawn by the latter for Waite's 1904 paper. From Waite 1904.

Intended to work the reef at low tide but rough weather suggested the rock pools at Stephens Look out. We limed a large pool and worked in a heavy thunderstorm & drenching rain. Our best catch was a large Aplodactylus etheridgii, new to our collection. (14.1.03)

By the time they returned from Lord Howe, McCulloch was playing an important part in Waite's ichthyological work. He was dispatched on errands that required some judgment, consulting ichthyology texts in the Linnean Society's library, and finding specimens in the Macleay Museum. Waite was developing some dependence on the young man: 'Wrote description of genus & species of the fish & Allan made me an outline drawing and reduced it by pantograph.' (WD, 27.3.03) And later, 'Allan is now drawing plates for me under my eye, and thus materially relieving the strain on my sight.' (2.5.03) A valuable paper on Lord Howe Island fishes (Waite 1904a) resulted from the study of their collections, and was partly McCulloch's work, as Waite acknowledged in the publication; in his diary he wrote, 'Handed in eight plates two by myself, the rest by Allan. They contain 24 species and are in illustration of L.H.I Fishes ...' (WD, 8.12.03)

Waite was delighted with McCulloch's work, but hardly distinguished it from his own; this is demonstrated not too subtly by another diary entry. Waite kept an album of his zoological drawings, or 'plates', and indexed it with a 'contents' page:

> Made 'contents' of the volume of 'plates' in which I paste copies of all figures drawn by me, and am now including those by Allan. (WD, 16.2.04)

Their relationship outside working hours remained cordial, Allan bringing photographs (taken with the new camera which Waite had chosen for him) to be developed and staying to tea. Waite had been amusing himself by making an incubator, complete with thermostat, for hatching the eggs of fowls and ducks; he made one for McCulloch's mother, who then lived at Paramatta, at her request:

> Left at noon with Allan and went to Paramatta. Fixed & set the incubator going, returned 8.25 train. (WD, 9.1.04)

Independent work: 1904

Waite must have had paternal, or at least rather proprietorial, feelings towards Allan. He was certainly taken aback when his young assistant showed signs of working independently: Allan was drawing eleotrids and making notes with a view to having them published independently. Then, 'Allan handed in his first work. (on Hypsypops) [White Ear, now *Parma microlepis*].' (WD, 31.3.04). Waite used his compliant Curator, Etheridge, to block McCulloch:

> The Museum board declined to publish the little paper of Allans regarding him as too young & also expressing opinion that one subject should not be studied by two individuals in the institution. (WD, 15.4.04)

There were few zoologists at the Australian Museum and it was rational to have them dealing with different subjects. However, McCulloch had received direct instruction in ichthyology from Waite and was deeply involved in the subject. His attempt to produce his own work is easily understood and his disappointment inevitable. He seemed to cope well enough with the situation at the time but later in life his antipathy for Waite and Etheridge was frequently manifest, and perhaps dates back to this early injury.

McCulloch's contribution to Waite's last publications in Sydney: 1903–1906
Judging from Waite's diaries and the continuing cooperation evident in his publications, the relationship between Waite and McCulloch seemed to be unimpaired. Waite continued to acknowledge at least part of McCulloch's contributions to his papers but still appeared as the sole author. They worked together on local collections, and those from Lord Howe Island and Western Australia.

McCulloch's drawings featured constantly in Waite's publications until Waite left for New Zealand. From 1902 to 1906 Edgar Waite acknowledged his assistant's work, specifically in the collection and illustration of many of the species of fish that formed the subject of his papers. The first of McCulloch's illustrations published was that of a shark from Lake Macquarie, described by Waite (1903).

Bronze Whaler, *Carcharhinus brachyurus* (Günther, 1870) drawn by McCulloch for Waite's 1903 paper. From Waite 1903.

In 1904, Waite published several papers in the *Records of the Australian Museum* to which McCulloch made significant contributions. He drew 18 of the 23 illustrations for *Additions to the Fish-Fauna of Lord Howe Island, No. 4* (1904a); this was primarily the report of their own collections and Waite mentioned McCulloch as his assistant, referring to the illustrations:

> A few of the accompanying illustrations are, as on previous occasions, my own productions, the greater number were however prepared by my assistant, Mr. A.R. McCulloch, whom I have specially trained for the work. The drawings were made under my immediate supervision and may therefore be relied upon for accuracy of detail and general proportions. (Waite 1904a)

Waite also acknowledged McCulloch's study of some members of the Pomacentridae (Demoiselles); and, in discussing the colour of an eel, wrote:

> The green coloration is entirely due to the mucous [sic] which covers the body, an interesting fact discovered by my assistant, Mr. A.R. McCulloch, when preserving a recent example. (op. cit.)

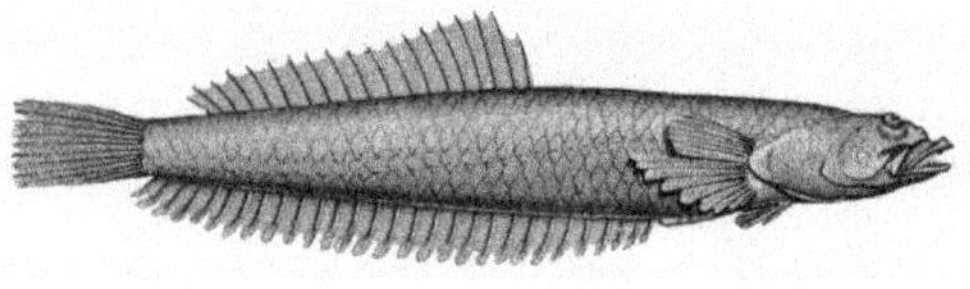

Two specimens of the Brokenfin Sandburrower, *Schizochirus insolens*, the largest 6 cm long, were collected at Maroubra Beach, Sydney by Thomas Whitelegge, drawn by McCulloch and described by Waite. From Waite 1904.

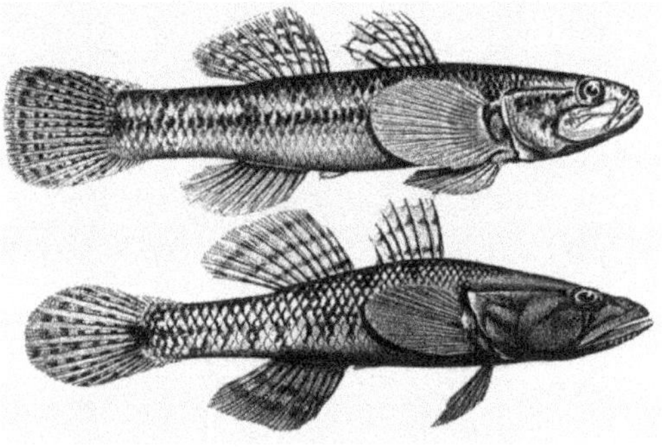

Cox's gudgeon, *Krefftius* [now *Gobiomorphus*] *coxii* (Krefft, 1864) and the Flathead Gudgeon, *Philypnodon grandiceps* (Krefft, 1864). McCulloch drew these Gudgeons hoping to publish a paper on them under his own name but Waite used the drawings in 1904. Both are common in eastern Australian coastal streams. From Waite 1904.

In *New Records and Recurrences of Rare Fishes from Eastern Australia, No. 3* (1904b) McCulloch was responsible for the six illustrations, as Waite acknowledged, 'the plates being reproduced from drawings by Mr. A.R. McCulloch.' They published a revision of the gudgeons found in the freshwater streams near Sydney (Waite 1904c). Waite had been collecting these fishes for several years, mostly as aquarium specimens. The stimulus for the preparation and publication of this paper was at least in part McCulloch's drawings and his attempt to have them published independently. Waite acknowledged the work: 'The figures, of most of our species, will be of distinct value, and are reproduced from the work of Mr. A.R. McCulloch.' He also mentioned that McCulloch had donated some of the specimens.

Waite received collections of fishes from Western Australia and his third report on these (Waite 1905b) was illustrated by McCulloch. He acknowledged the source of the illustrations and named one of the fish, supposedly a new species, after McCulloch. The subject of one of the notes in this work was taken from 'the little paper of Allans' (WD, 15.4.04), the publication of which the Museum Board had prohibited. Waite misidentified the species from the Western Australian collection as a demoiselle, *Hypsipops microlepis*, and stated that it was also common in Sydney Harbour. He wrote:

> The changes which take place, during the life of the species to maturity, have been lately studied by Mr. Allan R. McCulloch, and the following notes are based upon his observations. (Waite 1905b)

The Western Australian fish was a different species to that which McCulloch had studied in Sydney and was named *Parma mccullochi* by Whitley (1929a).

The last paper that Waite published before leaving Sydney was a collection of bits and pieces: *Descriptions of and Notes on some Australian and Tasmanian Fishes* (Waite 1906). Some of the specimens upon which this paper was based were from his recent trip to Melbourne, others forwarded from Tasmania and two new species had been found locally by McCulloch. Waite credited McCulloch as the collector and acknowledged his assistant's preparation of nine illustrations for this paper in a footnote, '... the drawings being made by Mr. A.R. McCulloch.'

At the least, Waite had been rather ungenerous to McCulloch in regard to recognition for his contributions to these last papers, published before leaving for New Zealand. Despite his age, there was a solid case for McCulloch having been formally recognised as co-author.

McCulloch studied the White Ear, *Hypsipops* [now *Parma*] *microlepis* in Sydney Harbour and drew its alterations in shape and colour with growth. Waite published McCulloch's findings, with suitable attribution, in his 1905 paper on Western Australian fishes. From Waite 1905.

Field work and Waite's departure: 1904–1906.

In October 1904, McCulloch accompanied a party from the Australian Museum, led by Charles Hedley, the Museum conchologist, on a trip to the Barrier Reef. This was the former's first experience on the reef, and he was in good company: Hedley was an experienced marine biologist with a special interest in the Barrier Reef fauna. Hedley's own experience included several trips to South Pacific islands, including Funafuti and New Caledonia, and he had previously brought back large collections to the Museum.

Hedley's group spent eight days on Masthead Island. McCulloch was captivated by the marine life of the reef and collected fishes that he brought back for study. From this trip, Waite recorded about 130 fish in the Museum register. This was the first of many such trips that McCulloch undertook, the Great Barrier Reef being second only to Lord Howe Island among his favourite places. McCulloch illustrated part of Hedley's publication on the molluscs of Masthead Island's reef (Hedley 1906).

Waite and McCulloch continued to pursue field work together around Sydney and, when Waite was away from the Museum, his cadet kept him informed and received instructions through an exchange of letters. McCulloch was living with his mother at Paramatta and collected at nearby Duck Creek, 'Allan brought me Utricularia [carnivorous water plants] & Najas [brown snakes] from Duck Creek'. (WD, 31.1.05) A joint excursion followed:

> Took day on Service [paid by the Museum]. Met Allan at Clyde & we worked Duck Creek with the seine net & got some large Pseudomugils in fresh water, also Philypnodon grandiceps. Returning to the Station I killed a 5 ft. brown snake. Went

to Paramatta & was joined by Rose & Claude & Mrs. McCulloch [Allan's mother]. (WD, 9.2.05)

Waite had participated in some of the dredging parties undertaken by Sydney University, and by E.P. Ramsay of the Museum. He modified a small dredge and conducted a modest excursion of his own:

> Had a boat out 3/- and started at 8.30 with Allan and Goldfinch. We first worked the rocks at low tide and got Anemones etc. then worked dredge got Hippocampus, Syngnathus, Stigmatopora [Seahorse and Pipefish] etc. Allan went home with me to tea when we put animals into the aquarium. Whitelegge also came to tea. (WD, 24.8.05)

Their last joint excursion included a friend of Waite's, E.K. Scott; they camped for a week on the Nepean River in December 1905. Extracts from Waite's record indicate a memorable field trip:

> Left at 1.25 with Scott, Allan joining the train at Parramatta. Arriving at Penrith paid grocer bought fresh meat & fruit thence to the boatshed ... and pulled up to Glenbrook Creek where we got fresh water. Camped a little higher on the opposite bank of the river raising the tent. (WD, 9.12.05)
>
> Rowed back to creek for water, and after breakfast Allan shot a small bird & on going to look for it came across a young diamond snake. He called me & then had his foot on it but on attempting to pick it up received a bite, the lower jaw made two punctures like those of a venomous snake. I caught it & we put it into a mosquito net bag. Got small Potamalosa [Freshwater Herring, *Potamalosa richmondia*] with the fly. (WD, 10.12.05)
>
> Regaining the boat we pulled up to junction of the Nepean and the Warragamba & had an exciting time through the rapids. Allan waded & I punted to avoid the rocks, we once got broadside on & nearly lost control. We camped on the Warragamba [River] sleeping under a huge rock shelter some distance above the water level. Had a lovely row by moonlight. The reflexions of the rocks & trees in the calm water having weird effect. We had a beautiful bed of sand & many visitors who left their footprints. I recognised Bettongia, Mus and Petrogale while a sinuous track which could scarcely be anything but a snake put Scott ill at ease next day. (WD, 11.12.05)
>
> Lyre birds around. Scott shot his first & only Wallaby. At night we camped under the same rock & I shot a rat with the pea rifle. (WD, 12.12.05)
>
> I drew a strap across Scott who was lying next to me, he said 'Waite,

The Freshwater Herring, *Potamalosa richmondia* (Macleay ,1879) is common in the coastal rivers of New South Wales. McCulloch drew this fish from the Hastings River. From McCulloch 1917.

> Waite, here's a d-d snake crawling over me'. I promptly secured it for him. (WD, 13.12.05)
>
> Moved camp again going back to the Nepean above its juncture with the Warragamba ... Rain at night. Got a good lot of herrings, some eels ... (WD, 14.12.05)
>
> Pulled down the river and did the rapids easily. Owing to rain we drew the boat under a large rock overhanging the water and were quite out of the wet. Camped at Glenbrook Creek using the tent. The rain was somewhat heavy result of a thunder storm. (WD, 15.12.05)
>
> Allan & I pulled down to Penrith in desultory rain. We got the 2.57 train and I reached home about 5. o'clock. I had done cooking throughout & was voted a 'good-hand'. (WD, 16.12.05)

Although this gives only Waite's view of the trip, there is an impression of pleasure in one another's company. Shortly after this excursion Waite, probably disappointed at his career stagnating at the Australian Museum, took an appointment in Christchurch as Director of the Canterbury Museum.

In April 1906 McCulloch was among the crowd at the dock to see the Waites off to New Zealand on the *Waikare*. McCulloch had helped with the packing and had undertaken the care of Waite's pet magpies. He also inherited the weightier responsibility for mammals, reptiles, amphibians, fishes and crustaceans at the Australian Museum.

Independence, the *Woy Woy* and Torres Strait: 1906–1907

McCulloch must have seen Waite's departure as fortuitous: a remarkable opportunity. He had already begun to make Crustacea his own specialty, although he had shown that he would prefer to study fish. He was 20 and the *ad hoc* rules introduced by the Museum Board presented the prospect of publication at age 21. However, for as long as Waite occupied the position of ichthyologist, it is likely that McCulloch would have had at most a junior author's recognition for his work on fishes. Now, he had inherited the very position he wanted and, as the new 'Assistant in Charge of Vertebrates', he was quick to demonstrate his competence and enthusiasm. McCulloch had prepared a brief paper on *Crustacea new to Australia* and this, his first publication, appeared in 1906.

Shortly after Waite's departure there was an opportunity to sample the deep sea fauna off Sydney. William Haswell, Professor of Biology at Sydney University since 1889 and a member of the Board of Trustees of the Australian Museum, maintained an active interest in the Museum and its staff. He had invited Waite on dredging and trawling excursions on Sydney Harbour, although eventually Waite lost interest, noting wryly, 'I have had enough of University Students en fête.' (WD, 3.11.97) In 1906, Haswell had funds from the Royal Society of London to engage in limited investigations of the deep sea off Sydney. Two short expeditions were undertaken. Initially, in June 1906, dredging for invertebrates was performed from the *H.M.C.S. Miner* in depths to about 500 metres.

In October 1906, from the *Woy Woy*, Haswell conducted a trial of trawling, using a small net of his own design, 35 miles [60 km] due east of Sydney. The young Assistant in Charge of Vertebrates was delighted to be included:

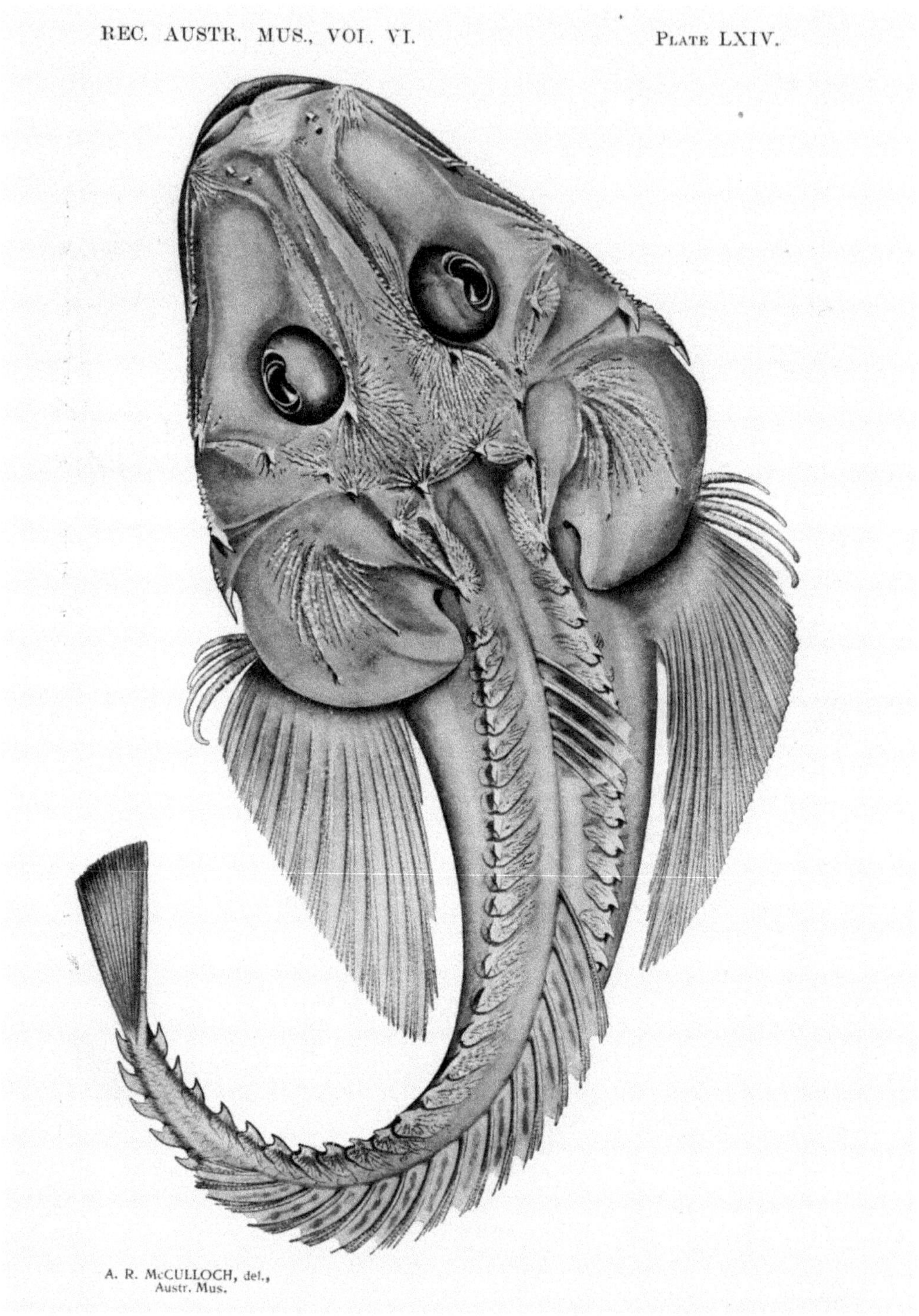

Deepsea Flathead, *Hoplichthys haswelli*. McCulloch collected the trawled fish from the deck and described its colours as '… pinkish yellow above, white beneath. Dorsals, caudal and pectorals pinkish, the latter edged with white. The vertical fins and tail with darker markings.' From McCulloch 1907.

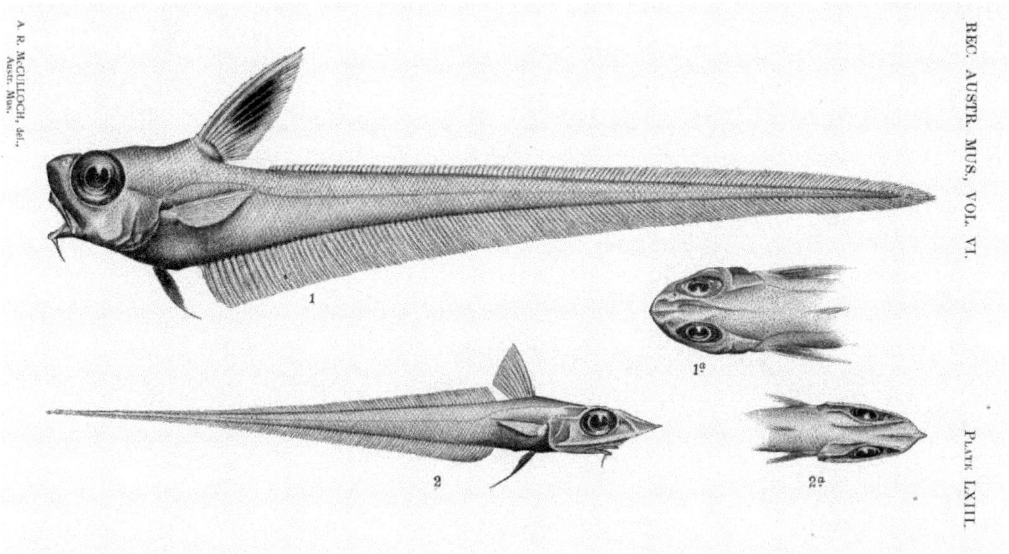

Deepwater Whiptails trawled by the *Woy Woy* in 1500 metres and described by McCulloch: he had five specimens of the Blackspot Whiptail, *Macrourus* [now *Lucigadus*] *nigromaculatus* (top) and two of the Notable Whiptail, *Coelorhynchus* [now *Coelorinchus*] *innotabilis* (bottom) from which he described the species. From McCulloch 1907.

> Professor Haswell ... kindly invited me to accompany him and has further honoured me by placing the Fishes and Crustacea in my hands for description. (McCulloch 1907)

A single cast of the trawl in a depth of 1500 metres produced specimens that pleased Haswell and McCulloch, and enabled the latter to describe several new species.

Fishes and Crustaceans from 800 fathoms (McCulloch 1907) was a worthy first publication on ichthyology. There was some luck involved in that although only 18 fishes were taken, they belonged to seven different species, three of which McCulloch described as new (and they are still regarded as valid). He was handicapped by several of the species from the haul being represented only by juveniles and by some damage to the fishes from sea urchins:

> ... a hundred living specimens of *Porocidaris elegans* which choked the net, and with their long spines scraped most of the scales off the fishes that lay near them. (McCulloch 1907)

McCulloch named for Haswell the most spectacular of the new species, a Deepsea Flathead, *Hoplichthys haswelli*.

McCulloch accompanied Charles Hedley on another small expedition in 1907, this time to Murray Island in Torres Strait. With the assistance of the inhabitants they sampled the marine and terrestrial fauna of the island, and made an extensive ethnological collection. Molluscs were Hedley's main interest and his party collected from the reefs at low tide and dredged the inshore waters. McCulloch's chief expertise lay in crustaceans and fish and he was responsible for collecting a great many. He collected by hand, by hook and line, by nets and with poison; the Islanders helped by spearing some that usually evaded these means.

Crustaceans: 1906–1908

McCulloch had been working with Frederick E. Grant at the Australian Museum, preparing reports on decapod crustaceans [crabs, lobsters, prawns]. These were published with McCulloch as junior author: *On a Collection of Crustacea from the Port Curtis District, Queensland* (Grant and McCulloch 1906) and *Decapod Crustacea from Norfolk Island* (Grant and McCulloch 1907). Grant was to have received the crustaceans from the deepsea trawl of the *Woy Woy* as well but he died in 1906; the specimens naturally came to McCulloch who included them in his report on the fishes.

For several years after Waite's departure, it seemed reasonable for McCulloch to handle both crustaceans and fish. McCulloch published the first of his *Studies in Australian Crustacea* (1908b):

> Under this heading I contemplate a series of short papers dealing with small collections as they come to hand.

Perhaps he reckoned without Ogilby, who was soon accumulating large collections of crustaceans through the activity of his AFAQ supporters. In the first *Studies* McCulloch dealt with crabs from Moreton Bay (from Ogilby) and from Port Phillip, Victoria. He named a small Spider Crab for Ogilby. Later workers (Griffin and Tranter 1986) have decided that it is probably synonymous with a species described from the Gulf of Martaban, Burma in 1893 but they remarked that '[t]he original description and figures of *Hoplophrys ogilbyi* [a small spider crab] given by McCulloch are extremely accurate and detailed.' He was as meticulous with his crustaceans as with his fishes.

Long term aims in ichthyology

McCulloch's studies of ichthyology took the form of a series of major projects which he developed according to the opportunities afforded by collections. He mainly worked on the temperate fishes of eastern, southern and western Australia and on Queensland fishes. There was a large component of fishes from deeper shelf waters, thanks to the government's trawling experiments, and this was the first time that an ichthyologist had had access to these fishes in numbers. He published on all of the collections that came his way but his chief ambition was to produce a comprehensive catalogue of the fishes of Australia. To this end he worked to resolve the difficult synonymies that arose from the confused early work of his predecessors and, whenever sufficient material was available, he revised genera and families.

As the work proceeded, McCulloch took advantage of the relatively advanced state of knowledge of the fishes of New South Wales to prepare a handbook. Work towards this manual was an integral part of the greater work needed for a text on the fishes of the whole of Australia.

Publication of McCulloch's work

The results of his researches were published in various journals. Work on specimens from New South Wales, Victoria and Lord Howe Island usually appeared in the *Records of the Australian Museum*, in the *Proceedings of the Linnean Society of New South Wales* and in the

Australian Zoologist; on Queensland fishes, usually in the *Memoirs of the Queensland Museum*; on South Australian fishes, in the *Transactions of the Royal Society of South Australia* and later in the *Records of the South Australian Museum*; on Western Australian fishes, in the *Records of the Western Australian Museum*. His trawling reports were published by the Board of Trade and Customs. He also contributed popular articles to the *Australian Museum Magazine*.

Studies in Australian Fishes: from 1908

McCulloch prepared eight *Studies in Australian Fishes* over nearly 20 years. He introduced the first in 1908: 'It is my intention to contribute under this heading a series of short papers dealing with material as it comes to hand.' The first paper (McCulloch 1908a) was much more circumscribed than this statement would lead one to expect, consisting of a thorough review of a group of clinid fishes from New South Wales and Victoria. In this paper he described no new species but brought preliminary order to a difficult group. He recognised eight species, each of which he described and illustrated, placing them in three existing genera that he defined with precision.

The Australian Clinidae, or Weedfishes, are small sedentary carnivores that bear live young and are found in shallow rocky and seagrass habitats. The group that McCulloch revised is now known to include in excess of 30 species from temperate Australian waters. All but one of the eight species that formed the subject of McCulloch's paper is still recognised; his conservative approach was appropriate for this early review and was a reasonable basis that facilitated the work of later ichthyologists.

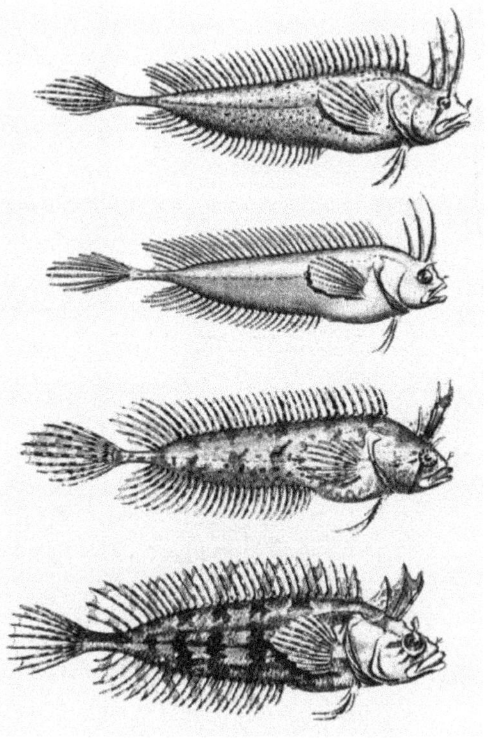

Weedfish (Clinidae) drawn by McCulloch for his *Studies in Australian Fishes. No. 1.* He correctly identified these (top to bottom) as *Cristiceps aurantiacus* Castelnau, 1879, *Cristiceps argyropleura* Kner, 1865, *Cristiceps australis* Valenciennes, 1836 and *Petraites* [now *Heteroclinus*] *roseus* (Günther, 1861). From McCulloch 1908.

The 'Studies' that followed involved the description of new Australian species, new state records with synonymies and revisions of groups whenever appropriate. For instance, the third of the series (McCulloch 1913a) consisted, in part, of a revision of a group of temperate wrasse (Labridae). For this revision, McCulloch assembled a large collection of specimens and acknowledged loans from Kershaw (Melbourne), Hamlyn-Harris (Brisbane), Haswell and Edgeworth David (Macleay Museum, Sydney) and Woodward (Perth). The taxonomic history of this group of fishes is complex and McCulloch observed:

> Count Castelnau, Sir William Macleay and Mr. de Vis relied largely upon their colour and colour-markings to distinguish them, and often regarded important structural details as being of only secondary value. But if a series of fresh specimens of any one species of this genus be examined it will be found that the actual colouration of different individuals differs considerably, and that even the colour-marking varies greatly in degree of development, according to the age of the specimens. Again, as I have found in P. tetricus, the young may bear no resemblance whatever to the adults ... (McCulloch 1913a)

McCulloch dealt with the nomenclature of the Bluethroat Wrasse, *Notolabrus tetricus* (Richardson, 1840) and noted that '[a] well graduated series of twenty-six specimens, 180-420 mm. long, shows that the colour-markings of this species vary remarkably, the variations apparently depending upon both age and sex.' Much later it was discovered that these fish are protogynous hermaphrodites (beginning as female and altering to male with growth).

McCulloch's problems with Australian authors

The reference above to Castelnau, Macleay and de Vis is rather restrained. McCulloch sometimes expressed himself more forcefully. A small leatherjacket from temperate Australia was named *Monacanthus* [now *Scobinichthys*] *granulatus* by John White, surgeon of the First Fleet, but poorly described (see figure page 22). The same species was described and named from different specimens in four separate publications, between 1872 and 1878, by Count Castelnau. The specimens came from Fremantle (Western Australia), Gulf St Vincent (South Australia) and Hobson's Bay (Victoria). Not only did Castelnau fail to identify the fish from White's original description but he gave it five new names, and described it inaccurately on each occasion. In this case McCulloch, who worked on the Leatherjackets in 1912, found that the synonymy had been largely worked out by Klunzinger of Stuttgart who had been sent specimens by the Melbourne botanist, Ferdinand von Mueller. As usual, McCulloch confirmed the identifications wherever possible by referring to the type specimens. He described this work in a letter to Ogilby, concluding:

> What solid pleasure one could get by murdering the man who could make such a mess of things as Castelnau did of this. (McCulloch to Ogilby, 9.4.12, QMA)

This was typical of the problems that needed to be sorted out to put the taxonomy of Australian fishes on a sound basis.

To a lesser extent, McCulloch had similar problems with the work of Robert Johnston in Tasmania although Johnston was a less prolific writer on ichthyology. He named about 20 Tasmanian fishes as new species, over the course of 20 years, most of which proved to be synonymous with known species. Ramsay, Saville-Kent, McCoy and a handful of others contributed a little extra confusion, but their work was much easier to deal with since their ambitions in ichthyology were limited.

The work of J. Douglas Ogilby was a prime cause for concern. McCulloch found that his old friend had contributed more than a little to the confusion. In some of his early papers, and those that he co-authored with E.P. Ramsay, Ogilby had described new species of fish without adequate reference to earlier literature. However, his descriptions were generally sufficiently detailed and accurate for his errors to be corrected with some confidence later, by himself or by McCulloch. Much the same could be said for Waite's work, although he did not practise taxonomy on the scale undertaken by Ogilby.

Once Ogilby began to resume publication in Queensland, his work became of great concern to McCulloch. Fortunately, the latter was able to guide and advise, and to a remarkable extent control Ogilby's output of new species descriptions subsequent to 1911. Through the last decade of Ogilby's work, he described relatively few new species as he concentrated, largely at McCulloch's behest, on working out synonymies and gathering biological information. Ogilby was an asset, albeit one whose maintenance demanded the expenditure of large amounts of McCulloch's time and energy.

United States ichthyologists

McCulloch, like Ogilby and Waite before him, had help from David Jordan of Stanford University in California, and from his associates and students. Jordan soon developed respect for McCulloch's work and the former's wide experience was at McCulloch's disposal. Jordan had become interested in the fishes of Australia, having exchanged collections with both Ogilby and Waite, but he did not interfere directly with the taxonomy of Australian fishes.

Jordan's own work was focussed on the fauna of North and Central America, and the Pacific. The rich fauna of Japan had attracted his attention; he and his associates worked on Japanese fishes and taught Japanese zoologists Jordan's approach to taxonomy. The American influence in Japan was much more direct than in Australia. Jordan and his associates undertook voyages to Japan and some resided there for lengthy periods; Japanese ichthyologists were also employed at Stanford at times.

David Starr Jordan: a portrait painted in 1921 when he was 70. He had recently completed *The Genera of Fishes*. From Jordan 1922.

McCulloch frequently sought, and usually took, Jordan's advice with regard to difficult taxonomic problems. For instance, when Ogilby asked McCulloch which was the appropriate generic name for a genus of electric rays, the latter said that he had 'followed Jordan whose work covering such an enormous field places him in what is probably the best position as a judge.' (McCulloch to Ogilby, 16.1.20, QMA)

Sometimes Jordan's information was almost an embarrassment, as when he sent the fourth part of his *Genera of Fishes* to McCulloch for comment:

> Have just card-indexed friend Jordan's papers on new genera (P. Ac. Philad. 1918 p. 341) and am unable to decide whether I like or dislike him for the trouble he has given me. (McCulloch to Ogilby, probably 26.6.19, QMA)

Jordan's works on fish genera had a great rationalising influence on the taxonomic studies of the ichthyologists of his day, and on that of subsequent workers. He relied on his American associates for most of the information collated in these publications, but also solicited contributions and comment from ichthyologists outside the United States.

McCulloch's studies, organised and strengthened by his card indices, allowed him to make useful additions to, and criticisms of, Jordan's work. He wrote to Waite:

> Jordan was good enough to forward me a rough list of his classified genera of the world, which I compared with ... my card-index, and so discovered a number of errors. These included one or two of his own American genera, whereat he moaned largely, though he was big enough to be quite glad to have the corrections. He has been very lavish with Families, and not a few will ~~not be accepted~~ be very difficult to define, but he is risking this for the sake of simplification. (McCulloch to Waite, 14.8.22, SAMA)

Jordan's strategy was to make taxonomic units above the species level smaller than did most other workers, so-called 'splitting', necessitating the definition of many new genera and families. McCulloch had some well-founded reservations about this approach and did not fully implement the changes advocated by Jordan.

One of the few problems caused directly by an American ichthyologist, that McCulloch was obliged to consider, was due to E.D. Cope (see page 157). Vertebrate Palaeontology was Cope's principal study; he was a highly respected zoologist who had begun his career in the 1860s with studies of reptiles, amphibians and fish. He briefly described two new species of leatherjacket (Monacanthidae), suggesting that they may have come from Australia (Cope 1871). McCulloch mentioned the problem in a letter to Ogilby:

> Monacanthus homopterus & M. hypargyreus, Cope are described from specimens 'said to have come from Australia'. They may therefore be Q^{ld} [Queensland] species or may not. Bust him! (McCulloch to Ogilby, 19.4.12, QMA)

He was not able to resolve this problem and, in the check-list of Australian fishes published posthumously (McCulloch 1929–1930), the two species bear the locality 'Australia.' Later research showed that Cope's *Monacanthus homopterus* was a synonym of a well-known Atlantic species, the Tail-light Filefish, *Cantherhines pullus* (Ranzani, 1842). *Monacanthus hypargyreus* was a West Pacific fish, the Yellow-spotted Leatherjacket, *Thamnaconus hypargyreus* (Cope, 1871): Cope's specimen may or may not have come from Australia.

British and European ichthyologists

By the time McCulloch began publishing on fishes Albert Günther, then in his late 70s, had retired from the British Museum. George Boulenger was in charge of the Department of Zoology and it was to him that McCulloch initially appealed for help with type specimens and for more general advice. C.T. Regan had been appointed assistant to Boulenger in the Zoology Department in 1901 and specialised in ichthyology. McCulloch found Regan responsive and helpful, and they corresponded frequently.

Waite sent McCulloch a small, unidentified fish among a collection from the Kermadec Islands. McCulloch found that it belonged to the family Girellidae and mentioned it to Ogilby:

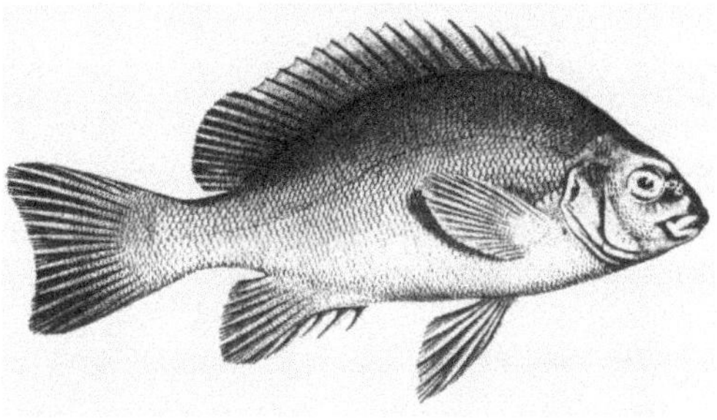

Girellops [now *Girella*] *fimbriata* from the Kermadec Islands, drawn and described by McCulloch. From McCulloch 1920.

> My new Girellid from the Kermadecs apparently belongs to Regan's Girellops, but he has failed to note the real generic characters in the dentition. I shall have to wait till I hear from him before I can definitely settle its identity. Meanwhile I have described and figured it for future use. (McCulloch to Ogilby, 26.6.19, QMA)

Regan had established the new genus *Girellops* almost as an aside: 'Related to *Tephraeops*, but with uniserial teeth.' (Regan 1913a) He specified *Girella nebulosa*, a fish described from Easter Island the previous year, as the genotype. McCulloch's problem was that his Kermadec fish had a second, irregular row of teeth behind the main series. After receiving Regan's advice, McCulloch decided to place his fish in Regan's new genus, describing it as *Girellops fimbriatus* (McCulloch 1920).

Regan's most significant papers were revisions of the taxonomy of groups of fishes based upon his own dissections and osteological studies. The results of these researches became available during the time that McCulloch was engaged with the taxonomy of Australian fishes, and Regan's work was of great importance to him. He and Ogilby usually accepted Regan's taxonomic decisions; his conclusions about the relationships of groups of fishes provided a modern framework for their papers.

Contact with the ichthyologists of the European museums was mainly concerned with reference to type specimens. McCulloch found that he could still get some help from Franz Steindachner of Vienna, and several times addressed queries to him, although this correspondence was inhibited during and immediately following the war. McCulloch wrote to Ogilby about a 'Demoiselle' which he considered could be:

> ... the long lost Pomacentrus unifasciatus Stdr, [Steindachner] but as I suppose it is bad form to write to Vienna these days, I can hardly prove this. Maybe I'll let it wait a while. (McCulloch to Ogilby, 16.1.20, QMA)

Castelnau had sent types to the Paris Museum in the 1870s and McCulloch asked for help with these. Photographs were taken of the relevant type specimens in Paris and sent to McCulloch in Sydney: he mounted them in a small folio, labelled 'Fish Types in the Paris Museum. 1914. McCulloch'.

Card indices

Waite had been an advocate for card index systems and had begun several databases of this sort at the Australian Museum, advocating the technique as a key to the collections (Waite 1900b). McCulloch modified and expanded this system with a view to having separate card entries for all references to Australian fish in a massive card index. He occasionally mentioned the work to Ogilby:

> My card-index (new-pattern, 1912-1914) is growing steadily, I having just put away 32,000 reference slips. Have ordered another 40,000 to go on with which should see me through the Zoological Record. (McCulloch to Ogilby, 30.11.14, QMA)

The index became a powerful tool for the study of fish taxonomy and McCulloch, as its designer, knew its strengths and limitations. He kept it up to date as new information became available from published work and as his own labour clarified taxonomic issues.

Small additions and modifications to the index were required from day to day and there were occasional upheavals, such as that caused by David Jordan's publications on the Genera of Fishes. McCulloch criticised Jordan's logic in his treatment of some generic names, concluding grumpily, '... all things considered, I fancy Jordan must be suffering from his age.' (McCulloch to Ogilby, 26.6.19, QMA). Ogilby, too, produced some papers which had seismic consequences for McCulloch's index. On receipt of Ogilby's *Edible Fishes of Queensland, Parts X – XXIV*, McCulloch remarked that he would need 'the devil of a shuffle to bring my cards into line with all the conclusions arrived at' (McCulloch to Ogilby, 11.8.18, QMA) and later, 'I'm fairly itching to rearrange the cards tomorrow.' (McCulloch to Ogilby, 11.8.18, QMA)

The day to day value of the card indices, which were maintained as a key to the collections, is clear from McCulloch's references. For instance, he was able quickly to discover which Grunters were available to send Ogilby to assist in some of his work:

> ... am very sorry to find they are mere babies. When I turned up the catalogue cards to see what we had, I did not notice their small size. (McCulloch to Ogilby, 2.6.15, QMA)

On another occasion:

> Am at a loss to understand the inclusion of the Endeavour River among the localities of Conger vulgaris by Macleay since there is no specimen so labelled in his Museum. I have a very complete card-index of the Macl. Collection, and the only eels from the Endeavour River are Gymnothorax tesselatus and a small 'Muraena sp' which is only 4 inches long. (McCulloch to Ogilby, 26.2.20, QMA)

These references to collection card indices, some of which were started by Ogilby, or Waite, indicate how important such databases had become in keeping track of specimens. They were not an alternative to the accession registers, large books in which the details of new specimens were entered, but they provided a rapid means to locate references and specimens.

Collaboration with Ogilby: 1908–1921

The active collaboration between McCulloch and Ogilby lasted for over a decade. During this period only three of their published works were formally co-authored, but they contributed to one another's papers in many ways. In the first of the co-authored papers, a review of the Wobbegongs (Orectolobidae), Ogilby was the senior author. At the beginning of 1908 Allan McCulloch, although only 22, was in charge of a significant portion of the zoology of the Australian Museum. In that year his publications were extensive: the first of his major works on Australian fishes (McCulloch 1908a), the first of a series on Australian Crustacea (McCulloch 1908b), description of a new genus and species of turtle (McCulloch 1908c) and the results of his first collaboration with Ogilby (Ogilby and McCulloch 1908).

Ogilby was still struggling to re-establish his scientific reputation after his disastrous move to Brisbane. They were obliged to modify their paper on the Wobbegongs when, just prior to publication, C.T. Regan's review of the same group appeared in September 1908:

> When our paper was almost ready for press we received Mr. C. Tate Regan's 'Revision of the Sharks of the family Orectolobidae' ... as he has taken into consideration all the species of the family, we have curtailed our first intentions, and have confined ourselves to the Australian species only. (Ogilby and McCulloch 1908)

Although Regan had to some extent stolen their thunder, the paper by Ogilby and McCulloch comprised a full and able coverage of the Australian species then known.

McCulloch found Ogilby an energetic if somewhat eccentric ally in the work of resolving synonymies. This was the most difficult task facing Australian ichthyologists and only Ogilby and McCulloch were willing to engage fully with it. Their first tentative steps in collaboration appear to have been taken in 1907–08. The process was threatened when Ogilby ignored McCulloch's right to work on the Queensland trawl collections (see pages 219–221) in 1910, but by 1912 they were corresponding frequently. From this time until 1920, when Ogilby's work was stopped by illness, he was McCulloch's most effective colleague. They shared the same view of the task, each appreciating the importance of sorting out the taxonomy of the fish fauna and bringing the information together in descriptive works. The significance of their relationship can hardly be overestimated. McCulloch's letters to Ogilby are illuminating (and much quoted below).

As their relationship matured, McCulloch asserted a benign dominance over Ogilby: the quality of his papers was greatly enhanced by McCulloch's encouragement and intervention. The advice was sound and Ogilby generally acted upon it. The following five headings are examples of McCulloch's advice on as many topics.

1. Importance of publication:

Ogilby had a paper on trevally in preparation for his *Edible Fishes of Queensland* series but wished to have McCulloch's opinion about the dentition of specimens in the Australian Museum. McCulloch was worried that much of Ogilby's work might be wasted through procrastination and wrote:

> Do not think it better to hold your paper back under any circumstances. Publish what you have and do the rest later. (McCulloch to Ogilby, 14.9.14, QMA)

This was a recurrent theme throughout the correspondence; Ogilby mentioned that he had a specimen of the Comb Wrasse, *Coris picta* and McCulloch urged him to publish the record:

> And so Coris picta turns up in your wonderful Moreton Bay. Well its not surprising, but don't fail to put the record down on paper now and at once. Why do I say that? Just because your letters for years back have told me of new records time & time again for which I have looked in each succeeding volume of your Memoirs [of the Queensland Museum] in vain. Its very human to loose [sic] interest in a beast after we have run it down and satisfied ourselves, but it is damned bad science. (McCulloch to Ogilby, 2.8.19, QMA)

2. *Nomina nuda*:

William Saville-Kent (1893) published a large number of new scientific names for Queensland fishes. Many of these names were associated with neither descriptions nor illustrations, although some were based upon Queensland Museum specimens that Charles de Vis had labelled. It seems that Saville-Kent had taken at least some of the names on his list from the labels. These names had not been published prior to their appearance in Saville-Kent's book and subsequent identification of the species was difficult or impossible. Ogilby, McCulloch and Waite had agreed that these scientific names should be discounted as *nomina nuda* ('bare names', having no status in formal nomenclature).

In 1915 Ogilby proposed to quote Saville-Kent's names in synonymies and McCulloch was horrified:

> Surely there is no question of the status of nomina nuda. If a name has nothing to substantiate it (technically speaking, you cannot prove that de Vis' label really applies to the specimen it is associated with) then it has no value or interest to science. There can be no question in my mind, and Mr. Hedley agrees with me, that should Lethrinus regius prove to be a valid species, and redescribed by you, it would certainly be L. regius, Ogilby, and De Vis and Kent would require no mention. Consider what would be the case in the reverse. Suppose you credit this name to either de Vis or Kent, then what is to happen to the 50 or so other similar names which are based on specimens which will never be found? I fail to see how you can have any doubts upon the matter, and sincerely hope you will leave the names in their grave of rubbish. (McCulloch to Ogilby, 2.6.15, QMA)

Ogilby argued that this particular case was different, that the fish could be identified. This provoked McCulloch to a further defence of his policy:

> Surely you are not going back on your decision to ignore Kent's list at the end of his Gt. Barrier Reef. You will remember that we agreed to disregard it, as

> it is full of MS. names, while Waite also agreed upon this detail. Acting upon this, I have consistently discarded it ever since, and sincerely hope you will do likewise. If you deal with one name, you must dispose of the rest. Further, surely the synonyms of Lethrinus spp., are numerous enough without you deliberately adding another. (McCulloch to Ogilby, 2.5.15, QMA)

Ogilby was persuaded: he dealt with *Lethrinus* [Emperors] without mentioning *L. regius* (Ogilby 1916c).

3. Uniformity of plan.

McCulloch, in having Ogilby join him in a paper on the gobies, hoped that Ogilby would conform to a suggested plan:

> On the back hereof is the skeleton plan of all my descriptions for the paper which, with a complete one in hand, you should have no difficulty in deciphering. Of course I don't insist it is the best, but if you can arrange your descriptions on the same lines, it will improve the harmony of our paper. (McCulloch to Ogilby, 15.2.18, QMA)

A practical if undemocratic approach. McCulloch was confident of the value of his own methods and imposed them, fairly gently, on others. He praised a paper (Ogilby 1918) on Queensland fishes, then offered a series of mild and constructive criticisms:

> The confusion in your lay out of Nemipterus guntheri [Threadfin Bream, *N. furcosus* (Valenciennes, 1830)] nearly had me beaten, but I unravelled it after a while. For why does the synonymy go at the tail-end of your notes on the species instead of in the usual place, & hopelessly confused with the text of the succeeding species. (McCulloch to Ogilby, probably 11.8.18, QMA)

4. Specifying types:

The importance of type specimens of new species, and the specification of types for new genera and subgenera, were topics to which McCulloch paid special attention. He scolded Ogilby:

> Its a dangerous thing to propose new subgenera in a footnote – Acanthogonia [proposed by Ogilby as a subgenus of *Paraplesiops*] p. 46 – since they can be so easily overlooked, but its a real crime not to fix the type when one does so. (McCulloch to Ogilby, probably 11.8.18, QMA)

5. Preparation of keys.

McCulloch advocated the use of keys in identifying the species of a genus, or the genera of a family. Ogilby had sent useful notes on the genus *Plectorhinchus* [Sweetlips] and McCulloch advised him to prepare a key to them for publication. He added:

> I had arrived at the determination to prepare generico-specific (some term?) keys for publication after the descriptions of any odd species I should write upon, as I found the identification of large collections greatly facilitated when one could turn to such quick guides. We are sadly lacking in such keys as regards our Australian species, and I submit the idea to you for consideration. A key to the Siganus spp [Rabbitfishes] would be all to the good. I know nothing of the brutes, but congratulate you on your notes on S. lineatus, C & V. (McCulloch to Ogilby, 6.10.18, QMA)

Following their paper on Wobbegongs (Ogilby and McCulloch 1908) a second paper, also with Ogilby as the senior author, appeared in 1916. This was a review of the freshwater Grunters (see page 231). Their last collaboration, in which McCulloch was the senior author, was a large paper on the Gobies of Australia (McCulloch and Ogilby 1919).

Some Australian Fishes of the Family Gobiidae was intended to be a full review of all Australian species but several factors limited its eventual scope. Most important was the fact that the group was simply too big for effective review within the time that the authors envisaged. Hoese *et al.* (2006) estimated that there were 344 Australian species out of 1500 worldwide; many remain to be named and indeed discovered. Also, McCulloch and Ogilby included the related eleotrids (Gudgeons) in their paper. They mentioned the perennial difficulty in obtaining some of the pertinent literature. Lastly, McCulloch's enlistment in the army in the latter part of 1918 abbreviated his researches on the group.

The authors explained their conservative approach:

> It was the original intention of the authors to revise all the Australian species of the family Gobiidae, but circumstances have prevented us from carrying out our design. We therefore submit descriptions and figures of such species as we have been able to deal with, and give references to the others. We have been unable to allocate some of the species dealt with to any genera known to us, but as we lack several important papers on the classification of the Gobiidae, we have preferred to leave them under the broader headings of *Gobius* and *Eleotris* rather than create unnecessary additions to the already long list of Gobioid genera. (McCulloch and Ogilby, 1919)

This was typical of the manner in which McCulloch attempted to simplify the taxonomy of the groups he studied, usually preferring to leave a dubious specimen imperfectly determined, though well defined by thorough description and illustration. He might almost be following the caution attributed to Hippocrates: 'First, do no harm.'

The goby paper was based upon specimens from several main sources:

> We have had the advantage of examining the very large collections contained in the Australian Museum, the Queensland Museum, the Macleay Museum, and the South Australian Museum. (op. cit.).

These sources, particularly the first three, were repositories for large numbers of the type specimens of Castelnau, Macleay and de Vis, representing species already described and named elsewhere. McCulloch and Ogilby defined nearly 70 species, almost half of which were illustrated by Phyllis Clarke in this paper, most for the first time. The majority of these were correctly determined although the generic allocations have often altered since that time, and are still being altered. Useful keys and extensive synonymies were a feature of the paper. The partial nature of the success of this study was recognised by McCulloch, but he was nevertheless pleased with it, and incidentally pleased to have it off his hands:

> Rejoice with me. Our Goby Paper is to go to press this morning, so I've spent the whole week-end scribbling as hard as I have been able to bring it into shape for publication. It will of course be terribly incomplete as a revision, but will make a wholesome little contribution nevertheless. (McCulloch to Ogilby, probably 3.3.19, QMA)

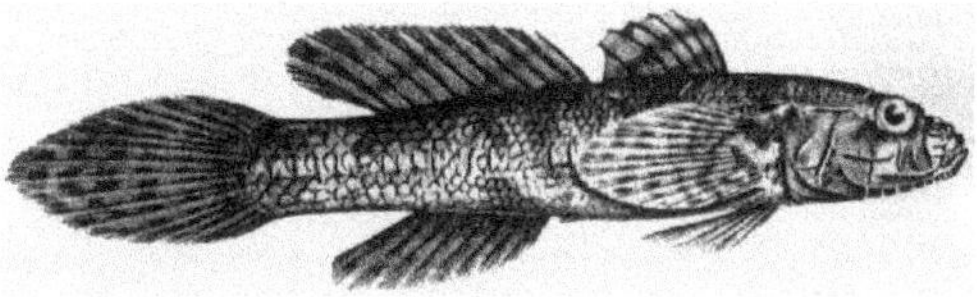

Drawings of gobies were executed by Phyllis Clarke under McCulloch's supervision. Over 100 specimens of the Sculptured Goby, *Callogobius mucosus* (Günther, 1872), a southern Australian species, were to hand. From McCulloch and Ogilby 1919.

The Redhead Goby, *Paragobiodon echinocephalus* (Rüppell, 1830) lives among branching corals and is distributed from East Africa to the Marshall Islands. McCulloch had 62 specimens in all, from several localities on the Great Barrier Reef and from Papua; he had collected the fish at Murray, Green and Masthead Islands on various field excursions. McCulloch and Ogilby provided an extensive synonymy, in particular synonymising two of Macleay's puzzling species, collected 40 years earlier at Cooktown, with this species. From McCulloch and Ogilby 1919.

McCulloch and Ogilby resolved many of the taxonomic problems which they encountered and their paper had a good reception:

> A pleasant note from David Starr [Jordan] approves muchly of our Gobiidae effort. (McCulloch to Ogilby, probably 10.10.19, QMA)

Jordan questioned one of their taxonomic decisions but McCulloch was able to defend their opinion successfully, boasting cheerfully to Ogilby, 'Thats one up for us.' In his discussion of gobioid osteology and classification, Victor Springer (1983) described the problem of goby taxonomy as an 'infamous and unwieldy morass.' At present approximately 350 species of the family Gobiidae are recorded from Australian waters. McCulloch and Ogilby (1919) considered only a small proportion of the Australian fauna but their contribution was useful.

McCulloch's other collaborations

As the most junior assistant at the Australian Museum, McCulloch began by working with Waite on fishes and, later, with Grant on crustaceans. His first contributions were primarily through preparing illustrations for the publications of Waite, Grant, Charles Hedley and other museum staff members. In his capacity as Mechanical Assistant, he also carried out photographic work, some of which was used to illustrate papers for the *Records of the Australian Museum*.

McCulloch conducted collaborations at distance with great energy. His work with Ogilby is the best example of this but, once independent, he manipulated his work with Waite effectively. The resultant papers, although McCulloch may not have had as much credit for his contributions as he deserved, were well organised and valuable. Waite knew and respected McCulloch's skill as a taxonomist: he sought and deferred to his opinion on many occasions.

In 1916, Robert Etheridge had been Curator of the Australian Museum for 15 years. McCulloch had provided drawings to illustrate Etheridge's papers on palaeontology since Waite had left the Museum and he joined Etheridge in 1916 in writing and illustrating a paper on sub-fossil crustaceans (Etheridge and McCulloch 1916).

An effective instructor, McCulloch conducted joint projects with his assistants. These sometimes led to useful publications, in particular with Frank McNeill on fishes (McCulloch and McNeill 1918) and crustaceans (McCulloch and McNeill 1923), and with Gilbert Whitley on fishes (McCulloch and Whitley 1925a and 1925b). McNeill ultimately made crustaceans his specialty but, during his first few years at the Museum, he undertook some useful projects on fishes with McCulloch's guidance. Under instruction from McCulloch, McNeill made competent drawings of Blennies for their paper.

McNeill also conducted a brief but useful project on the Luderick (or Blackfish),

McCulloch and his assistant, Frank McNeill, collaborated to produce a study of blennies (1918). McNeill drew the eight species they considered. They examined specimens in the Macleay and Australian Museums; Longman of the Queensland Museum sent specimens for which Ogilby provided notes. McCulloch had collected blennies from Murray Island, Cairns and Masthead Island and the museum held collections from India and from the New Hebrides. No new species were described, the object of the study being to reduce the disorder caused by previous workers. The fish above are (from top to bottom) the Lined Rockskipper, *Salarias* [now *Istiblennius*] *lineatus* (Valenciennes, 1836); Dussumier's Rockskipper, *Salarias* [now *Istiblennius*] *dussumieri* (Valenciennes, 1836); *Salarias geminatus* of which they wrote 'It is not improbable that *S. geminatus* is merely the male form of *S. dussumieri*' (and this has proven to be so); the Whitedotted Blenny, *Cirripectes alboapicalis* (Ogilby, 1899). From McCulloch and McNeill 1918.

Girella tricuspidata, a common herbivorous fish of some commercial importance. Two forms of Blackfish were thought to occur in temperate eastern Australia: they were known as *Girella tricuspidata* and *Girella simplex*. The main difference between them was their dentition, the teeth of the former were tricuspid whereas those of the latter were simple incisors. David Stead had declared, without offering supporting evidence, that these were simply sexual differences in a single species but Ogilby and McCulloch were sceptical. McCulloch set McNeill to investigate and informed Ogilby of the result in a postscript:

> Assistant McNeill has proved Stead correct in his unproved dogmatic statement that Girella simplex & tricuspidata are sexual forms of the same species. Extracting roes and milts together with the jaws, he has found they coincide without exception. Interesting, eh? (McCulloch to Ogilby, 5.7.19, QMA)

McCulloch included McNeill's results in *Studies of Australian Fishes. No. 6* (McCulloch 1920), acknowledging McNeill's work.

The publication of McCulloch's check-list of the fishes of New South Wales (McCulloch 1922a and 1922b) proved to be an effective stimulant to similar work elsewhere. In 1923 Clive Lord of the Tasmanian Museum published a list of Tasmanian fishes in the *Proceedings of the Royal Society of Tasmania*. McCulloch mentioned this to Waite:

> Lord of Tasmania recently forwarded me a list of Tasmanian Fishes for revision which is largely based upon my list, while Griffin and Phillipps of New Zealand have found it quite helpful. (McCulloch to Waite, 14.8.22, SAMA)

Although not collaborating in the complete sense, McCulloch was asked for help and gave it cheerfully.

New Zealand ichthyology has held a special interest for many Australian ichthyologists and McCulloch was no exception. Waite had caused some disappointment, particularly among the New Zealand fisheries interests, as he had not provided a catalogue of New Zealand fishes. After he moved to Adelaide, he gave up all thought of this project. McCulloch was approached in 1920:

> A letter from G.M. Thomson of N. Zealand asks what I would be prepared to go over there for to write an illustrated catalogue of the N. Zealand fishes. This is very tempting, especially as he insists that I should run around the various museums there and do all necessary collecting. So its just a matter of an arrangement between the N. Zealand Govt. & our Trustees which is not likely to present any difficulties as things stand at present. (McCulloch to Ogilby, 8.1.20, QMA)

He may have been particularly attracted to the proposition as he was apprehensive that Waite might be appointed to the directorship of the Australian Museum; McCulloch had no wish to work under Waite again. Waite did not get the job and McCulloch did not go to New Zealand.

William J. Phillipps brought Waite's list of New Zealand fishes up to date and sought help from McCulloch. McCulloch and Phillipps published a brief paper 'based upon specimens of several fishes from New Zealand waters, belonging to various families, which have been examined by us jointly' (McCulloch and Phillipps 1923). Later Phillipps, then of the Dominion Museum, Wellington, published an extensive *Bibliography of New Zealand Fishes* (1927), elaborating Waite's basic list after consulting both Waite and McCulloch.

The Blue-eyed Triplefin, *Notoclinops segmentatus*, common in rocky areas in New Zealand, was described by Phillipps and McCulloch. McCulloch illustrated the paper. From McCulloch and Phillipps 1923.

Endeavour collections (1909–1914) and reports (1911–1926)

The *Woy Woy* haul from 1500 metres was an intimation of the rudimentary state of knowledge of the deep water fishes of eastern Australian waters. Of more immediate significance to fisheries science and to ichthyology was an investigation of the continental shelf (waters to 200 metres in depth) and the adjacent part of the continental slope. These waters showed promise for developing fisheries and were populated by a diversity of fish, many new to science. Systematic trawling experiments were required and for this purpose a vessel, the *Endeavour*, was built at the Government Dockyard in Sydney in 1908.

The *Endeavour* investigations were planned and supervised by the new Commonwealth Director of Fisheries, Harald Christian Dannevig (1871–1914). Dannevig had worked in his native Norway and in Scotland in ocean fisheries and in fish hatcheries. He was chosen by the New South Wales Government in 1902 to manage their fisheries research and attempted to introduce some North Sea species to Australia (ADB). He brought with him several hundred Plaice (*Pleuronectes platessa*), Turbot (*Scophthalmus maximus*) and other marine species, but (fortunately) they did not become established. His trawling experiments extended to southern Queensland, New South Wales, Victoria, Tasmania, South Australia and south Western Australia. The feasibility of commercial trawl fisheries was demonstrated and productive grounds were located and charted. Dannevig published several papers on New South Wales fisheries, aquaculture and acclimatisation, and latterly a series of reports on the trawling experiments. The last of these (Dannevig 1913) included an optimistic summary of the potential of Australia's trawl fisheries. His work ceased prematurely when he and the vessel, with the 18 men on board, were lost; they had set out in a gale, returning from Macquarie Island in December 1914, and were heard from no more.

Marine organisms taken by the trawl were sent to specialists for study and the results published through the Department of Trade and Customs. Fishes from the *Endeavour* were sent to Allan McCulloch at the Australian Museum and he contributed five reports on these specimens between 1911 and 1926. McCulloch was aboard the *Endeavour* from 12th March to 6th April 1914 and 'saw the trawl worked at various depths between sixty and two hundred fathoms'. [110 to 370 metres] In being the sole recipient of the *Endeavour* fish collections, McCulloch had the responsibility for working out their taxonomy. This included the description of new species and the identification of many species whose taxonomy was in a confused state.

The *Endeavour* trawling experiments were initiated in 1909–1910 in depths to 170 metres. A portion of the catch was retained for scientific purposes and forwarded to McCulloch. The initial consignment consisted of 1070 specimens, from which he identified 94 species. He published a lengthy account of them in his first *Endeavour* report (McCulloch 1911).

First *Endeavour* report: 1911

In this report, McCulloch dealt with the fish in systematic order. He began with the sharks and rays, proceeded to those groups of bony fish that were supposedly more primitive, and ceased part way through the more advanced forms. About 50 families were covered in this way and approximately 15 left for the next part of the report. This systematic method could not be sustained, not least because the enormous collections that were delivered for his attention came in anything but systematic order. In subsequent reports McCulloch modified his approach, leaving some of the troublesome deepsea forms until later reports and concentrating on specimens and groups that promised material for useful revisions and the solution of taxonomic problems.

Up to this time, ichthyologists in Australia usually had few specimens at their disposal when preparing descriptions. Alterations with growth and variations in form and coloration among specimens of a single species could be easily misinterpreted but McCulloch found that, together with the collections at the Australian Museum, the *Endeavour* material allowed the resolution of many problems:

> With a series of forty or fifty specimens from wide-spread localities before one, it is often a simple matter to recognise the identity of fishes which have been regarded as distinct when isolated examples afforded no clue. (McCulloch 1911)

He particularly acknowledged Harald Dannevig's efforts:

> Many of the more striking variations were specially selected by Mr. Dannevig on board the ship, and it is largely due to his discrimination in this direction that so many extreme forms are here dealt with. When individuals were lacking to complete a series he has gone to considerable trouble to obtain them, and in the case of fishes in which the colours were of importance, he has sent me fresh material packed in ice. (op. cit.)

An extreme example of variation was afforded by the '[f]ifty-two specimens, of all sizes between 75 and 300 mm. long' of the 'Short Boar Fish'. McCulloch was impressed that the changes in proportion did not occur strictly according to size, and illustrated this with a series of text figures.

The provision of illustrations of species that had not been adequately represented in the literature had a high priority throughout McCulloch's work. He followed Waite's policy with this. In the initial *Endeavour* report, nine species were described as new and all were drawn. In all, 26 species were illustrated in the plates and several others in text figures. A young recruit to the Museum staff helped: 'I have to acknowledge the very willing assistance of Mr. Dene Fry, of the Australian Museum, in the preparation of the plates.' (op. cit.)

In his first *Endeavour* report McCulloch prepared a series of outline drawings to show irregular changes in the form of the Short Boarfish, *Zanclistius elevatus* (Ramsay and Ogilby, 1888) with growth. His specimens were trawled in about 90 metres in various parts of southern Australia. From McCulloch 1914.

Fry drew under McCulloch's supervision and his work appeared in other papers, always scrupulously acknowledged.

Fry enlisted in the army and Ogilby later wrote:

> Heard through Longman the sad news of Fry's death at the front; it is most regrettable, not only because of the loss of a young & valuable life, that is too common now-a-days to excite much comment, but because he was our most promising young herpetologist; his death will be a great loss to the staff of the Australian Museum ... (Ogilby diary fragment, 3.5.17, QMA)

The *Endeavour* collections generated an enormous volume of work, much of it in the form of communication with fellow workers. Correspondence and the exchange of specimens with authorities in other institutions was an integral part of McCulloch's research. For instance, he thanked Edward Stirling of the South Australian Museum for sending a specimen of a catshark, and himself sent a photo of a seven-gill shark to C.T. Regan of the British Museum for his opinion as to its relationships. He sought extra information from Waite and Ogilby as a matter of course.

This portion of the *Endeavour* material included specimens of a deepwater fish, *Apogonops anomalus*, first described by Ogilby 15 years earlier (see page 189). McCulloch re-described the species as he found some errors in the description of the scales and teeth in the original work. He was gentle with Ogilby:

> All of these characters are much more easily detected in my largest specimens, 122 mm long, than in the type which is only 54 mm long, and, considering the care which Mr. Ogilby usually lavished on his descriptions, the differences are doubtless due to the want of sufficient microscopical appliances on his part. (op. cit.)

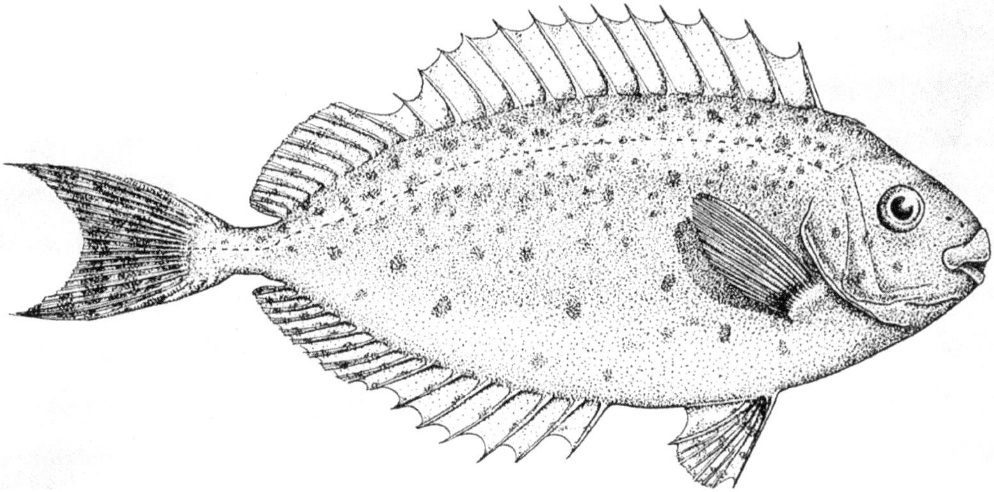

The Black Rabbitfish, *Siganus consobrinus* was described by Ogilby (1912) and drawn by Dene Fry under McCulloch's guidance from the type specimen sent by Ogilby. It has proven to be a synonym of *Siganus fuscescens* (Houttuyn, 1782). From Ogilby 1912.

McCulloch, quite reasonably, showed less forbearance in dealing with some of the lapses in Ogilby's hurried 1910 paper on the Queensland *Endeavour* fishes.

Endeavour: Queensland in 1910

In 1910 the *Endeavour* was in Queensland waters with Ogilby aboard and he seized the opportunity to describe a host of new species (see page 219). McCulloch found Ogilby's behaviour extraordinary and referred to it in his introduction to the second report (McCulloch 1914b). In re-describing Ogilby's new species, McCulloch several times refers to the origin of Ogilby's inaccuracies.

Ogilby described an Anglerfish, *Chirolophius laticeps* (synonym of *Lophiomus setigerus*) while on board the *Endeavour* from a single specimen 18 cm long. He subsequently forwarded the specimen to the Australian Museum and McCulloch, in re-describing and illustrating the fish, commented:

> [The fish was sent] without any label accompanying it. Ogilby gave the length as 196 mm., but this measurement includes the lower jaw when the mouth is wide open. His description is unintelligible in parts and misleading in others, errors which are doubtless the result of its absurdly hurried preparation and publication. (McCulloch 1914b)

McCulloch also questioned, giving his reasons, Ogilby's placement of the fish in the genus *Chirolophius*.

In his initial report, McCulloch began to deal with the fishes that had been taken by the *Endeavour* in Queensland waters, between Moreton Bay and Bowen, in the winter of 1910. The Anglerfish mentioned above was just one of 30 species which Ogilby had

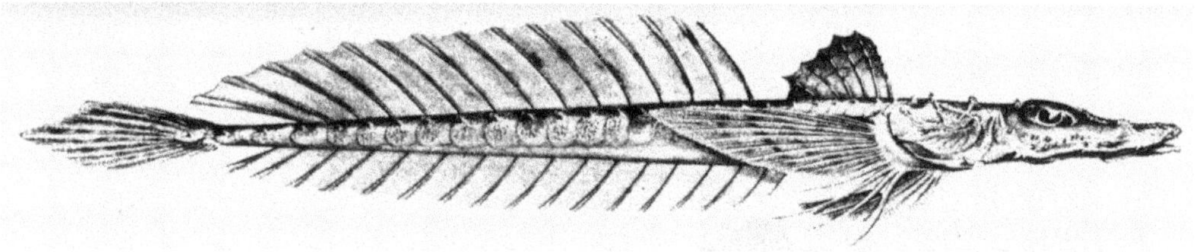

Ogilby's Ghost Flathead, *Hoplichthys ogilbyi* was trawled by the *Endeavour* off Cape Moreton, Queensland and described by McCulloch in his second *Endeavour* report. From McCulloch 1914b.

described as new. Ogilby had not labelled the type specimens so McCulloch, who decided to treat Ogilby's work as though it had been properly published, began by attempting to recognise Ogilby's types among the specimens. He then re-described the fish in more detail and with greater accuracy than Ogilby. In later publications, *Endeavour* Reports II (1914b), III (1915a) and IV (1916a), McCulloch continued to deal with Ogilby's new species, eventually re-describing many of them. McCulloch paid tribute to Ogilby when naming a deepwater flathead, taken north of Cape Moreton: 'I have much pleasure in associating the name of Mr. J. Douglas Ogilby with this pretty Queensland species'. (McCulloch 1914b)

Endeavour: Great Australian Bight, 1912–1913

The second of McCulloch's *Endeavour* reports was published in 1914 by which time the trawling trials had been extended into the Great Australian Bight. Preliminary work was carried out in the Bight in 1912 and more extensive investigations in 1913. Deepsea fishes were a conspicuous component of the second report. McCulloch noted:

> The 'Endeavour' ... has lowered a large commercial trawl into 450 fathoms [830 metres] and several lesser depths in the Great Australian Bight, and, as might be expected, has captured a wonderfully rich collection of new or little known fishes. (McCulloch 1914b)

He was intrigued to discover that:

> ... such of these deep-sea fishes as have been examined bear a marked resemblance to those described from Japan, Hawaii and the Indian Ocean, while a few are apparently identical with species occurring in the Mediterranean and the neighbouring Atlantic. (op. cit.)

Jordan and his associates had published the results of their deep water surveys around Hawaii and Japan. Major Alfred Alcock of the Indian Medical Service, Superintendent of the Indian Museum at Calcutta, had given an account of the deepsea fishes around India (Alcock 1899); and several authors had published on those of the north-east Atlantic and Mediterranean. McCulloch consulted these works, and many others.

In recognising the wide distribution of some of the deep water fishes, McCulloch avoided the trap into which many of his predecessors had fallen. Unfamiliar fishes were not automatically described as new, but were compared with those from other parts of the world. In this, he had the advantage of Goode and Bean's encyclopaedic *Oceanic Ichthyology* (1896) and John Gilchrist's work on South African fishes. Gilchrist had been collecting and studying deep sea fishes since 1902 and had published many papers relevant to the taxonomy of the Australian fauna.

The *Endeavour* fishes were impressive and McCulloch mentioned the latest arrivals, which he had assessed only superficially, to Ogilby:

> A new lot of Endeavour material just to hand, including collections from 100 faths off S. Australian Coast. Spotted a striking Hippocampus [Seahorse] which looks new – black cross bands on a pure white ground. Hoplichthys haswelli by the score etc. etc. My next part [of the *Endeavour* reports] will consist of descriptions of new species only, as they are rather getting ahead of me. Do you know a Tetraodon – Spheroides? about 3 inches long, greyish with indefinite bands, and a striking black ocellus-like mark round the base of the pectoral and the gill-opening. I have several from the Southern Australian Coast. The last consignment also includes several beautifully marked sharks which are new to me. (McCulloch to Ogilby, 19.4.12, QMA)

The plan of the *Endeavour* reports was changing. Initially, McCulloch had intended a systematic presentation, family by family; now he contemplated a report consisting of new species descriptions. The second report (McCulloch 1914b) was a mixture of these: 14 of

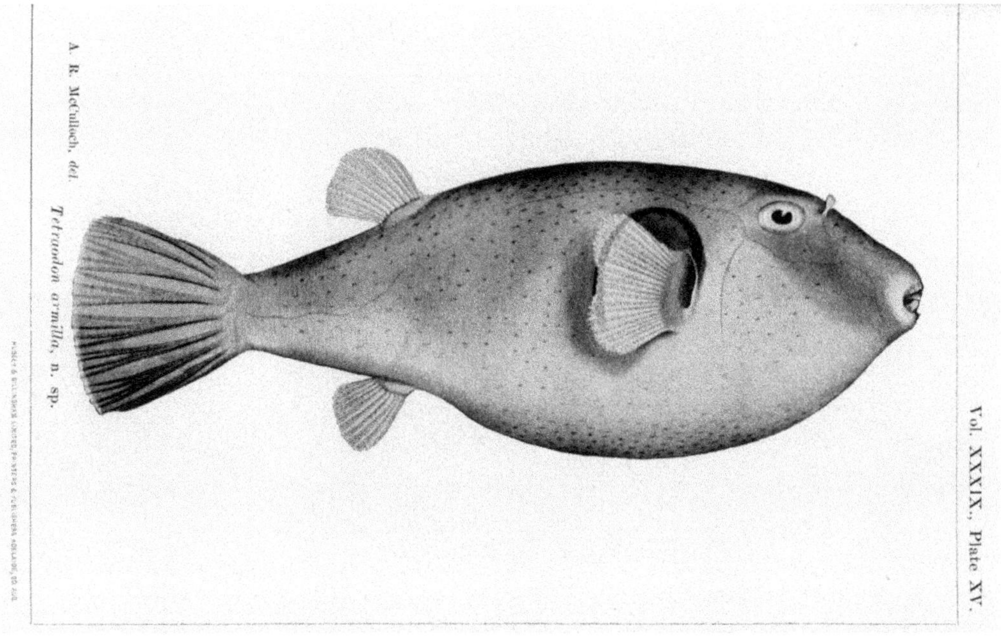

A Toadfish had appeared in the *Endeavour* hauls off South Australia and Waite later collected it from the *Simplon* in the Great Australian Bight. Fish from both collections were used in describing it as *Tetraodon* [now *Omegophora*] *armilla* (Waite and McCulloch, 1915). McCulloch drew it for the *Simplon* publication. From Waite and McCulloch 1915.

the 58 species in this report were described as new and he presented some of the families that had been omitted from the first report.

The catch from the Bight was recorded initially in McCulloch's second report. Those specimens that were more difficult to determine were held over for later publication. The fish were taken between April and June 1913 and included several new species.

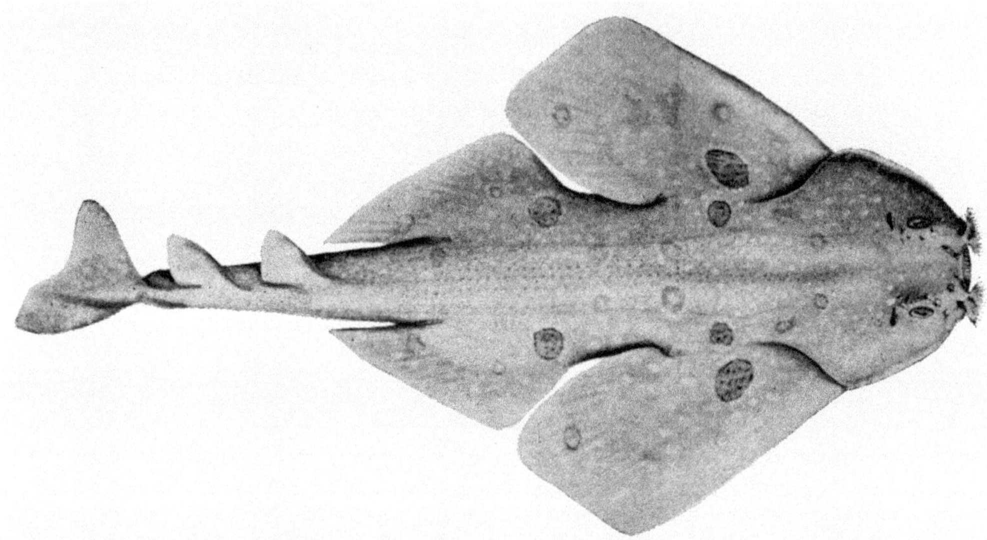

McCulloch began his second *Endeavour* report (1914b) by describing some of the small sharks that had been trawled in the Great Australian Bight in the winter of 1913. He named the Australian Angelshark, *Squatina tergocellata* from a single specimen 42 cm in length, trawled in 300 metres. From McCulloch 1914.

The Prickly Dogfish, *Centrina* [now *Oxynotus*] *bruniensis* (Ogilby, 1893) was described from Bruny Island, Tasmania. The *Endeavour* captured several specimens in the Bight in about 200 metres. Regan suggested that the Australian species might be identical with the Mediterranean *Oxynotus centrina* so McCulloch asked Jules Richard, Director of the Musée Océanographique, for Mediterranean specimens for comparison; it proved easy to distinguish the two species. From McCulloch 1914.

McCulloch's work in 1914

Even a summary of McCulloch's work, in any year subsequent to Waite's departure, is impressive. 1914 proved to be productive but frustrating. For the first time, the volume of work facing McCulloch appeared truly excessive. His publications in this year give a reasonable indication of his activity, and neither the number nor length of the publications was exceptional:

Notes on some Western Australian Fishes, No. 2	16 pages	4 species drawn
Revision of Fresh-water Crayfishes of south-western Aust	8 pages	4 species drawn
'Endeavour' Report, Part II	84 pages	22 species drawn
Notes on some Australian Pipe-fishes	3 pages	2 species drawn
Fishes of New South Wales (brief summary)	8 pages	
Note on *Jordanidia solandri*	1 page	

'Species drawn' in the table indicates the drawings executed by McCulloch, and published, as distinct from those he commissioned and supervised, but which were drawn by Phyllis Clarke.

In 1914, the *Endeavour* collections could easily have been a full time job but McCulloch found it almost impossible to decline work. Early that year, just before Waite left Christchurch for Adelaide, he received a letter from McCulloch, 'Letter from Allan, he wishes to relinquish the "Endeavour" fishes.' (WD, 4.2.14) There was no hint of any such wish in his letters to Ogilby at this time. Perhaps his comment was a hint that he had more than enough to do without odd jobs for Waite.

McCulloch also undertook work on both of his 'Studies' series, on Australian fishes and crustaceans. Ogilby offered more crustacean specimens but McCulloch declined:

> Thanks all the same but crustacea are off with me. I am hoping to send our interesting stuff to Miss Rathbun for examination, and you might perhaps care to do the same. (McCulloch to Ogilby, 22.11.14, QMA)

Mary Jane Rathbun was a remarkable character who had worked on crustaceans at the Smithsonian Institution in Washington for many years. She accepted McCulloch's collections and, with her protégé Waldo Schmitt, contributed to the results of the *Endeavour* work.

Two years earlier McCulloch had worked on Western Australian fishes and now new collections from the west demanded attention. Ogilby wrote asking for information on the sharks trawled by the *Endeavour* in Queensland, but was told he would have to wait:

> The specimens you mention are somewhere in a large tank mixed up with all the other sharks and rays taken by the Endeavour, and I very much regret that I cannot spare the time to hunt them out and identify them for you. (McCulloch to Ogilby, 3.6.14, QMA)

Ogilby wanted to know why McCulloch had not drawn some fish that seemed to require illustration; this provoked a small outburst from McCulloch:

> Why do I not figure Atypichthys, Chiloscyllium, etc.? Simply because I can do no more than is possible, and I am absolutely at the limit of my pace at present. I have lost an assistant in Fry, and have had his department [herpetology] thrust back on my shoulders, while my own branch is growing faster than I can cope with it. I could give you a list of 100 species I should like to figure, but they will have to remain unfigured at present, at any rate. (McCulloch to Ogilby, 3.6.14, QMA)

McCulloch had been drawing most of his illustrations himself with some help from Dene Fry. At about this time, the Curator recommended the provision of limited funds for the employment of Phyllis Clarke (see page 378) to draw some fishes for McCulloch's publications.

From the same letter to Ogilby (3.6.14), there emerges an impression of some of McCulloch's day to day activity in ichthyology. He was attempting to sort out the species of Latridae (Trumpeters), a family of large fishes of commercial importance; he spoke to David Stead, then General Manager of the infant New South Wales trawling industry:

> I have been down to the fisheries Dept. today discussing the matter with Stead who was very positive he had examples which agreed with Richardson's figure. [John Richardson had illustrated a latrid from the Erebus and Terror expedition in the 1840s] When I went down to have a look at them, however, I found them to be Dactylophora nigricans, [fish of a related family] so I have rather lost faith in his dogmatic statements concerning various species. (McCulloch to Ogilby, 3.6.14, QMA)

Stead had published *Fishes of Australia* (1906) and *Edible Fishes of New South Wales* (1908a), interesting popular works preoccupied with angling and fisheries information. He had little direct interest in taxonomic problems. McCulloch resented the fact that Stead and other popular authors profited by his work, generally without attribution, and made little effort to support him. He was happy to find fault with Stead's work: Ogilby told McCulloch that he was publishing descriptions to make clear the difference between two species of trevally and McCulloch responded, 'hope you will rub it in to those (including Stead) who have stated that they are identical.' (McCulloch to Ogilby, 3.6.14, QMA)

Part of McCulloch's museum work was concerned with exhibitions in the public galleries. He took opportunities for having interesting specimens cast. One such appears to have impressed him greatly; he mentioned it to Ogilby:

> Have just identified A. fish. It is about three feet [a metre] long and was caught on a rod and line off Tweed Heads. (McCulloch to Ogilby, 3.6.14, QMA)

The fish was doubly interesting: a spectacular animal in itself and the apparent solution to a taxonomic problem. This was the Shark Mackerel, *Grammatorcynus bicarinatus*, originally described by Quoy and Gaimard, the naturalists who had accompanied Louis de Freycinet in 1817–20. A specimen was taken by them in Shark Bay, Western Australia 'and apparently nothing more than a drawing was preserved (see colour plate 3). I have had the [Tweed Heads] specimen cast, and shall shortly describe and figure it.' (McCulloch to Ogilby, 3.6.14, QMA) He did so (McCulloch 1915b), providing a full description and an accurate figure.

British Association for the Advancement of Science: 1914

In the latter part of 1914, McCulloch's work was interrupted by a meeting of the British Association for the Advancement of Science (B.A.A.S.) in Sydney. He grumbled to Ogilby:

> Several [of the B.A.A.S. visitors] have made our place their headquarters, and I can scarcely find space to sit down in my own room. Work has been practically out of the question as we have had to spend our time showing them round ... (McCulloch to Ogilby, 24.8.14, QMA)

However, he appeared pleased to feel that he could speak to his distinguished visitors as an equal:

> Professor Jungersen, Director of the Copenhagen Museum, called on Friday to see such Pegasidae [Sea Moths] as I could show him. My recent work on these beasts stood me in good stead ... (McCulloch to Ogilby, 24.8.14, QMA)

McCulloch followed the rules of scientific nomenclature and felt confident enough to argue with Jungersen, who was apparently impatient with some of them. The Professor was presenting a paper to the B.A.A.S. (Jungersen 1915) and planned to use a generic name that was not strictly available:

> When I protested that this was not consistent with the rules etc., he explained that he did not like all these rules. (McCulloch to Ogilby, 24.8.14, QMA)

McCulloch could share the joke with Ogilby, who also had a history of bending the rules of nomenclature to suit himself; McCulloch was keeping Ogilby in order.

For the 1914 conference of the British Association for the Advancement of Science, McCulloch contributed a short article summarising the ichthyology of New South Wales. It was included in the *New South Wales Handbook*, published by the B.A.A.S. (McCulloch 1914d). In this, McCulloch drew attention to the opportunities for ichthyological and fisheries investigations that existed in New South Wales. He discussed some of the biogeographic factors that influenced the composition of the fish fauna and outlined the chief fisheries of the day.

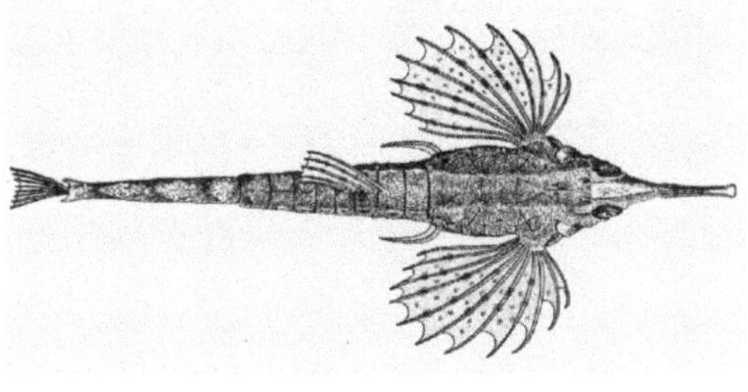

The Sculptured Seamoth, *Pegasus lancifer* Kaup, 1861 was collected by the *Endeavour* at the entrance to Oyster Bay, Tasmania in 1909. McCulloch established the genus *Acanthopegasus* for this species but subsequent studies have not supported this generic distinction. From McCulloch 1915.

Western Australian fishes: Records of the WA Museum, 1912 and 1914

Work on the *Endeavour* collections was interspersed with many other pressing tasks. In particular, McCulloch wished to deal with the sharks and rays from southern Australia, well represented in the 1913 collections, 'but I must first dispose of two large collections from W. Australia.' (McCulloch to Ogilby, 3.6.14, QMA). McCulloch (1912) had published a paper on these collections, sent by Bernard Woodward of the Western Australian Museum and Mr A. Abjornssen, the Chief Inspector of Fisheries in Western Australia. Most of the Western Australian specimens had been collected in the south-west, 'but a few from Port Hedland add a little to our knowledge of the great North-western Coast, which is, at present, largely unknown.' (McCulloch 1914a) In 1914 McCulloch completed publication of the results of his work on the Western Australian collections, the two papers together comprising the most significant contributions to the ichthyology of the state to that time.

The papers were typical of McCulloch's work, exhibiting his usual procedure when faced with new material. His method did not differ substantially from that of his contemporaries, but he was more thorough than Waite and more accurate than Ogilby. Where possible, McCulloch identified the specimens with known species, preparing synonymies and paying special attention to the ill-defined species named by Castelnau from the west. For instance, the Blue Devil, *Paraplesiops meleagris* was readily identified but proved, on comparison with an obscure description published by Castelnau (1873), to be identical with his *Ruppelia prolongata*. Until this time, Castelnau's *Ruppelia* was an unwelcome mystery.

In preparing descriptions, McCulloch preferred to work with a large number of specimens of differing sizes and from different locations. He used the Australian Museum and Macleay Museum collections to supplement those from the west. Where there was sufficient material to make the revision of groups feasible, he included known species from outside Western Australia and presented descriptions and keys. In this way, he revised the

The Black Jawfish, *Opistognathus inornatus* was described in 1887 by Ramsay and Ogilby who had received two small specimens from Derby, Western Australia. McCulloch secured a large specimen from Port Hedland and identified it from the original description. He located the type specimens and drew one of them for his paper, copying the spots on the body of the fresh specimen onto his drawing. McCulloch used the genus *Gnathypops* for the fish but it has since been placed in *Opistognathus*. From McCulloch 1914.

The Fringefin Trevally, *Caranx* [now *Pantolabus*] *radiatus* was described by William Macleay in 1881. It was taken at Port Hedland by the *Endeavour* and McCulloch identified it by comparison with the long-preserved material in the Macleay Museum. He drew it for his third *Endeavour* report. Only the male has extended fin rays. From McCulloch 1915.

Australian species of the Scorpion Fish genus, *Scorpaena*. He was able to extend the recorded distribution of several species and named two new species. He provided illustrations of the new species and of others that had not previously been satisfactorily represented. In all, he illustrated 14 species in the two papers.

Linnean Society of New South Wales

Since Waite first brought him along to a meeting in 1899, McCulloch had been associated with the Linnean Society of New South Wales. He conformed to the custom followed by his museum colleagues in exhibiting specimens at the meetings and submitted occasional papers. These included a description of larval sunfish which had been 'taken from the stomach of a king-fish caught swimming near the surface, during the passage of H.M.S. "Torch," between Ellice and Union Islands ...' (McCulloch 1913b) The tiny fish were only a centimetre in length but McCulloch, who identified them as larval sunfish, drew and described them. Although they differed in form from those he found described in the literature, he resisted the temptation to name them as a new species.

McCulloch (1915b) published *Notes on, and Descriptions of Australian Fishes* in the *Proceedings of the Linnean Society of New South Wales*. The paper was similar to his series of *Studies on Australian Fishes* being 'based on specimens which have been forwarded to the Australian Museum from various sources.' In this single publication, McCulloch dealt with 13 fishes, each forming a topic for resolution of a taxonomic problem or description of a new species. This was the paper in which he dealt with *Grammatorcynus* (see page 354). Other examples are outlined below.

He had been sent specimens of the Tasmanian Whitebait from the Derwent River and used them to prepare an accurate description and illustration, neither of which were available in the literature up to that time. He erected a new genus, *Lovettia* for the species:

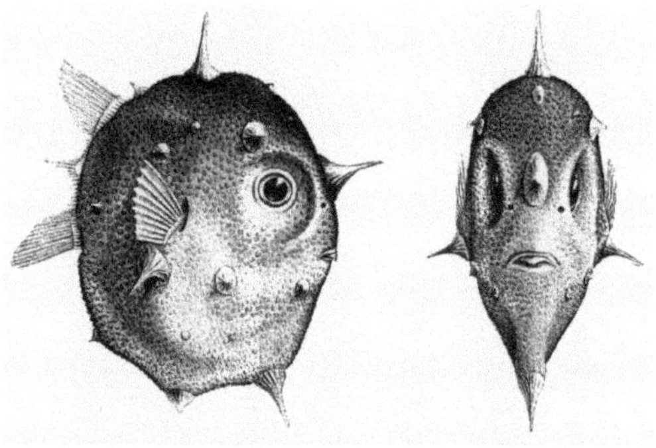

Planktonic larvae of a species of Sunfish, *Mola* sp., about 1 cm in length. McCulloch's drawings show a view of the right side with the larva facing to the right; and an anterior view. From McCulloch 1913.

> I associate with this genus the name of Mr. E.F. Lovett, of Hobart, to whom I am indebted for the beautifully preserved specimens described below, as well as many other interesting Tasmanian fishes. (McCulloch 1915b)

The Tasmanian Whitebait, *Lovettia sealii* (Johnston, 1883) is an anadromous fish that migrates from the sea in shoals to spawn in the rivers of Tasmania. McCulloch received specimens collected by Mr Lovett in the Derwent River and established the genus *Lovettia* for the fish. From McCulloch 1915.

McCulloch's most active donor of freshwater fishes, who also contributed fishes and information to Ogilby, was Fisheries Inspector H.K. Anderson. Among some specimens he sent from the Murrumbidgee River, was a fish just 95 mm long which appeared to be of special interest; McCulloch was initially delighted with it:

> Have a ridiculously new Therapon from the Murrumbidgee, and in the same lot, a set of Nannoperca australis [Southern Pygmy Perch] which enables me to correct the several striking errors in the original description & figure of that species. (McCulloch to Ogilby, 17.12.14, QMA)

McCulloch examined the Therapon more closely and wrote to Ogilby again:

> My 'ridiculously new' Therapon of my last letter proves to be nothing more exciting than the young of T. bidyana [Silver Perch, *Bidyanus bidyanus*], though it is a wonderfully different-looking fish. I have therefore figured it and described the differences, so that nobody else need have as much trouble with it again. (McCulloch to Ogilby, 22.12.14, QMA)

McCulloch thus avoided a pitfall but his experience illustrates the potential problems for the taxonomist, arising from the alteration in shape and coloration with growth.

Royal Zoological Society of New South Wales

McCulloch became a council-member of the Society and edited their journal, the *Australian Zoologist*, for several years, contributing occasional articles. Although he was not an experimental scientist McCulloch made frequent ecological observations and published an account of a series of simple but effective experiments on *Galaxias* (1915d).

Common Galaxias, *Galaxias maculatus*, live in freshwater but lay eggs on vegetation in the tidal regions of coastal streams. The larvae hatch and are carried out to sea. They return to the streams after about six months and ascend in enormous shoals. Before these habits were understood, McCulloch observed the tiny fish in a shallow creek near Freshwater Beach in Sydney. He erected cheesecloth barriers across the stream and, with the 'assistance of several sharp-eyed boys' (op. cit.), proved that the fish were migrating upstream, against the current. He captured some of the larval fish and tried them in aquaria of differing salinities, keeping some in fresh water for several months and remarking on their metamorphosis:

> Soon after transference to fresh water the larvae lost their transparency, and rapidly assumed the characters of the adult fish. (op. cit.)

He mentioned the find to Ogilby:

> Made the interesting discovery of the larval Galaxias (attenuatus?) migrating from the sea into fresh-water last Saturday. Kept specimens in salt-water for several hours and then transferred them to my fresh-water aquarium, and they liked it. Proved the point conclusively and will write it up shortly. (McCulloch to Ogilby, 1.9.14, QMA)

McCulloch's special interest in this phenomenon lay in its biogeographic significance. This little fish is found in South America and New Zealand, as well as southern Australia.

A collection of fishes made by the Field Naturalists' Club of Tasmania provided material for another of McCulloch's contributions to the *Australian Zoologist*. The club had collected some pipefish by dredging in about 20 metres of water in Wineglass Bay, Tasmania and forwarded their catch to the Australian Museum. From these specimens McCulloch was able to describe the Crested Pipefish, *Histiogamphelus briggsii* named for the collector; and the remarkable female of the Widebody Pipefish, *Stigmatopora nigra*:

> The male is subcylindrical and almost without ornamentation, while his mate is broad and depressed, with rich pink and black markings. They are so dissimilar in their appearance that one can only be sure of their specific identity after a critical comparison of all their characters. (McCulloch 1914c)

This form of sexual dimorphism, in which the female is the conspicuous sex, is common in pipefish. It is presumably associated with the fact that the female transfers her eggs to the male who carries them in a brood pouch until they hatch.

The Royal Zoological Society of New South Wales published McCulloch's major work on the fishes of that state in three parts (McCulloch 1919; 1921a; 1922a). *The Fishes and Fish-like Animals of New South Wales*, also published as a handbook (McCulloch 1922b), was much more effective than any of the works on New South Wales fishes that had been published to that time. McCulloch listed 588 species and expressed the hope that 'the accompanying keys and illustrations will indicate the identity of any [fish] recorded from the waters of this state.' Previous authors had provided only incomplete accounts, generally concentrating on the commercial species.

McCulloch regarded illustration of the check-list as essential. Finding published figures of good quality was not easy:

> Having to hunt up figures for reproduction in my illustrated check-list, I'm just staggered to not[e] the rottenity of the average illustration! Not a figure of Heterodontus or Gyropleurodus [Port Jackson sharks] anywhere that look like the beasts. (McCulloch to Ogilby, 2.8.19, QMA)

The illustrations were copied from a variety of sources and assembled as miniature reproductions. Theodore Roughley (see page 422), then Economic Zoologist at the Technological Museum in Sydney, carried out the photographic work:

> For the preparation of the excellent photographs which compose the plates I am indebted to Mr. T.C. Roughley, who has spared no pains to overcome numerous difficulties in their production; stains in the paper of the older plates, folds and other surface irregularities, and conflicting colours, have each presented their troubles, which he has consistently disposed of. (McCulloch 1922)

373 of the 588 species were illustrated in the photographic plates, usually at least one member of each genus.

Pleased to have Ogilby's criticism of the first part of his new catalogue, McCulloch was nevertheless disconcerted when Ogilby sent him a list of suggested alterations in the nomenclature. The response to another of Ogilby's criticisms was easier:

> The reason that *Rhizoprion crenidens* was left out of my N.S.Wales list is because the compiler thereof is a silly ass. (McCulloch to Ogilby, 16.1.20, QMA)

As a consequence, the Milk Shark, now *Rhizoprionodon acutus*, appeared in a footnote.

McCulloch had a realistic appreciation of the general reader's lack of enthusiasm for scientific keys: he suggested that identifications should be carried out by leafing through the photographs until a near match was found. This was intended to bring the user to the correct genus, after which a short key might need to be negotiated to reach the correct species. The handbook was a success and was reprinted in 1927 and 1934. For his handbook McCulloch published 43 pages of plates, page 30 of which showed 11 species of wrasse. The table opposite shows the origin of each: 3 of his own, 2 from Waite, 1 from Roughley and the remainder from overseas authors. In the introduction to his handbook McCulloch praised the work of T.C. Roughley in preparing the reduced illustrations as plates:

> I have further to gratefully acknowledge the loan of numerous originals of the figures quoted, with permission to republish them in reduced form. The greater number of these have been lent by the Trustees and the Director of the

> Australian Museum, and others by the Director of the Queensland Museum, and the Director of the South Australian Museum [Waite]; I am similarly indebted to the Councils of the Linnean and Royal Societies of New South Wales. Likewise equally grateful acknowledgement is here rendered to the numerous authors of papers quoted, whose figures have been copied on the accompanying plates.

Plate number	McCulloch's nomenclature	Present identification and nomenclature	Origin of plate	Publication date of plate
256a	*Pseudolabrus guntheri*	*Pseudolabrus guentheri*	McCulloch	1913
256d	*Pseudolabrus tetricus*	*Notolabrus tetricus*	McCulloch	1911
257a	*Pictilabrus laticlavius*	*Pictilabrus laticlavius*	Richardson	1848
258a	*Eupetrichthys angustipes*	*Eupetrichthys angustipes*	McCulloch	1913
259a	*Novaculichthys jacksonensis*	*Iniistius jacksonensis*	Waite	1899
262a	*Ophthalmolepis lineolatus*	*Ophthalmolepis lineolatus*	Kner	1865
263a	*Coris picta*	*Coris picta*	Waite	1903
264a	*Choerodon venustus*	*Choerodon venustus*	Roughley	1916
265a	*Achoerodus gouldii*	*Achoerodus viridis*	Klunzinger	1879
266a	*Lepidaplois vulpinus*	*Bodianus frenchii*	Waite	1902
267a	*Verreo oxycephalus*	*Bodianus unimaculatus*	Jordan & Snyder	1902

Cooperation with Christchurch Waite: 1906–1914

From Edgar Waite's New Zealand diaries there emerges an impression of the extent of communication with his one-time protégé. The control which Waite had exercised over McCulloch's work since 1898 was naturally lost at Waite's abrupt departure for New Zealand. Their relationship was altered but it was natural, and expedient, to maintain contact. Waite was soon writing to his Sydney colleagues for assistance in obtaining specimens for the New Zealand International Exhibition. McCulloch sent some aquatic plants and received crustacean specimens in return, '[w]rote to Allan & sent him Crawfish.' (WD, 19.10.06)

The improvement in McCulloch's situation at the Australian Museum was soon clear:

> Letter from Allan, he has been appointed to succeed me at £150 to start with. Replied, congratulations. (WD, 10.1.07)

With the improvement of his situation at the Australian Museum, McCulloch presumably felt more secure in dealing with his former mentor.

There was a consistent exchange of letters, generally initiated by Waite who asked for information that he could not obtain in New Zealand. He used McCulloch to consult the literature and fish collections held in Sydney:

> Wrote to Allan for dates etc. for fish list. [Waite required references for his basic list of New Zealand fishes] (WD, 12.3.07)

> Wrote to Allan for outline drawing of Astrape japonica. [an electric ray] (WD, 18.10.07)

> I wrote to Allan for details of Scyllium [i]sabella, Broussonet. [a small shark of the genus *Cephaloscyllium*] (WD, 19.10.08)

Waite became aware that McCulloch did not want him to return to the Australian Museum; when the assistant curator's position was created in 1908, Waite observed:

> Letter from Allan. Neither he nor Rainbow [another Sydney colleague with whom Waite corresponded] inform me of the new position. (WD, 29.10.08)

McCulloch, pleased with his new independence, was prepared to cooperate with Waite while the latter was at a distance, but had no wish to work as a subordinate again.

McCulloch provided Waite with information but asserted himself when he felt that he was being exploited. Waite '[s]ent some Kermadec Is. fishes to Allan & asked him to look over my Norfolk Is. list.' (WD, 25.11.09) The response suggested that Waite had asked for too much:

> Letter from Allan, he was going for a holiday & therefore my requirements as to Norfolk Is. fishes will have to wait. (WD, 9.12.09)

and then:

> Finished my paper on Kermadec Island fishes Allan having written to say he is too busy to make the comparisons I asked for. (WD, 30.12.09)

When Waite proposed including a note of his own in a major article of McCulloch's, the latter declined.

While Waite was working on the *Nora Niven* trawl fish, he frequently required help from McCulloch, and the help was generally given promptly. By this time (1909 to 1911) McCulloch was working on the enormous trawled collections from the *Endeavour*. They sometimes found themselves dealing with the same, or similar, species. Both ichthyologists saw illustration as an important part of their work and Waite proposed a rationalisation of this process. He wrote to McCulloch in June 1910 listing the species he was illustrating, intending to avoid duplication, but McCulloch refused to suppress any of his own illustrations.

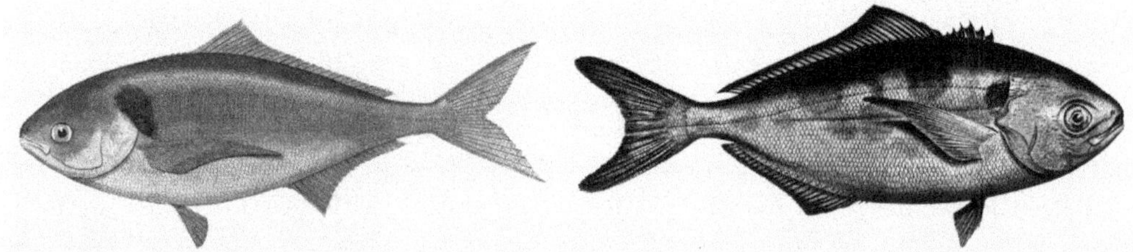

The Warehou, *Seriolella brama* (Günther, 1860) is found in both southern Australian and New Zealand waters. It was taken by both the *Endeavour* and the *Nora Niven*. Waite suggested that McCulloch's illustration (right) should be used and his own (left) suppressed but both were published in 1911. From McCulloch 1911 and Waite 1911.

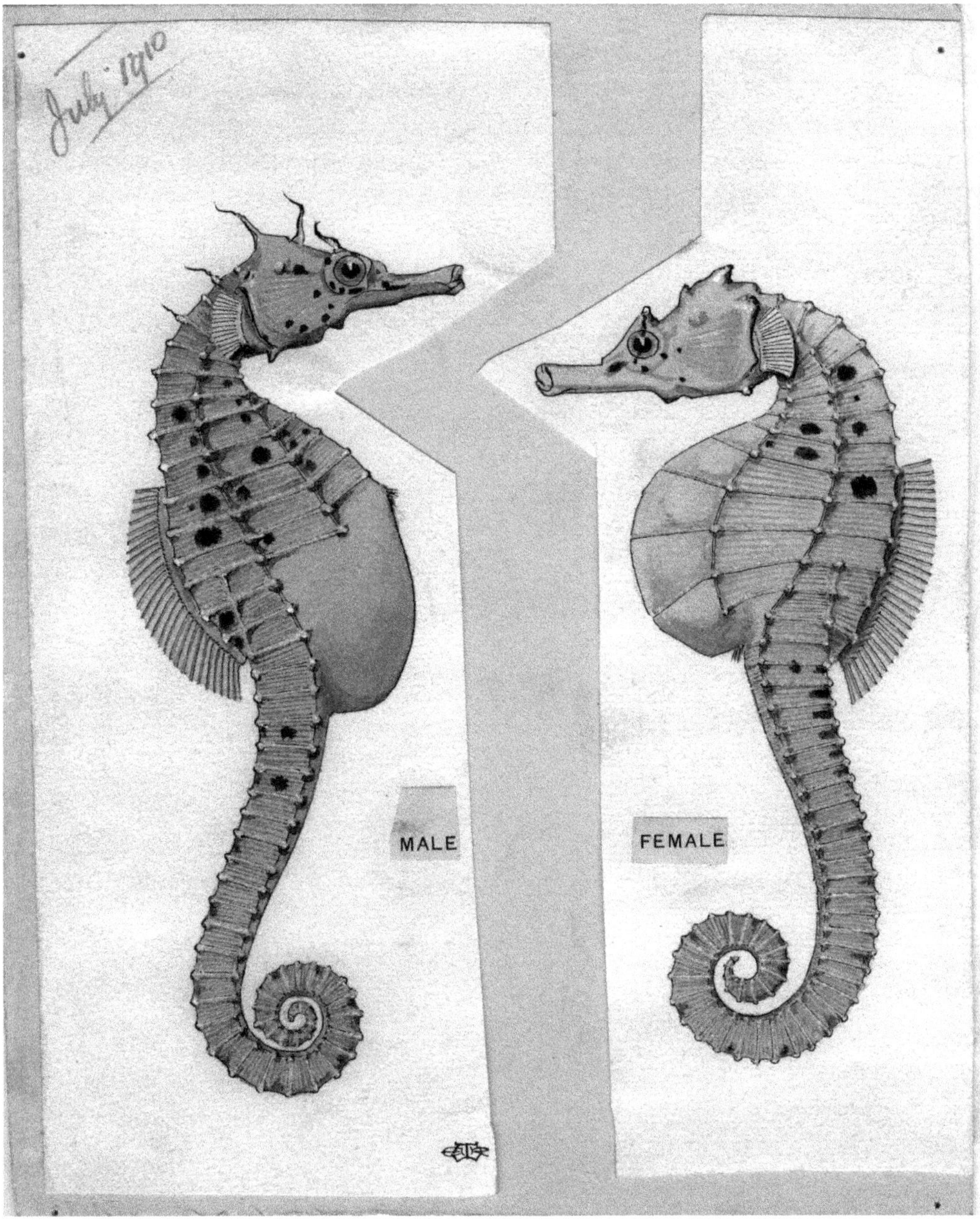

Waite's seahorses (*Hippocampus abdominalis*) in preparation in 1910 (see page 276). From AA356 Waite Collection, courtesy of the South Australian Museum Archives.

The correspondence between Waite and McCulloch was sustained and each acknowledged help from the other in publications. Waite, writing on the fishes of the Kermadec Islands in 1910, quoted McCulloch's opinion repeatedly. He sent specimens frequently for his colleague's examination; for example, a leatherjacket obtained at Sunday Island in the Kermadecs:

> The type being preserved in the Australian Museum, I forwarded the specimen to Mr. McCulloch, who kindly writes as follows: 'The specimen is undoubtedly *Pseudomonacanthus analis* [Darkvent Leatherjacket, now *Thamnaconus analis*]. The type, as you will remember, is imperfect and beach-worn, and the position of the gill opening and pectoral are consequently not quite correctly shown in the figure.' (Waite 1910a)

This was typical of the sort of help that Waite sought. The process of consultation was reciprocal: McCulloch was presented with some Dories which had been trawled by the *Endeavour* on the outer continental slope in Bass Strait. He had 56 specimens. He identified them tentatively as Silver Dory, *Cyttus novaezealandiae*, first described by Arthur (1885) in New Zealand. He noticed some discrepancies between the specimens before him and the original description, and reported the result of consulting Waite:

> Mr. Waite ... has kindly compared some [specimens] that I have sent him with other New Zealand examples which he identifies as *C. novae-zealandiae*, and has further forwarded me a half-grown example for examination. This last is undoubtedly identical with my specimens, the largest of which is 135 mm. long, and we are agreed that they are really Arthur's species. (McCulloch 1911)

McCulloch's occasional self-assertion should have been a salutary reminder to Waite that he was dealing with a professional equal. He noted that McCulloch wrote 'criticising points in my trawling paper.' (WD, 3-8-11) However, the tone of McCulloch's letters was generally polite and deferential, perhaps making it natural and easy for Waite to persist in seeing him as an assistant.

Whenever Waite visited Sydney (or any of the capitals) he gravitated to the natural history museum; in 1911 he spent a month in Sydney, and renewed contact with McCulloch:

> 11. Wed. ... had lunch with Mrs. McCulloch ... [Allan McCulloch's mother; she and Allan lived at Paramatta.]
>
> 13. Fri. Spent the whole day at the Museum. Read proofs of Allan's fishes of the Endeavour Exp.
>
> 19. Thurs. Spent morning at the Museum ... Also looked into the fishes of Kermadec Islands.
>
> 20. Fri. Had a demonstration on the Aerograph [spray system for fine work in painting] with Allan.
>
> 27. Fri. ... journeyed to Paramatta, stayed the night with Allan.
>
> 28. Sat. Met Allan after lunch & we went to Maroubra ...
> (WD, excerpts from October 1911)

An article *In Memoriam* was published in 1915 for the *Endeavour* and its people, lost on the return journey from Macquarie Island in 1914. It included obituaries for the biologist, Charles Harrisson and the Director of Fisheries, Harald Dannevig. From Lockyer 1915.

When Waite returned to Sydney seven months later, ready to embark in the *Aurora* for Macquarie Island, he and McCulloch worked together again:

> 6. Mon. ... In afternoon oscillated between the Museum (where Allan and I looked through our respective fish lists) and the dentists.
>
> 8. Wed. ... thence to Museum & borrowed lantern slides from Allan.
>
> 11. Sat. Finished comparison of fish lists with Allan.
>
> 16. Thurs. joined ... Allan & others in launch & went to Aurora at Pyrmont. (WD, excerpts from May 1912)

They continued to exchange information on fishes and, when Waite was successful in applying for the directorship of the Museum in Adelaide, McCulloch was quick to congratulate him. (WD, 2.1.14)

Illness

Allan McCulloch suffered from a form of depressive illness that sometimes interfered seriously with his work, particularly towards the end of his life. In his letters to Ogilby, symptoms of the illness were exhibited: episodes of feeling gloomy and listless, poor concentration, erratic behaviour and paranoia. He occasionally showed some insight into his problems but was largely at the mercy of the illness. The communication with Ogilby appears to have had some cathartic function and it is clear that he valued the correspondence highly.

The loss of the government trawler, with all of its crew, had a profound effect on McCulloch:

> Am feeling utterly wretched over the fate of the 'Endeavour'. It seems narrow old jossers, all suffering more or less from senile decay should hang on here till the place fairly rots with them. The idea that there are people upstairs who govern the things of this earth may be right, but if so they have certainly got the reins mixed up or have gone to sleep or something of that sort. Yours in deep depression ... (McCulloch to Ogilby, 22.12.14, QMA)

He recovered more rapidly from these feelings at 30 than he was able to do a few years later, but tones of dissatisfaction are frequent:

> Have just returned from a trip to Port Macquarie where I have been surfing & sunning for a week. Life seemed almost worth while for a time, but I managed to bring back an awful cold with me which rather takes the cream off things. (McCulloch to Ogilby, 30.3.15, QMA)

His third *Endeavour* report was about to be published, but he was far from enthusiastic:

> It contains the usual stuff I write, but seems to lack interest somehow or other. However, I shall look forward to your criticism of it. (McCulloch to Ogilby, 30.3.15, QMA)

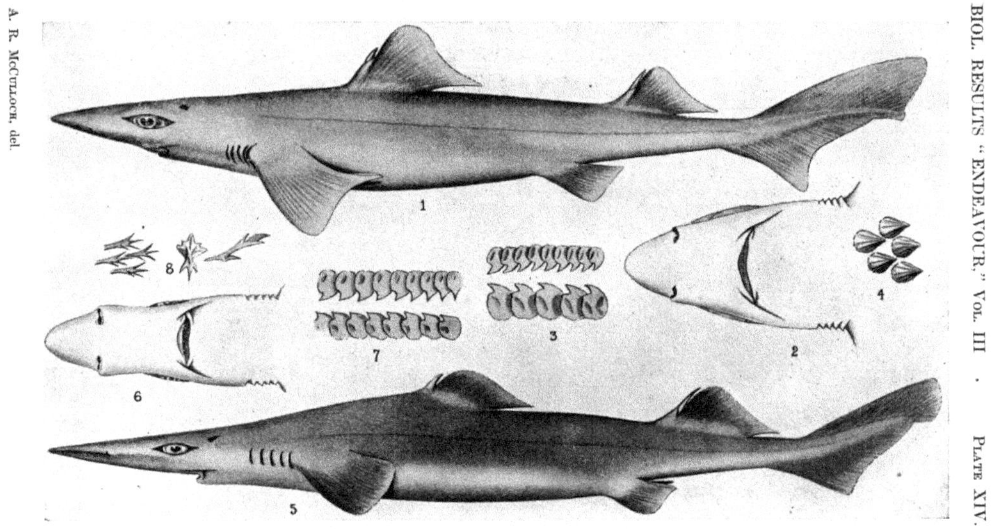

Small benthic sharks trawled in from 200 to 800 metres off Gabo Island, Victoria by the *Endeavour* and described as new species by McCulloch: Harrisson's Dogfish, *Centrophorus harrissoni* (top), named for biologist Charles Harrisson; the Longsnout Dogfish, *Acanthidium quadrispinosum* [now *Deania quadrispinosa*] (bottom). From McCulloch 1915.

The Viper Fish, named *Chauliodus dannevigi* by McCulloch but since synonymised with the widely distributed *Chauliodus sloani*. This bathypelagic predator was trawled in 400 to 500 metres by the *Endeavour* off Cape Everard, Victoria. McCulloch wrote 'The specimen described above was one of the last fishes preserved on board the "Endeavour" before she left on her ill-fated voyage to Macquarie Island. I associate with it the name of my friend, the late Mr. Harald C. Dannevig, who collected it, and whose untimely loss terminated a grand chapter in the fisheries investigation in Australia.' From McCulloch 1916.

Ogilby wrote complaining that there was a delay in the publication of his paper on sharks and McCulloch sympathised, expressing dissatisfaction with Etheridge's editorial policy:

> Your note about the postponement of your Selachian paper in favour of a foreigner's bugs is enough to make one weep. We suffer something similar from a blighter who writes on myriapods [centipedes and millipedes], & quotes pages of german from other authors. Its a mystery to me that any editor should give preference to outsiders over his own staff. (McCulloch to Ogilby, 10.12.14, QMA)

McCulloch frequently interpreted Etheridge's behaviour as overbearing and unfair:

> ... [Etheridge] has hung up my Endeavour Report pending further red-tapeism with the Commw. Authorities, notwithstanding we have every right and reason to go right ahead. (McCulloch to Ogilby, 18.2.15, QMA)

On occasion, McCulloch's complaints have a flavour of paranoia:

> Was hauled over the coals by R.E. [Etheridge] yesterday for my work on the Girellidae, because it entailed neglect of the preparation of a popular guide to the fish gallery which is a routine job. The fact that they have enough guides in hand to keep our snail-printer going for twelve-months is disallowed as an argument – but then, when one wants to hang a dog, one doesn't use gold chain to carry out the deed. I'm not worrying overmuch, anyhow. (McCulloch to Ogilby, 5.7.19, QMA)

Following one of his several disappointments at being rejected for enlistment, McCulloch wrote a touchy letter to Ogilby. He complained about Ogilby's obtuseness and Waite's lack of courtesy and concluded

> Not having looked at a fish for a fortnight, I have nothing to tell you, and feel too beastly blue to settle down to work yet. (McCulloch to Ogilby, 2.6.15, QMA)

Mood swings were often manifest. In a letter of March 1919, after discussing a variety of taxonomic issues and grumbling about Waite, he concluded:

> I don't think there is anything to excite you in the foregoing, nor matters requiring answers. All this is merely the outcome of a feeling that I'd like to chat with you over things generally, which, being impossible, causes me to substitute the next best thing. (McCulloch to Ogilby, 25.3.19, QMA).

There was a degree of reliance on Ogilby, which was soon compromised by Ogilby's own illness.

Among the letters written by McCulloch to Ogilby, preserved in the Queensland Museum Archives, is apparently the last before Ogilby's incapacitating stroke. The mood is of depression with some puzzlement; McCulloch did not understand why he felt so listless:

> My dear Mr. Ogilby,
> You will put me down for a low sort of cow for having left your several letters unanswered, but I have been so unwell lately that things have gone all awry ... If only I could feel a bit brighter, things would go a little more easily, but I'm as dopey as an owl nine days out of ten. And there has been a little lassie who has been filling up a lot of my thoughts lately – I might even have fallen in love with her if she hadn't emphatically stated that she hadn't fallen in love with me. Of course that proves her good sense and all that sort of thing, but its annoying just the same. Which makes me realise I'm writing rot. (McCulloch to Ogilby, 5.4.20, QMA)

The reference to McCulloch's preoccupation with 'lassies' was a long-standing joke that he shared with Ogilby.

Support for the diagnosis of endogenous depression can be derived from the letters mentioned above, from some fragments of draft letters written in 1924 (ML) and from a diary he kept near the end of his life, the latter discussed briefly below. His eventual suicide in 1925, at the age of 40, is also consistent with a depressive illness.

Enlistment: 1914–1918

McCulloch made many efforts to enlist in the army between 1914 and 1918 and was dismayed at repeated rejections:

> I went up to the barracks on Monday, and had no end of a busy time chasing around after the various people and sections I wished to see. I had some useful letters with me which apparently served their purpose, as I have been instructed to wait until the formation of the fourth hospital is announced in the papers, when I am to go up for enrollment. So here I am disposing of various odd jobs, and praying that I shall not be rejected when the time comes. I am nervous of the last, as so many strong and apparently healthy chaps have been turned down for one cause or another. Even Fry [McCulloch's young assistant at the Museum] was rejected at first, and had to improve his condition before applying a second time ... So as soon as I learn whether I am accepted or not, I shall let you know, and provided that luck favours me, will send you my address. All letters from you will be greatly appreciated as always. (McCulloch to Ogilby, 21.5.15, QMA)

Disappointment followed:

> Here I am still writing from the old address, and I am likely to remain so for a while at any rate, having been turned down twice on account of insufficient chest expansion. I don't remember ever having felt so small before. I learnt that the next hospital is not likely to be formed before next August, so being unable to wait about so long, I went up with a couple of friends to enlist in the ordinary way. But I am evidently not sturdy enough for field work as two different doctors both rejected me on the chest measurement, so I shall have to wait patiently for the hospital as I originally intended. Hope you are not bored

> by all these details, but its a relief to tell ones friends of ones disappointment. (McCulloch to Ogilby, 2.6.15, QMA)

In 1917, after his mother's death, McCulloch tried again to enlist. He was upset and anxious; his frequent letters to Waite around this time suggest that he still felt some dependence upon his old mentor.

Eventually, in 1918, McCulloch was accepted for enlistment and he wrote to Ogilby from the 'University Company Camp', buoyant with his success. He answered Ogilby's taxonomic queries from memory as best he could and concluded:

> Well old chap, I think that's as much as I can answer of your letter, and it only remains to tell you something of my present manner of living. Circumstances demanded that I should go into camp a week earlier than I anticipated ... Consequently, I've had to put in Sundays [at the Museum] endeavouring to clear up things a little. The Curator and Trustees have been most considerate, and I received an official letter this afternoon which makes my hat feel distinctly tight around the brim. As a soldier I'm a huge joke, and am learning that there are many things to be done in camp life other than drilling and firing rifles. I'm a specialist on fatigue duties by now, and can clean up a w.c with the best of them ... I've been on guard in the early hours of the morning, all dressed up in pack and fixed-bayonet and the nowhere to go feeling. Its a vast change from the pleasant work of matters ichthyological, and very hard just at first, but no doubt I'll be used to it all in good time ... I'm almost afraid we shall not be wanted over there, for which I feel very selfishly regretful, but I suppose its a bit early to be forming conclusions on the matter. (McCulloch to Ogilby, 6.10.18, QMA)

The armistice was signed on 11.11.18, so McCulloch's military career was suspended almost before it was begun. He was distressed, returning to the Museum, initially with little enthusiasm:

> Back to the old job, Mr. Ogilby, having come considerable of a guttser. There is so much on hand requiring attention that I shall take a week or two to get into the groove, so don't start firing in difficult problems too soon. (McCulloch to Ogilby, 25.11.18, QMA)

Queensland fishes: *Memoirs of the Queensland Museum* (and other journals)

The fishes of Queensland presented a great and immediate challenge to Australian fish taxonomists in McCulloch's time. There were several layers of complication. First, the fauna was much larger than that of the temperate regions. Second, the distribution of the northern fishes was frequently much wider than that of those from the south so that many of the species had been described from the Red Sea, India, Indonesia, Japan or elsewhere. Lastly, the terrible triad of Macleay, de Vis and Castelnau had all dealt with collections of Queensland fishes in the 1870s and 1880s. The taxonomy of these fishes was thus in a particularly bad state.

Waite had had little to do with the region. Ogilby began working actively on this fauna once his base of operations was reasonably secured at the Amateur Fishermen's Association of Queensland. McCulloch's first real experience of these fishes, aside from the trip with Waite to Lord Howe Island, was with Charles Hedley in Torres Strait in 1907 (see page 327). Hedley, although a conchologist, had a long-standing interest in matters ethnological and McCulloch seems to have shared the interest, making notes on the people they saw. They watched pearl shell and bêche-de-mer fishing, and collected invertebrates and fish on the reefs of Murray Island. They returned with large collections of shells and fish.

McCulloch had many visits to Queensland and the islands of the Great Barrier Reef, particularly Masthead Island. He also collected in the New Hebrides in September 1910, gaining more experience of the tropical Indo-Pacific reef fishes and again increasing the Australian Museum collections of these fishes.

In 1911, the *Endeavour* collections that reached Sydney included those from Queensland waters, and most of the ichthyological results were published in McCulloch's second and third *Endeavour* reports. By 1912, Ogilby and McCulloch were immersed in the problems associated with this fauna. All of Ogilby's publications for the greater part of the next decade were concerned with these fishes and McCulloch contributed advice and research to Ogilby's efforts. In return, Ogilby searched for de Vis' old type material, buried among the collections in the Queensland Museum, and helped McCulloch with his vast experience.

Aside from his contributions to Ogilby's many publications on Queensland fishes, McCulloch published three papers of his own in the *Memoirs of the Queensland Museum*, comprising a series entitled *Notes and Illustrations of Queensland Fishes*. In the first of these, McCulloch (1915c) described and illustrated a few fish that he had collected himself at Masthead or Murray Islands, and others that had been presented to the Australian Museum.

The identification of Dottybacks (subfamily Pseudochrominae), small coral reef fishes, shows McCulloch's use of information gathered in the field. Several of these fishes which he collected from Masthead Reef exhibited marked differences in colour, and he commented:

> Their remarkable colour variation suggests that they represent more than one species, but as I found them living together in the same pool, and there being no important structural differences between them, I prefer to regard them as one. (op. cit.).

The second and third in this series on Queensland fishes were published after Ogilby's illness and concerned fishes presented initially to Ogilby. In 1919 Ogilby asked McCulloch for information about a Messmate Fish (family Carapidae) that had been found in the viscera of a bêche-de-mer (holothurian) trawled in 80 metres. The fish was in a collection given to Ogilby by Captain Hoult, then in charge of the Queensland Government's trawler. McCulloch advised:

> Please don't hope for information about Fierasfer from me, because I don't know one damn thing about the genus. Have never collected it myself, and have scarcely looked at any specimens. But if that Capt. Hoult is a little sandy

> haired chap who was on our trawlers and fell out with Stead, you freeze on to him, as he is a good sort, and will do his best for you. (McCulloch to Ogilby, 10.10.19, QMA)

Ogilby named the Messmate Fish *Fierasfer houlti* [later synonymised with *Encheliophis gracilis* (Bleeker, 1856)]. Ogilby suffered a stroke few weeks later and the Queensland collections were sent to McCulloch, who reported on them in two papers. He introduced the initial paper (McCulloch 1921b):

> The specimens dealt with in the following pages were secured by Captain Hoult during the trawling operations of the Queensland State trawler *Bar-ea-mul.* This vessel worked along the entire coast between the southern boundary of the State and Cairns, principally in the ship channels inside the Barrier Reef, but occasionally in the passages through the reef to its outer edge. Captain Hoult found the area between Rockhampton and the Whitsunday Passage to be swarming with small fish of many kinds, but too small to be of commercial value.

This diversity of small fry with a lack of marketable fishes repeated the findings of the *Endeavour* trawling, carried out in 1910. Similarly, line fishing 'produced an abundance of large edible fishes.' The experiment was terminated in 1921.

After Ogilby's retirement through illness, ichthyological problems in Queensland generally came first to his protégé, Tom Marshall. Marshall could deal with most of the enquiries but not the fishes that presented taxonomic problems: these he sent to McCulloch. E.H. Rainford of Bowen in North Queensland had been donating fish to the Queensland Museum for many years; on Ogilby's retirement, the Director of the Queensland Museum, Heber Longman, sent Rainford's collections to McCulloch. In 1923, McCulloch included an account of fishes from one of Rainford's collections in two papers in which he dealt also with fishes from Lord Howe Island (McCulloch 1923b and 1923c).

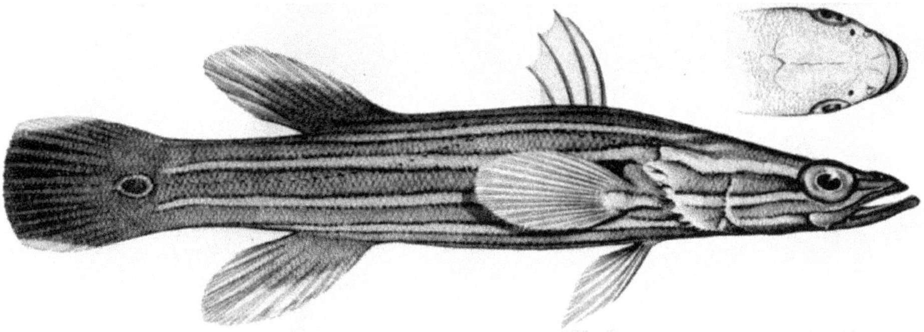

3

A. R. McCulloch and F. A. McNeill, del.

E.H. Rainford collected a single specimen of the Flathead Perch which McCulloch described and illustrated, naming it *Rainfordia opercularis*. It is endemic to northern Australia, living in caves on coral reefs. From McCulloch 1923.

Rainford's collection included several species of Butterflyfishes (Chaetodontidae); McCulloch took the opportunity to present a key to the family, re-described several species and named a new species *Chaetodon rainfordi* (McCulloch 1923b). A remarkable and colourful little fish was included in the collection. McCulloch consulted David Jordan and implicated him in the decision to establish not only a new genus (*Rainfordia*) but also a new family (Rainfordiidae) for the fish:

> I therefore follow the advice of Professor D.S. Jordan, to whom I have submitted sketches and notes, and establish a new family for it to be ranged near the *Grammistiidae.* (McCulloch 1923c)

McCulloch's last work on the fishes of Queensland was undertaken with the help of his assistant, Gilbert Whitley, and consisted of a modification of the former's embryonic Fishes of Australia to produce a list of the fishes of Queensland. McCulloch was too ill to complete the work and the finishing touches were by Whitley (McCulloch and Whitley 1925b).

Lord Howe Island

The 1902–03 excursion with Waite (see page 261) was McCulloch's introduction to Lord Howe Island. He returned to the island many times and wrote several popular articles for the *Australian Museum Magazine*, derived from his experiences there. His feeling for the island was profound:

> Dreams of sunlight, palms and open spaces, and also of freedom from the cares of this troublous world. And yet, for those whose tastes run to more substantial things, likewise an island of fascinating interest. There are delightful homes with gardens and charming people, and all those things that go to make civilisation worth while. (McCulloch 1921d)

It was his principal escape from the problems he found with life in Sydney.

Increasing knowledge of the fishes of Lord Howe Island owed much to McCulloch. The initial papers were produced by Ogilby, beginning in 1889 and Waite inherited this work, publishing five papers on the fishes of Lord Howe Island between 1900 and 1904. In the 1920s the residents of the island were still forwarding specimens to the Australian Museum. Beachcombing could be rewarding, oceanic species frequently stranding. McCulloch reported on these fish in two papers (1923b and 1923c). At this time McCulloch was upset with Waite and this seems to have coloured some of his comments. He referred to a drawing of the Blue Razorfish, *Iniistius pavoninus*, a synonym of *Iniistius pavo* (Valenciennes, 1840), which Waite had published in 1901:

> [t]he position of the anterior dorsal spine ... is not so far back as is illustrated in Waite's rather crude figure, which is inaccurate in other details such as the backward extension of the mouth and the relative length and depth of the head. (McCulloch 1923b)

When dealing with the synonymy of one of the deep water fishes, McCulloch found some evidence of Waite being careless:

> When examining specimens of *Maurolicus australis* Hector, Waite, by some strange circumstance, counted only three instead of four gill-arches. (McCulloch 1923c)

Waite (1910b) established a new genus, *Triarcus* (= three arches), for the Pennant Pearlside (*Maurolicus australis* Hector, 1875) but McCulloch had the two specimens which Waite had used sent from New Zealand and found that they had four gill-arches. He also noticed that a gill-arch had been removed from one side of one specimen, leaving three only on that side; the inference is that Waite established a new genus after counting the gill-arches of only one side in one of the two available specimens.

Popular work

McCulloch was judged an effective speaker, his lectures at the Museum and elsewhere being well attended and widely praised. The notes he made for some of his lectures (ML) show great care in preparation and bear markers to indicate where slides are to be shown.

A new Australian Museum publication, the *Australian Museum Magazine*, was initiated in 1921. McCulloch contributed to the first issue, and thereafter until his death. Many of his articles were inspired by his field trips to New Guinea, Lord Howe Island, the Great Barrier Reef and Torres Strait. He wrote on some of the more spectacular fishes that he encountered: *Electric Rays* (1921c), *Sea Dragons* (1923a) and *Stone Fishes* (1925b). For the *Illustrated Australian Encyclopaedia* he contributed articles on individual species of fish, and a summary of fish and fisheries in Australia (1925a). He also wrote on game fishes for the New South Wales Anglers' Casting Club and on marine topics for the American *Mid-Pacific Magazine.*

Photography had been of great interest to McCulloch since his first lessons from Waite. He had tried stereoscopic techniques and sent some of his results to Waite: 'Letter from McCulloch with photos of Lord Howe Is. (stereoscopic).' (WD, 26.11.21) He also experimented with the bulky movie cameras of his day, filming nesting sea-birds on the cliffs and fishes in the reef pools. He wrote to Frank Hurley of some early success:

> Expect to cause a splash with a lecture in June for which I have exposed 2000 feet of film of fishes under water. Pterois volitans ('Butterfly Cod') with enormous fins of staggering beauty, and deadly poisonous needle-like spines was my star performer, so I allowed him 200 feet. (McCulloch to Frank Hurley, 5.5.24, ML)

For the *Australian Museum Magazine*, he submitted an illustrated article, *Fishes and the Movies* (McCulloch 1924b), describing some of his problems in making the 'cinematographic pictures':

> The camera, tripod, spare film-boxes and other appurtenances which I used weighed in all about 70 pounds. [32 kilos] Its transport to scenes of operation, often over exceedingly rough country, was a matter of great difficulty, and taxed the good temper of several friends who assisted me... Over the roughest of mountain tracks, stepping from boulder to boulder, climbing upon steep cliff-

Lion Fish (or Butterfly Cod), *Pterois volitans* (Linnaeus, 1758) was McCulloch's best behaved fish for the 'cinematograph' while he was experimenting with the gear on Lord Howe Island in 1924. From McCulloch 1924.

> faces in search of nesting sea-birds, through dense forests where tangled vines caught and held the load, or out in a cranky dinghy to reach the coral reef ...

Some of the fish were filmed in the pools, others placed in an 'improvised aquarium'. They captured a 'Butterfly Cod' but its performance was postponed:

> We dallied so greatly admiring its slow evolutions and trailing fins, that the light of the late afternoon became too weak to admit of its portraiture, and it had to be kept alive in a bird-cage, sunken in the lagoon, until the next day. (op. cit.)

The fishes did not cooperate readily:

> The fish perhaps moves when the machine is out of action, or remains studiously still when it is required to be active, and hides so successfully that special tricks have to be adopted to bring it within range of the camera. (op. cit.)

The purpose of the movies was to illustrate a series of public lectures that McCulloch was giving at the Australian Museum.

McCulloch prepared exhibits at the museum with great care. He often painted the backgrounds himself and sometimes described their preparation in articles in the *Australian Museum Magazine*. He wrote guides to the galleries and was always alert for specimens suitable for the preparation of casts.

Cooperation and disagreements with Waite: 1915–1923

McCulloch was uneasy working with Waite. There was an unresolvable ambivalence in his attitude to his one-time mentor, now a co-worker. However, their joint work on the fishes of South Australia provided the basis for an understanding of the composition of that part of the Australian fauna. Much of the background for this work is described by Waite, so his diaries once again provide a framework for the following account.

As Edgar Waite and his family passed through Sydney in March 1914 on the way to Adelaide, McCulloch was away with the *Endeavour*. He wrote to Waite about the trip shortly after his return to the Museum, '[l]etter from Allan re his recent trip in the "Endeavour", (WD, 17.6.14) and a few weeks later, 'Received another part of "Endeavour" fishes from Allan.' (WD, 8.7.14) They continued to exchange information, Waite still being the main beneficiary of this cooperation, as he still needed access to the collections and library of the Australian Museum.

In January 1915, Waite visited McCulloch in Sydney. He brought some of the specimens from the catch of the *Simplon*. It was natural to pool the information from the only two collections taken in the Great Australian Bight: those of the *Endeavour* and the *Simplon*. Waite and McCulloch spent the greater part of four days together at the Australian Museum, working on the *Simplon* fishes and a South Australian fish catalogue. This marked the initiation of work on the catalogue, which began as Waite's project: 'To Museum, cataloguing' he recorded in his diary. (WD, 27.1.15) However, McCulloch had most of the collections and literature on hand in Sydney, so the greater part of the work devolved upon him when Waite returned to Adelaide.

McCulloch sent Waite an instalment of the basic list for the catalogue in March 1915:

> Received a first batch of an epitome of Allan's list of fishes for our joint list of S.A. species. (WD, 15.3.15)

A few months later, Waite received a request from McCulloch suggesting that, since he was doing most of the work for the *Simplon* fishes, he should appear as the senior author. Waite appeared to regard this as fair and agreed. A month later, McCulloch altered the proposition, suggesting that Waite formally author the *Simplon* fishes and he, McCulloch, author the allied work on Boxfish. Waite replied with a reasonable compromise:

> Replied to Allan. Agreed to splitting the paper but suggested that we should still be associated. The Aracaninae [Boxfish] by McC. & W. & the Simplon by W & McC. (WD, 31.8.15)

McCulloch accepted this and Waite negotiated with the Royal Society of South Australia to have the work published:

> Received letter, M.S. and plates from Allan, he having fallen in with my suggestion (31st). Sent title of the two papers to Roy. Soc. S.A. (WD, 6.9.15)

The process of publication of the two papers, the *Simplon* fishes and the Boxfish, proceeded rapidly; Waite soon tried to involve McCulloch in more of his projects.

In his diary, Waite gave a very different impression of the relationship between himself and McCulloch to that derived from McCulloch's letters to Ogilby. McCulloch wrote:

> Tickled to note that you too find difficulty in squeezing courtesy out of Mr. Waite. I recognize that he is a very busy man, but strongly object to his complete failure to recognise my existence when it does not happen to be convenient for him to do so. When he wanted something done recently I put in more than a whole day making diagrams and hunting out catalogues for him, which he even failed to acknowledge the receipt of. When I ask for a few pounds to have illustrations prepared for a paper we are writing together (that is to say he has provided half the material to be identified), he answers that he is pleased to learn that I am going to draw them, in the coolest manner possible. I'll see him 14-times damned first. (McCulloch to Ogilby, 2.6.15, QMA)

This brief but unrestrained deluge of complaint to a third party showed vehement feelings that McCulloch generally concealed from Waite.

Partial fish lists for Norfolk Island had been prepared by Waite while he was in New Zealand, and McCulloch had also worked on this project. Arthur Lea, then entomologist at the South Australian Museum, had sent Waite collections from both Norfolk and Lord Howe Islands, and Waite hoped to enlist McCulloch's aid in preparing publications that would include the new collections. He wrote to McCulloch, but initially received no encouragement:

> Letter from Allan re the Norfolk Island fishes 'too busy to look into the matter, at present.' (WD, 18.1.16)

Waite had sent part of Lea's collections for McCulloch's consideration, and the latter eventually responded:

> Letter from Allan with notes on Norfolk Island fishes I had asked. (WD, 3.8.16)

A week after receiving this information, Waite had modified his list of Norfolk Island fishes and sent it back to McCulloch for his consideration.

Waite showed no sign of appreciating his one-time assistant's difficulties. McCulloch's work load was very heavy: aside from Waite's projects, he was deeply involved with Ogilby's Queensland work, revising flatfish and other groups for publication, coping with the *Endeavour* trawl collections, and he had a host of other duties at the Australian Museum. McCulloch, in turn, saw Waite as something of an opportunist who took unfair advantage of their relationship. He did not seem to realise that Waite himself carried out an enormous amount of work, albeit that its focus was not on publication so much as the effective administration of his Museum and the various societies to which he belonged.

The fruits of cooperation between Waite and McCulloch in 1916 took the form of two publications. *Additions to the Fauna of Lord Howe Island, No. 5* (McCulloch and Waite 1916) was fully illustrated by McCulloch and he had done most of the taxonomic work as well. The second paper was essentially a list of the fishes of Norfolk Island. Waite (1910a) had published the first such list, including 33 fish; it was, as he acknowledged, a crude arrangement, some of the fish identified only to genus with some inevitable misidentifications. In this original paper, Waite compared the known Norfolk fish fauna, at a generic level, with that recorded from Lord Howe and Kermadec Islands. The 1916 paper extended the Norfolk list to 72 species, more satisfactorily identified thanks largely to McCulloch, and once again compared with the records from Lord Howe and the Kermadec Islands. The biogeographic comparisons were here extended to Australia and New Zealand. The paper (Waite 1916b) included the description of two new species, McCulloch and Waite each illustrating one of these. In spite of McCulloch's significant contribution, acknowledged in the introduction to the paper, Waite appeared as its sole author.

Field trips were McCulloch's delight, a taste he shared with Waite. McCulloch had hoped to go to New Guinea in a 'war boat' early in 1917, (WD, 30.1.17) but the trip was cancelled. At much the same time Waite led a Museum expedition to Cooper's Creek. The ichthyological results were naturally sparse (see page 302), but he caught a species of Grunter (Therapon Perch); McCulloch had recently studied the group so Waite sent it to him and was gratified when he had a '[l]etter from Allan saying the Cooper's Creek Therapon is new.' (WD, 14.3.17) He also sent 'an "O.C." [old collection] Grunter from Innamincka' and this, too, proved to be a new species. McCulloch prepared descriptions and sketches of these fish and the results were duly included in the brief *Pisces* part (McCulloch and Waite 1917) of the expedition report. Waite's contribution appeared to be limited to supplying the specimens and choosing their names.

Waite's enthusiasm for a South Australian fish catalogue was unabated and he proposed to extend the scheme:

> Wrote to Allan & suggested that we include the whole of the Bight in our S.A. Fish lists. (WD, 13.4.17)

This proposition was acceded to by McCulloch, thereby entailing the inclusion of many of his *Endeavour* fishes. This was a logical move, explained by Waite in the introduction to the catalogue:

> The western border of the State [of South Australia] is the 129th meridian of East longitude, but as it would scarcely be practicable to extend this limit to the ocean, the fishes recorded from the whole of the Bight (the littoral of Western Australia excepted) have been included. (Waite 1921)

They exchanged letters over the catalogue and, when Waite visited Sydney in July 1917, they worked on it together. Waite had collected a number of fishes from Kangaroo Island and elsewhere in South Australia, and had unearthed some specimens from the old collections that Zietz had shelved. He now proposed that these should form the basis of yet another joint paper with McCulloch. The specimens were of great interest and McCulloch complied once again. Waite was delighted with the prospect of having McCulloch's cooperation for two more important papers: the catalogue and a work on select groups of South Australian fishes. Having decided to include in the catalogue illustrations of as many South Australian species as possible, Waite set to work to raise interest in the project and money for the

Charles Hedley (left), conchologist, with colleagues at the Australian Museum in about 1914. From left to right: Phyllis Clarke, Joyce Allan and R. Bretnall. Phyllis Flockton Clarke (1891–1989) was born in Charters Towers, Queensland and worked as an illustrator at the Australian Museum from about 1910. Her work adorns many Australian fish publications of the time. Her water colours of Australian fungi have recently been reproduced in modern books (as in Grgurinovic 1997). Joyce Allan (1896 to 1966) became an authority on molluscs; she wrote and illustrated excellent books on this group and also illustrated *The Seahorse and its Relatives* (Whitley and Allan 1958). Courtesy of the Australian Museum Archives 1897-1m.

illustrations. The publication of such a catalogue presented great difficulties due in part to wartime shortages.

The illustrations were culled from various published sources, some originally drawn by McCulloch or by Waite himself. When Waite raised money, he sent it to McCulloch who was having extra illustrations prepared by Phyllis Clarke, a talented artist who worked on contract at the Australian Museum.

Work on the catalogue extended over several years and McCulloch found Waite expecting him to resume the project at short notice, without reference to McCulloch's own occupations. He became thoroughly frustrated by what he saw as Waite's peremptory demands but was still essentially cooperative as he saw the intrinsic worth of the project. Aside from having done much of the basic work on the list, and from providing many of the illustrations, McCulloch made a great contribution to the taxonomic work for the catalogue.

The second half of 1917 was an unsettled time for McCulloch. He wrote to Waite, telling him of the death in action of Dene Fry, who had been McCulloch's young assistant before the war. (WD, 15.5.17) He corresponded frequently with Waite at this time, complained of the excessive amount of work on hand and apparently considered applying for the Curator's position in Brisbane, suddenly vacant due to the abrupt retirement of Hamlyn-Harris:

> Received 3 fish-drawings from Allan, part of Miss Clarke's contract. Letter from Allan, he says that as the Trustees are treating him badly in the matter of pay, £225, he will become an applicant for the Brisbane vacancy. (WD, 13.9.17)

A few months later, Mrs McCulloch died; Allan McCulloch had lived with his mother and was evidently distressed. Waite wrote:

> Letter from Allan. In consequence of disruption of home life on death of his Mother, he proposes to offer for enlistment. (WD, 1.11.17)

And a few days later:

> Letter from Allan: the Museum Trustees have given him permission to enlist. (WD, 6.11.17)

Over the next few weeks, Waite received more of the drawings for their joint publications and then a letter from McCulloch:

> Letter from Allan, he is finally rejected for enlistment. (WD, 12.12.17)

The Brisbane Museum and the army failing to provide an escape from his routine, McCulloch planned a trip to north Queensland, telling Waite that he was going to spend a month in Cooktown. (WD, 11.5.18) That the trip eventuated is shown in a note to Ogilby, in which McCulloch mentions having collected 'a baby [Bony Bream] in a beach-pool near Cooktown.' (McCulloch to Ogilby, 5.4.20, QMA). The Museum register for July 1918 includes about 90 fish collected from the Endeavour River estuary, Cooktown by McCulloch.

Eventually McCulloch was accepted for enlistment and Waite was informed of this by Etheridge. Waite recorded:

> Rumour of surrender of Germany, unconfirmed. Business letter from Etheridge in which he mentions that Allan 'has enlisted and passed for the time being from the service of the Trust.' (WD, 18.10.18)

The timing was unfortunate as McCulloch's life was complicated further by the army episode. The paper which included Waite's Kangaroo Island fishes was completed, largely by McCulloch, and published in the first number of the new *Records of the South Australian Museum* (McCulloch and Waite 1918a). Waite was asked to 'contribute [an] article to Angus & Robertson's *Encyclopaedia of Australia* on Fishes in the absence of McCulloch, who has enlisted.' (WD, 24.10.18) He no sooner agreed than the armistice was signed and he wrote:

> As men in camp will be disbanding McCulloch will be able to take up the fish article, so I wrote to Carter, declining to proceed further in respect to Encyclopaedia. (WD, 16.11.18)

Within a week or so, Waite and McCulloch were exchanging letters on ichthyological topics once again.

Waite was in Sydney again in January 1919, keen to resume work on the South Australian fish catalogue. He and McCulloch spent several days together at the Museum, bringing the list closer to publication. From Adelaide, Waite kept McCulloch up to date with his progress in securing finance for the book:

> SA Museum,
> Adelaide,
> 21.ii.19.
>
> My dear Allan,
> I have, today, received a letter from Mr. Etheridge informing me that the Trustees [of the Australian Museum] have granted use of the drawings of Fishes for which we asked.
> At my Committee meeting [South Australian Museum] yesterday, the first since I left Sydney, the publication of the Catalogue was approved & my request that the Fisheries Dept. be asked to pay half the cost, supported. I will personally see the Chief Inspector of Fisheries: we are friends, and he will do what he can.
> I am confident of being able to publish the catalogue so we may go ahead. I will assemble what original drawings we have here and though, apart from finance, I have done what you consider my part of the work I shall be glad to know if I can still further the project. Do you think it possible to be able to go to press in April 1920?
> Yours very sincerely,
> Edgar R. Waite.
> (Waite to McCulloch, 21.2.19, AMA)

This letter was consistent with Waite's diary entry of the same date:

> Wrote to Allan <u>re</u> Catalogue of Fishes of S.A. informing him that Committee had adopted my recommendation to ask the Fisheries Dpt. to share ½ cost of publication. The total estimated at £200. (WD, 21.2.19)

This appears clear, but the phrase 'so we may go ahead' was taken literally by McCulloch. In particular, he committed himself to the expense involved in the preparation of illustrations. His reply to the letter of 21-2-19 alerted Waite to misunderstandings and he tried to correct them:

> S. Australian Museum,
> Adelaide, 3 March, 1919.
>
> My dear Allan,
> I have just received your letter of Feb. 27th. and hasten to correct any impression that my former letter may have given you as to finality of arrangement in respect to the Fish Catalogue. The matter is still sub-judice, inasmuch as we have not yet had reply from the Minister of the Fisheries

Dept. I intended to convey to you that my Committee approved my recommendation & decided to approach the Dept. ... You appear to have forgotten that I intimated to you that the photographs would possibly have to be made here and that I applied for the loan of the original drawings to this end. At least I judge so from your remark that you 'will look around for a photographer.' ... Reverting to the catalogue, I would say go on with the M.S. and I will do my part in respect to illustrations and engineering finance etc.
Yours sincerely,
Edgar R. Waite. (Waite to McCulloch, 3.3.19, AMA)

The caution arrived too late; McCulloch had proceeded with the illustrations. He informed Ogilby of his progress:

Have just submitted to Waite dinkyiest little photos to serve as the figures for our illustrated S.A. list. I had not hoped for such results, and trust he will find no difficulty in securing the necessary funds to proceed with them. They are to be taken from the original drawings mostly, but when such are lacking, we shall fall back upon any good published figures. (McCulloch to Ogilby, 4.3.19, QMA)

The photographs were being prepared, at McCulloch's request, by Theodore Roughley of the Technological Museum, Sydney (see page 422). The phrase 'and trust he will find no difficulty in securing the necessary funds' shows that McCulloch was already aware that finance was not finally arranged.

And a few weeks later, Waite was still trying to organise the money. McCulloch had sent him some of Roughley's photographs as samples of the work. Waite wrote again:

S.A. Museum,
Adelaide,
10 Mch. 1919.

My dear Allan,
Though I am writing today I will hold over the letter until I have definite information from the Minister controlling the Fisheries Dept. respecting joint payment, (with my Board) of the catalogue of Fishes ... You ask me to send the drawings at once, so that they may be photographed forthwith: I should like to explain that even if the Minister should put a line on the estimates, the money would not be available until they are passed: would this suit Mr Roughley?, usually they are passed in October. As the money may <u>not</u> be voted it would not do to presume that it will be, and so incur expenses which could not be met.

13th March
I have had a communication from the Secretary to the Minister of Industry and he tells me that, owing to the financial condition of the country he would not take any responsibility in the matter, but would refer it to Cabinet next

> week. You will see, therefore, that though non-monetary work may be gone on with I cannot at present authorize expenditure ... I will write to you as soon as I have definite information. Even should this scheme fail, I do not intend to let the matter drop and will endeavour to pull it off by hook or by crook.
> Yours sincerely,
> Edgar R. Waite.
> (Waite to McCulloch, 13.3.19, AMA)

In his diary, Waite recorded:

> Minister for Industry informed me that the question of voting £100 towards cost of printing the Fish Catalogue will be brought before the Council next week. Wrote Allan & informed him. (WD, 13.3.19)

McCulloch was upset by Waite's delays, but there was also an element of willful misunderstanding. He was genuinely upset that he had misled Roughley, but had done so knowing that the finance was not definitely available. McCulloch wrote to Ogilby, to Roughley and to Waite.

> Friend Waite continues to cause me much annoyance with his vacillating ways. Giving me authority to go ahead, I had the most beautiful illustrations prepared to be submitted as samples. Ere he had received them he decided we wouldn't go ahead, and now he has decided we'll go ahead, but not at present. That I and the photographer should wait on his pleasure is of course a matter of no concern ... (McCulloch to Ogilby, 25.3.19, QMA)

McCulloch bundled up three of Waite's letters (21-2, 3-3 and 13-3: quoted above) and sent them to Roughley by way of explaining the confusion and delay in payment for the photographs. In his covering letter, he blamed Waite:

> Enclosed are three very conflicting letters from Waite relative to the preparation of The S. Australian Catalogue, which will better explain the present circumstances than any amount of talk on my part. I feel more annoyed that you should be let down than I can express, but am afraid that his position cannot be assailed. (McCulloch to Roughley, 29.3.19, AMA)

Roughley's response was muted and reasonable:

> I have received the correspondence relative to the preparation of photographs for the Adelaide Museum and agree with you that it is very contradictory. Nevertheless, you hinted to me that, although Mr. Waite told you to go straight ahead, you were not a bit satisfied that the financial arrangements were settled, and in consequence of this I did not build too much upon it.
> However, I am grateful to you for the efforts you have made to finalise the matter and, well – it may eventuate some day.
> (Roughley to McCulloch, 3-4-19, AMA)

Despite his statements to the contrary, McCulloch had realised that Waite still had reservations regarding the finances.

To Waite, McCulloch expressed his disappointment and his sense of having been misled:

> Of course if the money is not available we cannot proceed with the work, but your letter of Feb. 21st so clearly states that 'we may go ahead', that I felt safe in arranging with Roughley to undertake the preparation of the photographs. If you could compare your three letters of Feb. & March to me re. the Catalogue, you would realise that they contain enough conflicting instructions to puzzle the conventional Philadelphian law[y]er. (McCulloch to Waite, copy, 29-3-19, AMA)

He goes on to say that he will resume work on the *Endeavour* fishes, implying that Waite has chosen to delay the South Australian catalogue.

In reality Waite was active, casting around for means to publish. Seven months later:

> Wrote to Allan re the S.A. list of Fishes, after ascertaining that the Fisheries Dept. had put £100 on the estimates. Saw Howchin [Chairman of the South Australian Museum Board] who says that Museum must otherwise stand the lot. (WD, 10.10.19)

Waite was jubilant and promptly notified McCulloch of his success in obtaining funding:

> Preparation of the Fish Catalogue was approved by my Committee at its meeting on Wednesday, and the necessary funds voted, so now you can have 'no possible doubt, no possible shadow of doubt whatever', or whatever Gilbert said.[30] ... I am pleased to have your letter of the 3rd. inst. informing me that you are going on with your part of the work, and agreeing to the inclusion of the whole of the 'Bight' species. [This had been agreed between them in 1917] (Waite to McCulloch, 7.11.19, AMA)

Two weeks later McCulloch had another of his life's props threatened:

> Letter from Allan re the Fish Cat. he says Ogilby has had a stroke ... (WD, 24.11.19)

The letter in which he mentioned this must have had frank criticism of Waite who, in his reply, suggested that details of procedure in preparation of the catalogue and its illustrations should be discussed:

> ... It is advisable that all adjustments in this matter should be discussed between you and me in the first place, for some overlapping seems inevitable, situated as we are, at some distance apart ...
> Your concluding remarks (Letter Dec 5) are scarcely up to your usual standard of courtesy, but I will conclude that you were not well when writing.

30 Of that there is no possible doubt – No probable, possible shadow of doubt – No possible doubt whatever. Gilbert, W.S.: (1889) *The Gondoliers* Act 1.

Waite had heard that Ellis Troughton, one of McCulloch's assistants, was soon to accompany a visiting American zoologist to Adelaide and thence to northern South Australia. Waite concluded the letter:

> I wish you were coming instead, though I should not desire to consign you to Mt. Lyndhurst and Mt. Hopeless. (Waite to McCulloch, 9.12.19, AMA)

McCulloch replied promptly, quite dissatisfied with Waite's suggestions as to the illustrations, and presenting a virtual ultimatum:

> Glancing over our correspondence re this check-list, I find it is not as free from misunderstandings as might be desired. In order to avoid further confusion I submit the accompanying plan of work which, if it meets with your approval, you might initial and return to me ... If on the other hand you prefer any other arrangement, it will be best for us to work separately, in which case I shall return you your M.S. and retain the illustrations for my future Australian list, and will settle with Roughley myself ... Please do no suppose that I wish to be unduly dictatorial. We evidently regard some of the matters concerned from rather different view-points, and it is only to avoid further misunderstanding that I venture to put the whole matter so plainly. (McCulloch to Waite, draft, 12.12.19, AMA)

Among other points, the plan that Waite received from McCulloch stipulated:

> 241 photographic prints to be paid for by S. Austr. Museum as per Roughley's account, but duplicate prints from the same negatives shall be available to either author for future use in any other publication. (McCulloch to Waite, draft, 12-12-19, AMA)

McCulloch intended to use some of the photographs in his projected work on the fishes of Australia. Waite agreed to McCulloch's proposals but said:

> When you forward the photographs, please also send the negatives for those for which we are paying in full ... It is not in my power to use illustrations printed in our records elsewhere, nor to grant the privilege to others, but I have little doubt but that permission will be granted to you if applied for officially. (Waite to McCulloch, 15.12.19, AMA)

McCulloch planned to keep control of the photographs, even at his own expense. He promptly returned the manuscript:

> I regret that you insist upon having both the negatives and prints of the S. Australian fishes while being unable to authorise their publication in any other work, since this is quite fatal to our agreement. (McCulloch to Waite, undated draft, AMA)

He wrote that he had expected Waite to be satisfied with a set of prints, that he would not require the negatives. He continued, the draft letter clearly revealing his feelings:

> As your letter makes it clear that we cannot agree upon these matters, I am returning your M.S. by registered post. I am honestly sorry that you should have had all the difficulty of assaying about the money matters etc., ~~but after reviewing all our correspondence, cannot think you have given me a wholly square deal~~ and shall be only too glad if I can lend a hand in case you propose carrying on the list yourself. (McCulloch to Waite, undated draft, AMA)

He may have thought Waite incapable of completing the preparations for publication; if so, he was wrong. Waite recorded only:

> Letter from Allan; he withdraws from joint authorship of Cat. of Fishes, wanting to get on with the 'Endeavour' fishes. (WD, 24.12.19)

Waite acknowledged the return of the manuscript, writing:

> Your somewhat frequent hint of retiring from collaboration, coupled with your expressed desire to return to work on the 'Endeavour' Fishes, in a measure, prepared me. As to your stated reason 'least said soonest mended'.
> I also have thank you for part 1 of your 'Check-list' [of New South Wales fishes] which is a decided contribution to our literature. The notes are interesting and instructive, I did not know, for example, that the egg-cases of H. philippi [the common Port Jackson Shark] were provided with tendrils, I have unfortunately stated otherwise. [This was obnoxious of Waite: he was well aware that these egg-cases had no tendrils; and he knew that McCulloch knew he knew.] ... It is kind of you to offer to lend me a hand, should I find it possible to carry on the catalogue myself, an offer I will gratefully bear in mind. (Waite to McCulloch, 31.12.19)

McCulloch wrote to Ogilby:

> The proposed S. Australian check-list in conjunction with Waite has gone fut, and 'his' M.S. (written from my notes by his assistant) has been returned. I have had to personally pay about £17 for the illustrations which have been prepared for it, but this was preferable to working with a chap who does his best to work points upon one all the time. He was to have had 77 illustrations free of charge, and all the figures (241) were complete, but he insisted upon retaining a sort of copyright over them so that I could not use them again in my future Australian list – and this notwithstanding that they were all either lent by this museum or cribbed from other books. And so I returned him his so-called M.S., & we agreed to differ. (McCulloch to Ogilby, 8.1.20, QMA)

Again, McCulloch was out of patience with Waite and exaggerating his injuries. He represented his withdrawal from the South Australian catalogue to Ogilby in a rather selective light.

Waite and his young assistant, Herbert Hale, continued to work on the catalogue through 1920 and McCulloch helped by providing literature and references from time to time. The Director of the Australian Museum, Robert Etheridge, died early in 1920 and Waite promptly applied for the position. McCulloch, intimidated by Etheridge for many years, anticipated a more agreeable regime; he hoped that his friend, Charles Hedley, would be appointed:

> You will doubtless have heard of the death of Etheridge last Monday while at his farm at Colo Vale. I look forward to far more reasonable conditions here in the future, and in company with most of the staff, hope Charles H. will be appointed Curator as he surely deserves. (McCulloch to Ogilby, 8.1.20)

He made it clear to Ogilby that he did not want Waite back at the Australian Museum:

> I'm only praying he [Waite] may not be appointed to the Directorship over C.H's [Charles Hedley] head. (McCulloch to Ogilby, 16.1.20, QMA).

And in the next letter:

> We are endeavouring to make a good show during the six months of Mr. Hedley's probation, as we don't want our worthy friend E.R.W. or others of his kind stepping into the Curatorial shoes. (McCulloch to Ogilby, 8.2.20, QMA)

This feeling against Waite had been noticeable in McCulloch's letters to Ogilby for years.

McCulloch asked Waite for a photograph of himself (Waite) and of Ogilby to use in his New South Wales fish catalogue (WD, 15.6.20); Waite sent the negatives but they were not used. The first part of McCulloch's New South Wales check-list was published in 1919 and by the time Waite's illustrated catalogue was ready for publication in 1921, the second part of McCulloch's list was also in print. In April 1921, Waite was at last able to send his *Illustrated Catalogue of Fishes of South Australia* to McCulloch. His contribution to the South Australian catalogue was largely overlooked and McCulloch was not happy. Ogilby, who had suffered a more serious stroke, was no longer the recipient of McCulloch's confidences.

Cooperation was maintained between McCulloch and Waite. In 1922, there was a typical letter from McCulloch, responding to one from Waite; McCulloch was happy and about to participate in a trip to Papua with Frank Hurley:

> Dear Mr. Waite,
> Very pleased to have your cheery note. Had heard you were possibly away on a tour around the north-west coast, so had not hoped to hear from you before I leave for Papua at the end of the month. You may have seen Hurley's picture Pearls and Savages on the screen? Having made quite a nice sum out of it, he is off again to do even better so as to make the picture suitable to tour America and Europe, and he has invited me to accompany him. He is taking two seaplanes, so I expect to have a pleasant trip. We will not be back before the end of the year if all goes well.

He then proceeded to answer an ichthyological question that Waite has posed, and gave Waite news of the success of the check-list of the fishes of New South Wales, of recent

Allan McCulloch (left) on a Museum field trip to Lord Howe Island in 1921. With him is Ellis Troughton, later curator of mammals at the Australian Museum. Courtesy of the Australian Museum V15643 Large Format Negatives Collection.

contact with David Starr Jordan and David Stead. McCulloch described some of his recent work on Queensland fishes and returned to the subject of his trip:

> ... lessons in wireless operating at every spare moment. We are taking up complete outfit with us to Papua, so as to keep in touch ... My only regret in going away just now is that I am, for the first time, in a position to finish off the mass of the 'Endeavour' material, having not only completed the card index but also in having another assistant who can lend me a hand with the illustrations. But Papua has been my objective for many years, and I would not miss it for anything ... A cheery word to Mrs Waite for me.
> Sincerely,
> Allan.
> (McCulloch to Waite, 14.8.22, AMA)

Waite and McCulloch seemed to have resumed a normal and pleasant correspondence. This was not to endure.

Waite modified his catalogue, republishing it as a handbook in 1923. He omitted the synonymies and added fin formulae, colour descriptions and some notes of general interest. He also omitted all acknowledgement of McCulloch's contribution. As usual, Waite sent

McCulloch a copy of the work, naively expecting congratulations. He was very proud of the handbook and was unprepared for McCulloch's response:

> The Australian Museum,
> 28th May 1923.
> Dear Mr. Waite,
> I have to acknowledge the receipt of your handbook to the Fishes of South Australia, though its appearance is naturally the reverse of pleasing to me. I am one of those unfortunates quoted by Professor Jordan who, impelled by a love of this kind of work, asks only for recognition of his labours. And this you have studiously denied me.
> You may remember that your first catalogue of the South Australian Fishes, based upon Zietz's lists, was made concrete by Troughton [Ellis Troughton, then assisting McCulloch] and myself when we worked through my card-index and prepared a basic list for you. This you expanded by the inclusion of many references, and then you handed it to me for preparation for publication under our joint authorship. Your request to the Trustees for the loan of my original drawings was acceded to, and following your instructions, I had them photographed, but your later contradiction led to our disagreement, and a charge upon my pocket of £17 odd.
> You then published your catalogue, copying many of my figures from the published plates though you claimed in the introductory part that you had had the originals. You so arranged your reference quotations that none might know which figures were so copied without turning to the works quoted ...
> And now appears your Handbook, illustrated with 93 of my own drawings and without even a reference to them. You acknowledge in your preface that your Board of Governors has granted permission to reproduce them, but I wonder what that Board would think if its members knew the facts, and whether they would appreciate being held indirectly responsible for the copyrighting of these very drawings you have been able to utilise through that courtesy of science which prohibits copyright of its labours.
> My dear Mr. Waite, you have done more to shatter my enthusiasm as a student of ichthyological research than you will guess, even though you owe more than a little to my drawings that have served you so well. I have just that very natural desire for the recognition of my labour, which you could have satisfied without detracting one iota from the value of your own.
> Have you chanced on Jordan's flattering reference to my work in Mem. Carnegie Museum x.1, p.44, under Maccullochina. [now *Synagrops*, a genus of fishes named by Jordan in compliment to McCulloch] It, at least, takes some of the sting from your treatment of my efforts.
> Allan R. McCulloch.
> (McCulloch to Waite, copy, 28-5-23, AMA)

The point regarding copyright had been taken up by Waite previously; he had protested against the General Secretary of his Museum Board, with whom he had frequent battles, from applying it to issues of the *Records of the South Australian Museum*.

Waite did not feel at fault and responded with sarcasm, and without sympathy:

> My dear Allan,
> I have received your kind letter of May 28th. in which you so graciously thank me for a copy of the Handbook to the Fishes of S. Aust.
> I must, for the present, content myself with a bare acknowledgement, but I may write to you more fully later ... I wonder whether Virgil was referring to you or me:- 'Quantum mutatus ab illo'[31].
> With kind regards,
> Yours sincerely,
> Edgar R. Waite.
> (Waite to McCulloch, 7-6-23, AMA)

Waite commented on McCulloch's letter in his diary:

> A remarkable letter from McCulloch in which he says I have 'studiously ignored him in the Handbook of fishes'. (WD, 5.6.23)

Waite had treated McCulloch poorly, through neglect rather than malice. He had been using McCulloch's work for 20 years; frequently, but not infallibly, with appropriate acknowledgement. He took McCulloch for granted but on this occasion McCulloch's long-smothered resentment had become manifest.

The affair rapidly became more serious as McCulloch appealed to officialdom. His Director wrote to the Secretary of the South Australian Branch of the British Science Guild, who had published the Handbook, complaining of the lack of acknowledgement of McCulloch's drawings and the application of copyright to the publication. Waite noted:

> Cleland as Sec. of the Brit. Sci. Guild received letter from Australian Museum, evidently inspired by McCulloch drawing attention to absence of acknowledgement in the Fishes Handbook. (WD, 9.6.23)

Waite was editor of the Handbooks and took part in the discussion of the complaint. Waite's diary account is succinct:

> Meeting Handbooks' Committee to consider complaint from Aust. Mus. decided to express regret (see 9th.) and to print a slip acknowledging source of illustrations. Wrote and took slip to Government Printer. (WD, 12.6.23)

The 'slip' mentioned was a minimal response, not at all generous to McCulloch:

> ... a general acknowledgment is here made. The blocks obtained ... were prepared from illustrations made by, or accompanying, the writings of ...

31 *Quantum mutatus ab illo. Hectore qui redit exuvias indutus Achilli.* How greatly changed from that Hector who came back arrayed in the armour of Achilles! Virgil: *Aeneid* book 2, l. 274.

Then followed a list of 24 authors, most of whom were dead, McCulloch's name undistinguished among the list. Not very conciliatory.

There were faults on both sides. McCulloch had acted precipitately in engaging Roughley, willfully misunderstood Waite, acted the victim, and rationalised his own position by a contortion of logic. Waite took McCulloch's work for granted and his flagrant neglect of proper acknowledgement was indefensible. The professional relationship between McCulloch and Waite did not function for 12 months after this crisis.

Contact resumed

The first move to resume contact appears to have been a letter from McCulloch, commenting on a recent publication of Waite's (WD, 2.6.24). A meeting of the Australian Association for the Advancement of Science was due to be held in Adelaide shortly after this, and McCulloch was expected to attend. Waite took the opportunity for peace-making:

> Wrote to Allan & invited him to stay with us during the visit of the A.A.A.S.
> (WD, 13-6-24)

But McCulloch declined:

> Letter from McCulloch, he cannot come to Adelaide for A.A.A.S., having charge of Museum during Anderson's [Director] absence.

They resumed the exchange of information on ichthyological matters; for instance:

> Letter from Allan confirming my identification of <u>Crapatalus</u> <u>arenarius</u> [the Flathead Sandfish, now identified as *Lesueurina platycephala* Fowler, 1908].
> (WD, 9.3.25)

McCulloch then wrote a friendly letter from Lord Howe Island, on the eve of his own birthday (WD, 20.6.25), when he was about to attend a Fisheries Congress in Honolulu. Waite responded pleasantly:

> My dear Allan,
> I was very pleased to receive your letter written from L.H.I. and, writing on this date, I take the opportunity of wishing you 'very many happy returns of the day.'
> I envy the opportunities you have of visiting the island, for though I have been on many, both in the Subantarctic and in the Bight, none has the charm of L.H.I. and when the warm drizzling rain falls I am somehow reminded of the palm trees. I have often wished to see Honolulu, if for nothing more than the Aquarium and now you are going there.

Waite chatted about some minor matters and mentioned that he hoped to 'revisit England and the Continent after an absence of 32 years'. He realised that McCulloch had been unwell and closed:

> Shall be glad to have a line from you when you find time and opportunity and hoping that you may soon pull round.
> Yours ever,
> Edgar R. Waite. (Waite to McCulloch, 20.6.25, ML)

The relationship appeared to be restored.

Field work

McCulloch seemed unable to resist an opportunity for field work and had a wide experience of collection and observation in both marine and freshwater environments. An examination of the Australian Museum accession registers shows much of his activity. Reflecting his exertions within New South Wales, McCulloch registered fishes that he had taken around Sydney, from Port Macquarie, the Nepean River, the Hastings River and Port Stephens.

He worked widely in Queensland, the origins of these collections including the Brisbane River, Port Curtis, Cairns, Torres Strait and several islands on the Great Barrier Reef, to which he returned many times. His trips to Lord Howe Island frequently involved the collection of fishes.

McCulloch had seen trawling aboard the *Woy Woy* off Sydney (1906), the *Endeavour* in Bass Strait (1914), and on a New South Wales State Government trawler (1916). From each of these excursions, McCulloch registered specimens.

Among the Museum's more enthusiastic donors were a pair whose contributions, generally from the New Hebrides (but in December 1911 from Cornwall, England!), were indicated by 'Paymaster Stevens & Dr Cummins' of HMS *Pegasus*. These donations were first recorded by McCulloch in the register in 1910, and comprised 15 to 50 fish at a time, from once to several times a year, for more than a decade. McCulloch joined them on passages to the New Hebrides in September 1910 and again in July 1922, about 100 fish being registered from the latter trip.

McCulloch's New Guinea trip with Frank Hurley in 1922–23 was intended to be primarily an ethnological investigation, but McCulloch also collected zoological specimens. On his return, he registered fishes from Thursday Island, from a reef off Port Moresby and from Lake Murray, Papua.

McCulloch's taxonomy

McCulloch's taxonomy could be characterised as conservative for his time. His approach to the description of new species was generally circumspect, although there were exceptions. For instance, in 1909, one of his regular donors, Mr A.T. Basset Hull, found a small fish stranded on a beach near Sydney and presented it to McCulloch. He drew the specimen and prepared a careful description, writing:

> This is doubtless the young form of some large species of Caranx or allied genus, though just which of the several closely-related divisions of the Carangidae I am unable to decide. The great development of the fins and the pronounced colour markings at once distinguish this species from all others that are known to me. (McCulloch 1909)

Although realising that the fish was a juvenile, he described it as new, naming it *Caranx hullianus.* The specimen was later identified as the young of a species that was not uncommon in northern Queensland. McCulloch usually described new species only from adequate adult material.

Caranx hullianus, named by McCulloch from a tiny specimen, proved to be the young of the Cottonmouth Trevally, *Uraspis secunda* Poey, 1860. From McCulloch 1909.

Well aware of the problems posed by changes with growth, sexual dimorphism and sexual dichromatism, McCulloch usually avoided these pitfalls in establishing new species. Once again, there were exceptions: the southern Australian Cowfish (Aracaninae) were the best example (see page 297).

Customarily, McCulloch's care in literature searches and his ready consultation with colleagues both within Australia and overseas, together with his familiarity with Museum collections in Australia, imparted a high degree of reliability to his work. His conservatism led him astray at times, particularly in dealing with some of the closely related groups of freshwater species. For instance, he observed the large variation in form present in freshwater fish from different drainages but underestimated the number of species present. This was particularly so in groups of freshwater Gobies (Gobiidae) and Rainbowfish (Melanotaeniidae).

Of those fish McCulloch described and named as new, approximately two thirds of the 100 odd species are now regarded as valid. This compares favourably with Waite (1 in 2); Ogilby (1 in 3); Castelnau and Macleay (each 1 in 5) and de Vis (1 in 13).

Much the same could be said for his generic descriptions, most of which have proven acceptable. Over three quarters of the 21 genera established by McCulloch survive in modern fish taxonomy and his definitions still serve in most cases. His success, measured in this way, was higher than that of his near contemporaries in Australia (Waite and Ogilby), and still higher than that of the earlier ichthyologists (Macleay, Castelnau and de Vis) and of his immediate successor (Whitley).

The immediate focus for much of McCulloch's work was the resolution of past errors, the ultimate intention being the production of a major text on Australian fishes. For this reason, the description of new species comprised a relatively minor part of his work. He was perennially keen to establish the identity of species that had been described by his predecessors, and found Ogilby very useful in this pursuit. McCulloch asked Ogilby to try to track down a species sketchily described by de Vis in the *Proceedings of the Linnean Society of New South Wales* (1884a):

> Now I should particularly like to know what Homodemus cavifrons (P.L.S.N.S.W., IX, 1884, p. 396) is, or at any rate, where to put it in my index. (McCulloch to Ogilby, 18.2.15, QMA)

Ogilby found that the fish was a Murray Cod which de Vis had not recognised, describing it as new. Similarly, Ogilby helped with another de Vis conundrum, *Herops munda* (De Vis, 1884a):

> I am indebted to Mr. Ogilby for the suggestion as to the identity of De Vis' puzzling genus. (McCulloch 1920)

This proved to be a common Queensland fish, the Silver Flagtail, *Kuhlia munda.*

McCulloch made his views of the importance of this sort of work clear. On one occasion, Ogilby wrote to say that he had a Grunter from a Queensland stream which might prove to be *Therapon hillii*, which Castelnau had described from the Dawson River in Queensland (Castelnau, 1878a); when he wrote again to say that the descriptions did not tally, McCulloch responded:

> Am very sorry to learn that it is not T. hillii, Castelnau, since the rediscovery of this species would be worth a dozen new ones.
> (McCulloch to Ogilby, 22.12.14, QMA)

Among Australian ichthyologists, this sentiment could have been shared only with Ogilby.

One of McCulloch's strengths as a taxonomist was his appreciation of the importance of type specimens. He repeatedly referred to this in his letters to Ogilby and published comments on the topic, pleased with the growing appreciation of its significance:

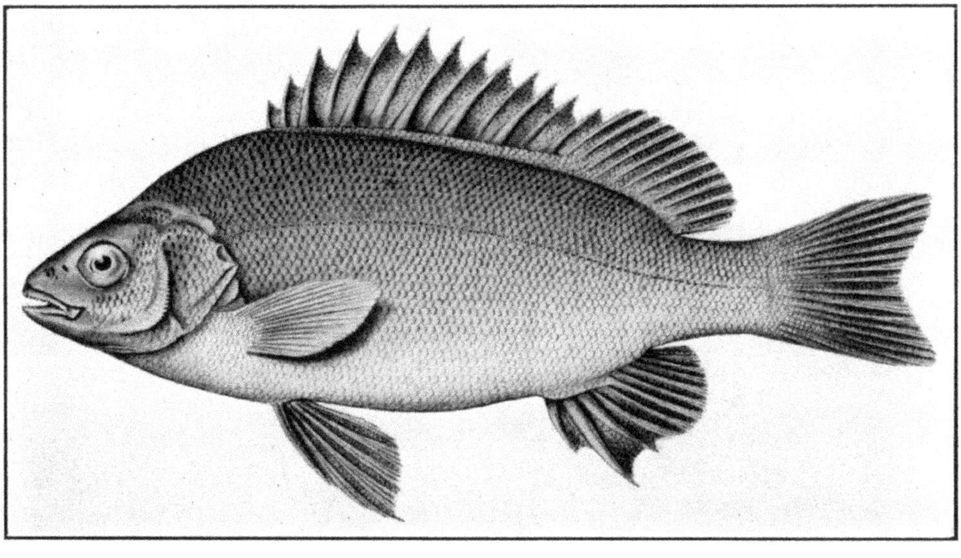

Fig. 2.—THERAPON HILLII Castelnau, imm. ⅝ Nat. Size.

A. R. McCulloch, del. Page 121.

The Leathery Grunter, *Therapon* [now *Scortum*] *hillii* was first described by Count Castelnau (1878). Ogilby tried to rediscover it through his contacts in Queensland and was successful thanks to Dr Thomas Bancroft, an enthusiastic biologist who supplied Ogilby with many freshwater fish, and some notes: 'Dr Bancroft writes that this fish is uneatable; no matter how long it may be boiled it retains the consistency of shoe-leather ...' McCulloch drew the fish from a freshly preserved specimen. From McCulloch 1916.

> An appreciation of the extreme importance of typical specimens and of others that have been described or illustrated in literature has been generally developed, and invaluable material, hitherto largely neglected or lost in a maze of unsorted miscellanea, has been rediscovered and valued at its proper worth ... This improvement has come none too soon, since not a few types and other specimens of equal value have either been lost or are decayed beyond redemption. It is fortunate for Australian ichthyology that the specimens on which Castelnau, Macleay, de Vis, and Johnston founded their numerous species are now being diligently sorted out of the various collections in which they are contained, and made available for examination. (McCulloch 1916b)

McCulloch extended these comments some years later:

> ... improved conditions of these collections is steadily reducing the number of lost types, and the elimination of the comparatively few fishes described by Australian authors which remain known only by their names and ludicrously meagre descriptions is gradually approaching completion. (McCulloch 1924a)

In this paper McCulloch acknowledged the Macleay Museum Committee of Management for the loan of types; and Tom Marshall for rearranging and classifying the Queensland Museum collection, in the course of which he had located some misplaced types of de Vis and Ogilby.

Last works in progress: 1924–1925

Some of McCulloch's work early in 1924 can be deduced from the subjects of the letters he sent to overseas correspondents. For instance, he sent Professor Max Weber in the Netherlands sketches and notes of Dottybacks, *Pseudochromis* spp., as he hoped to have his opinion for the fifth *Endeavour* report, in particular since some of Weber's work had been published in a journal to which McCulloch had no access (McCulloch to Weber, draft, 14.1.24, ML). He wrote to von Bonde, a South African ichthyologist, 'very interested in the wonderfully close relationship between your species and ours', suggesting an exchange of specimens. McCulloch was interested in the possible identity of a species, recently described from South Africa, of the oceanic genus *Cubiceps* (McCulloch to von Bonde, draft, 16.1.24, ML). To Doctor Georg Duncker of the Hamburg Museum of Natural History, McCulloch sent specimens of Günther's Pipehorse, *Solegnathus guntheri*, [now *S. lettiensis*] asking for comment that he wished to include in a paper. (McCulloch to Duncker, draft, 12.2.24, ML).

Correspondence with David Jordan: throughout 1924, McCulloch attempted to make progress with his many projects, but often appeared disheartened. He maintained a correspondence with David Jordan but it assumed a pessimistic and melancholy aspect. McCulloch had hoped to attend a conference in Hawaii and was resentful because the Museum administration had not approved his plans, although David Stead had joined the conference:

> I should have enjoyed meeting yourself, Dr. Evermann, and all the others gathered at Honolulu more than I can tell you, but my friend Stead understands the art of wrangling these things better than I do. He even claimed to represent the Australian Museum and the Fisheries Department in the Pan-Pacific official publications, but this little exaggeration has been officially corrected by both departments.

His hopes, his accomplishment to this date, and his difficulties with the Museum were all mentioned:

> I am seriously contemplating a catalogue of the Australian fishes on the lines of your 1905 'Fishes of Hawaii'. Most of Macleay's and Castelnau's species, which you spoke of when last in Australia as a source of much trouble, are now more or less placed, so that the list can be reasonably accurate. The principle difficulty lies in arousing interest to supply the necessary cash, because I must have someone to help me with the figures.
>
> I have still an enormous collection of undescribed fishes, mostly deep-water forms, from the 'Endeavour' collection, but have pretty well lost hope of completing their investigation. If Stead and Waite would do their bit as did old Ogilby, instead of maintaining a sort of policy of piracy by which they strive to acquire much kudos with the smallest amount of work possible, it would be better fun. But now that old Ogilby is out of it (I believe he still exists in a mental hospital), it is only the magnificent standard set by yourself and some of your associates that sustains my ambition. (McCulloch to Jordan, draft, 14.7.24, ML)

Waite and Stead were 'pirates'. Ogilby was 'out of it'. His old friend Charles Hedley had left the Museum in March 1924, after working there for 35 years, to take up the position of Director of the Great Barrier Reef Committee in Brisbane. Four months later, McCulloch wrote to Jordan again, appearing even more disturbed:

> We are blessed with a very strange assortment of trustees at present, who have no interest in science matters whatever, and even our publication, the Records, is threatened with suppression ... My work upon fishes has been largely stifled, and unless some change occurs in the near future, my aim to produce a Catalogue of Australian Fishes, will be permanently squashed. (ML)

The curator and trustees of the Museum did not see the full importance of McCulloch's work. He felt isolated, his ambitions of concern to no one but himself.

At about this time, judging from its position among his papers and on the content, McCulloch appears to have received a proof of an article about himself from 'Who's Who,' Soho Square, London. He presumably supplied the details, including the fact that he was 'engaged on a Catalogue of Australian Fishes.' There was reference to his positions on the councils of the Linnean and Zoological Societies of New South Wales and the Great Barrier Reef Committee.

Flounders, Sole and their relatives are 'Flatfish' (order Pleuronectiformes). A great variety of them is found in the marine waters of Australia, and a few inhabit the northern rivers. They proved a difficult group to work with in Australia, partly because many of the species were infrequently encountered and consequently poorly represented in collections.

John R. Norman, Regan's assistant in the British Museum, was specialising in the study of flatfish and McCulloch had been sending him specimens. He had also prepared a manuscript work on Australian flatfish, entitled *List of the Heterosomata recorded from Australia with notes*, which included synonymies and distribution records. This work was to be sent to Norman and the whole was punctuated by McCulloch's notes and opinions on the validity of species. For instance, under *Pseudorhombus multimaculatus*:

> Mr Regan compared specimens of novae-cambriae from New South Wales with Günther's type of multimaculatus for me and declared them identical. Ogilby informed me, however, that he could not accept Regan's dictum, so I will be glad if you will publish your opinion of the matter.
> (McCulloch, undated MS, ML)

Norman found *Pseudorhombus multimaculatus* to be a synonym of *Pseudorhombus jenynsii* (Bleeker, 1855). McCulloch sent his manuscript, together with the *Endeavour* Flatfish, to Norman. In publishing his report on this collection (Norman 1926), and in his subsequent *Monograph of the Flatfishes* (Norman 1934), Norman acknowledged McCulloch's assistance.

In Gilbert Whitley's first publication he was the junior author, with McCulloch, of a paper on *Some Little Known Australian Flat-Fishes* (McCulloch and Whitley 1925a). This paper provided descriptions of species represented in the collections only by single specimens which could not reasonably have been sent to Norman:

> The risk of damage to, or even loss of these unique specimens forbids their transport to and from London ... (McCulloch and Whitley 1925)

The specimens were drawn and described in appropriate detail, synonymies prepared and identifications suggested.

McCulloch and William Paradice

Surgeon Lieutenant-Commander W.E.J. Paradice (1897–1927) of the Royal Australian Navy was interested in natural history and was encouraged by McCulloch to collect and study fishes. His first donation was a Unicorn Leatherjacket, *Aluterus monoceros*, collected off Montague Island, New South Wales. It was recorded in the Australian Museum register in June 1920.

Later, Paradice was cruising with the HMAS *Geranium* through the waters of northern Australia and had many opportunities for collecting fishes. McCulloch provided him with containers, preservative, and advice on collecting and preserving specimens. Their relationship was prospering, though at a time when McCulloch himself was in very poor health.

Paradice wrote:

> Dear Mr. McCulloch,
> In the course of our work we have discovered and named two new reefs after illustrious workers in the Zoological world.
> The new reefs are Hedley Reef and McCulloch Reef.
> Yesterday the Captain and myself fired a charge just over the edge of your reef and I collected quite a nice lot of fish.
> These will shortly be forwarded to you and I hope that they will prove of more than usual interest on account of the locality.

Paradice mentioned that he had collected a Black Kingfish (or Cobia), *Rachycentron canadum* and was sending the skin and skeleton.

> By the way we have again run out of containers so if you send back the cans they will again be used especially the large ones. And another jar of formalin would also be useful as I find that even 15% is not preserving the large fish this hot weather.

From a cruise to the Pellew Islands, Paradice sent McCulloch an account of the fishes:

> ... large hauls of fish were obtained by the seine net the cheif [sic] varieties being gar-fish, longtoms and trevally ... Large sharks were caught by line in the muddy zone ... Surrounding Pearce and Urquhart Islands where [sic] are outside the muddy zone are beautiful growing corals and among these swim government bream, parrot fishes, rock cod, and others all of which take the bait avidly. (Paradice MS, FDAM)

The fish collection from this cruise was acknowledged by McCulloch, who reported on part of it in 1924:

> A valuable collection of fishes from the Pellew Group in the Gulf of Carpentaria has been made by Surgeon-Lieutenant W.E.J. Paradice, R.A.N., of H.M.A.S. 'Geranium.' (McCulloch 1924a)

McCulloch was appreciative of Paradice's efforts and hoped to accompany the *Geranium* on a northern cruise himself. However, his depression was affecting him profoundly. In March 1925 he attempted to reply to Paradice and drafts of his letter are preserved. The handwriting is disordered and in parts hardly legible, words have been omitted and there are many cancelled words and phrases. Isolated phrases, set down as attempted draft replies, reflect McCulloch's distress:

> This, my dear sir, is the outcome of a severe nervous [blank]
>
> The worry and troubles associated with my work at the office [illegible] and I'm little more than a jangled set of nerves.

> Dear Dr. Paradice,
> The worry and difficulties of the last twelve months have got me down so completely that I am compelled to take a long rest as far away from office affairs and Trustees etc. as can be managed. I am hopeful that the old gents who run this institution will find it in their hearts to permit me to remain a member of the staff, as I greatly want to return in due course and carry on with my allotted job – the Catalogue of Australian Fishes. Should they forget my services in the past and decide to do without me in future, I shall have to break out somewhere else, just as circumstances determine. (undated MS drafts, ML)

Eventually McCulloch completed a more appropriate letter, congratulating Paradice on his Pellew Islands report, and concluding:

> I am sending the Report of the Australian Museum for 1923-4 which records your earlier collections in the Gulf district.
> Am off to the hospital tonight for a brief spell, and shall hope to write up some more of your collections, if they will allow me to get to work at them, in a few weeks time. (McCulloch to Paradice, draft, 6.3.25, ML)

On leaving the hospital McCulloch, on his doctors' advice, took extended leave from the Museum and set out to recuperate on Lord Howe Island.

McCulloch's last assistant, Gilbert Whitley: 1922–30

Gilbert Percy Whitley (1903–1975) came to Australia with his parents in 1921 and joined the staff of the Australian Museum as a cadet in 1922. He became McCulloch's assistant but the latter was absent in New Guinea during part of Whitley's first year at the Museum. He took a course in zoology at the Sydney Technology College and gained a distinction in the second year zoology examination at Sydney University in 1924 (ADB; Paxton and McGrouther 1997). This was the limit of his formal education but, as a cadet apprenticed to McCulloch, he began to understand Australian ichthyology.

McCulloch's depressive illness interfered with his ability to instruct Whitley and the young cadet had to undertake much more unsupervised work than was intended. He was enthusiastic and eager to learn, but McCulloch was aware that his supervision left much to be desired. In April 1925, McCulloch's illness became worse and he was granted 12 months leave from the Museum. He planned to spend time initially on Lord Howe Island, his favourite place, and gave Whitley instructions before he left.

From the island, McCulloch's letters gave an indication of his anxieties. He wrote a long letter to Whitley (2.5.25, ML) shortly after arriving on Lord Howe Island:

> Have thought of many things to write about while wandering around, my dear Whitley, but I am likely to forget most of them now that I have the 'portable' [typewriter] on my knee. How are things in the ichthyology department ...?

McCulloch was worried that his instructions on leaving had been inadequate:

Gilbert Whitley counting fin rays using an early stereo microscope. Courtesy of the Australian Museum V15796 Large Format Negatives Collection.

> I was so dreadfully ill when leaving that my memory would not act, and apart from the disconnected ideas I gave you, I forgot some written suggestions I had in my pocket.

He urged Whitley not to neglect the most important routines of the department:

> And the TANKS. [of preserved fish] Let their overhaul be your chief consideration, carrying on along the same lines as you did the earlier ones. Don't forget that everything you can do towards improving the arrangement of the collections will go a long way towards simplifying work upon the Australian Catalogue which I hope to write with you.

Whitley was still very new to his job and McCulloch seemed to be confident of his young assistant's enthusiasm rather than his competence:

> And now as to the Queensland list. I have thought much about this, and wonder if you have managed to complete it in time. Had I anticipated being away, I should never have started you on so difficult a task, and if you have been able to subdivide the species into families, and complete a bibliographical list, with cross references, etc., you will have done an excellent job, though one which I think is not entirely beyond you.

> Call upon Mr. Iredale[32] for assistance whenever you are in difficulties ... Only the fear of disappointing you after all your efforts causes me to allow the typescript to be submitted without my supervision before going to press. But make sure I get a proof. (McCulloch to Whitley, 2.5.25, ML)

A draft of the 'Queensland list' began as a list of the fishes of Australia, being modified to include only those recorded from Queensland. Whitley worked on it and typed a clean copy to which he referred in answer to McCulloch:

> I only finished typing the Queensland list yesterday ... I completed the first twenty-one pages of typescript and sent them to Brisbane [for the *Memoirs of the Queensland Museum*] last week, and I hope to send the remainder with your Introduction in a few days. (Whitley to McCulloch, 12.5.25, ML)

The list was published later in the year.

Whitley had been introduced to McCulloch's card index but must have appeared to rely on it excessively. McCulloch showed that he felt that Whitley's understanding of the index and its workings was imperfect:

> It is unlikely that you will fully appreciate it yet, but my decision to deprive you of the cards is quite the best thing for you as it will force you to take up the work for yourself. If we are going to work together in the future, we can as easily refer to three card indices as two, and if you select your authors carefully, you will never regret the work entailed. The very best work to begin with is Jordan's 'Classification', with a cross reference to the Genera of Fishes on each card. This will teach you a whole heap of things about which you have not dreamed so far, and you should index the fossil as well as the recent genera ... if you will follow it up with indexing Günther's Catalogue, you will have, in two works, a wonderfully good lay out of genera and species. Well, that's that. (McCulloch to Whitley, 2.5.25, ML)

Or so McCulloch hoped. He considered that Whitley needed to construct his own rudimentary card index system if he were to understand the possibilities and limitations of the major systems that McCulloch had prepared over many years. However, Whitley wrote: 'I have not yet done anything about the removal of your card index ...'; and later in the same letter, 'I still go to the Art Class, so do not get much time for card-indexing; still, I have done a little ...' (Whitley to McCulloch, 12.5.25, ML)

Beyond doubt, Whitley was busy; he remained so throughout his long professional life, and continued to expand and use his predecessor's card indices. The legacy was enormous but Whitley would surely have been a better ichthyologist if McCulloch had been able to instruct him for a longer period; insisting on the card index exercise would have been a good start.

The relationship between McCulloch and Whitley appears to have been friendly and rather informal. Whitley omitted part of his mentor's introduction to the Queensland

32 Tom Iredale (1880–1972), an ornithologist from the British Museum, came to Australia and was conchologist at the Australian Museum, 1924–44.

list, as he felt that it showed some lack of judgment. McCulloch had paid tribute to Ogilby's work on Queensland fishes but made a florid reference to his alcoholism and strokes (see page 238).

Whitley consulted another of McCulloch's protégés, Ellis Troughton, and wrote diffidently:

> Mr. Troughton thinks, and I am inclined to agree with him, that this sentence might cause offence; but if you specially desire its inclusion, I can add it in when I hear from you.

The easy relationship is reflected in the whole tone of Whitley's letter, in which there was a mixture of business and small talk:

> I went to Gunamatta Bay Fish Hatchery with Mr. Iredale and Boardman for a week-end recently. The wealth of life on the sand-flat was amazing. A stiff southerly was blowing and I quite expected to get pneumonia or something and die but, mirabile dictu, I am still alive, and it only remains for you to register still another bitter disappointment to your list.
> (Whitley to McCulloch, 12.5.25, ML)

Whitley attempted in this letter to reassure McCulloch; he showed that he had affairs of the ichthyology department well under control and that the work was progressing on several fronts.

In Whitley's obituary written for the Linnean Society of New South Wales, T.G. Vallance (1977) said:

> His pen scarcely rested. Were there truth in the dictum 'publish or perish' Gilbert Whitley ought to have been practically immortal!

He began well as junior author, with McCulloch, of the paper on flatfish, and then the Queensland list. By the end of 1930, Whitley had published 16 papers on Australian ichthyology, using six different journals. McCulloch had inspired a most energetic character and Whitley, seeing an enormous task before him, took up the instruments that McCulloch had left to hand. Not least among these was the card index which, in spite of McCulloch's caution, Whitley put to use without much inhibition.

Lord Howe Island and an invitation: 1925

McCulloch had spent two months of his 12 months leave, during which he hoped to be relieved of his depression, at Lord Howe Island. Charles Hedley, then Director of the Great Barrier Reef Committee and based in Brisbane, knew of McCulloch's illness and was concerned for him. McCulloch had known Hedley as long as he had known Waite. They had shared many field trips and a feeling for Lord Howe Island:

> My dear Allan,
> My last news of you through 'the boys' at the Museum [Whitley, McNeill, Troughton] was that already you were feeling the better for the change. Nothing ever agrees with you as well as L.H.I.

Hedley mentioned that he had taken the last part of the Queensland fish list manuscript to Brisbane. His conclusion shows his concern:

> I wish that I could spend a day or two with you loafing in the sun on a pleasant beach ... Meanwhile believe that I often think of you and hope that things go pleasantly for you. I wish that I could make them go still better.
> So with all good wishes from your old friend,
> Charles Hedley.
> (Hedley to McCulloch, 29.5.25, ML)

McCulloch answered from Sydney:

> Dear Mr. Hedley,
> ... I returned from Lord Howe a week ago on receipt of a wireless from Troughton informing me that Sir Joseph Carruthers [ex Premier of New South Wales] had squeezed £150 out of the Government towards a trip to Honolulu. I greatly regretted leaving Lord Howe as I was recovering at a fine pace, but it seemed folly to miss such an excellent opportunity of seeing Hawaii, and it is not likely that there will be overmuch work to do in connection with it ...
>
> I am delighted to hear that the Royal Society awarded you the Clarke medal. Oh how I wish these belated honours and appreciations of your work had come in time to keep you with us here, for apart from all else, I miss your friendly chats, helpful advice, and occasional rambles more than I can well tell.
> Sincerely
> (McCulloch to Hedley, draft, 22.6.25, ML)

McCulloch also wrote to another friend, Leighton Kesteven (see page 423), who was working on the skeletons of vertebrates. He gave information about a small isopod crustacean that inhabited the shallow waters around Lord Howe Island and could be used to clean soft tissues from bones. He incidentally discussed his situation and plans:

> I leave by the Aorangi with Sir Joseph Carruthers to attend the fisheries conference at Hawaii. Maybe you know that I am on twelve months leave of absence ...

He mentioned that the leave was due to his 27 years of service to the Museum and that he has consulted doctors who have advised rest.

> So I'm paying half my expenses and the Govt. the remainder. If luck holds, I hope to return by way of Fiji, New Caledonia, Malekula in the New Hebrides, Norfolk Island, and then a month or so at Lord Howe again before returning to the office. But Hurley may go to Macquarie Island and the Ice Barrier this summer to make a drama out of the penguins and another triangular love story for the movies, in which case I'm very hopeful of going with him ...
> (McCulloch to Kesteven, 27.6.25, ML)

Hurley had sent the proofs of his book on their travels in New Guinea; McCulloch was delighted with it and with the invitation to participate in another expedition. McCulloch hurried back to Sydney, where he stayed only briefly, and embarked for Honolulu via New Zealand, Fiji and the Phoenix Islands.

Brief diary and McCulloch's death: 1925

Allan McCulloch kept a personal diary from 7.7.25 to 3.8.25; this is held in the Mitchell Library, Sydney. He explained:

> ... keeping a diary proves, in my case at any rate, to be much more a matter of intent than execution ... the copious details recorded by Frank Hurley in his Papuan diaries ... taught me to value the written record of impressions made while the subjects they are based upon remain fresh in one's mind ... and remembering the fun the reading of his daily entries afforded ... during our long isolation in the lonely Fly River, I venture to write down things just as they drift into my mind in the belief that their re-reading may someday afford me renewed interests in the happenings of today.
> (McCulloch diary, 7.7.25, ML)

His ship was in New Zealand when he began the diary and he spent some time with the Director of the museum at Auckland.

An examination of the diary gives some insight into McCulloch's behaviour towards the end of his life. It shows some of the symptoms of his illness as it gained power over him, but also moments of humour and hints of his aspirations for the future.

In Suva, he observed some conflict between the Fijians and Indians and wondered about the influence of the west. He reflected upon his experiences in Papua and his observations in Fiji and questioned the separation inherent in the 'White Australia' policy:

> But looking the matter squarely in the face, I am led to wonder if such a scheme is fair, practicable, or even possible. (McCulloch diary, 12.7.25)

He made the acquaintance of a variety of people aboard the vessel, including an engineer who had been a fellow student with the German engineer, Rudolph Diesel. The man claimed that 'poor Diesel committed suicide by jumping off the Dover to Calais steamer because he foresaw naught but failure ahead for his invention.' (McCulloch diary, 14.7.25)

McCulloch seemed unable to relax and his diary entries show that he contributed to the entertainments on board on several occasions; this included giving a lecture on 'New Guinea and some coral islands'. He took the performance seriously, illustrating the talk with lantern slides:

> The usual lantern troubles occurred but I was able to talk and joke through them, having that family characteristic – a gift o' the gab, which proves so useful at times. (McCulloch diary, 15.7.25)

He was jubilant in success:

> I may smile at myself later when reading over these writings, but at the moment, after evading a positive storm of congratulations from all and sundry, I cannot help feeling grateful for that collusion of kinks which enables me to entertain my fellow men and women, young and old alike, to the extent that they are glad, genuinely glad to tell me afterwards of the pleasure I have afforded them. Well, thanks for that to such good friends as Frank Hurley, who have taught me the art of photography ... (McCulloch diary, 15.7.25)

The next day, as they neared Hawaii, McCulloch visited the engine room, and later attended a concert. He found that a New York singer, Kate Rooney, was aboard. They had been neighbours in 1896, she learning singing while the 11-year-old McCulloch and his sisters had piano practice 'which Mother insisted upon as a preliminary to breakfast.' He practised on the grand piano in the music room on board and eventually accompanied 'Madame Rooney' in Tosti's 'Goodbye.' Again, the diary entry is extravagant.

On arrival at Honolulu, McCulloch's optimistic mood was destroyed. The fisheries conference was being organised by Mr Alexander Hume Ford, Director of the Pan-Pacific Union. McCulloch had expected the conference, which he believed would be attended by the leading American ichthyologists of the day, to be imminent. He had been misled:

> I learnt that the Fisheries Conference, which I had supposed to be an event fixed for July, was largely 'in the air', and that no others interested in fishes were present save myself. Mr. Ford did not even know I was coming, and Professor Jordan and Dr. Evermann were not expected till September, if then. That settled it, and I readily decided to return upon the Aorangi, sailing in three weeks time. Ford assures me he will arrange for the others, including Dr. Nishikawa of Japan, to be here if I will only wait ... A sorry muddle, and much bad temper. (McCulloch diary, 17.7.25)

He saw the sights and attended official functions. He gave a public lecture on 'Head Hunters' at the Pan-Pacific Research Institution. The preparation for the lecture was complicated by his lantern slides being held up by the customs, then by the non-appearance of the 'lanternist'. A substitute was found and instructed, and the lecture was a great success.

Problems associated with the conference and the lecture upset McCulloch, 'and the strain and nervous tension generally was sufficient to keep me awake for many hours, and throughout today I have been far from well as a result thereof.' He was also disturbed by the inaccuracies of the press:

> 'Australian Explorer'. Could anything be more absurd? And again the title of Dr., which, despite all my disclaimers is thrust upon me whenever my name is either printed or spoken. I wish, most sincerely, that I could lay claim to it, which makes it all the more tantalising to have to deny the right. But I turn it down on every possible occasion, a circumstance which led a prominent journalist whom I so corrected to say 'Waal you ought to be in the museum you come from, for yure the firrst man of science I've met heere who ain't a

> perfessor or daacter of some kind or other.' Maybe the lack of a degree is a distinction? (McCulloch diary, 25.7.25)

He wandered about Honolulu, looked at the dance halls, watched the surfers, the hula dancers; but the sense of strain is evident frequently in his diary entries. He appears to have been loosely attached to Sir Joseph Carruthers' official party, but had some disagreements with them, and did not approve of their haughty treatment of the hotel staff. The diary has many unexplained blank pages. He was still sleeping poorly, 'Sleep, as a restful proceeding seems to have deserted me, and is the outcome of exhaustion only.' (McCulloch diary, 3.8.25)

The last entry in the diary recounts a riding excursion which McCulloch found interesting; and it had an amusing side:

> Two of the dames had just no idea whatever of even using the reins, and when their steeds stopped, as they frequently did without obvious reason, they simply waved their arms, implored the horses to move on, and usually remained helpless till one of us rode back to their assistance. (McCulloch diary, 3.8.25)

The remainder of the notebook is blank.

McCulloch's behaviour at the Conference, which finally took place in August 1925, attracted only favourable attention. He left articles for publication, posing 'Questions to be considered by the Pan-Pacific Fisheries Congress'. He referred to the depletion of Pacific fisheries and quoted David Starr Jordan who had written that the life histories of the animals of the region urgently required study, summarising Jordan's views: the 'results of such studies should form the basis of international treaties for animal conservation.' McCulloch promoted 'the need for a Fisheries Training Centre.' (McCulloch 1925c) He also left the description of a flight over Oahu, with comments on the appearance of the coral reefs, and this was published as *Allan R. McCulloch's Last Article* (McCulloch 1925d). His friends and colleagues were shocked when it was discovered, on the first of September, that he had shot himself in his hotel room.

Reactions to McCulloch's death

The Australian Museum's Director, C. Anderson (1926), prepared an obituary in which he attributed McCulloch's death to overwork in the cause of science. Although suitable for the obituary, this interpretation omits reference to the depressive illness which had affected McCulloch for many years. McCulloch's ashes were brought to Lord Howe Island in September 1925, and a memorial erected.

David Jordan had described McCulloch as 'one of the most accurate workers in systematic ichthyology now living'. In recommending McCulloch to Alexander Hume Ford of the Pan-Pacific Union, Jordan wrote:

> He is not only the best authority in Australia, but one of the six or eight best in the world, so far as fishes are concerned. He is a good friend of mine and I esteem him and his work very highly. (Jordan to Ford, 17.7.25, ML)

He wrote that McCulloch's premature death represented a loss to the study of ichthyology.

When McCulloch killed himself in Honolulu only a couple of months after their last friendly exchange of letters, Waite was distressed. He wrote:

> Geisler [aquarist friend] rang me and asked if I had seen the report of McCulloch's death in an hotel at Honolulu. As I had not it proved quite a shock, but it was not until I searched the newspaper that I learned what a tragic death it was. (WD, 3.9.25)

About six weeks later, presumably with an interesting mixture of motives, Waite explored the possibility of utilising McCulloch's preparations for what was to have been a major work:

> Wrote Anderson, Sydney re McCulloch's MS. Cat. of Australian fishes. (WD, 15.10.25)

Whitley was already involved with the work, and it would have been common knowledge at the Australian Museum that McCulloch would not have wanted Waite to have the benefit of his work.

Waite was upset when he found that Regan had published a confused note about McCulloch in *Nature*. His work and his age were wrong and Waite protested: 'Wrote to Tate Regan re his note in Nature ...' (WD, 9.12.25) He included a note of correction, which was duly published in March 1926.

Returning from an exhausting and only partially satisfactory trip to Europe and North America, Waite found himself in Honolulu for a few hours, and wrote in his diary:

> Several times I thought of Allan McCulloch who shot himself here, Aug. last year. (WD, 18.10.26)

Posthumous publications: 1926–1930

In the year following McCulloch's death, the last (eighth) of his *Studies in Australian Fishes* was published (McCulloch 1926a), as was the last (fifth) of his *Endeavour* Reports (McCulloch 1926b).

Gilbert Whitley played an important part in preparing McCulloch's last works for the press (see colour plate 14). The most important of these was *A Check-list of the Fishes Recorded from Australia* (McCulloch, 1929–30), published in 4 parts. McCulloch had used his card indices to prepare a list of the species of fish recorded from Australia's marine and fresh waters. He planned to include descriptions, synonymies and illustrations of every Australian fish. After McCulloch's death, Whitley completed the list so that it included each species recorded up to 1929 and added distribution records. The final publication provided the starting point for taxonomic research for decades and is still an essential reference.

McCulloch's *Check-list of the Fishes and Fish-like Animals of New South Wales* remained under Whitley's care through new editions in 1927 and 1934, new records for the state being included in each.

McCulloch's contributions to ichthyology

Allan McCulloch made a greater contribution to Australian ichthyology than either Ogilby or Waite. He was a dedicated and talented worker, and he had some remarkable advantages. Ogilby began to work on Australian fishes in 1885, when he was 32; Waite in about 1895, at 29. McCulloch's introduction to Australian ichthyology was in 1898: he was 13.

Ogilby pioneered the rational approach to Australian fish taxonomy and was the first to see just what needed to be done. His most useful guidance came from correspondence with David Jordan and he understood the necessity of familiarising himself with the latest literature on fish classification, particularly with the works of Theodore Gill and George Boulenger. Waite did not see taxonomy in quite the same light, not having any urgent sense of the need for a comprehensive work on Australian fishes.

McCulloch was fully engaged with the problem by the time Waite left for New Zealand. At 20, he had a variety of responsibilities at the Museum but his priority soon became, and remained, the production of a work on the fishes of Australia. He had some great advantages:

- Useful preliminary work had been done by Ogilby.
- McCulloch began his work at an early age and had excellent instruction from Waite.
- He had access to substantial collections, and often to large series of fresh specimens.
- He was a talented artist.
- Partly through McCulloch's own efforts, Australian Museum curators had begun to appreciate the significance of type specimens.
- Jordan and his colleagues were helping to formulate, and were applying, rational rules of nomenclature.
- He had assistance from Ogilby in Brisbane, the repository of many of the types of the de Vis species. Later, Waite was able to help from his base in Adelaide.

McCulloch also had disadvantages:

- The size and complexity of the problem.
- Distractions in the form of other Museum work.
- Professional isolation: he had only Ogilby upon whom he could rely; he perceived problems clearly, but could rarely delegate tasks to competent assistants.
- The impact of staff and materials shortages due to the war, and his own efforts to enlist.
- The common problems of limited access to the literature and type specimens in overseas institutions.
- His depressive illness.

McCulloch's relationship with Waite was enduring and complex. C.P. Snow described clearly:

> ... the strain that comes between a protégé and his patron - where all emotion is ambivalent, unless both parties are magnanimous beyond the human limits: if they are ordinary humans, there is the demand for gratitude on one side, resentment on the other, and those forces must drive them further apart. (Snow 1947)

There is abundant documentary evidence of the contradictory feelings at work in this relationship and of the truth of Snow's insight. Positive facts remain: Waite was an effective mentor and McCulloch an apt apprentice; they cooperated professionally for 25 years; and the published work of each was often complemented by the influence of the other.

McCulloch resolved large numbers of troublesome synonymies and made complete and accurate descriptions of many species, revising groups and constructing keys to facilitate identification and, wherever practicable, providing illustrations of the species that had previously been unfigured. This work allowed him to produce a handbook of New South Wales fishes and was essential to the production of a similar work on the fishes of South Australia; Tasmanian and Queensland lists were also facilitated by his work. The card indices were used to good effect by his last assistant, Gilbert Whitley, who published the references derived therefrom as a check-list of the fishes of Australia. Much later Whitley published a *List of the Fishes of Western Australia* (1948), based largely on McCulloch's hand-written database.

McCulloch's work was meticulous and his insight into the relationships within groups led to a rational nomenclature at generic and specific levels. The state of Australian ichthyology improved enormously in his hands from 1907 to 1925. His successor, Gilbert Whitley, would presumably have developed a less idiosyncratic approach if he had been subjected to McCulloch's discipline for longer. It is impossible to study the changes that McCulloch wrought without wishing that he had been able to apply himself to the problems for another decade or so.

Contemporaries of Ogilby, Waite and McCulloch

Morton

Alexander Morton (1854–1907) was born in Louisiana and came to Queensland when his father emigrated. He studied natural history and was employed by the Australian Museum in 1877, collecting at various times in north Queensland, around Darwin, New Guinea, the Solomon Islands and Lord Howe Island.

In 1884 Morton was appointed curator of the natural history museum in Hobart and became Director of the Tasmanian Museum and Botanic Gardens. He helped to establish the Queen Victoria Museum in Launceston where he was honorary curator from 1891 to 1896. He travelled between the two museums, which had few staff, organising both institutions and interesting himself in all aspects of acquisition, curation and administration.

Morton described several new species of fish, working with J. D. Ogilby, to whom he sent large numbers of specimens. Ogilby's description of the Prickly Dogfish, *Oxynotus bruniensis* was based upon a specimen sent to the Australian Museum by Morton (Ogilby 1893b). Morton later sent a specimen of the Spinycoat Anglerfish from Tasmania's east coast directly to Ogilby who described it as *Antennarius* [now *Trichophryne*] *mitchellii*, sending the description to Morton for him to publish (Morton 1897).

Saville-Kent

William Saville-Kent (1845–1908) was born William Savill Kent in Devon, England. He studied biology and laboratory techniques under Thomas Huxley and obtained work at the British Museum under Richard Owen, then superintendent of the natural history collections, in 1868. At this time he was set to work on the corals; many years later, in the preface to his monumental work on Australia's Great Barrier Reef, he wrote

> A day-dream of the author, when attached, in years gone by, to the Natural History Department of the British Museum, and for the nonce engaged in the arrangement and nomenclature of Madrepores or Stony-corals in that Institution, was to be afforded an opportunity of seeing those organisms growing in their native seas and in their wonderful living tints. (Saville-Kent 1893)

In the 1870s he had the management of large public aquaria, first at Brighton and later Manchester, where he made observations on the behaviour of a variety of marine animals. He hoped to establish in England a counterpart to the Naples marine research station to advance the study of marine biology in his own country, but was unable to accomplish this (Harrison 1997).

Among Saville-Kent's main interests in marine biology were corals, which he had studied in the British Museum collections, fish from his work with aquaria, and protozoa. He was an experienced microscopist and published the ambitious *Manual of the Infusoria* (1880–82). He had also studied marine fisheries and the culture of oysters.

William Saville-Kent (1845–1908) demonstrating his method of photography at Rocky Island on the Great Barrier Reef. From Saville-Kent 1893.

Through Thomas Huxley's recommendation, Saville-Kent was offered an appointment with the Tasmanian Government to establish aquaculture of oysters and fish. He was successful with the oysters but had an unfortunate enthusiasm for acclimatisation. Trout were already established but Saville-Kent improved the culture techniques and attempted the aquaculture of some native Tasmanian fishes as well. He also introduced more efficient fishing methods to the commercial fisheries in Tasmania, based on what he had learned in England.

During his time in Tasmania, from 1884 to 1886, Saville-Kent published a report on the fisheries of Tasmania (1884, 1885) and several papers on the native fishes, including the description of a Tasmanian Weedfish, *Clinus* [now *Heteroclinus*] *johnstoni* (Saville-Kent, 1886).

He was obliged to leave Tasmania due to a controversy that involved the so-called 'Salmon Commissioners.' This was an influential group who claimed to have successfully introduced the Atlantic salmon to Tasmanian waters. Their claim was ill-founded, as Saville-Kent pointed out, being based upon the misidentification of large trout (which *had* been successfully introduced).

After leaving Tasmania, Saville-Kent was employed for a few months by the government of Victoria to produce reports on the fisheries. He participated in the surveying cruise of HMS *Myrmidon* in north-western Australia in 1888 and wrote an account (Saville-Kent 1889a). This included the description of three new species of fish, all of which are still recognised.

Saville-Kent was appointed Commissioner of Fisheries for Queensland in 1889, a position he held for three years. During this time he accomplished a remarkable volume of work: he assessed the fisheries, continued his work on oyster culture and described many supposedly new species of fish. He published a preliminary report on the commercial fishes, *Food Fishes of Queensland* (1889b), illustrated by his own line drawings. He was also supplied with some specimens of 'Barramundi' by Frank Jardine of Somerset Station in Torres Strait and this proved to be a new species of Saratoga, *Osteoglossum* [now *Scleropages*] *jardinii* (Saville-Kent 1892).

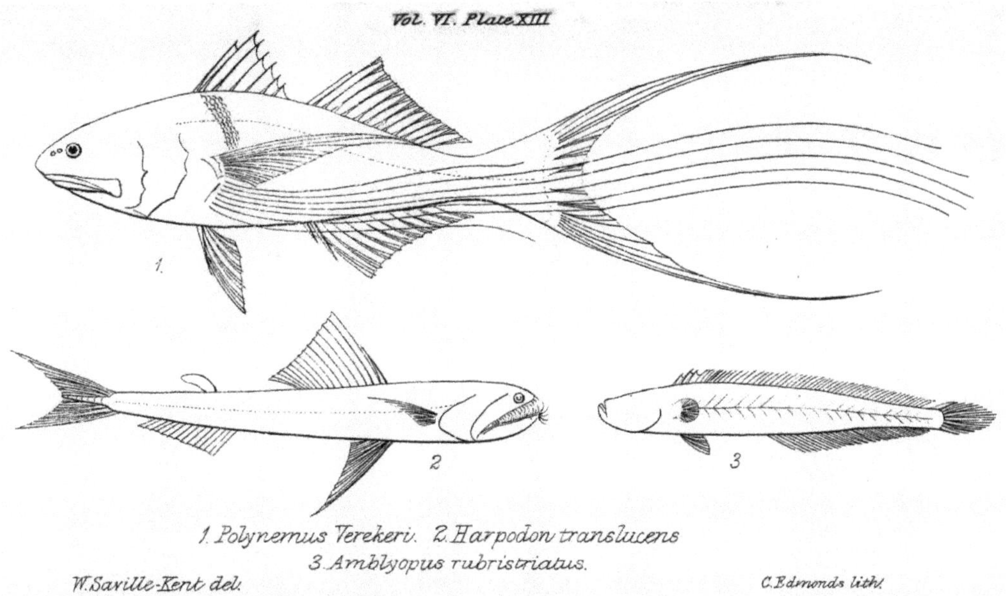

Saville-Kent described the Streamer Threadfin, *Polynemus* [now *Parapolynemus*] *verekeri*, Glassy Bombay Duck, *Harpadon translucens* and Red Eelgoby, *Amblyopus* [now *Caragobius*] *rubristriatus*. His drawings, although crude, were useful. From Saville-Kent 1889.

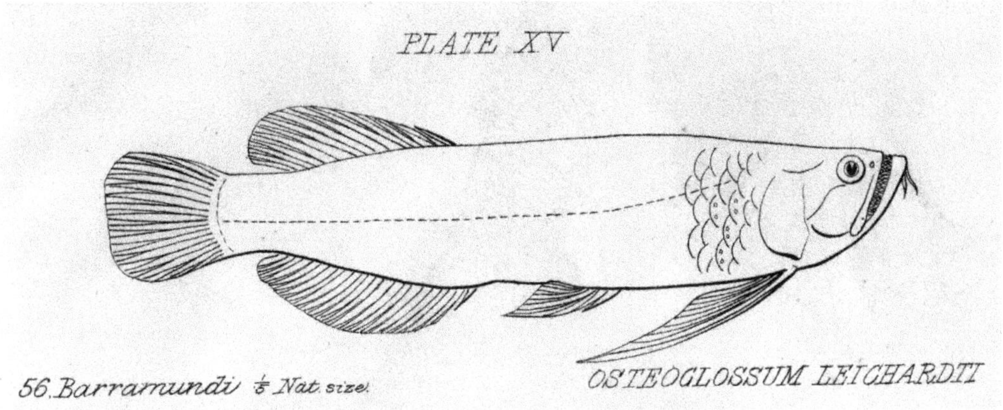

As Commissioner of Fisheries for Queensland Saville-Kent published a report on the commercial fishes, illustrated by his own line drawings, including a drawing of the Southern Saratoga, *Osteoglossum* [now *Scleropages*] *leichhardti*. Frank Jardine of Somerset Station in Torres Strait sent him a new Barramundi which he described in 1892 as *Osteoglossum* [now *Scleropages*] *jardinii*. From Saville-Kent 1889.

In Queensland, Saville-Kent had the opportunity to realise his 20-year daydream and he made the most of it. He conducted scientific explorations of Queensland's coral reefs, and produced a major illustrated work, *The Great Barrier Reef of Australia; its Products and Potentialities* (1893). Under the most trying circumstances he took large numbers of photographs (most originally as 'stereo pairs') and reproduced them in his book. He explored, collected and photographed at many sites on the reef, from Thursday Island in the north to Lady Elliot Island in the south. In this publication he discussed Queensland fishes in a chapter entitled 'Food and Fancy Fishes' and appended a 'classified list of Queensland food fishes'. The work was illustrated by many of his own photographs and he presented his own drawings of 20 species of the magnificently coloured reef fishes.

Saville-Kent (1893) published a number of new scientific names for Queensland fishes, some of which were associated with neither descriptions nor illustrations, most often based upon Queensland Museum specimens that Charles de Vis had labelled. Ogilby, McCulloch and Waite agreed that these scientific names should be regarded as *nomina nuda*, lacking status in formal nomenclature (see page 340). Those that have since been defined have proven to be synonymous with species that had been previously described.

At the termination of his contract with the Queensland Government, Saville-Kent moved to Western Australia where he again worked on fisheries assessments and acclimatisation of northern hemisphere and eastern Australian species in the west. He returned to England in 1895 and wrote a well-known popular account of Australian natural history (Saville-Kent 1897).

Saville-Kent contributed much to the establishment of fisheries science in Australia and presented information that was valuable to both the general public and marine biologists in his publications, particularly that on the Great Barrier Reef.

Tosh

James R. Tosh occupied the position of Government Marine Biologist in Queensland during the early part of Ogilby's residence there, and relied on Ogilby to identify the fishes that he studied. He sent large collections of tropical marine fishes to Ogilby.

Tosh published two papers on the fishes of Queensland. *On the Common Whiting of Moreton Bay* (Tosh 1902) included a study of the embryology of the species, nicely illustrated by the author. Ogilby, placing too much reliance on the silvery streak common to many species of whiting, misidentified the Moreton Bay Whiting as *Sillago bassensis* (it proved to be *S. ciliata*). In this paper there was also a lively account, part of which is quoted below, of the whiting's habit of sand diving to escape a threat:

> The dive can be executed with great rapidity and is a most serviceable accomplishment. When fishing for whiting with a seine net one can observe as the bunt of the net nears the shore here and there a small cloud of sand thrown up; the fisherman marks the place, and when his net is in, wades out and feels about in the sand with his feet; when a fish moves under his foot he stamps his

> foot down to hold it, and then picks it up with the haul. Often as many as a dozen fish are so taken which had otherwise escaped the foot-rope of the net. Very small whiting, an inch and a half long, have the trick. When burying the whiting throws up its tail, and actually takes a header into the sand using its tail fin vigorously ... (Tosh 1902)

Notes on the Habits, Development, etc., of the Common Food Fishes of Moreton Bay (Tosh 1903) was an account of a more ambitious project. The principal interest of this work lay in his account of the reproductive biology, and particularly the embryology, of the 'Tygum Mullet', *Liza argentea* and Bream, *Chrysophrys* [now *Acanthopagrus*] *australis*. Once again, Tosh's drawings of the fishes (30 species) and the developing embryos were carefully executed.

Tosh became Assistant Professor of Zoology at the St Andrews University in Scotland. He died in Mesopotamia during the First World War.

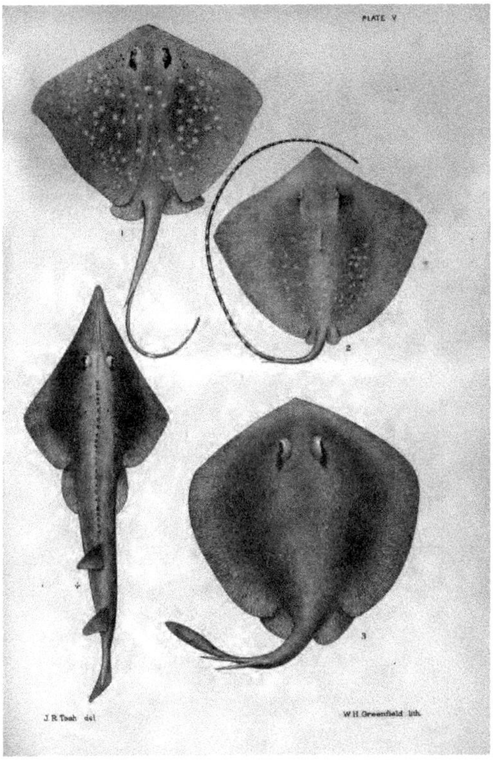

James Tosh drew some of the fishes of Queensland, including rays of Moreton Bay, relying on Ogilby to identify them: *Dasyatis* [now *Neotrygon*] *kuhlii* (Müller and Henle, 1841) (top left); *Dasyatis uarnak* (later described as a new species, *Himantura toshi* (Whitley, 1939) (top right); *Urolophus testaceus* [now *Trygonoptera testacea* Müller and Henle, 1841] (bottom right); *Rhinobatus rhinobatos* [now *Glaucostegus typus* (Bennett, 1830)] (bottom left). From Tosh 1903, courtesy of the Queensland Museum.

Marshall

Tom C. Marshall (1896–1976) joined the Queensland Museum as a cadet in 1910, where he had the advantage of contact with J.D. Ogilby. While working at the museum Marshall took a course in art at Brisbane's Central Technical College. He prepared fine casts of fish and other specimens, painting them himself, and assembled imaginative displays that proved very popular at the museum. In the late 1930s Marshall made films on marine life of the Great Barrier Reef.

Illness obliged Ogilby to cease work in about 1920 after which Marshall was the de facto ichthyologist for Queensland. He relied on first Allan McCulloch, and later Gilbert Whitley, for help with fish taxonomy. While still with the Queensland Museum he published several brief papers between 1925 and 1928, noting species that were new records for Queensland. His preoccupation with museum displays was evident in the notes on a two metre Ribbon-fish, *Trachipterus jacksonensis* washed ashore in 1921:

> When this Ribbon-fish was received at night by the Director (Mr. Heber Longman), its silvery surface was glowing with phosphorescence and it formed a remarkable object. In order to preserve it satisfactorily it was cut into two portions. Subsequently these were united and the specimen was placed on exhibition in formalin solution in a neatly-made cement tank with the exterior blackened and the interior enamelled white. As it was found impracticable to place a glass cover on this cement tank, the whole was enclosed in a large case with other specimens and casts. Except for the occasional need to replenish the formalin lost through evaporation, the specimen requires little attention and forms a popular exhibit. (Marshall 1925)

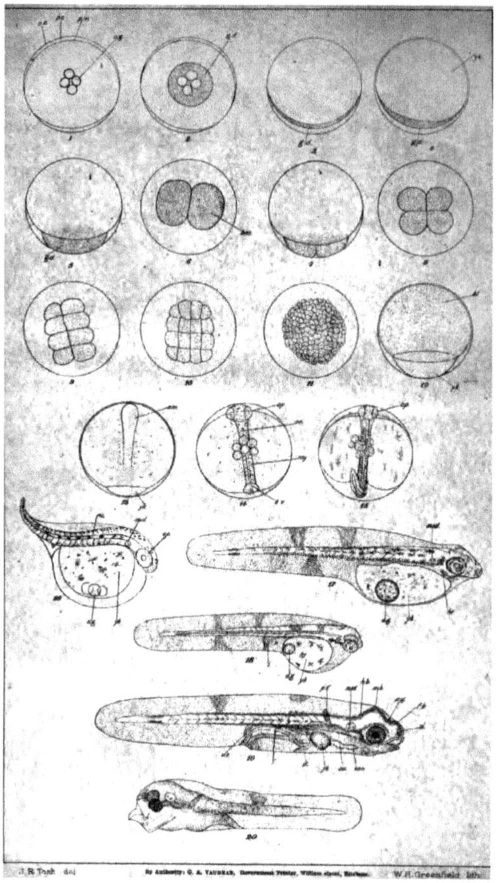
The development of the 'Tygum Mullet', *Liza peronii* [now *Liza argentea* (Quoy & Gaimard, 1825)] from the egg two hours after fertilisation to a larva one week old. From Tosh 1903, courtesy of the Queensland Museum.

In 1942 Marshall transferred to the Department of Harbours and Marine in Brisbane, working as an ichthyologist in the fisheries branch of that department. Between 1941 and 1957 he continued occasional publications recording many new records for Queensland. During this time specimens that came to the museum were referred to him and included in his publications. The last of this series of publications was the best organised and most significant and in it Marshall recorded many new records for Australia and Queensland, once again thanking Gilbert Whitley for his help. The best known of Marshall's works, *Fishes of the Great Barrier Reef and Coastal Waters of Queensland* (1964), was published two years after his retirement: a beautiful book, illustrated in part by paintings executed by his friend George Coates.

Lucas

Arthur Lucas (1853–1936) was born at Stratford-on-Avon (his given names were Arthur Henry Shakespeare). He studied at Oxford and London Universities and subsequently began a medical course. Lucas, an accomplished scholar, came to Australia in 1883 where he taught science in Melbourne and, as a member of the Senate of the University of Melbourne, promoted science in school and university curricula (ADB).

Lucas had an enduring interest in biology and wrote on botanical and zoological topics. He collected some rockpool fishes and, in 1889, also attempted a 'census' of the fishes recorded from Victoria up to that time (Lucas 1890). He had an exaggerated respect for the work of Count Castelnau, writing that 'The Count's labours have made the work of those who follow him mainly of a supplementary nature'. However, the 'census' established a basic list with some references, local names, and an indication of distribution. The following year he described new species, some of which are still recognised (Lucas 1891). In this paper he described the common Port Phillip Pipefish, *Syngnathus* [now *Vanacampus*] *phillipi.* James Sherrard[33], in compiling his account of the Melbourne 'Aquarium, Picture Salon, Cyclorama. Museum, and Technological Collections', reprinted and illustrated Lucas' 'census' of Victorian fish (Sherrard 1900).

Arthur Lucas. From Walkom 1925.

Lucas moved to Sydney to be headmaster at Newington College in 1892 and taught until his retirement at the age of 70. In retirement he occupied himself largely by working in the neglected field of Australian marine algae; his papers on these organisms appeared from 1909 to the year of his death, in 1936, when the first part of *The Seaweeds of South Australia* (dealing with the green and brown algae) was published. Part two of this work (on the red algae) was published posthumously after completion by his co-worker, Mrs Florence Perrin (Lucas and Perrin 1947).

Radiograph of an Old Wife, *Enoplosus armatus* included in Sherrard's *Illustrated Handbook*. From Sherrard 1900.

33 Many of the illustrations in Sherrard's pamphlet were labelled with hilarious inaccuracy: his 'hammerheaded shark' was an Elephant Fish, *Callorhinchus milii*; his 'ludrick' a Barber Perch, *Caesioperca rasor* and so on. However, the publication was made more interesting by the inclusion of illustrations 'taken by the Röntgen ray process'; the radiograph of an Old Wife, *Enoplosus armatus* was particularly well reproduced.

Spencer

Walter Baldwin Spencer in Alice Springs in 1901, taken during a 12-month anthropological expedition in Central Australia conducted with his friend, Frank Gillen. Courtesy of Museum Victoria, Indigenous Collections.

Walter Baldwin Spencer (1860–1929) was born in Lancashire, England. He studied at the Manchester School of Art and took science at Oxford. Professor H.N. Moseley was among his lecturers and apparently encouraged Spencer to do his research in Australia (ADB). He was primarily a biologist although he is now best known for his innovative research in anthropology. He came to Australia in 1887 to occupy the newly created chair of biology at Melbourne University.

In 1899 Spencer became Honorary Director of the National Museum of Victoria, bringing his administrative experience to bear on the museum's many problems and working on the ethnographical collection. In 1894 he led the Horn Expedition to the MacDonnell Ranges (see pages 246 and 292). He and his colleagues examined the geology, botany, zoology and anthropology of the region and Spencer personally edited the four volumes of reports. From his experiences with the aboriginal people whom he met, and from his association with Francis Gillen, which dates from the expedition, Spencer was led to study many aspects of aboriginal life and culture. Between them, Spencer and Gillen established a new approach to anthropology. Spencer retired in 1919 and returned to England in 1927. He died in 1929 while carrying out anthropological field work in Tierra del Fuego (Mulvaney *et al.* 1997).

Baldwin Spencer published several papers on fishes. His early work on the pineal eye of reptiles led him to examine its histological structure in the Australian Shorthead Lamprey, *Mordacia mordax* (Spencer 1890). He concluded:

> There can be little doubt that here, as I believe, in all other animals in which it is now found, the Pineal eye must, as an organ for the perception of rays of light, be regarded as rudimentary and functionless.

He demonstrated the presence of a retina but appeared to base his conclusion upon the presence of opaque tissues above the 'eye', perhaps misled by the state of preservation of the animal.

Spencer published notes on the habits of the Australian Lungfish, derived from his observations on a field trip to the Burnett River in 1881. He had hoped to obtain embryos but the season was too advanced. Spencer dissected the fish he collected, observing that '... *Ceratodus* seems to live principally, if not entirely, on vegetable matter. The alimentary canals of those which I opened in late September and early October, being filled with the fruit of the gum tree *Eucalyptus tereticornis* which overhangs the river banks.' (Spencer 1892) The Australian Lungfish had already been the subject of numerous papers, apart from the original description of Krefft (1870) and Günther's major work (1871). The latter included a description of the animal's circulatory system but Spencer carried out dissections which made some aspects clearer. He compared the arrangement of vessels with that found in amphibia and elasmobranchs, and drew a series of clear diagrams (Spencer 1893) (see colour plate 15).

Spencer's involvement with the Horn Expedition included making his own field notes on the habits of fishes which he dispatched, together with the expedition's collection of fishes, to A.H. Zietz at the South Australian Museum (see page 291).

Spencer (1899) published several brief studies of parasites, notably a re-description of a spectacular intestinal parasite from the Elephant Fish, *Callorhinchus milii.* This tapeworm, now identified as *Gyrocotyle rugosa*, was minutely described and beautifully illustrated by Spencer (Arundel *et al.* 2009).

University zoology courses generally included, and still include, some knowledge of the lancelet, a marine, fish-like chordate without vertebrae. In Melbourne, Spencer utilised a local species and encouraged Ethel Remfrey Morris and Janet Raff to study its structure (Morris and Raff 1909). Raff later described a second distinctive species of lancelet trawled by the *Endeavour* (Raff 1912).

Kershaw

James Andrew Kershaw (1866–1946) of Melbourne became an assistant at the National Museum of Victoria at the age of 17. Initially he worked with his father, a taxidermist and preparator, and later became assistant to Frederick McCoy, the Director of the museum, from whom he learned zoology and museum administration. When McCoy died in 1899, Baldwin Spencer was appointed Honorary Director and Kershaw became curator of zoology. In 1929 Kershaw himself became Director of the museum (ADB).

He had wide interests in zoology, was particularly attracted to entomology, and undertook field trips for the museum. He dealt with the housekeeping aspects of ichthyology, maintaining the collections, answering queries and reporting new records for the state. He used the *Victorian Naturalist* for a series entitled *Some Additions to the Fish Fauna of Victoria* (1906 to 1928). In the introduction to the first of these publications he summarised the earlier work on Victorian fishes. The museum collection clearly contained a large number of unidentified fishes at this time and Kershaw had difficulty with them:

> During the last few years I have, as opportunity permitted, devoted some of my time to the examination of Victorian fishes in the National Museum, with the view to the identification of many of the species not yet determined. Owing, however, to the fact that such opportunities have been few and far between, and even then have been much interrupted with other official duties, I have, so far, made but little progress. (Kershaw 1906).

He was hampered by the confusing literature, lack of advice, lack of experience with fish taxonomy and, as he said, little time to spare from his other duties.

Kershaw gave a rather circumstantial description of the collection of one of the more spectacular new fish:

> ... Mr. W.H. Baldwin, while riding along the shore about 20 miles east of Apollo Bay, noticed what appeared to be a strange fish floundering about in the shallow water. On dismounting, he found it to be a fish about 4 feet long, with an unusually square-shaped head, surmounted by a long, erect spine, and large and slightly prominent eyes, giving to it a rather fierce appearance.
>
> Being afraid to handle it, for fear, as he explained, of being poisoned, he endeavoured to land it by means of a stockwhip he was carrying. Finding the animal altogether too lively, however, he secured a net, by means of which he succeeded in capturing it without serious injury. (Kershaw 1909)

Duly identifying the fish as the Crested Bandfish, *Lophotes cristatus* [now *Lophotus guntheri*], he gave a careful description and added a more interesting comment from the collector:

> According to Mr. Baldwin, the fish, while being secured, ejected a dark, inky fluid from the vent, which discoloured the water all around it. (op. cit.)

Thus an early description of the presence of an ink sac, a characteristic of the family (Lophotidae) and apparently a defensive mechanism, comparable in function to that of squid and octopi.

Hale

Herbert M. Hale (1895–1963) trained as an accountant but his enthusiasm for natural history led him to apply for work at the South Australian Museum. He became Edgar Waite's assistant in 1914, shortly after Waite came to Adelaide from New Zealand (see page 299). From Waite, Hale received instruction in scientific illustration, photography and zoology in general. Hale was promoted to the curatorship of the South Australian Museum after Waite's death in 1928, then Director a few years later, the position he held until he retired in 1960.

He began publishing observations made while operating his aquaria, an enthusiasm he shared with Waite, in 1918. This included information on native marine and freshwater fishes, published in *Aquatic Life*, Philadelphia. By 1924 he was publishing notes on South Australian crustacea, and these animals remained the focus of his research throughout his life. While still working with Waite, he prepared *Crustaceans of South Australia* (Hale 1927–29), a work that remained the essential text on the subject for nearly 80 years, and which is still consulted regularly.

Under Waite's guidance, Hale contributed to the preparation of the *Catalogue of the Fishes of South Australia* (Waite 1923) and other publications, and co-authored the *Review of the Lophobranchiate Fishes of South Australia* (Waite and Hale 1921). He collected fish on a trip to Princess Charlotte Bay in North Queensland in 1927 and these were eventually reported by Whitley (1935). After Waite's death, Hale published occasional papers on fish. He described the egg-case of the Gulf Cat Shark, *Asymbolus vincenti* (Hale 1935); noted new distribution records for fishes in South Australia (Hale 1939, 1944, 1947); described mouth breeding in the Southern Cardinalfish, *Vincentia conspersa* (Hale 1947); and described two supposedly new species of pipefish at different times (Hale 1939 and 1941), both of which proved to be synonymous with *Histiogamphelus cristatus* (Macleay 1882).

Hale's history of the South Australian Museum is a very useful account of South Australian zoology and incidentally demonstrates the breadth of his interests in natural history (Hale 1956).

Cleland and Duhig

John Burton Cleland (1878–1971), born in Adelaide, studied medicine in Adelaide and Sydney. He worked as a microbiologist in Western Australia and New South Wales, returning to Adelaide in 1920. Although professionally he was a microbiologist and pathologist, Cleland was also active in studying anthropology, ornithology and mycology.

While employed by the Bureau of Microbiology in Sydney he wrote on animals harmful to man, including a generous section on fishes. This review included shark attacks, injuries from stingrays, envenomation by catfish and scorpionfish (including the stonefish) and poisoning by ingestion. His own researches included the preparation of tissue sections from near the base of the spine of the Estuary Catfish, *Cnidoglanis macrocephalus.* He described '... a racemose gland, doubtless providing the poison which enters the wound made by the spine.' (Cleland 1912) This was the first of three reports that appeared over 30 years.

Cleland also delivered a talk on doctor-naturalists to the Royal Australian College of Physicians in 1949, referring to a number who contributed to Australian ichthyology: John White, Benjamin Bynoe, George Bennett, Thomas Bancroft, and William Paradice (Cleland 1950). All have their place in the present history.

J.V. Duhig and Gwen Jones (1928a and 1928b) studied the venom apparatus and toxin of the Stonefish, *Synanceia horrida.* Duhig, a pathologist who founded the Department of Pathology at the University of Queensland, received stonefish from Heber Longman (Director of the Queensland Museum) and others. They described the fin spines, the glands and ducts associated therewith, and examined the nature of the toxin as seen by its effect on

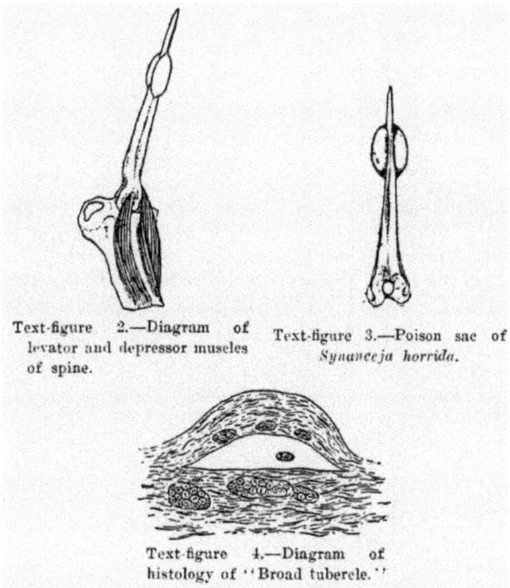

The 'venom apparatus' of the Estuarine Stonefish, *Synanceia horrida* (Linnaeus, 1776). From Duhig and Jones 1928.

guinea pigs. This was the first study of the toxicology of stonefish venom (Halstead 1970). They also described the histology of the mucus-secreting glands in the skin of *Synanceia* and were impressed by their efficacy:

> In every case we found the fish completely covered by a slime so dense that the fish rather resembled a rock ... after removal of this coating of slime the colour of the fish was shown to be rather beautiful - dappled grey and brown in general, with patches of deep violet about the pores of the tubercles over the dorsum, and rich saffron ventrally and especially under the opercular [pectoral] fins. Why the fish conceals this pretty natural appearance in exchange for its slimy coat of dirty blackish green is hard to explain.

They were clearly unaware that the excess mucus was a reaction to irritation: their colour description reflected the appearance of the fish when undisturbed.

Stead

David George Stead (1877–1957), born in Sydney, had an active interest in the fishes of Australia throughout his life. He was trained as a compositor but Harald Dannevig (see page 346) recruited him as a scientific assistant to the Fisheries Commission in 1902. He became Naturalist to the Board of Fisheries for New South Wales and in 1914–15 he was overseas examining European and American fisheries methods for the Australian government. On his return in 1915 Stead managed the New South Wales trawling industry but he was dismissed in 1920 when this was found to be unprofitable.

Stead was a populariser, most of his fish publications being aimed at the general populace, and these books and articles had a wide readership. *Fishes of Australia* (Stead 1906) and *Edible Fishes of New South Wales* (Stead 1908a) were illustrated, in part, 'by means of photography, directly from the fishes themselves; the latter having been previously specially preserved by me to enable this to be successfully carried out.' The

Stead chose the Yellow Crested Weedfish, *Cristiceps aurantiacus* Castelnau, 1879 for the frontispiece of his *Fishes of Australia* (1906). From Stead 1906.

former work included only a fraction of the fishes known from Australia at the time; the latter was a good representation of the fishes to be found in the Sydney fish market. Stead took much of his information from the recent publications of Ogilby (and to a lesser extent, Waite) but did not formally acknowledge the debt. His habit of using Australian scientific publications without making his sources known persisted and was resented, particularly by Allan McCulloch, whose regard for Stead as an ichthyologist was not high (see page 354). However, McCulloch did Stead justice in a letter to Ogilby (see page 345) when Stead's view that there was but one species of Blackfish (or Luderick), *Girella tricuspidata* proved to be correct.

Among Stead's more interesting publications was a small booklet *Notes on Eggs and Breeding-Habits of fishes* (Stead 1907b) which appeared as *Brooks's Nature-study Series, No.2.* This drew heavily on overseas publications, all duly cited, but also

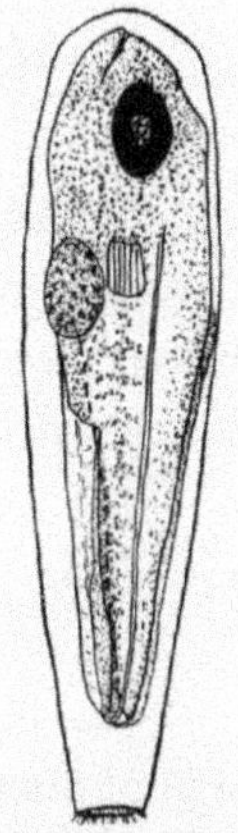

Fig. 10.—Egg (containing young) of LONG-JAWED OYSTER-GOBY (*Gillichthys microphthalmus*). Natural position.

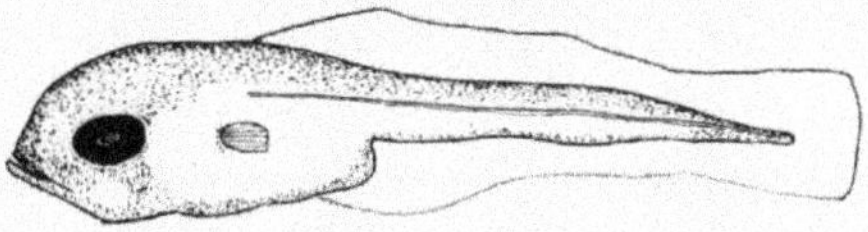

Fig. 11.—Newly-hatched Young of LONG-JAWED OYSTER-GOBY.

Note the beginning of the formation of separate fins.

Greatly magnified. (Drawn from Nature.)

Stead drew the developing embryo and 'newly-hatched young' of the 'Long-jawed Oyster Goby', *Gillichthys microphthalmus* [now *Redigobius macrostoma* (Günther, 1861)]. From Stead 1907.

NEW FISHES FROM NEW SOUTH WALES. PLATE III.

PLATYCEPHALUS MARMORATUS Stead.—*Marbled Flathead* (Dorsal aspect).

The Marbled Flathead, *Platycephalus marmoratus* Stead, 1908. From Stead 1908.

included a few of the author's own observations and sketches 'Drawn from Nature.' *On the Need for More Uniformity in the Vernacular Names of Australian Edible Fishes* (Stead 1911) was an early plea for some sensible regulation (a recent version of this has led to the hopefully doomed suggestion that the Tommy Rough of southern Australia be known as Herring). *Giants and Pigmies of the Deep* (Stead 1933) comprised essentially an account of the large and the fierce, the 'pigmies' not needing much space.

Stead also took upon himself the task of recording new distribution records (Stead 1907a) and a few new species for New South Wales (Stead 1908b). Two of his eight new species proved to be valid.

He prepared sketches of the fisheries of New South Wales (Stead 1910) and Northern Australia (Stead 1929). His popular work on sharks and rays (Stead 1963) was revised by Whitley and published posthumously.

Roughley

Theodore Cleveland Roughley (1888–1961) was born in Sydney. He began the medical course at Sydney University but left to work at the Sydney Technology Museum as the Economic Zoologist. His interests in art and microscopy were constant themes in his publications. While with the Technology Museum he published *Fishes of Australia. Their Utilisation in Applied Art* (Roughley 1913).

His publication of *Fishes of Australia and their Technology* (Roughley 1916) was a useful and very well illustrated guide to the commercial fishes of New South Wales and the fisheries that had been established to that time. The illustration of this work had been carried out under Roughley's supervision. Casts of the New South Wales fishes were prepared, coloured by an artist at the Technology Museum, and then photographed (see colour plate 16). The results were pleasing and the fishing public of New South Wales had been presented with an easy way to identify their common commercial fishes. Roughley acknowledged the artists:

> The casts from which the coloured plates are reproduced are the work of Mr. Alex. Murray, of the Sydney College Staff; the colouring is from the brush of Mr. Chas. Toms, the museum artist, and late teacher of decorative art at the aforementioned College. Both the casting and colouring were carried out under my close supervision, and I am greatly indebted to these officers for the interest and pains taken in their work. (op. cit.)

He also thanked Allan McCulloch and others for their help, and acknowledged a debt to the works of Ogilby and Stead. The book gave a good picture of the local fisheries and included a section on his pet topic, fishes in art. McCulloch in turn thanked Roughley for the work he had done in providing illustrations for McCulloch's own work on New South Wales fishes and for Waite's fishes of South Australia (see page 360).

Roughley's study of the Sydney rock oyster (Roughley 1933) attracted wide interest as he reported that oysters changed sex through their lives. He also published *Wonders of the Great Barrier Reef* (Roughley 1936), a sensitive and informative popular account that was well received.

Roughley joined the fisheries branch in New South Wales as a research officer in 1939, held the position of Deputy-Controller of Fisheries from 1943 to 1947, and thereafter Superintendent of Fisheries. Roughley revised his 1916 book and it was published as *Fish and Fisheries of Australia* (Roughley 1951), becoming very popular under that title and, as with his book on the Barrier Reef, going through many editions.

Glauert and Alexander

Ludwig Glauert (1879–1963) was born in Yorkshire, England and trained as a geologist. In 1908 he came to Western Australia and, as a palaeontologist, was employed in the Geological Survey. He curated the materials in the Western Australian Museum and became the scientific assistant to the Director of the Museum in 1910. He carried out field work as a geologist and palaeontologist and discovered a Pleistocene mammalian fauna related to that still present in eastern Australia. In 1914 Glauert became keeper of geology and ethnology at the museum.

After the First World War he studied Australian natural history specimens in the British Museum and, on his return to Perth in 1920, had the position of keeper of the biological collections at the Museum. Although an active field worker and researcher, the financial constraints on the Museum between the wars reduced the technical staff and for some years Glauert was the only scientist. He was provided with a single technical assistant and a taxidermist.

Most of Glauert's zoological work was with reptiles and scorpions but he also published on the fishes of Western Australia. These were all brief papers, most describing new records for Western Australia. His largest fish paper was a report on the fishes taken by the government trawler *Penguin* near Albany in south-western Australia (Glauert 1921). Whitley recognised his contributions by naming a small Anglerfish, *Echinophryne* [now *Allenichthys*] *glauerti* found in south-western Australia for Glauert (Whitley 1944).

W.B. Alexander, Keeper of Biology at the Western Australian Museum, carried out a little work on ichthyology. He described a new species of Scabbardfish, *Evoxymetopon* [now *Assurger*] *anzac* (1917) and published brief notes on Western Australian Lampreys (1920). He also produced a series of papers on the history of zoology in Western Australia (1915, 1916, 1918), quoting liberally from Vancouver, Péron and others.

Kesteven

H. Leighton Kesteven (1881–1964), born in Fiji, became a technical assistant at the Australian Museum (1903–1907) prior to undertaking a medical course. He worked as a general practitioner in Queensland but returned to Sydney in 1919 to take an M.D. in comparative anatomy. He moved to country New South Wales (Bulahdelah) where he pursued his interest in comparative anatomy and was appointed to the Australian Museum as an honorary zoologist.

His first contribution to ichthyology was a note *On the Venom of the Fish, Notesthes robusta* (the Bullrout of fresh waters in New South Wales) (Kesteven 1914). This was a description of the clinical signs of envenomation rather than a study of the venom.

The bones of the fish skull have been studied by most of the great comparative anatomists, generally with a view to identifying homologies with the skull elements of other classes of vertebrates. Cuvier's descriptions were the basis upon which Owen and Huxley built there own work. Theodore Gill, George Boulenger and C.T. Regan were among the ichthyologists who used osteological studies to elucidate relationships between groups of fishes. Kesteven entered this field with a paper challenging Huxley's conclusions as to the origins of the jaws and palate (Kesteven 1922). He proceeded to publish a series of *Contributions to the Cranial Osteology of the Fishes*, using chiefly Freshwater Catfish (1925); the Freshwater Eel (1926); Flathead, Scorpionfish, Gurnard and Stonefish (1926); Snapper, Queensland Grouper and Morwong (1928); and inevitably, the Australian Lungfish (1931). Kesteven's conclusions were politely but firmly treated by William K. Gregory, comparative anatomist and curator of the American Museum of Natural History, in his *Fish Skulls, a Study of the Evolution of Natural Mechanisms* (1933). Gregory found no virtue in Kesteven's deductions and drew attention to his unfortunate choice of such specialised species as catfish and the eel for his early work.

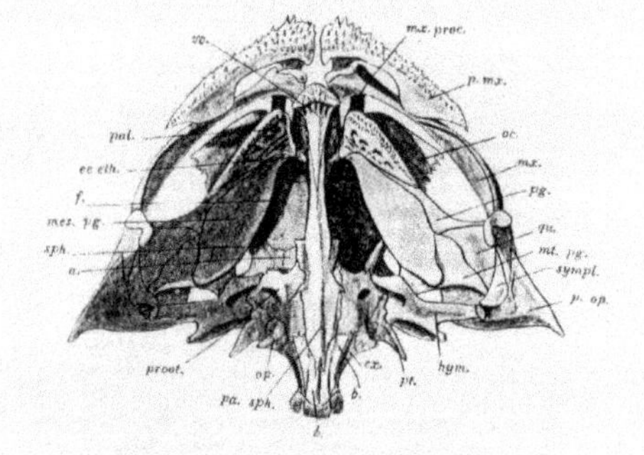

Fig. 1. *Promicrops itaiara*, palatal view of the complete skull.

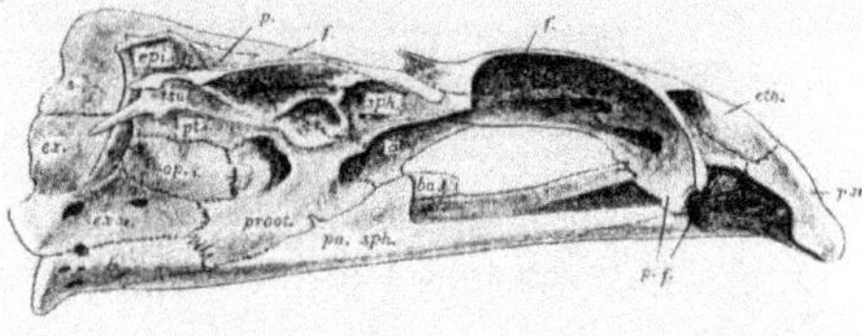

Fig. 2. *Promicrops itaiara*, side view of the cranium and bones derived from the chondrocranium.

The skull of the Queensland Grouper, *Epinephelus lanceolatus* (Bloch, 1790) from Kesteven's osteological studies. From Kesteven 1922.

Paradice

Surgeon Lieutenant-Commander William E. J. Paradice (1897–1927) of the Royal Australian Navy collected natural history specimens in the course of his cruises in northern Australia. McCulloch encouraged him to collect and study fishes.

Inspired by Cleland's account of injuries to man from animals (Cleland 1912), Paradice reported injuries of this sort that had come to his attention, including illness (now called ciguatera) from eating the Chinamanfish, *Symphorus nematophorus*. In an address to the Health Inspectors' Association (Paradice 1926) he discussed and illustrated fish that might help to destroy mosquito larvae, and poisonous and venomous fishes. He wrote 'It is a pleasure for me to speak about fish at any time, the study of these fascinating animals having been a hobby of mine for several years' and concluded his address with personal observations on Mud Skippers (*Periophthalmus*) and Archerfish (*Toxotes*). This paper was printed with a sprinkling of interspersed advertisements for 'pest experts' and an 'Automatic Glass Washer'.

While with HMAS *Geranium* Paradice made large collections of fishes and later wrote of his experiences. His official report on the Sir Edward Pellew group of islands, which

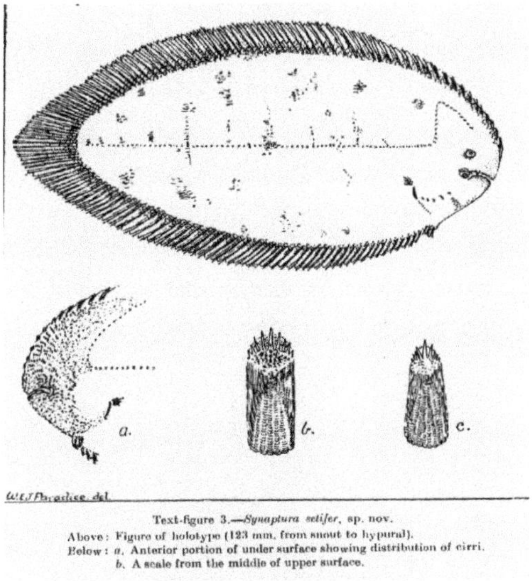

Paradice's Sole, *Synaptura* [now *Paradicula*] *setifer* (Paradice, 1927). From Paradice and Whitley1927.

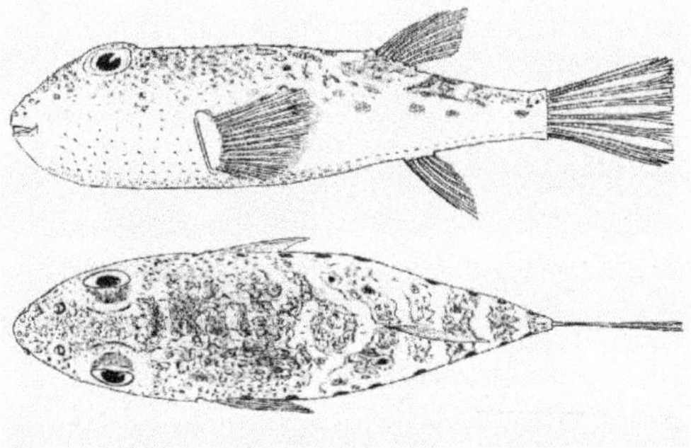

Whitley's Toadfish, *Spheroides* [now *Torquigener*] *whitleyi* (Paradice, 1927) from the cruise of the *Geranium*, drawn by Paradice. From Paradice and Whitley 1927

are located in the Gulf of Carpenteria, contained accounts of the fishes he encountered (Paradice 1924). McCulloch (1924a) reported on part of the *Geranium* collection and, after McCulloch's death, Paradice collaborated with Whitley to produce an annotated list of the fishes taken on the *Geranium* cruises (Paradice and Whitley 1927). This included some of Paradice's own fish drawings. In November 1927 Paradice was killed in a ferry collision on Sydney Harbour.

Lord and Scott

Clive Errol Lord (1889–1933), born in Tasmania and trained as an architect, became active in the Tasmanian Field Naturalists' Club. He was appointed assistant curator of the Tasmanian Museum in 1917 and curator a few years later. Lord was an active voice for conservation and was instrumental in the establishment of national parks in Tasmania (Reynolds 1934).

Ichthyology was peripheral to his main interests but, following the publication of regional guides by McCulloch for New South Wales and Waite for South Australia, Lord brought Johnston's checklist up to date (Lord 1923). For this task he had the active assistance of McCulloch, acknowledged in the introductory remarks: '... it is largely due to his kindly personal assistance that the compilation of the present list has been made possible.'

The Barracouta, *Thyrsites atun* (Euphrasén, 1791). Some of Mrs. Meredith's drawings were photographed to illustrate *The Vertebrate Animals of Tasmania*. From Lord and Scott 1924.

Herbert Hedley Scott (1841–1922) succeeded Alexander Morton as curator of Launceston's Queen Victoria Museum in 1897. He held this position until his death when his son, Eric[34], took over the curatorship. Scott senior collaborated with Lord in writing *A Synopsis of the Vertebrate Animals of Tasmania* (1924), including an annotated list of fishes, illustrated chiefly with line drawings. This work also reproduced some of Louisa Meredith's fine drawings. Louisa Ann Meredith (1812–1895), well known as a writer for her social commentary, came from England to live in Tasmania in 1840. She was an accomplished artist who represented the natural world of Tasmania and whose interests included the fishes of Tasmania's east coast; her drawings of them were accurate and attractive.

T.H. Johnston

Thomas Harvey Johnston (1881–1951), born in Sydney, was in charge of the new Department of Biology at the University of Queensland from 1911, and became Professor of Zoology at the Adelaide University in 1922. He had wide-ranging interests but was best known for his work in parasitology (ADB).

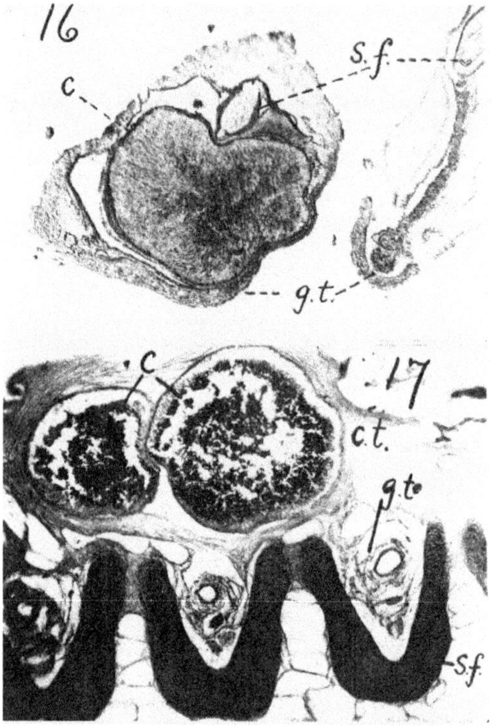

Sporozoan parasites of the gill filaments and gill arch of Callop, *Plectroplites ambiguus* [now *Macquaria ambigua*]. From Johnston and Bancroft 1918.

Harvey's parasitological publications occasionally included work on freshwater fishes. The first paper of this kind was an account of an epidemic of *Saprolegnia* in the fishes of the Thomson River in Queensland (Johnston 1917). Ogilby provided some of the fish identifications and Johnston quoted his account of the infection near Sydney (Ogilby 1888b). Later Johnston, working with Thomas Bancroft's daughter Jo, described several new species of sporozoan parasite found in Queensland freshwater fishes (Johnston and Bancroft 1918). The authors remarked that '... the specimens have been kindly identified for us by Mr. J. Douglas Ogilby of the Queensland Museum' and named one of the parasites, *Myxosoma ogilbyi*, for him. The same authors summarised their experience of episodes of mass mortality in the fishes of Queensland rivers and speculated upon the possible causes (Johnston and Bancroft 1922).

Much later, during his time in Adelaide, Johnston and Patricia Mawson (daughter of Douglas Mawson) published descriptions of many new species of nematodes found in Australian freshwater fishes (Johnston and Mawson 1940).

34 E.O.G. Scott, working at the Queen Victoria Museum in Launceston, published a series of *Observations on some Tasmanian Fishes*, beginning in 1934 and, over the next forty years, described many supposedly new fish, most of which have proven to be synonymous with known species. He co-authored *Fishes of Tasmania* (Last *et al.* 1983).

A glance forward from 1930

Whitley and others after 1930

With the death of Ogilby, Waite and McCulloch, Australian ichthyology was in the hands of Gilbert Whitley. Over the next 40 years his publications, several hundred of them, were the most visible manifestation of ichthyological work in Australia. He appears to have leant heavily on McCulloch's card index in spite of McCulloch's warning that he needed to develop this system on his own if he were to understand its value and its limitations. Among his more useful early tasks was the provision of supplements to subsequent editions of McCulloch's check-list of New South Wales fishes; in the third edition (McCulloch 1934) Whitley added 52 species to the original list.

Whitley practised an eccentric form of taxonomy that has created some confusion and attracted more than a little criticism. He described many new species and defined new genera for existing species, often neglecting precise definitions of his taxa. In all, he described over 150 valid species of fish from Australia between 1927 and 1971.

A biogeographic scheme was proposed by Tom Iredale, then conchologist at the Australian Museum, and modified by Whitley who prepared a map of biogeographic zones applicable to fish (Whitley 1932). These were superseded when Briggs (1974) published *Marine Zoogeography*, utilising knowledge of the distributions of a variety of marine organisms to propose new biogeographic boundaries for marine faunas around the world. Iredale and Whitley later applied their knowledge of fish and mollusc distributions to delineate freshwater zones that they termed fluvifaunulae (Iredale and Whitley 1938).

Whitley's awareness of historical perspectives and near encyclopaedic knowledge of the history of Australian zoology enabled him to produce interesting biographic sketches and obituaries, of which he wrote many, and useful summaries of Australian zoology. He established and maintained an extensive card index of biographical references to people relevant in the history of natural science. He wrote a history of the Australian Museum[35], a simple chronological account without pretensions to analysis, but it apparently found no favour with the administration; it is dated 1963 but was not published.

Whitley's major work in progress was a Fishes of Australia. Part One was published as *Sharks, Rays, Devil-Fish, and other Primitive Fishes of Australia and New Zealand* (Whitley 1940). His nearest approach to success in the larger scheme was a modest, pocket-sized publication in two parts entitled *Marine Fishes of Australia* (Whitley 1962), a popular summary dealing with 'representatives of the principal native families.' On its back cover he was quoted: '… [he is] currently engaged on a large work on all the fishes of Australia which, at his present rate of progress, should be published on his 99th birthday, in A.D. 2002.' A companion volume on Australian freshwater fishes (Whitley 1960) had preceded this.

35 Whitley's history of the Australian Museum is held in the Basser Library, Canberra.

A Survey of Australian Ichthyology (Whitley 1964) comprised a list of Australian fishes with a treasury of references to their literature and a summary of significant workers in the field, the latter derived from his card index. This was perhaps Whitley's single most useful contribution to Australian ichthyology.

Theodore Roughley and David Stead continued to produce popular literature on Australian fishes. Roughley became Superintendent of New South Wales Fisheries and, on his retirement, published *Fish and Fisheries of Australia* (1951), a much revised version of his *Fishes of Australia and their Technology* (1916). Whitley revised Stead's manuscript on Australian elasmobranchs and published *Sharks and Rays in Australian Seas* (1963).

A determined effort to produce a Fishes of Australia was made by Ian S. Munro (1919–1994), fish taxonomist for CSIR (later CSIRO) in New South Wales from 1943. This appeared as removable pages in the *Fisheries Newsletter* from 1956 to 1961. This useful publication appeared in 42 parts but was suspended, through no fault of Munro's, when about two thirds of the fauna had been covered.

Changes in the study and perception of the fish fauna since 1930

Advances in Australian ichthyology since 1930 have been manifest in many disciplines: cladistic taxonomy, molecular biology, ecology, fisheries science and aquaculture, larval studies, biogeography, parasitology, physiology, behaviour, and many others. From about 1890 to 1930 ichthyological research within Australia was conducted almost exclusively in state museums. Since that time it has diffused through the zoology departments of universities, state fisheries organisations and institutions specialising in marine research. Purpose-built research vessels have literally brought to light a fauna from depths that had been only superficially sampled. An influx of overseas-trained ichthyologists, from the 1970s, brought Australian ichthyology into closer contact with the science as it is practised overseas, particularly in the United States.

McCulloch's *Check-list* (1929–1930) comprised approximately 2000 nominal species but Whitley, in preparing it from McCulloch's card index, was obliged to include many names that represented unresolved problems of taxonomy; even McCulloch's *bêtes noires*, Saville-Kent's *nomina nuda*, found a place in the list. In his *Survey of Australian Ichthyology* (1964) Whitley listed nearly 2500 fishes and most of these were 'good' species. In excess of 4500 species have now been recorded from Australia. The greater number of additions to the list since 1964 represent range extensions from other parts of the Indo-Pacific although many new species have been described from Australia over this period.

Trawling deeper waters has added many species to the Australian lists. A host of small midwater species, such as Lanternfish (Myctophidae), are recorded and benthic trawls have located substantial numbers of Whiptails (Macrouridae) and Brotulas (Ophidiidae and Bythitidae). From the continental slope of southern Australia, between about 500 and 1500 metres, more than 30 species of Snailfish (Liparidae) have been trawled over the last 20 years: a significant component of the fauna that was previously unsuspected. Among the most remarkable additions are the many new species of sharks and rays which have been described in the last few years, particularly from the North-West Shelf. These discoveries are presented in the magnificent *Sharks and Rays of Australia* (Last *et al.* 2009).

Since 1930 the fishes of Australian fresh waters have provided some surprises. Near North West Cape fish have been found in the groundwater: Whitley (1945) described a Blind Gudgeon, *Milyeringa veritas* and Mees (1962) a Blind Cave Eel, *Ophisternon candidum*. In the south-west, the location of the little Salamanderfish (*Lepidogalaxias salamandroides* Mees, 1961) established an endemic family (or perhaps subfamily) of fishes. The complexity of the fauna of the northern streams and rivers has been elucidated, confirming its relationship with southern New Guinea. Many new species have been found, particularly among the Grunters (Terapontidae), Rainbowfish (Melanotaeniidae) and Tasmanian galaxiids.

A few puzzling species, found long ago, were later placed in their appropriate groups: Richardson included the Shore Eel, *Alabes dorsalis* (Richardson, 1845) with the Swamp Eels (Synbranchidae) as it has a single gill opening but its affinities with the Clingfish (Gobiesocidae) were demonstrated 130 years later (Springer and Fraser 1976); the Moonlighter, *Tilodon sexfasciatus*, long placed with the Butterflyfishes (Chaetodontidae) was recognised as a Sweep of the family Microcanthidae; and the Rainbow Fish, *Heteroscarus acrotpilus*, supposedly a Parrotfish (Scaridae), was found to be a Weed Whiting (Odacidae).

Fishes of Australia

It seems appropriate to finish this account with a reference to what seemed to be the ultimate goal of Australian ichthyologists for many years. The *Check-List of the Fishes Recorded from Australia* (McCulloch 1929–30), prepared for publication by Whitley, led to the three splendid *Fishes* books, comprising volume 35 of the *Zoological Catalogue of Australia* (Hoese *et al.* 2006). Swainston (2011) has presented a magnificently illustrated summary of Australian fishes, derived from this. These works are stalked by a fine legion of ghosts: Richardson, Castelnau, Macleay, Ogilby, Waite, McCulloch, Whitley and Munro, to name only those who made the greatest contributions to the discovery of Australia's fishes.

Appendix: Classification of fish species

This appendix provides a classified list of the species of fish mentioned in the text with the scientific name, author and date of original description. The levels of classification used are restricted to the general group (agnathan, holocephalan, elasmobranch, sarcopterygian, actinopterygian), order (ending 'iformes'), suborders but only for the Perciformes (ending 'oidei') and family (ending 'idae). The usual convention is followed whereby the author and date of the original description are enclosed in parentheses when the species has later been placed in a different genus.

Agnathans (lampreys and hagfish)

Myxiniformes

Myxinidae
Eptatretus cirrhatus (Forster, 1801)

Petromyzontiformes

Geotriidae
Geotria australis Gray, 1851

Petromyzontidae
Mordacia mordax (Richardson, 1846)

Holocephalans (chimaeras)

Chimaeriformes

Callorhinchidae
Callorhinchus milii (Bory de Saint-Vincent, 1823)

Chimaeridae
Hydrolagus ogilbyi (Waite, 1898)

Elasmobranchs (sharks and rays)

Heterodontiformes

Heterodontidae
Heterodontus galeatus (Günther, 1870)
Heterodontus portusjacksoni (Meyer, 1793)

Hexanchiformes

Hexanchidae
Notorynchus cepedianus (Péron, 1807)

Lamniformes

Mitsukurinidae
Mitsukurina owstoni Jordan, 1898

Cetorhinidae
Cetorhinus maximus (Gunnerus, 1765)

Lamnidae
Carcharodon carcharias (Linnaeus, 1758)

Carcharhiniformes

Scyliorhinidae
Asymbolus vincenti (Zietz, 1908)
Cephaloscyllium isabellum (Bonnaterre, 1788)

Triakidae
Galeorhinus galeus (Linnaeus, 1758)
Mustelus antarcticus Günther, 1870

Carcharhinidae
Carcharhinus brachyurus (Günther, 1870)
Carcharhinus leucas (Müller & Henle, 1839)
Carcharhinus limbatus (Müller & Henle, 1839)
Galeocerdo cuvier (Péron & Lesueur, 1822)
Loxodon macrorhinus Müller & Henle, 1839
Rhizoprionodon acutus (Rüppell, 1837)

Orectolobiformes

Parascylliidae
Parascyllium collare Ramsay & Ogilby, 1888
Parascyllium variolatum (Duméril, 1853)

Brachaeluridae
Brachaelurus colcloughi Ogilby, 1908
Brachaelurus waddi (Bloch & Schneider, 1801)

Orectolobidae
Eucrossorhinus dasypogon (Bleeker, 1867)
Orectolobus maculatus (Bonnaterre, 1788)
Orectolobus ornatus (De Vis, 1883)
Sutorectus tentaculatus (Peters, 1864)

Hemiscylliidae
Hemiscyllium ocellatum (Bonnaterre, 1788)
Hemiscyllium trispeculare Richardson, 1843

Stegostomatidae
Stegostoma fasciatum (Hermann, 1783)

Squaliformes

Dalatiidae
Isistius brasiliensis (Quoy & Gaimard, 1824)

Oxynotidae
Oxynotus bruniensis (Ogilby, 1893)
Oxynotus centrina (Linnaeus, 1758)

Somniosidae
Proscymnodon plunketi (Waite, 1911)

Centrophoridae
Centrophorus harrissoni McCulloch, 1915
Deania quadrispinosa (McCulloch, 1915)

Squalidae
Squalus megalops (Macleay, 1881)

Squatiniformes

Squatinidae
Squatina tergocellata McCulloch, 1914

Pristiophoriformes

Pristiophoridae
Pristiophorus cirratus (Latham, 1794)

Torpediniformes

Hypnidae
Hypnos monopterygium (Shaw, 1795)

Narcinidae
Narcine tasmaniensis Richardson, 1841
Typhlonarke aysoni (Hamilton, 1902)

Pristiformes

Pristidae
Pristis clavata Garman, 1906

Rajiformes

Rhinobatidae
Aptychotrema rostrata (Shaw, 1794)
Aptychotrema vincentiana (Haacke, 1885)
Glaucostegus typus (Bennett, 1830)
Trygonorrhina dumerilii (Castelnau, 1873)
Trygonorrhina fasciata (Müller & Henle, 1841)

Rajidae
Dentiraja lemprieri (Richardson, 1845)
Dipturus australis (Macleay, 1884)
Dipturus polyommata (Ogilby, 1910)

Arhynchobatidae
Pavoraja nitida (Günther, 1880)

Myliobatiformes

Dasyatidae
Dasyatis fluviorum Ogilby, 1908
Himantura toshi Whitley, 1939
Neotrygon kuhlii (Müller & Henle, 1841)

Urolophidae
Trygonoptera testacea Müller & Henle, 1841
Urolophus cruciatus (Lacépède, 1804)

Myliobatidae
Myliobatis australis Macleay, 1881
Myliobatis hamlyni Ogilby, 1911

Mobulidae
Manta birostris (Walbaum, 1792)

Rhinopteridae
Rhinoptera neglecta Ogilby, 1912

Sarcopterygians (lungfish)

Ceratodontiformes

Ceratodontidae
Neoceratodus forsteri (Krefft, 1870)

Lepidosirenidae
Lepidosiren paradoxa Fitzinger, 1837

Actinopterygians (bony fish)

Osteoglossiformes

Osteoglossidae
Scleropages jardinii (Saville-Kent, 1892)
Scleropages leichhardti, Günther, 1864

Elopiformes

Megalopidae
Megalops cyprinoides (Broussonet, 1782)

Notacanthiformes

Notacanthidae
Notacanthus sexspinis Richardson, 1846

Anguilliformes

Anguillidae
Anguilla anguilla (Linnaeus, 1758)

Muraenidae
Gymnothorax prasinus (Richardson, 1848)
Strophidon sathete (Hamilton-Buchanan, 1822)

Ophichthidae
Leiuranus semicinctus (Lay & Bennett, 1839)
Scolecenchelys breviceps (Günther 1876)

Clupeiformes

Engraulidae
Stolephorus waitei Jordan and Seale, 1926

Clupeidae
Etrumeus teres
Nematalosa erebi (Günther, 1868)
Potamalosa richmondia (Macleay, 1879)

Gonorynchiformes

Gonorynchidae
Gonorynchus greyi (Richardson, 1845)

Cypriniformes

Cyprinidae
Cyprinus carpio Linnaeus, 1758

Siluriformes

Ariidae
Ariopsis graeffei (Kner & Steindachner, 1867)
Cinetodus froggatti, (Ramsay & Ogilby, 1886)
Nemapteryx armiger (De Vis, 1884)

Pimelodidae
Phractocephalus hemiliopterus (Bloch, 1801)

Plotosidae
Cnidoglanis macrocephalus (Valenciennes, 1840)
Plotosus lineatus (Thunberg, 1791)
Porochilus argenteus (Zietz, 1896)
Tandanus tandanus (Mitchell, 1838)

Salmoniformes

Salmonidae
Oncorhynchus tshawytscha (Walbaum, 1792)
Salmo salar Linnaeus, 1758

Retropinnidae
Prototroctes maraena Günther, 1864
Retropinna retropinna (Richardson, 1848)
Retropinna semoni (Weber, 1895)

Galaxiidae
Galaxias brevipinnis Günther, 1866
Galaxias maculatus (Jenyns, 1842)
Galaxias occidentalis Ogilby, 1899
Galaxias olidus Günther, 1866
Galaxias rostratus Klunzinger, 1872
Lepidogalaxias salamandroides Mees, 1961
Lovettia sealii (Johnston, 1883)

Stomiiformes

Sternoptychidae
Maurolicus australis Hector, 1875

Stomiidae
Chauliodus sloani Bloch & Schneider, 1801
Echiostoma barbatum Lowe, 1843
Opostomias micripnus (Günther, 1878)
Pachystomias microdon (Günther, 1878)

Aulopiformes

Aulopidae
Latropiscis purpurissatus (Richardson, 1843)

Synodontidae
Harpadon translucens Saville-Kent, 1889
Saurida undosquamis (Richardson, 1848)

Myctophiformes

Myctophidae
Myctophum phengodes (Lütken, 1892)

Lampridiformes

Lamprididae
Lampris guttatus (Brunnich, 1788)

Lophotidae
Lophotus guntheri Johnston, 1883

Trachipteridae
Trachipterus jacksonensis (Ramsay, 1881)

Regalecidae
Agrostichthys parkeri (Benham, 1904)
Regalecus glesne Ascanius, 1772

Ophidiiformes

Carapidae
Encheliophis gracilis (Bleeker, 1856)

Bythitidae
Dermatopsis multiradiatus McCulloch & Waite, 1918
Monothrix polylepis Ogilby, 1897

Gadiformes

Macrouridae
Coelorinchus innotabilis McCulloch, 1907
Lucigadus nigromaculatus (McCulloch, 1907)

Moridae
Pseudophycis barbata Günther, 1863

Batrachoidiformes

Batrachoididae
Batrachomoeus dubius (White, 1790)
Halophryne diemensis (Lesueur, 1824)

Lophiiformes

Lophiidae
Lophiomus setigerus (Vahl, 1797)

Antennariidae
Allenichthys glauerti (Whitley, 1944)
Phyllophryne scortea (McCulloch & Waite, 1918)
Histrio histrio (Linnaeus, 1758)
Rhycherus filamentosus (Castelnau, 1872)
Trichophryne mitchellii (Morton, 1897)

Brachionichthyidae
Brachionichthys hirsutus (Lacépède, 1804)
Sympterichthys politus (Richardson, 1844)

Sympterichthys unipennis (Cuvier, 1817)
Sympterichthys verrucosus McCulloch & Waite, 1918

Mugiliformes

Mugilidae
Liza argentea (Quoy & Gaimard, 1825)
Mugil cephalus Linnaeus, 1758
Rhinomugil nasutus (De Vis, 1883)
Trachystoma petardi (Castelnau, 1875)

Atheriniformes

Notocheiridae
Iso rhothophilus (Ogilby, 1895)

Melanotaeniidae
Cairnsichthys rhombosomoides (Nichols & Raven, 1928)
Melanotaenia fluviatilis (Castelnau, 1878)
Melanotaenia nigrans (Richardson, 1843)
Melanotaenia splendida tatei (Zietz, 1896)

Pseudomugilidae
Pseudomugil signifer Kner, 1867

Atherinidae
Atherion maccullochi Jordan & Hubbs, 1919

Beloniformes

Belonidae
Belone belone (Linnaeus, 1761)

Exocoetidae
Cheilopogon cyanopterus (Valenciennes, 1847)
Cypselurus naresii (Günther, 1889)
Cypselurus ogilbyi Jordan & Snyder, 1908
Exocoetus volitans Linnaeus, 1758

Hemiramphidae
Hyporhamphus australis (Steindachner, 1866)
Hyporhamphus regularis (Günther, 1866)

Beryciformes

Anomalopidae
Photoblepharon palpebratum (Boddaert, 1781)

Monocentridae
Cleidopus gloriamaris De Vis, 1882

Trachichthyidae
Paratrachichthys macleayi (Johnston, 1881)
Trachichthys australis (Shaw, 1799)

Berycidae
Centroberyx gerrardi (Günther, 1887)
Centroberyx lineatus (Cuvier, 1829)

Zeiformes

Cyttidae
Cyttus novaezealandiae (Arthur 1885)

Oreosomatidae
Oreosoma atlanticum Cuvier, 1829

Zeidae
Zenopsis nebulosus (Temminck & Schlegel, 1845)

Caproidae
Antigonia rubicunda Ogilby, 1910

Gasterosteiformes

Pegasidae
Pegasus lancifer Kaup. 1861

Syngnathidae
Campichthys tryoni (Ogilby, 1890)
Corythoichthys intestinalis (Ramsay, 1881)
Filicampus tigris (Castelnau, 1879)
Haliichthys taeniophorus Gray, 1859
Hippichthys penicillus (Cantor, 1850)
Hippocampus abdominalis Lesson, 1827
Hippocampus bleekeri Fowler, 1908
Hippocampus breviceps Peters, 1869
Hippocampus planifrons Peters, 1877
Hippocampus whitei Bleeker, 1855
Histiogamphelus briggsii McCulloch, 1914
Hypselognathus rostratus (Waite & Hale, 1921)
Idiotropiscis australe (Waite & Hale, 1921)
Kaupus costatus (Waite & Hale, 1921)
Leptoichthys fistularius Kaup, 1853
Leptonotus elevatus (Hutton, 1872)
Leptonotus norae (Waite, 1911)
Phycodurus eques (Günther, 1865)
Phyllopteryx taeniolatus (Lacépède, 1804)
Solegnathus hardwickii (Gray, 1830)
Solegnathus lettiensis Bleeker, 1860
Solegnathus spinosissimus Günther, 1870
Stigmatopora nigra Kaup, 1856
Vanacampus phillipi (Lucas, 1891)

Fistulariidae
Fistularia petimba Lacépède, 1803

Macroramphosidae
Centriscops humerosus (Richardson, 1846)

Synbranchiformes

Synbranchidae
Monopterus albus (Zuiew, 1793)
Ophisternon candidum (Mees, 1962)

Scorpaeniformes

Neosebastidae
Neosebastes scorpaenoides (Guichenot, 1867)

Scorpaenidae
Parascorpaena picta (Kuhl and Hasselt, 1829)
Pterois antennata (Bloch, 1787)

Pterois volitans (Linnaeus, 1758)
Scorpaena cardinalis Richardson, 1842

Apistidae
Apistops caloundra (De Vis, 1886)

Tetrarogidae
Glyptauchen panduratus (Richardson, 1850)
Gymnapistes marmoratus (Cuvier, 1829)
Liocranium praepositum Ogilby, 1903
Notesthes robusta (Günther, 1860)
Paracentropogon vespa Ogilby, 1910
Richardsonichthys leucogaster (Richardson, 1848)

Synanceiidae
Synanceia horrida (Linnaeus, 1766)

Congiopodidae
Perryena leucometopon (Waite, 1922)
Zanclorhynchus spinifer Günther, 1880

Gnathanacanthidae
Gnathanacanthus goetzeei Bleeker, 1855

Aploactinidae
Paraploactis trachyderma Bleeker, 1865

Pataecidae
Neopataecus waterhousii (Castelnau, 1872)
Pataecus fronto Richardson, 1844

Caracanthidae
Caracanthus unipinna (Gray, 1831)

Triglidae
Chelidonichthys kumu (Cuvier, 1829)
Pterygotrigla polyommata (Richardson, 1839)

Platycephalidae
Platycephalus bassensis Cuvier, 1829
Platycephalus marmoratus Stead, 1908
Platycephalus speculator Klunzinger, 1872
Thysanophrys cirronasa (Richardson, 1848)

Hoplichthyidae
Hoplichthys haswelli McCulloch, 1907
Hoplichthys ogilbyi McCulloch, 1914

Perciformes

Percoidei

Latidae
Lates calcarifer (Bloch, 1790)

Ambassidae
Ambassis jacksoniensis (Macleay, 1881)
Ambassis marianus Günther, 1880

Acropomatidae
Apogonops anomalus Ogilby, 1896
Howella brodiei Ogilby, 1899
Polyprion oxygeneios (Forster, 1801)

Serranidae
Caesioperca rasor (Richardson, 1839)
Cephalopholis argus Bloch & Schneider, 1801
Cephalopholis boenak (Bloch, 1790)
Cephalopholis sonnerati (Valenciennes, 1828)
Epinephelides armatus (Castelnau, 1875)
Epinephelus daemelii (Günther, 1876)
Epinephelus fasciatus (Forsskål, 1775)
Epinephelus lanceolatus (Bloch, 1790)
Epinephelus radiatus (Day, 1868)
Epinephelus rivulatus (Valenciennes, 1830)
Hypoplectrodes jamesoni Ogilby, 1908
Hypoplectrodes maccullochi (Whitley, 1929)
Lepidoperca tasmanica Norman, 1937
Rainfordia opercularis McCulloch, 1923

Callanthiidae
Callanthias allporti Günther, 1876
Callanthias australis Ogilby, 1899
Callanthias platei Steindachner, 1898

Percichthyidae
Bostockia porosa Castelnau, 1873
Gadopsis marmoratus Richardson, 1848
Maccullochella macquariensis (Cuvier, 1829)
Maccullochella peelii (Mitchell, 1838)
Macquaria ambigua (Richardson, 1845)
Macquaria australasica Cuvier, 1830
Macquaria colonorum (Günther, 1863)
Nannatherina balstoni Regan, 1906
Nannoperca australis Günther, 1861
Nannoperca obscura (Klunzinger, 1872)

Plesiopidae
Belonepterygion fasciolatum (Ogilby, 1889)
Notograptus guttatus Günther, 1867
Paraplesiops meleagris (Peters, 1869)
Trachinops caudimaculatus McCoy, 1890

Pseudochromidae
Congrogadus spinifer (Borodin, 1933)
Congrogadus subducens (Richardson, 1843)

Opistognathidae
Opistognathus eximius (Ogilby, 1908)
Opistognathus inornatus Ramsay & Ogilby, 1887

Apogonidae
Archamia leai Waite, 1916
Vincentia conspersa (Klunzinger, 1872)
Vincentia punctata (Klunzinger, 1880)

Sillaginidae
Sillaginodes punctata (Cuvier, 1829)
Sillago bassensis Cuvier, 1829
Sillago ciliata Cuvier, 1829
Sillago maculata Quoy & Gaimard, 1824

Malacanthidae
Malacanthus lattovittatus (Lacépède), 1801

Dinolestidae
Dinolestes lewini (Griffith, 1834)

Pomatomidae
Pomatomus saltatrix (Linnaeus, 1766)

Echeneidae
Remora remora (Linnaeus, 1758)

Glaucosomatidae
Glaucosoma buergeri Richardson, 1845
Glaucosoma hebraicum Richardson, 1845
Glaucosoma magnificum (Ogilby, 1915)
Glaucosoma scapulare Macleay, 1881

Kuhliidae
Kuhlia munda (De Vis, 1884)
Kuhlia rupestris (Lacépède 1802)

Rachycentridae
Rachycentron canadum (Linnaeus, 1776)

Coryphaenidae
Coryphaena hippurus Linnaeus, 1758

Carangidae
Alectis ciliaris (Bloch, 1787)
Alectis indica (Rüppell, 1830)
Caranx bucculentus Alleyne & Macleay, 1877
Caranx ignobilis (Forsskål, 1775)
Caranx papuensis Alleyne & Macleay, 1877
Caranx sexfasciatus Quoy & Gaimard, 1825
Elagatis bipinnulata (Quoy & Gaimard, 1825)
Naucrates ductor (Linnaeus, 1758)
Pantolabus radiatus (Macleay, 1881)
Pseudocaranx georgianus (Cuvier, 1833)
Seriola lalandi Valenciennes, 1833
Trachinotus anak Ogilby, 1909
Trachinotus botla (Shaw, 1803)
Trachinotus coppingeri (Günther, 1884)
Trachurus declivis (Jenyns, 1841)
Trachurus novaezelandiae Richardson, 1843
Uraspis secunda Poey, 1860

Menidae
Mene maculata (Bloch & Schneider, 1801)

Bramidae
Brama brama (Bonnaterre, 1788)

Emmelichthyidae
Emmelichthys nitidus Richardson, 1845

Lutjanidae
Lutjanus argentimaculatus (Forsskål, 1775)
Lutjanus bitaeniatus (Valenciennes, 1830)
Lutjanus carponotatus (Richardson, 1842)
Symphorus nematophorus (Bleeker, 1860)

Lobotidae
Lobotes surinamensis (Bloch, 1790)

Sparidae
Acanthopagrus australis (Günther, 1859)
Chrysophrys auratus (Forster, 1801)
Rhabdosargus sarba (Forsskål, 1775)

Lethrinidae
Gymnocranius griseus (Temminck & Schlegel, 1843)

Nemipteridae
Nemipterus furcosus (Valenciennes, 1830)
Pentapodus emeryii (Richardson, 1843)
Pentapodus paradiseus (Günther, 1859)

Polynemidae
Parapolynemus verekeri (Saville-Kent, 1889)

Sciaenidae
Argyrosomus japonicus (Temminck & Schlegel, 1844)
Nibea leptolepis (Ogilby, 1918)

Mullidae
Upeneichthys vlamingii (Cuvier, 1829)

Leptobramidae
Leptobrama muelleri Steindachner, 1878

Monodactylidae
Schuettea scalaripinnis Steindachner, 1866
Schuettea woodwardi Waite, 1905

Toxotidae
Toxotes chatareus (Hamilton-Buchanan, 1822)

Chaetodontidae
Chaetodon aureofasciatus Macleay, 1878
Chaetodon rainfordi McCulloch, 1923
Chaetodon tricinctus Waite, 1901
Heniochus diphreutes Jordan, 1903

Pomacanthidae
Chaetodontoplus duboulayi (Günther, 1867)

Enoplosidae
Enoplosus armatus (White, 1790)

Pentacerotidae
Pentaceropsis recurvirostris (Richardson, 1845)
Zanclistius elevatus (Ramsay & Ogilby, 1888)

Kyphosidae
Atypichthys strigatus (Günther, 1860)
Girella cyanea Macleay, 1881
Girella elevata Macleay, 1881
Girella fimbriata (McCulloch, 1920)
Girella nebulosa Kendall & Radcliffe, 1912
Girella tricuspidata (Quoy & Gaimard, 1824)
Girella zebra (Richardson, 1846)
Kyphosus sydneyanus (Günther, 1886)

Scorpis lineolata Kner, 1865
Tilodon sexfasciatus (Richardson, 1842)

Arripidae
Arripis georgianus (Valenciennes, 1831)
Arripis trutta (Forster, 1801)
Arripis truttaceus (Cuvier, 1829)

Terapontidae
Amniataba caudavittata (Richardson, 1845)
Amniataba percoides (Günther 1864)
Bidyanus bidyanus (Mitchell, 1838)
Bidyanus welchi (McCulloch & Waite, 1917)
Hephaestus carbo (Ogilby & McCulloch, 1916)
Hephaestus fuliginosus (Macleay, 1883)
Pelates octolineatus (Jenyns, 1840)
Pelates quadrilineatus (Bloch, 1790)
Pelates sexlineatus (Quoy & Gaimard, 1824)
Pelsartia humeralis (Ogilby, 1899)
Scortum barcoo (McCulloch & Waite, 1917)
Scortum hillii (Castelnau, 1878)

Cirrhitidae
Neocirrhites armatus Castelnau, 1873
Notocirrhitus splendens (Ogilby, 1889)

Oplegnathidae
Oplegnathus conwaii Richardson, 1840
Oplegnathus woodwardi (Waite, 1900)

Chironemidae
Chironemus microlepis Waite, 1916
Threpterius maculosus Richardson, 1850

Aplodactylidae
Aplodactylus arctidens Richardson, 1839
Aplodactylus etheridgii (Ogilby, 1889)
Aplodactylus punctatus Valenciennes, 1832

Cheilodactylidae
Cheilodactylus spectabilis Hutton, 1872
Dactylophora nigricans (Richardson, 1850)
Nemadactylus macropterus (Bloch & Schneider, 1801)

Cepolidae
Cepola australis Ogilby, 1899

Latridae
Latridopsis forsteri (Castelnau, 1872)

Labroidei

Labridae
Achoerodus gouldii (Richardson, 1843)
Achoerodus viridis (Steindachner, 1866)
Bodianus frenchii (Klunzinger, 1880)
Bodianus unimaculatus (Günther 1862)
Cheilinus undulatus Rüppell, 1835
Choerodon cyanodus (Richardson, 1843)
Choerodon fasciatus (Günther, 1867)
Choerodon venustus (De Vis, 1884)
Coris auricularis (Valenciennes, 1839)
Coris aygula Lacépède, 1801
Coris picta (Bloch & Schneider, 1801)
Epibulus insidiator (Pallas, 1770)
Eupetrichthys angustipes Ramsay & Ogilby, 1888
Gomphosus caeruleus Lacépède, 1801
Hemigymnus melapterus (Bloch, 1791)
Hologymnosus annulatus (Lacépède, 1801)
Iniistius jacksonensis (Ramsay, 1881)
Iniistius pavo (Valenciennes, 1840)
Notolabrus tetricus (Richardson, 1840)
Ophthalmolepis lineolatus (Valenciennes, 1839)
Pictilabrus laticlavius (Richardson, 1839)
Pseudolabrus guentheri Bleeker, 1862
Thalassoma hardwicke (Bennett, 1829)
Thalassoma lunare (Linnaeus, 1758)
Thalassoma quinquevittatum (Lay & Bennett, 1839)

Odacidae
Haletta semifasciata (Valenciennes, 1840)
Heteroscarus acroptilus (Richardson, 1846)
Siphonognathus argyrophanes (Richardson, 1858)
Siphonognathus beddomei (Johnston, 1885)
Siphonognathus radiatus (Quoy & Gaimard, 1834)

Scaridae
Chlorurus microrhinos (Bleeker, 1854)

Pomacentridae
Parma mccullochi Whitley, 1929
Parma microlepis Günther, 1862
Parma victoriae (Günther, 1863)
Pomacentrus coelestis Jordan and Starks, 1901

Zoarcoidei

Zoarcidae
Pachycara brachycephalum (Pappenheim, 1912)

Notothenioidei

Bovichtidae
Bovichtus angustifrons Regan, 1913
Bovichtus psychrolutes Günther, 1860
Pseudaphritis urvillii (Valenciennes, 1832)

Nototheniidae
Notothenia coriiceps Richardson, 1844
Paranotothenia microlepidota (Hutton, 1875)

Harpagiferidae
Harpagifer georgianus Nybelin, 1947

Artedidraconidae
Artedidraco shackletoni Waite, 1911
Dolloidraco longedorsalis Roule, 1913

Bathydraconidae
Akarotaxis nudiceps (Waite, 1916)

Trachinoidei

Pinguipedidae
Parapercis allporti (Günther, 1876)
Parapercis colias (Forster, 1801)
Parapercis haackei (Steindachner, 1884)

Cheimarrichthyidae
Cheimarrichthys fosteri Haast, 1874

Creediidae
Creedia haswelli (Ramsay, 1882)
Schizochirus insolens Waite, 1904

Percophidae
Percophis brasiliensis Quoy & Gaimard, 1825
Squamicreedia obtusa Rendahl, 1921

Leptoscopidae
Leptoscopus macropygus (Richardson, 1846)
Lesueurina platycephala Fowler, 1908

Uranoscopidae
Kathetostoma laeve (Bloch & Schneider, 1801)

Blennioidei

Tripterygiidae
Lepidoblennius haplodactylus Steindachner, 1867
Lepidoblennius marmoratus (Macleay, 1878)
Notoclinops segmentatus (McCulloch & Phillipps, 1923)

Clinidae
Cristiceps argyropleura Kner, 1865
Cristiceps aurantiacus Castelnau, 1879
Cristiceps australis Valenciennes, 1836
Heteroclinus johnstoni (Saville-Kent, 1886)
Heteroclinus roseus (Günther, 1861)
Ophiclinus gracilis Waite, 1906
Peronedys anguillaris Steindachner, 1884

Blenniidae
Aspidontus taeniatus Quoy & Gaimard, 1834
Cirripectes alboapicalis (Ogilby, 1899)
Istiblennius dussumieri (Valenciennes, 1836)
Istiblennius lineatus (Valenciennes, 1836)
Lipophrys pholis (Linnaeus, 1758)
Xiphasia setifer Swainson, 1839

Gobiesocoidei

Gobiesocidae
Alabes dorsalis (Richardson, 1845)
Creocele cardinalis (Ramsay, 1882)

Callionymoidei

Callionymidae
Callionymus lyra Linnaeus, 1758
Foetorepus calauropomus (Richardson, 1844)
Repomucenus calcaratus (Macleay, 1881)
Repomucenus limiceps (Ogilby, 1908)

Gobioidei

Eleotridae
Gobiomorphus australis (Krefft, 1864)
Gobiomorphus coxii (Krefft, 1864)
Hypseleotris klunzingeri (Ogilby, 1898)
Milyeringa veritas Whitley, 1945
Mogurnda adspersa (Castelnau, 1878)
Mogurnda larapintae (Zietz, 1896)
Oxyeleotris lineolata (Steindachner, 1867)
Philypnodon grandiceps (Krefft, 1864)

Gobiidae
Amblygobius bynoensis (Richardson, 1844)
Arenigobius bifrenatus (Kner, 1865)
Callogobius mucosus (Günther, 1872)
Caragobius rubristriatus (Saville-Kent, 1889)
Chlamydogobius eremius (Zietz, 1896)
Paragobiodon echinocephalus (Rüppell, 1830)
Paragobiodon xanthosoma (Bleeker, 1852)
Redigobius macrostoma (Günther, 1861)
Taenioides mordax (De Vis, 1883)
Valenciennea longipinnis (Lay & Bennett, 1839)

Kurtoidei

Kurtidae
Kurtus gulliveri Castelnau, 1878

Acanthuroidei

Ephippidae
Drepane punctata (Linnaeus, 1758)

Scatophagidae
Scatophagus argus (Linnaeus, 1766)

Siganidae
Siganus fuscescens (Houttuyn, 1782)

Luvaridae
Luvarus imperialis Rafinesque, 1810

Zanclidae
Zanclus cornutus (Linnaeus, 1758)

Acanthuridae
Acanthurus nigricans (Linnaeus, 1758)

Scombroidei

Sphyraenidae
Sphyraena novaehollandiae Günther, 1860

Gempylidae
Lepidocybium flavobrunneum (Smith, 1849)
Nealotus tripes Johnson, 1865
Rexea solandri (Cuvier, 1832)
Ruvettus pretiosus (Cocco,1833)
Thyrsites atun (Euphrasén, 1791)

Trichiuridae
Assurger anzac (Alexander, 1917)
Lepidopus caudatus (Euphrasén, 1788)

Scombridae
Acanthocybium solandri (Cuvier, 1831)
Gasterochisma melampus Richardson, 1845
Grammatorcynus bicarinatus (Quoy & Gaimard, 1825)
Scomber australasicus Cuvier, 1832
Scomberomorus commerson (Lacépède, 1800)

Xiphiidae
Xiphias gladius Linnaeus, 1758

Stromateoidei

Centrolophidae
Seriolella brama (Günther, 1860)

Nomeidae
Cubiceps whiteleggii (Waite, 1894)

Tetragonuridae
Tetragonurus atlanticus Lowe, 1839

Anabantoidei

Osphronemidae
Betta pugnax (Cantor, 1849)
Macropodus opercularis (Linnaeus 1758)

Pleuronectiformes

Bothidae
Arnoglossus muelleri (Klunzinger, 1872)
Lophonectes gallus Günther, 1880

Scophthalmidae
Scophthalmus maximus (Linnaeus, 1758)

Paralichthyidae
Pseudorhombus argus Weber, 1913
Pseudorhombus arsius (Hamilton-Buchanan, 1822)
Pseudorhombus jenynsii (Bleeker 1855)

Pleuronectidae
Ammotretis rostratus Günther, 1862
Azygopus pinnifasciatus Norman, 1926
Pelotretis flavilatus Waite, 1911
Pleuronectes platessa Linnaeus, 1758

Soleidae
Aserragodes haackeanus (Steindachner, 1883)
Brachirus selheimi (Macleay, 1882)
Paradicula setifer (Paradice, 1927)

Tetraodontiformes

Triacanthidae
Pseudotriacanthus strigilifer (Cantor, 1850)

Balistidae
Balistoides conspicillum (Bloch & Schneider, 1801)

Monacanthidae
Acanthaluteres brownii (Richardson, 1846)
Acanthaluteres spilomelanurus (Quoy & Gaimard, 1824)
Aluterus monoceros (Linnaeus, 1758)
Aluterus scriptus (Osbeck, 1765)
Brachaluteres jacksonianus (Quoy & Gaimard, 1824)
Cantherhines fronticinctus (Günther, 1866)
Cantherhines pullus (Ranzani, 1842)
Chaetodermis penicilligera (Cuvier, 1816)
Meuschenia australis (Donovan, 1824)
Meuschenia freycineti (Quoy and Gaimard, 1824)
Meuschenia galii (Waite, 1905)
Paraluteres prionurus (Bleeker, 1851)
Scobinichthys granulatus (White, 1790)
Thamnaconus analis (Waite, 1904)
Thamnaconus hypargyreus (Cope, 1871)

Ostraciidae
Anoplocapros lenticularis (Richardson, 1841)
Aracana aurita (Shaw, 1798)
Aracana ornata (Gray, 1838)
Capropygia unistriata (Kaup, 1855)

Tetraodontidae
Arothron firmamentum (Temminck & Schlegel, 1847)
Canthigaster solandri (Richardson, 1844)
Contusus richei (Fréminville, 1813)
Lagocephalus sceleratus (Gmelin, 1789)
Omegophora armilla (Waite & McCulloch, 1915)
Tetractenos hamiltoni (Gray & Richardson, 1843)
Torquigener whitleyi (Paradice, 1927)

Molidae
Mola mola (Linnaeus, 1758)
Mola ramsayi (Giglioli, 1883)

Glossary

The definitions are simplified and some are intended to apply only to this text.

anatomical characters features of structure revealed by dissection.

benthic living close to the seafloor.

binomial nomenclature Linnaean system of naming species, allotting each species a generic name and a specific name which are used together to designate the species.

biogeography study of the distribution of plants and animals.

checklist simple classified list of the species found in a specified region, with or without annotation.

chimaera Ghost Sharks and Elephant Fish: cartilaginous fishes with the gill openings concealed by a single gillcover.

Chordates vertebrates and related invertebrates (that at some stage in development have a rod-like structure supporting the body).

cladistic taxonomy classification of species by ordering them strictly on the grounds of recency of descent from other species.

dichotomous key means of identifying a specimen of a species by making a series of choices, each intended to eliminate one or more species from which that under examination needs to be separated.

endemic occurring naturally only in a specific geographic area.

endogenous depression depressive illness largely independent of external causes.

family level of classification; each family contains a single genus or all of a group of closely related genera.

filariasis parasitic disease of humans, and animals, caused by a 'round worm'.

first reviser a rule of the International Code of Zoological Nomenclature alluding to the first author to publish a reasoned choice among possible names for a species.

fluviatile inhabiting rivers.

ganoid fishes a group of primitive fishes, named for their enamel-like scales.

genotype in this context used for a species to be regarded as the type for a genus.

genus (plural: genera) level of classification; each genus contains a single species or a number of species that are more closely related to one another than to any other species.

histology study of the structure of organs at a cellular level.

ichthyofauna the diversity of fish species of a specified region or habitat.

international code refers to the International Code of Zoological Nomenclature.

lamprey eel-like fish with a cartilaginous skeleton, lacking jaws and paired fins.

lateral line sensory structure consisting of a noticeable row of modified scales on each side of a fish, usually extending from the base of the tail to the gill cover.

littoral relating to the shore.

monotreme egg-laying mammals, the platypus and echidna.

nomina nuda scientific names rejected as, although published, they do not satisfy the requirements for availability and validity.

ocellus in this context, a marking on the side of a fish resembling an eye.

ontogenetic relating to ontogeny, early development prior to birth.

osteology study of the bones and skeleton.

pelagic living in open water, not in close relation to the seafloor.

plankton plants and animals, most microscopic, living in open water and moved more by currents than by their own powers of swimming.

pylorus that part of the stomach leading into the intestine; often in fish associated with a cluster of tube-like outgrowths known as pyloric caeca.

sexual dichromatism different colours in males and females.

sexual dimorphism different body shapes in males and females, often with different fin shapes as well.

species one or more populations of an organism, the individuals of which are capable of interbreeding.

synonym each of two or more names for the same species.

synonymy chronological list of the scientific names that have been applied to a species, noting the author and date of each name; here often implying that a name has been relegated to such a list once an earlier name has been found.

systematics study of evolutionary relationships among species, genera, families and higher categories of a system of classification.

taxonomy procedure for classifying living organisms.

type locality geographic locality from which a type specimen was collected.

type specimen designated as pertaining to the original description of a species.

viviparous bearing the young alive as distinct from releasing eggs.

References

Agassiz L (1833–1844) *Recherches sur les poissons fossiles*. Neuchâtel.

Åhlander E, Kullander SO and Fernholm B (1997) Ichthyological Collection Building at the Swedish Museum of Natural History, Stockholm. In *Collection Building in Ichthyology and Herpetology*. (Eds TW Pietsch and WD Anderson) pp. 13–25. American Society of Ichthyologists and Herpetologists.

Alcock AA (1899) *A Descriptive Catalogue of the Indian Deep Sea Fishes in the Indian Museum*. Calcutta.

Alexander WB (1915) The History of Zoology in Western Australia. Part 1. Discoveries in the 17th Century. *Journal of the Natural History and Science Society of Western Australia* **5**, 49–64.

Alexander WB (1916) The History of Zoology in Western Australia. Part 2. 1791–1829. *Journal of the Royal Society of Western Australia* **1**, 83–149.

Alexander WB (1917) Description of new species of fish of the genus *Evoxymetopon* Poey. *Journal of the Royal Society of Western Australia* **2**, 104–105.

Alexander WB (1918) The History of Zoology in Western Australia. Part 3. 1829–1840. *Journal of the Royal Society of Western Australia* **3**, 37–69.

Alexander WB (1920) The History of Zoology in Western Australia. Part 3. 1829–1840. *Journal of the Royal Society of Western Australia* **3**, 37–69.

Allen GR (2009) *Field Guide to Marine Fishes of Tropical Australia and South-east Asia*. Western Australian Museum, Perth.

Allen GR, Midgely SH and Allen M (2002) *Field Guide to the Freshwater Fishes of Australia*. Western Australian Museum, Perth.

Alleyne HG and Macleay WJ (1877) Ichthyology of the Chevert Expedition. *Proceedings of the Linnean Society of New South Wales* **1(1)**, 261–281, 321–359.

Anderson B (1960) *Surveyor of the Sea. The Life and Voyages of Captain George Vancouver*. University of Washington Press, Seattle.

Anderson C (1926) Obituaries. Allan Riverstone McCulloch, 1885–1925. Bibliography. *Records of the Australian Museum* **15 (2)**, 141–143.

Anon. (undated) *Amateur Fishermen's Association of Queensland, Official History, 1904–1997*. Amateur Fishermen's Association of Queensland.

Anon. (1879) Rules of the Sydney Zoological Station. *Proceedings of the Linnean Society of New South Wales* **3(3)**, 163.

Artedi P (1738) *Ichthyologia sive Opera Omnia de Piscibus*. Lugduni Batavorum.

Arthur W (1885) Notes on New Zealand fishes. *Transactions of the New Zealand Institute* **17**, 160–172.

Arundel JH, Beveridge I and Sandeman M (2009) Parasitology in Victoria. In *A History of Parasitology in Australia and Papua New Guinea*. (Eds I Beveridge and PJ O'Donoghue) pp. 259–274. Australian Society for Parasitology.

Bancroft TL (1916) Some Preliminary Notes on the Habits of the Dawson River Barramundi, *Scleropages leichhardtii*. *Proceedings of the Royal Society of Queensland* **28**, 93–95.

Bancroft TL (1924) Some Further Observations on the Dawson River Barramundi: *Scleropages leichhardtii*. *Proceedings of the Royal Society of Queensland* **35**, 46–47.

Bancroft TL (1928) On the life-history of *Ceratodus*. *Proceedings of the Linnean Society of New South Wales* **53** (3), 315–317.

Banfield EJ (1908) *The Confessions of a Beachcomber*. Fisher Unwin, London.

Barrett PH and Freeman RB (1986) Introduction to Volume Six. In *Works of Charles Darwin, Volume 6, Zoology of the Voyage of H. M. S. Beagle* (Eds PH Barrett and RB Freeman) pp. 5–6. William Pickering, London.

Baulch W (1961) Ronald Campbell Gunn, F.R.S. F.L.S. A Biographical Note. In *Van Diemen's Land Correspondents. Letters from R.C. Gunn, R.W. Lawrence, Jorgen Jorgensen, Sir John Franklin and others to Sir William Hooker. 1827–1849*. (Eds TE Burns and JR Skemp) pp. xiii–xix. Queen Victoria Museum, Launceston.

Benham WB (1904) An apparently new species of *Regalecus (R. parkeri). Transactions of the New Zealand Institute* **36**, 198–200.

Bennett FD (1840) *Narrative of a Whaling Voyage Round the Globe.* 2 volumes. Bentley, London.

Bennett GA (1834) *Wanderings in New South Wales, Batavia, Pedir Coast, Singapore and China; being the Journal of a Naturalist in those Countries, during 1832, 1833 and 1834.* 2 volumes. Richard Bentley, London.

Bennett GA (1837) *A Catalogue of the Specimens of Natural History and Miscellaneous Curiosities deposited in the Australian Museum, Sydney.* Tegg, Sydney.

Bennett GA (1859) On the Fish called *Glyphisodon biocellatus. Proceedings of the Zoological Society of London, 1859*, 222–223.

Bennett GA (1860) *Gatherings of a Naturalist in Australasia; being observations principally upon the animal and vegetable productions of New South Wales, New Zealand and some of the Austral Islands.* J. van Voorst, London.

Bennett GA (1871) On the Toad Fish, *Tetraodon hamiltoni,* of New South Wales. *New South Wales Medical Gazette* **1**, 176–181.

Bennett GA, Gibson A and Sherrin W (1835) '*Report on the Epidemic Catarrh or Influenza Prevailing among the Sheep in the Colony of New South Wales in the Year 1835*'. Sydney.

Bennett JW (1828–1830) *Selection from the most remarkable and interesting fishes found on the coast of Ceylon from drawings made in the southern part of that island from living specimens.* Longman, London.

Bennett JW (1994) *Fishes of Ceylon.* Facsimile edition, Wildlife Heritage Trust, Sri Lanka.

Bicheno JE (1851) On a Specimen of *Pristis cirrhatus. Papers and Proceedings of the Royal Society of Van Diemen's Land* **1**, 223–225.

Blandowski W (1858) Recent discoveries in natural history on the lower Murray. *Transactions of the Philosophical Institute of Victoria* **2**, 124–130, 135–137.

Bleeker P (1855) Over eenige visschen van Van Diemensland. *Verhandelingen van het der Koninklijke. Akademie van Wetenschappen te Amsterdam* **2**, 1–3.

Bleeker P (1860) Derde bijdrage tot de Kennis der Vischfauna van Singapoera. *Natuurkundig Tidschrift voor Nederlandsch Indië* **22**, 101–102.

Bleeker P (1862) *Atlas Ichthyologique des Indes Orientales Néêrlandaises. Tome I. Scaroïdes et Labroïdes.* Frédéric Muller, Amsterdam.

Bleeker P (1863) Notices sur une collection de poissons de la Nouvelle Hollande faite à Port-Jackson. *Verslagenen Mededeenlingen der Kononklyke Akademie Amsterdam* **15**, 442–451.

Bleeker P (1983) *Atlas Ichthyologique des Indes Orientales Néêrlandaises. Plates Originally Prepared for Planned Tomes XI-XIV Published Here for the First Time.* Smithsonian Institution Press.

Blewett CF (1929) Habits of some Australian freshwater fishes. *South Australian Naturalist* **10 (2)**, 21–29.

Bloch ME (1785–1795) *Naturgeschichte der ausländischen Fische.* Berlin.

Bloch ME and Schneider JGT (1801) *Systema Ichthyologiae Iconibus.* 2 volumes. Berlin.

Blunt W (1971) *The Compleat Naturalist: A Life of Linnaeus.* London.

Boeseman M (1983) Introduction. In *Atlas Ichthyologique des Indes Orientales Néêrlandaises. Plates Originally Prepared for Planned Tomes XI-XIV Published Here for the First Time.* (P. Bleeker) pp. 1–12. Smithsonian Institution Press.

Bonnaterre JP (1782–1832) *Dictionnaire encyclopédie méthodique.* Paris.

Bonnaterre JP (1788) *Tableau encyclopédique et méthodique … Ichthyologie*. Paris.

Bonnemains J (1988) Biography of Charles-Alexandre Lesueur. In *Baudin in Australian Waters. The Artwork of the French Voyage of Discovery to the Southern Lands 1800–1804* (Eds J Bonnemains, JE Forsyth and B Smith) pp. 19–26. Oxford University Press, Melbourne.

Borodin NA (1932) Scientific Results of the Yacht "Alva" World Cruise, July, 1931 to March, 1932, in Command of William K. Vanderbilt. Fishes. *Bulletin of the Vanderbilt Marine Museum* **1 (3)**, 65–101.

Borodin NA (1933) A new Australian fish. *Copeia 1933* **3**, 141–142.

Bory de Saint-Vincent JBGM (1822–31) *Dictionnaire Classique d'Histoire Naturelle.* Volumes 1–17. Paris.

Botros GA (1971) Fishes of the Red Sea. Oceanography and Marine Biology. *Annals and Review of Oceanography and Marine Biology* **9**, 221–348.

Boulenger GA (1895) *Catalogue of the Fishes in the British Museum.* Second edition, volume 1. British Museum (Natural History) London.

Boulenger GA (1904) Teleostei (Systematic Part). In *Cambridge Natural History.* (Eds SF Harmer and JE Shipley) pp. 541–727. Macmillan, England.

Briggs JC (1974) *Marine Zoogeography.* McGraw Hill Book Company, USA.

Brittan MR (1997) The Stanford School of Ichthyology: Eighty Years (1891–1970) from Jordan (1851–1931) to Myers (1905–1985) In *Collection Building in Ichthyology and Herpetology.* (Eds TW Pietsch and WD Anderson) pp. 233–263. American Society of Ichthyologists and Herpetologists.

Broussonet PMA (1782) *Ichthyologia sistens Piscium descriptiones et icones, decas I.* Elmsly, London.

Brunton P (Ed) (1998) *The Endeavour Journal of Joseph Banks. The Australian Journey.* Angus & Robertson.

Buffon GL, Daubenton LJM, de Montbeillard PG, Bexon GLCA and Lacépède BGE (1749–1804) *Histoire naturelle, générale et particulière, avec la description du cabinet du roi.* Paris.

Calaby J (Ed) (1989) *The Hunter Sketchbook. Birds & Flowers of New South Wales drawn on the Spot in 1788, 89 & 90 By Captain John Hunter RN of the First Fleet.* National Library of Australia, Canberra.

Canestrini G (1869) Sopra alcuni Pesci dell'Australia. *Arch. Zool. Anat. Fisiol. (Genova) (series 2)* **1**, 151–155.

Cantor T (1850) Catalogue of Malayan Fishes. *Journal of the Asiatic Society of Bengal* **18**, xii + 983–1443.

Castelnau FL (1855) Poissons. In *Animaux nouveaux or rares recueillis pendant l'expédition dans les parties centrales de l'Amérique du Sud, de Rio de Janeiro à Lima, et de Lima au Para: exécuté par ordre du gouvernement Français pendant les années 1843 à 1847.* Volume **2**, pp. i–xii + 1–112.

Castelnau FL (1872a) Contributions to the Ichthyology of Australia. No. i - The Melbourne Fish Market. *Proceedings of the Royal Zoological Acclimatization Society of Victoria* **1**, 29–242.

Castelnau FL (1872b) Contributions to the Ichthyology of Australia. No. ii – Note on some South Australian Fishes. *Proceedings of the Royal Zoological Acclimatization Society of Victoria* **1**, 243–248.

Castelnau FL (1873a) Notes on the edible fishes of Victoria. *Intercolonial Exhibition Essays, Victorian Exhibition, 1872-73* **5**, 1–17.

Castelnau FL (1873b) Contributions to the Ichthyology of Australia. No. iii – Supplement to the Fishes of Victoria. No. iv – Fishes of South Australia. No. v – Notes on Fishes from North Australia. No. vi – Notes on Fishes from Knob Island. No. vii – Fishes of New Caledonia. No. viii – Fishes of Western Australia. No. ix – New Sorts for the Victorian Fauna, and List of Australian Fishes. *Proceedings of the Royal Zoological Acclimatization Society of Victoria* **2**, 37–158.

Castelnau FL (1875) Researches on the Fishes of Australia. Philadelphia Centenary Exhibition 1873 (Melbourne, 1875): Official Record. *Intercolonial Exhibition Essays no. 2*, 1–52.

Castelnau FL (1878a) Australian Fishes. New or little known species. *Proceedings of the Linnean Society of New South Wales* **2**, 225–248.

Castelnau FL (1878b) Notes on the Fishes of the Norman River. *Proceedings of the Linnean Society of New South Wales* **3**, 41–51.

Castelnau FL (1878c) On some new Australian (chiefly) Freshwater Fishes. *Proceedings of the Linnean Society of New South Wales* **3**, 140–144.

Castelnau FL (1878d) On a new Ganoid Fish from Queensland. *Proceedings of the Linnean Society of New South Wales* **3**, 164.

Castelnau FL (1879) Essay on the Ichthyology of Port Jackson. *Proceedings of the Linnean Society of New South Wales* **3**, 347–402.

Cleland JB (1912) Injuries and Diseases of Man in Australia Attributable to Animals (Except Insects). *Australasian Medical Gazette* **32**, 269–274.

Cleland JB (1950) The Archibald Watson Memorial Lecture. The Naturalist in Medicine with Particular Reference to Australia. *Medical Journal of Australia* **1 (17)**, 549–563.

Cockerell TDA (1913) The Scales of Some Queensland Fishes. *Memoirs of the Queensland Museum* **2**, 51–59.

Cockerell TDA (1915) The Scales of Some Australian Fishes. *Memoirs of the Queensland Museum* **3**, 35–46.

Cockerell TDA (1916) Some Australian Fish-scales. *Memoirs of the Queensland Museum* **5**, 52–57.

Cooper JG (1864) On new genera and species of Californian fishes – No. III. *Proceedings of the California Academy of Sciences* **(1) 3**, 108–114.
Cope ED (1871) Contribution to the ichthyology of the Lesser Antilles. *Transactions of the American Philosophical Society (new series)* **14**, 445–483.
Cotton BC (1964) *South Australian Mollusca. Chitons.* Government Printer, Adelaide.
Cox C (1905) *An Alphabetical List of the Fishes protected under the Fisheries Act of 1902, with Remarks on each Species.* Government Printer, Sydney.
Cunningham GG (1853) *History of England in the Lives of Englishmen.* London & Edinburgh.
Cuvier GL (1816) *Règne Animal distribué d'après son organization,* ... Edition 1, volume 2. Paris.
Cuvier GL (1817) Sur le genre *Chironectes* Cuv. (*Antennarius* Commerson), *Mémoires du Muséum National d'Histoire Naturelle, Paris* **3**, 418–435.
Cuvier GL (1818) Sur les diodons, vulgairement orbes épineux. *Mémoires du Muséum National d'Histoire Naturelle, Paris* **4**, 121–138.
Cuvier GL (1995) *Historical Portrait of the Progress of Ichthyology, from its Origins to Our Own Time.* (Ed TW Pietsch) Johns Hopkins University Press, Baltimore and London.
Cuvier GL and Valenciennes A (1828–1849) *Histoire Naturelle des Poissons,* volumes 1–22. Levrault, Paris.
Dampier W (1703) *Voyage to New Holland in the Year 1699.* Volume 3. James Knapton, London.
Dannevig HC (1913) *Fisheries. Notes on Australia's Fisheries with a Summary of the Results obtained by the F.I.S. "Endeavour".* Melbourne.
Darwin CR (1836) (Ed R Keynes) (2000) Charles Darwin's Zoology Notes & Specimen Lists from H.M.S. *Beagle.* Cambridge University Press.
Darwin CR (1859) *On the Origin of Species by Means of Natural Selection.* John Murray, London.
Day F (1865) *The Fishes of Malabar.* Quaritch, London.
Day F (1875–78) *The Fishes of India: being a natural history of the fishes known to inhabit the seas and fresh waters of India, Burma, and Ceylon.* William Dawson & Sons, London.
De Freycinet L (1824–25) *Voyage autour de Monde, entrepris par ordre du Roi, exécuté sur les corvettes de S.M. l'Uranie et la Physicienne, pendant les années 1817, 1818, 1819 et 1820.* Paris
De Freycinet R (1996) *The Journal of Rose de Freycinet on her Voyage around the World 1817-1820.* As *A Woman of Courage.* (Ed. MS Rivière) National Library of Australia, Canberra.
De Vis CW (1882a) Description of three new fishes of Queensland. *Proceedings of the Linnean Society of New South Wales* 7, 318–320.
De Vis CW (1882b) Descriptions of some new Queensland fishes. *Proceedings of the Linnean Society of New South Wales* 7, 367–371.
De Vis CW (1883a) Descriptions of New Genera and Species of Australian Fishes. *Proceedings of the Linnean Society of New South Wales* **8**, 283–289.
De Vis CW (1883b) Description of two new Queensland Fishes. *Proceedings of the Linnean Society of New South Wales* **7**, 620–621.
De Vis CW (1884a) New Fishes in the Queensland Museum. *Proceedings of the Linnean Society of New South Wales* **9**, 389–400.
De Vis CW (1884b) New Fishes in the Queensland Museum, Part II. *Proceedings of the Linnean Society of New South Wales* **9**, 453–462.
De Vis CW (1884c) New Fishes in the Queensland Museum, No. 3. *Proceedings of the Linnean Society of New South Wales* **9**, 537–547.
De Vis CW (1884d) New Fishes in the Queensland Museum, No. 4. *Proceedings of the Linnean Society of New South Wales* **9**, 685–698.
De Vis CW (1884e) Fishes from South Sea Islands. *Proceedings of the Linnean Society of New South Wales* **8**, 445–456.
De Vis CW (1884f) On new fishes from Moreton Bay. *Proceedings of the Royal Society of Queensland* **1 (3)**, 144–147.
De Vis CW (1885) New Fishes in the Queensland Museum, No. 5. *Proceedings of the Linnean Society of New South Wales* **9**, 869–887.
Dean BD (1906) Notes on the Living Specimens of the Australian Lungfish, *Ceratodus forsteri,* in the Zoological Society's Collection. *Proceedings of the Zoological Society of London 1906* **1 (12)**, 168–178.

Dean BD (1916–23) *A Bibliography of Fishes.* **1**, 1–718; **2**, 1–702; **3**, 1–707. American Museum of Natural History, New York.

Dodge ES (1971) *Beyond the Capes. Pacific Exploration from Captain Cook to the Challenger (1776–1877).* Gollancz, London.

Dollo L (1904) *Poissons. Expédition antarctique Belge. Zoologie. Résultats du voyage du S. Y. Belgica en 1897–1898–1899 sous le commandement de A. de Gerlache de Gomery.* Rapports Scientifique. Anvers.

Donovan EO (1824) *The Naturalist's Repository; or monthly miscellany of exotic natural history* **3**, unnumbered pages.

Duhig JV and Jones G (1928a) The Venom Apparatus of the Stone Fish (*Synanceja horrida*). *Memoirs of the Queensland Museum* **9 (2)**, 136–150.

Duhig JV and Jones G (1928b) Haemotoxin of the Venom of *Synanceja horrida. Australian Journal of Experimental Biology and Medical Science* **5 (2)**, 173–179.

Duke-Elder S (1958) *System of Ophthalmology, volume 1, The Eye in Evolution.* Henry Kimpton, London.

Duméril AHA (1853) Monographie de la tribu des Scylliens ou Roussettes (poissons plagiostomes) comprenant deux espèces nouvelles. *Revue Magasin Zoologie* (series 2) **5**, 8–25; 73–87; 119–130.

Duyker E (1998) *Nature's Argonaut, Daniel Solander 1733–1782.* Miegunyah Press, Melbourne.

Duyker E (2006) *François Péron. An Impetuous Life. Naturalist and Voyager.* Miegunyah Press, Melbourne.

Eschmeyer WN (Ed) (1998) *Catalog of Fishes.* 3 volumes. California Academy of Sciences, San Francisco.

Etheridge R (1916) The Australian Museum: fragments of its early history. *Records of the Australian Museum* **11**, 67–78.

Etheridge R (1917) Obituary, Edward Pierson Ramsay, LL.D. *Records of the Australian Museum* **11**, 204–217.

Etheridge R (1919) The Australian Museum: fragments of its early history. *Records of the Australian Museum* **12**, 339–399.

Etheridge R and McCulloch AR (1916) Sub-fossil Crustaceans from the Coasts of Australia. *Records of the Australian Museum* **11**, 1–14.

Eydoux JFT and Souleyet LFA (1850) Poissons. In *Voyage autour du monde exécuté pendant les années 1836 et 1837 sur la corvette La Bonite, commandée par M. Vaillant. Zoologie*, Volume 1 (part 2). pp. 155–216.

Finney C (1993) *Paradise Revealed. Natural History in nineteenth-century Australia.* Museum of Victoria, Melbourne.

Fisher C and Calaby J (2009) The Top End of the Top End: John Gilbert's Manuscript Notes for John Gould on Vertebrates from Port Essington and Cobourg Peninsula (Northern Territory, Australia); with Comments on Specimens Collected during the Settlement Period 1838 to 1849, and Subsequently. *The Beagle, Records of the Museums and Art Galleries of the Northern Territory, Supplement* **4**, 1-239.

Flannagan R (2001) *Gould's Book of Fish. A Novel in Twelve Fish.* Pan Macmillan, Australia.

Fletcher JJ (1893) *The Hon. Sir William Macleay, Kt., F.L.S., M.L.C.* Macleay Memorial Volume, Linnean Society of New South Wales, vii–li.

Forsskål P (1775) *Descriptiones animalium avium, amphibiorum, piscium, insectorum, vermium; quae in itinere orientali observavit ...* Post mortem auctoris edidit Carsten Niebuhr. Hauniae.

Forster G (1777) *Voyage round the World in His Britannic Majesty's Sloop, Resolution, commanded by Capt. James Cook, during the years 1772, 3, 4 and 5.* 2 volumes. London.

Forster JR (1788) *Enchiridion historiae naturali inserviens, quo, termini et delineationes ad avium, piscium, insectorum et plantarum adumbrationes intelligendas et concinnandas, secundum methodum systematis Linnaeani continentur.* Halae.

Forster JR (1844) *Descriptiones animalium quae in itinere ad maris Australis terras per annos 1772 1773 et 1774 suscepto collegit observavit et delineavit Ioannes Reinoldus Forster ... nunc demum editae ... Henrico Lichtenstein.* Berlin.

Fortuine R (1969) Sir John Richardson, M.D., Arctic naturalist and explorer. *Canadian Medical Association Journal* **100 (11)**, 531-534.

Fowler HW (1908) A Collection of Fishes from Victoria, Australia. *Proceedings of the Academy of Natural Sciences of Philadelphia* **59**, 419–444.

Fowler HW (1928) The Fishes of Oceania. *Memoirs of the Bernice P. Bishop Museum* **10**, i–iii and 1–540.

Fowler HW (1931) The Fishes of Oceania - Supplement 1. *Memoirs of the Bernice P. Bishop Museum* **11**, 313–381.

Fowler HW (1934) The Fishes of Oceania - Supplement 2. *Memoirs of the Bernice P. Bishop Museum* **11**, 386–466.

Fowler HW (1938) The Marine Fishes of West Africa Based on the Collection of the American Museum Congo Expedition, 1909–1915. *Bulletin of the American Museum of Natural History* **70**, 1–1493.

Fowler HW (1949) The Fishes of Oceania - Supplement 3. *Memoirs of the Bernice P. Bishop Museum* **12**, 37–186.

Fowler HW (1953) Australian Fishes Obtained or Observed by the United States Exploring Expedition, 1838–1842. *Ichthyological Notes (Brisbane)* **2**, 11–20. Department of Harbours and Marine, Queensland.

Gale A (1914) Notes on the breeding habits of the Purple-striped Gudgeon, *Krefftius adspersus. Australian Zoologist* **1 (1)**, 25–26.

Gale A (1915) *Aquarian Nature Studies and Economic Fish Farming.* Government Printer, Sydney.

Garman S (1899) Reports on an Exploration ... by the U.S. fish commission steamer "Albatross" during 1891, The Fishes. *Memoirs of the Museum of Comparative Zoology, Harvard* **24**, 1–431.

Garman S (1903) Some Fishes from Australasia. *Bulletin of the Museum of Comparative Zoology, Harvard* **39**, 229–241.

Garman S (1904) The Chimaeroids (Chismopnea Raf., 1815; Holocephala Mull., 1834), especially *Rhinochimaera* and its allies. *Bulletin of the Museum of Comparative Zoology, Harvard* **41**, 245–272.

Garman S (1906) New Plagiostomia. *Bulletin of the Museum of Comparative Zoology, Harvard* **46**, 203–208.

Garman S (1913) The Plagiostomia (sharks, skates, and rays). *Memoirs of the Museum of Comparative Zoology, Harvard* **36**, i–xiii and 1–515.

George AS (1999) *William Dampier in New Holland: Australia's First Natural Historian.* Bloomings Books, Victoria.

Giglioli EH (1883) Zoology at the Fishery Exhibition. II. Notes on the Vertebrata. *Nature* **28**, 313–316.

Gill TN (1862a) Synopsis of the family of Cirrhitoids. *Proceedings of the Academy of Natural Sciences, Philadelphia* **14 (3–4)**, 102–124.

Gill TN (1862b) On the limits and arrangement of the family of Scombroids. *Proceedings of the Academy of Natural Sciences, Philadelphia* **14 (3–4)**, 124–127.

Gill TN (1864) Synopsis of the pleuronectoids of the eastern coast of North America. *Proceedings of the Academy of Natural Sciences, Philadelphia* **16**, 214–220.

Gill TN (1874) On the Identity of *Esox Lewini* with the *Dinolestes Mülleri* of Klunzinger. *Annals and magazine of natural History* **(4) 14**, 159–160.

Gill TN (1881) A review of Dr Günther's Introduction to the Study of Fishes. *Forest and Stream, Sep. 1, 1881*, 86–88.

Gill TN (1891a) On the Genus *Gnathanacanthus* of Bleeker. *Proceedings of the United States National Museum* **14**, 701–704.

Gill TN (1891b) On the Genera *Labrichthys* and *Pseudolabrus. Proceedings of the United States National Museum* **14**, 395–404.

Gill TN (1893) A comparison of antipodal faunas. *Memoirs of the National Academy of Sciences* **6**, 91–124.

Gill TN (1905) Life History of the Angler. *Smithsonian Miscellaneous Collections* **52 (2)**, 500–516.

Gill TN (1909) The Archer-Fish and its Feats. *Smithsonian Miscellaneous Collections* **52 (2)**, 155–180.

Glauert L (1921) Fish Collected by the Government Trawler "Penguin" near Albany. *Journal of the Royal Society of Western Australia* **7**, 44–47.

Glover CJM and Olsen AM (1985) Fish and Major Fisheries. In *Natural History of Eyre Peninsula* (Eds CR Twidale, MJ Tyler and M Davies) pp. 169–181. Royal Society of South Australia.

Glover M (2007) Robert Neill. *Dictionary of Australian Artists Online.* <http://www.daao.org.au.>

Gmelin JF (1789) *Systema Naturae (Linné).* Edition 13. Delamollière, Leiden.

Gomon MF, Bray D and Kuiter RH (2008) *Fishes of Australia's Southern Coast.* Reed New Holland.

Gon O and Heemstra PC (1990) *Fishes of the Southern Ocean.* J.L.B. Smith Institute of Ichthyology, Grahamstown.

Goode GB and Bean TH (1896) Oceanic Ichthyology, a treatise on the deep-sea and pelagic fishes of the world, based chiefly upon the collections made by the steamers Blake, Albatross, and Fish Hawk in the northwestern Atlantic. *Special Bulletin of the United States National Museum* **2**, i–xxxv + 1–26 + 1–553.

Graham DH (1953) *A Treasury of New Zealand Fishes.* Reed, Wellington.

Grant FE and McCulloch AR (1906) On a Collection of Crustacea from the Port Curtis District, Queensland. *Proceedings of the Linnean Society of New South Wales* **31 (1)**, 2–52.

Grant FE and McCulloch AR (1907) Decapod Crustacea from Norfolk Island. *Proceedings of the Linnean Society of New South Wales* **32 (1)**, 151–156.

Gray JE (1827) Pisces. In King, P.P., *Narrative of a Survey of the Intertropical and Western Coasts of Australia performed between the years 1818 and 1822.* 2: appendix. pp. 435–437.

Gray JE (1830–1835) *Illustrations of Indian Zoology, chiefly selected from the collection of Major-General Hardwicke, F.R.S.* 20 parts in 2 volumes. Treutell, Wurtz, Treutell Jun. and Richter, London.

Gray JE (1838) Notes on Fish. *Annals of Natural History* **1**, 109–111.

Gray JE (1851a) Description of a new form of Lamprey from Australia, with a Synopsis of the Family. *Proceedings of the Zoological Society of London* **19**, 235–241.

Gray JE (1851b) *List of the Specimens of Fish in the Collection of the British Museum. Part 1. Chondropterygii.* London.

Gray JE (1859) Description of a New Genus of Lophobranchiate Fishes from Western Australia. *Proceedings of the Zoological Society of London* **27**, 38–39.

Greenop FS (1944) *Who Travels Alone.* K.G. Murray, Sydney.

Gregory KW (1933) Fish Skulls, a Study of the Evolution of Natural Mechanisms. *Transactions of the American Philosophical Society* **23**, 75–481.

Grgurinovic CA (1997) *Larger Fungi of South Australia.* Adelaide.

Griffin DJG and Tranter HA (1986) *The Decapoda Brachyura of the Siboga Expedition, part VIII, Majidae.* E.J. Brill, Leiden.

Griffith E and Smith CH (1834) *The class Pisces, arranged by the Baron Cuvier, with supplementary additions, by Edward Griffith, F.R.S., &c. and Lieut.-Col. Charles Hamilton Smith* ... London.

Grove DI (1990) *A History of Human Helminthology.* C.A.B.I., Wallingford, U.K.

Guichenot AA (1867) Notice sur le Néosébaste, nouveau genre de poissons de la famille des Scorpènoides, et description d'une nouvelle espèce. *Mémoires de la Société Impériale des Sciences Naturelles de Cherbourg* **13**, 83–89.

Günther ACLG (1859–1870) *Catalogue of the Acanthopterygian Fishes in the Collection of the British Museum*, 8 volumes. British Museum, London.

Günther ACLG (1863) On new species of fishes from Victoria, South Australia. *Annals and Magazine of Natural History* **11**, 114–117.

Günther ACLG (1864a) On a new Generic Type of Fishes discovered by the late Dr. Leichhardt in Queensland. *Annals and Magazine of Natural History* **14**, 195–197.

Günther ACLG (1864b) Descriptions of three new species of fishes in the collection of the British Museum. *Annals and Magazine of Natural History* **14**, 374–376.

Günther ACLG (1864c) *Zoological Record.* London.

Günther ACLG (1865a) *Zoological Record.* London.

Günther ACLG (1865b) On the Pipe-fishes belonging to the genus *Phyllopteryx. Proceedings of the Zoological Society of London 1865* **2**, 327–328.

Günther ACLG (1867a) Descriptions of some new or little-known Species of Fishes in the Collection of the British Museum. *Proceedings of the Zoological Society of London 1867*, 99–104.

Günther ACLG (1867b) Additions to the Knowledge of Australian Reptiles and Fishes. *Annals and Magazine of Natural History* **20**, 45–68.

Günther ACLG (1871a) Description of *Ceratodus,* a Genus of Ganoid Fishes, recently discovered in Rivers of Queensland, Australia. *Proceedings of the Royal Society 1871*, 377–379.

Günther ACLG (1871b) *Ceratodus*, and its place in the system. *Annals and Magazine of Natural History* **(4)** 7, 222–227.

Günther ACLG (1872a) *Zoological Record.* London.

Günther ACLG (1872b) Description of two new fishes from Tasmania. *Annals and Magazine of Natural History* **10**, 183–184.

Günther ACLG (1873–1910) Andrew Garrett's Fische der Südsee. *Journal of the Museum Godeffroy* 1–515.

Günther ACLG (1876a) Remarks on Fishes, with Description of New Species in the British Museum, chiefly from Southern Seas. *Annals and Magazine of Natural History* **17**, 389–402.

Günther ACLG (1876b) On the Urogenital Apparatus of a Blennioid Fish from Tasmania. *Annals and Magazine of Natural History* **17**, 403–404.

Günther ACLG (1878) Preliminary Notices of Deep-Sea Fishes collected during the Voyage of H.M.S. 'Challenger'. *Annals and Magazine of Natural History* **(5) 2**, 17–28, 179–187, 248–251.

Günther ACLG (1880a) *An Introduction to the Study of Fishes.* Black, Edinburgh.

Günther ACLG (1880b) Report on the Shore Fishes collected by H.M.S. 'Challenger' during the years 1873–1876. In *Voyage of H.M.S. 'Challenger'. Zoology* **(1) 6**, 1–74.

Günther ACLG (1884) Reptilia, Batrachia and Pisces. In *Report of the Zoological Collections made in the Indo-Pacific Ocean during the Voyage of H.M.S. 'Alert' 1881–2*, 29–33. London.

Günther ACLG (1886) Note on *Pachymetopon* and the Australian species of *Pimelepterus. Annals and Magazine of Natural History* **(5) 18**, 367–368.

Günther ACLG (1887) Report on the Deep-Sea Fishes collected by H.M.S. 'Challenger' during the years 1873–1876. In *Voyage of H.M.S. 'Challenger'. Zoology* **22**, 1–268.

Günther ACLG (1889) Report on the Pelagic Fishes. In *Voyage of H.M.S. 'Challenger'. Zoology* **31** (78), 1–47.

Gunther AE (1980) *The Founders of Science at the British Museum 1753–1900.* Halesworth Press, England.

Haacke W (1885a) Diagnosen zweier bemerkenswerthes südaustralischer Fische. *Zoologischer Anzeiger* **8**, 508–509.

Haacke W (1885b) Über Helotes scotus und Eimer's Theorie der Thierzeichnungen. *Zoologischer Anzeiger* **8**, 507–508.

Haast J (1874) On *Cheimarrichthys fosteri,* a new genus belonging to the New Zealand freshwater fishes. *Transactions and Proceedings of the New Zealand Institute* **6**, 103–104.

Hale HM (1927–29) *Crustaceans of South Australia.* Parts 1 and 2. Government Printer, Adelaide.

Hale HM (1928) Obituary and Bibliography of Mr. Edgar R. Waite. *Records of the South Australian Museum* **3**, 345–359.

Hale HM (1935) The Egg-case of a Cat Shark, *Scyliorhinus vincenti* (Zietz). *Records of the South Australian Museum* **5 (3)**, 367.

Hale HM (1939) Some Fishes Hitherto Unknown from South Australian Waters. *South Australian Naturalist* **19 (4)**, 1–4.

Hale HM (1941) A New South Australian Pipe-Fish. *South Australian Naturalist* **22**, 10.

Hale HM (1944) Record of the Oblong Sunfish (*Triurus laevis,* Pennant) from South Australia. *South Australian Naturalist* **22** (4), 1–2.

Hale HM (1947) Evidence of the Habit of Oral Gestation in a South Australian Marine Fish (*Apogon conspersus* Klunzinger). *South Australian Naturalist* **24 (3)**, 1–3.

Hale HM (1956) First Hundred Years of the Museum - 1856–1956, *Records of the South Australian Museum* **12**, 1–225.

Halstead BW (1970) *Poisonous and Venomous Marine Animals of the World. Volume 3. Vertebrates.* United States Government Printing Office, Washington.

Hamilton F (1822) *An Account of the Fishes found in the River Ganges and its branches.* Archibald Constable and Company. Edinburgh.

Hamilton H (1916) Notes on the Occurrence of the Genus Trachipterus in New Zealand. *Transactions of the New Zealand Institute* **48**, 371.

Hamilton J and Bruce J (1998) *The Flower Chain. The Early Discovery of Australian Plants.* Kangaroo Press, Australia.

Harrison AJ (1997) *Savant of the Australian Seas. William Saville-Kent (1845–1908) and Australian Fisheries.* Tasmanian Historical Research Association, Hobart.

Haswell WA (1882a) *Catalogue of the Australian stalk- and sessile-eyed Crustacea.* Australian Museum, Sydney.

Haswell WA (1882b) On the structure of the paired fins of *Ceratodus,* with remarks on the general theory of the vertebrate limb. *Proceedings of the Linnean Society of New South Wales* **7 (1)**, 2–11.

Haswell WA (1884a) Studies on the Elasmobranch Skeleton. *Proceedings of the Linnean Society of New South Wales* **9 (1)**, 71–119.

Haswell WA (1884b) Note on the young of the Saw-Fish Shark, *Pristiophorus cirratus. Proceedings of the Linnean Society of New South Wales* **9 (3)**, 680–681.

Haswell WA (1890) On the development of *Chilobranchus rufus. Report of the Australian Association for the Advancement of Science* **2**, 482–487.

Haswell WA (1897) On the Development of *Heterodontus* (*Cestracion*) *phillipi. Proceedings of the Linnean Society of New South Wales* **22 (1)**, 96–103.

Hector J (1872) *Notes on the Edible Fishes.* In FW Hutton and J Hector, *Fishes of New Zealand. Catalogue with Diagnoses of the Species.* James Hughes, Wellington.

Hedley C (1906) The Mollusca of Mast Head Reef, Capricorn Group, Queensland. *Proceedings of the Linnean Society of New South Wales* **31**, 453–479.

Heemstra PC and Randall JE (1993) *Groupers of the World (Family Serranidae, Subfamily Epinephelinae) An annotated and illustrated catalogue of the grouper, rockcod, hind, coral grouper, and lyretail species known to date.* FAO Species Catalogue, volume 16. Food and Agricultural Organization of the United Nations, Rome.

Henderson GC (1907) *Sir George Grey: Pioneer of Empire in Southern Lands.* J.M. Dent, London.

Henle FGJ (1834) Ueber *Narcine,* eine neue Gattung electrischer Rochen nebst einer Synopsis der electrischer Rochen. 1–43. Berlin.

Herzig-Straschil B (1997) *Franz Steindachner (1834–1919) and Other Prime Contributors to the Ichthyological Collection of the Naturhistorisches Museum Wien,* In *Collection Building in Ichthyology and Herpetology.* (Eds TW Pietsch and WD Anderson) pp. 101–108. American Society of Ichthyologists and Herpetologists.

Hobson EC (1842) On the *Callorhynchus Australis. Tasmanian Journal of Natural Science* **1 (1)**, 14–20.

Hoese DF, Bray DJ, Paxton JR and Allen GR (2006) Fishes. In *Zoological Catalogue of Australia. Volume 35.* ABRS and CSIRO Publishing, Australia.

Hora SL (1927) Hamilton-Buchanan's Fish-drawings. *Journal of the Asiatic Society of Bengal* (N.S.) **21,** 99–115.

Hordern M (1989) *Mariners are Warned! John Lort Stokes and H.M.S. Beagle in Australia 1837–1843.* Melbourne University Press.

Hordern M (1997) *King of the Australian Coast. The Work of Phillip Parker King in the* Mermaid *and* Bathurst *1817-1822.* Miegunyah Press.

Houston CS (Ed) (1984) *Arctic Ordeal. The Journal of John Richardson Surgeon-Naturalist with Franklin 1820–1822.* McGill's University Press, Kingston and Montreal.

Humphries P (2009) Wilhelm Blandowski's Contribution to Ichthyology of the Murray-Darling Basin, Australia. *Proceedings of the Royal Society of Victoria* **121 (1)**, 90–108.

Hutton FW (1872) *Fishes of New Zealand. Catalogue with Diagnoses of the Species.* James Hughes, Wellington.

Hutton FW (1873) Supplement to the Fishes of New Zealand. *Transactions of the New Zealand Institute* **5**, 1–12.

Hutton FW (1889) List of the New Zealand Fishes. *Transactions of the New Zealand Institute* **22**, 276–285.

Illidge T (1894) On *Ceratodus forsteri. Proceedings of the Royal Society of Queensland* **10**, 40–44.

Ingram, GJ (1990) The Work of Charles Walter de Vis, alias 'Devis', alias 'Thickthorn'. *Memoirs of the Queensland Museum* **28 (1)**, 1-34.

Iredale T and Whitley GP (1938) The Fluvifaunulae of Australia. *South Australian Naturalist* **18 (4)**, 64–68.

Jacquinot H and Guichenot A (1853) Poissons. In *Voyage au Pôle Sud et dans l'Oceanie sur les corvettes "L'Astrolabe" et "La Zélée"* (J Dumont d'Urville).

Jenyns LT (1842) Fish. In *Zoology of the Voyage of H.M.S. Beagle, part 4* (Ed CR Darwin) pp. v–xvi; 1–172. Smith, Elder, and Co., London.

Johnston RM (1881) Descriptions of two species of fishes (*Trachichthys macleayi* and *Mendosoma allporti*) caught in the estuary of the Derwent. *Proceedings of the Royal Society of Tasmania 1880*, 54–57.

Johnston RM (1883) General and Critical Observations on the Fishes of Tasmania; with a Classified Catalogue of all the known Species. *Proceedings of the Royal Society of Tasmania 1882*, 53–144.

Johnston SJ (undated) *On the Dissection of the Stingray, Trygonoptera testacea.* University of Sydney, Department of Zoology. F.W. White, Sydney.

Johnston TH (1916) Presidential Address. *Proceedings of the Royal Society of Queensland* **28**, 1–30.

Johnston TH (1917) Notes on a *Saprolegnia* epidemic amongst Queensland Fish. *Proceedings of the Royal Society of Queensland* **29**, 125–131.

Johnston TH and Bancroft MJ (1918) Some new Sporozoan parasites of Queensland Freshwater Fish. *Proceedings of the Royal Society of New South Wales* **52**, 520–528.

Johnston TH and Bancroft MJ (1922) The Freshwater Fish Epidemics in Queensland Rivers. *Proceedings of the Royal Society of Queensland* **33 (10)**, 174–210.

Johnston TH and Mawson P (1940) Some Nematodes parasitic in Australian freshwater fish. *Transactions of the Royal Society of South Australia* **64 (2)**, 340–352.

Jones P (1996) The Horn Expedition's place among nineteenth century inland expeditions. In *Exploring Central Australia. Society, the Environment and the 1894 Horn Expedition.* (Eds SR Morton and DJ Mulvaney) pp. 19–28. Surrey Beatty and Sons, New South Wales.

Jordan DS and Evermann JB (1896–1900) The Fishes of North and Middle America: a descriptive catalogue of the species of fish-like vertebrates found in the waters of North America, north of the Isthmus of Panama. *Bulletin of the United States National Museum* **47**, 4 parts.

Jordan DS (1898) Description of a species of fish *Mitsukurina owstonii* from Japan, the type of a distinct family of lamnoid sharks. *Proceedings of the California Academy of Sciences* **(3) 1 (6)**, 199–202.

Jordan DS (1900) Waite's Fishes of the Thetis Expedition. *American Naturalist* **34**, 438.

Jordan DS (1905) *A Guide to the Study of Fishes.* 2 volumes. Henry Holt, New York.

Jordan DS (1907) *Fishes.* Henry Holt, New York.

Jordan DS (1917–1920) *The Genera of Fishes.* 4 parts. Leland Stanford Junior University Publications.

Jordan DS (1922) *Days of a Man.* 2 volumes. World Book Company, New York.

Jordan DS (1923) A Classification of Fishes including families and genera as far as known. *Stanford University Publications, University Series, Biological Sciences* **3 (2)**, 79–244.

Jordan DS (1925) *Fishes.* Appleton & Company, New York, London.

Jordan DS and Evermann BW (1898) Descriptions of new or little-known genera and species of fishes from the United States. *Bulletin of the United States Fish Commission* **17**, 125–133.

Jordan DS and Evermann BW (1905) The Aquatic Resources of the Hawaiian Islands, with a general account of the fish fauna. *Bulletin of the United States Fish Commission* **23**, 1–574.

Jordan DS and Hubbs CL (1919) Studies in Ichthyology. A Monographic review of the family Atherinidae or Silversides. *Leland Stanford Junior University Publications, University Series*, 1–87.

Jordan DS and Seale A (1926) Review of the Engraulidae, with Descriptions of New and rare Species. *Bulletin of the Museum of Comparative Zoology, Harvard* **(5) 46 (11)**, 355–418.

Jordan DS and Snyder JO (1908) On a collection of fishes from Fiji, with notes on certain Hawaiian fishes. *Proceedings of the United States National Museum* **34 (1625)**, 603–617.

Jordan DS, Tanaka S and Snyder JO (1913) A Catalogue of the Fishes of Japan. *Journal of the College of Science, Imperial University of Tokyo* **33 (1)**, 1–497.

Jungersen HFE (1915) Some Facts regarding the Anatomy of the genus *Pegasus. Report of the British Association for the Advancement of Science 1914,* section D **6**, 420–422.

Kaup JJ (1856a) *Catalogue of Lophobranchiate Fish in the Collection of the British Museum*, 1–80. British Museum (Natural History), London.

Kaup JJ (1856b) *Catalogue of Apodal Fish in the Collection of the British Museum.* i–viii and 1–164. British Museum (Natural History), London.

Kaup JJ (1858) Uebersicht der Familie Gadidae. *Archiv für Naturgeschichte* **24 (1)**, 85–93.

Kershaw JA (1906) On some additions to the Fish Fauna of Victoria. *Victorian Naturalist* **23 (6)**, 121–127.

Kershaw JA (1909) Additions to the fish fauna of Victoria, No. 2. *Victorian Naturalist* **26 (6)**, 78–79.

Kesteven HL (1914) On the Venom of the Fish, *Notesthes robusta. Proceedings of the Linnean Society of New South Wales* **39 (1)**, 91–92.

Kesteven HL (1922) A New Interpretation of the Bones in the Palate and Upper Jaw of Fishes. *Journal of Anatomy* **56 (3–4)**, 307–324.

Kesteven HL (1925) Contributions to the Cranial Osteology of the Fishes, No. I. *Tandanus tandanus* Mitchell. *Records of the Australian Museum* **14 (4)**, 271–289.

Kesteven HL (1926) Contributions to the Cranial Osteology of the Fishes, No. II. The Maxillae in the Eels and the Identification of These Bones in the Fishes Generally. *Records of the Australian Museum* **15 (1)**, 132–140.

Kesteven HL (1928) Contributions to the Cranial Osteology of the Fishes, No. VI. Some Percomorph Skulls. *Records of the Australian Museum* **16 (7)**, 316–345.

Kesteven HL (1931) Contributions to the Cranial Osteology of the Fishes, No. VII. The Skull of *Neoceratodus forsteri.* A Study in Phylogeny. *Records of the Australian Museum* **18 (4)**, 236–165.

Klunzinger CB (1871) *Synopsis der Fische des Rothen Meeres.* Wien.

Klunzinger CB (1872) Zur Fischfauna Süd-Australien. *Archiv für Naturgeschichte* **38 (1)**, 17–47.

Klunzinger CB (1879) Die v. Müller'sche Sammlung australischer Fische in Stuttgart. *Sitzungsberichte der kaiserlichen Akademie der Wissenschaften Wien* **80**, 325–340.

Klunzinger CB (1964) *Synopsis der Fische des Rothen Meeres.* Cramer, Weinheim.

Kner R (1865–1867) Fische. Reise der österreichischen Fregatte "Novara" um die Erde in den Jahren 1857–1859, unter den Befehlen des Commodore B. von Wüllerstorf-Urbain. *Wien Zoologischer Theil* **5 (1)**, 1–109; **(2)**, 111–272; **(3)**, 275–433.

Kohlstedt SG (1983) Australian Museums of Natural History: Public Priorities and Scientific Initiatives in the 19th Century. *Historical Records of Australian Science* **5**, 1–19.

Krefft G (1864) Notes on Australian Freshwater Fishes and Descriptions of Four New Species. *Proceedings of the Zoological Society of London 1864*, 182–184.

Krefft G (1868a) Descriptions of some New Australian Freshwater Fishes. *Proceedings of the Zoological Society of London* 1867, 942–944.

Krefft G (1868b) *Notes on the Fauna of Tasmania. Fishes.* F. White, Sydney.

Krefft G (1869) *Snakes of Australia; An Illustrated and Descriptive Catalogue of all the Known Species.* Thomas Richards, Sydney.

Krefft G (1870a) *Letter to the Editor.* Sydney Morning Herald, Jan. 18, 1870: 65, column 5.

Krefft G (1870b) Description of a gigantic Amphibian allied to the genus *Lepidosiren*, from the Wide-Bay district, Queensland. *Proceedings of the Zoological Society of London 1870*, 221–224.

Krefft G (1871) *Mammals of Australia.* Government Printer, Sydney.

Kuiter RH (1993) *Coastal Fishes of South-eastern Australia.* University of Hawaii Press.

Lacépède BGE (1798–1803) *Histoire Naturelle des Poissons, par le Citoyen La Cépède ...* 5 volumes. Plassan, Paris.

Lacépède BGE (1804) Mémoire sur plusieurs animaux de la Nouvelle Hollande dont la description n'a pas encore été publiée. *Annales du Musée d' Histoire Naturelle, Paris* **4**, 184–211.

Laissus Y (1988) François Péron: A Biographical Note. In *Baudin in Australian Waters. The Artwork of the French voyage of Discovery to the Southern Lands 1800-1804.* (Eds J Bonnemains, E Forsyth and B Smith) pp. 31–34. Oxford University Press, Melbourne.

Last PR and Stevens JD (2009) *Sharks and Rays of Australia.* CSIRO Publishing.

Last PR, Scott EOG and Talbot FH (1983) *Fishes of Tasmania.* Tasmanian Fish Development Authority, Hobart.

Lawson P (1993) *The East India Company. A History.* Longman.

Lay GT and Bennett ET (1839) *Fishes.* In *Zoology of Captain Beechey's voyage ... to the Pacific and Behring's Straits ... in 1825–28*, 41–75. London.

Lesson RP (1828) Espèce nouvelle d'Hippocampe. *Bulletin des Sciences Naturelle et de Géologie (Férussac)* **11**, 127–8.

Lesson RP (1830–1831) Poissons. In *Voyage autour du Monde, exécuté sur la Corvette de la Majesté La "Coquille", pendant les anneés 1822, 1823, 1824 et 1825.* (LI Duperrey) Bertrand, Paris.

Lesueur CA (1822) Description of a *Squalus*, of a very large size, which was taken on the coast of New-Jersey. *Journal of the Academy of Natural Science, Philadelphia* **2**, 343–352.

Lesueur CA (1824) Description of two new species of the genus Batrachoid of Lacépède. *Journal of the Academy of natural Science Philadelphia* **3 (2)**, 395–403.

Lines P (1997) *Sir John Richardson and the Royal Hospital Haslar circa 1838–1855.*

Linnaeus C (1758) *Systema Naturae*, Edition X, Holmiae.

Linnaeus C (1766) *Systema Naturae*, Edition XII, Holmiae.

Lockyer N (1915) Harald Christian Dannevig, Director, and the Work and Loss of the F.I.S. "Endeavour". *Biological Results of the Fishing Experiments carried on by the F.I.S. "Endeavour" 1909–1914* **3**, i–vi.

Lönnberg E (1905) *The fishes of the Swedish South Polar expedition.* Wiss. Erg. Schwed. Südpolar-Expedition 1901–03, unter Leitung von Dr. Otto Nordenskjöld **5 (6)**, 1–69.

Lord CE (1923) A List of the Fishes of Tasmania. *Proceedings of the Royal Society of Tasmania 1922*, 60–73.

Lord CE and Scott HH (1924) *A Synopsis of the Vertebrate Animals of Tasmania.* Oldham, Beddome & Meredith, Hobart.

Lowie RH (1968) Amalie Dietrich. *Australian Zoologist* **14 (3)**, 236–245.

Lucas AHS (1890) Systematic Census of Indigenous Fish, hitherto recorded from Victorian Waters. *Proceedings of the Royal Society of Victoria (new series)* **2**, 15–47.

Lucas AHS (1891) On the occurrence of certain fish in Victorian Seas, with descriptions of some new species. *Proceedings of the Royal Society of Victoria* **2 (3)**, 8–14.

Lucas AHS (1936) *The Seaweeds of South Australia. Part I. Introduction, and the Green and Brown Seaweeds.* Government Printer, Adelaide.

Lucas AHS and Perrin F (1947) *The Seaweeds of South Australia. Part II. The Red Seaweeds.* Government Printer, Adelaide.

Mack G (1956) The Queensland Museum 1855–1955. *Memoirs of the Queensland Museum* **13 (2)**, 106–124.

Macleay WJ (1878) The Fishes of Port Darwin. *Proceedings of the Linnean Society of New South Wales* **2**, 344–367.

Macleay WJ (1879a) Descriptions of some new Fishes from Port Jackson and King George's Sound. *Proceedings of the Linnean Society of New South Wales* **3 (1)**, 33–37.

Macleay WJ (1879b) On the Clupeidae of Australia. *Proceedings of the Linnean Society of New South Wales* **4**, 363–385.

Macleay WJ (1880) On the Mugilidae of Australia. *Proceedings of the Linnean Society of New South Wales* **4**, 410–427.

Macleay WJ (1881–82) Descriptive Catalogue of the Fishes of Australia. *Proceedings of the Linnean Society of New South Wales* **5 (3)**, 302–444; **5 (4)**, 510–629; **6 (1)**, 1–138; **6 (2)**, 202–387.

Macleay WJ (1882a) Notes on the Pleuronectidae of Port Jackson, with descriptions of two hitherto unobserved species. *Proceedings of the Linnean Society of New South Wales* **7**, 11–15.

Macleay WJ (1882b) The Fishes of the Palmer River. *Proceedings of the Linnean Society of New South Wales* **7 (1)**, 69–71.

Macleay WJ (1882c) Description of two fishes lately taken in or near Port Jackson. *Proceedings of the Linnean Society of New South Wales* **7 (3)**, 366–367.

Macleay WJ (1882d) Contribution to a knowledge of the fishes of New Guinea - No. I. *Proceedings of the Linnean Society of New South Wales* **7 (2)**, 224–250.

Macleay WJ (1882e) Contribution to a knowledge of the fishes of New Guinea - No. II. *Proceedings of the Linnean Society of New South Wales* **7 (3)**, 351–366.

Macleay WJ (1883a) Contribution to a knowledge of the fishes of New Guinea - No. III. *Proceedings of the Linnean Society of New South Wales* **7 (4)**, 585–598.

Macleay WJ (1883b) Contribution to a knowledge of the fishes of New Guinea - No. IV. *Proceedings of the Linnean Society of New South Wales* **8 (2)**, 252–280.

Macleay WJ (1884a) Notes on a Collection of Fishes from the Burdekin and Mary Rivers, Queensland. *Proceedings of the Linnean Society of New South Wales* **8**, 199–213.

Macleay WJ (1884b) On a New Genus of Fishes from Port Jackson. *Proceedings of the Linnean Society of New South Wales* **8**, 439–441.

Macleay WJ (1884c) Some results of Trawl fishing outside Port Jackson. *Proceedings of the Linnean Society of New South Wales* **8**, 457–462.

Macleay WJ (1884d) Supplement to the Descriptive Catalogue of the Fishes of Australia. *Proceedings of the Linnean Society of New South Wales* **9 (1)**, 2–64.

Macleay WJ (1884e) Notices of New Fishes. *Proceedings of the Linnean Society of New South Wales* **9**, 170–172.

Macleay WJ (1885) New Fishes from the Upper Murrumbidgee. *Proceedings of the Linnean Society of New South Wales* **10**, 267–269.

Macleay WJ (1886a) A Remarkable Fish from Lord Howe Island. *Proceedings of the Linnean Society of New South Wales* **10 (4)**, 1–3.

Macleay WJ (1886b) Note on *Ctenodax wilkinsoni*, Macl. *Proceedings of the Linnean Society of New South Wales* **1 (2)**, 511–512.

Marshall TC (1925) New Fish Records for Queensland. *Memoirs of the Queensland Museum* **8 (2)**, 123–124.

Marshall TC (1964) *Fishes of the Great Barrier Reef and Coastal Waters of Queensland.* Angus & Robertson, Sydney.

Mather P (1986) Time for a Museum: the history of the Queensland Museum 1862-1986. *Memoirs of the Queensland Museum* **24**, 1–365.

McCoy F (1865) Remarks on the Australian fish of the genus *Arripis. Proceedings of the Royal Society of Victoria* **6**, 158–159.

McCoy F (1867) *On the Recent Zoology and Palaeontology of Victoria.* Intercolonial Exhibition Essays **7**, 1–24.

McCoy F (1878–1890) *Prodromus of the Zoology of Victoria: or figures and descriptions of Victorian indigenous animals.* 20 decades in 2 volumes. Government Printer, Melbourne.

McCulloch AR (1906) Crustacea New to Australia. *Records of the Australian Museum* **6 (3)**, 231.

McCulloch AR (1907) The Results of Deep-Sea Investigation in the Tasman Sea. ii. The Expedition of the "Woy Woy". Fishes and Crustaceans from Eight Hundred Fathoms. *Records of the Australian Museum* **6 (5)**, 345–355.

McCulloch AR (1908a) Studies in Australian Fishes. No. 1. *Records of the Australian Museum* **7 (1)**, 36–43.

McCulloch AR (1908b) Studies in Australian Crustacea. No. 1. *Records of the Australian Museum* **7 (1)**, 51–59.

McCulloch AR (1908c) A new Genus and species of Turtle, from north Australia. *Records of the Australian Museum* **7 (2)**, 126–128.

McCulloch AR (1909) Studies in Australian Fishes. No. 2. *Records of the Australian Museum* **7 (4)**, 315–321.

McCulloch AR (1911) Report on the Fishes obtained by the F.I.S. "Endeavour" on the Coasts of New South Wales, Victoria, South Australia, and Tasmania. Part i. *Zoological Results of the Fishing Experiments carried out by F.I.S. "Endeavour" 1909–1910,* **1 (1)**, 1–87.

McCulloch AR (1912) Notes on some Western Australian Fishes. *Records of the Western Australian Museum* **1 (2)**, 78–97.

McCulloch AR (1913a) Studies in Australian Fishes. No. 3. *Records of the Australian Museum* **9 (3)**, 355–389.

McCulloch AR (1913b) A Description and Figures of Three Specimens of *Molacanthus* from the Central Pacific Ocean. *Proceedings of the Linnean Society of New South Wales* **37 (3)**, 553–555.

McCulloch AR (1914a) Notes on some Western Australian Fishes, No. 2. *Records of the Western Australian Museum* **1 (3)**, 211–227.

McCulloch AR (1914b) Report on some Fishes obtained by the F.I.S. "Endeavour" on the Coasts of Queensland, New South Wales, Victoria, Tasmania, South and South-Western Australia. Part ii. *Zoological Results of the Fishing Experiments carried out by F.I.S. "Endeavour" 1909–1910* **2 (3)**, 77–165.

McCulloch AR (1914c) *Notes on some Australian Pipe-fishes. Australian Zoologist* **1 (1)**, 29–31.

McCulloch AR (1914d) *The Fishes of New South Wales.* New South Wales Handbook, British Association for the Advancement of Science.

McCulloch AR (1915a) Report on some Fishes obtained by the F.I.S. "Endeavour" on the Coasts of Queensland, New South Wales, Victoria, Tasmania, South and Southwestern Australia. Part iii. *Zoological Results of the Fishing Experiments carried out by F.I.S. "Endeavour" 1909–1910* **3 (3)**, 97–170.

McCulloch AR (1915b) Notes on, and Descriptions of Australian Fishes. *Proceedings of the Linnean Society of New South Wales* **40 (2)**, 259–277.

McCulloch AR (1915c) Notes and illustrations of Queensland Fishes. *Memoirs of the Queensland Museum* **3**, 47–56.

McCulloch AR (1915d) The Migration of the Jolly-tail or Eel-gudgeon, *Galaxias attenuatus*, from the Sea to Fresh-water. *Australian Zoologist* **1 (2)**, 47–49.

McCulloch AR (1916a) Report on some Fishes obtained by the F.I.S. "Endeavour" on the Coasts of Queensland, New South Wales, Victoria, Tasmania, South and South-western Australia. Part iv. *Zoological Results of the Fishing Experiments carried out by F.I.S. "Endeavour" 1909–1910* **4 (4)**, 169–199.

McCulloch AR (1916b) Ichthyological Items. *Memoirs of the Queensland Museum* **5**, 58–69.

McCulloch AR (1919) The Fishes and Fish-like Animals of New South Wales, part i. *Australian Zoologist* **1 (7)**, 217–227.

McCulloch AR (1920) Studies in Australian Fishes. No. 6. With a description of a new *Girellops* from the Kermadec Islands. *Records of the Australian Museum* **13 (2)**, 41–71.

McCulloch AR (1921a) The Fishes and Fish-like Animals of New South Wales, part ii. *Australian Zoologist* **2 (2)**, 14–58.

McCulloch AR (1921b) Notes and Illustrations of Queensland Fishes No. 2. *Memoirs of the Queensland Museum* **7 (3)**, 164–178.

McCulloch AR (1921c) Electric Rays. *Australian Museum Magazine* **1 (3)**, 89–90.

McCulloch AR (1921d) Lord Howe Island - A Naturalist's Paradise. *Australian Museum Magazine* **1 (2)**, 31–47.

McCulloch AR (1922a) The Fishes and Fish-like Animals of New South Wales, part iii. *Australian Zoologist* **2 (3)**, 86–130.

McCulloch AR (1922b) *The Fishes and Fish-like Animals of New South Wales. Australian Zoologist Handbook, No. 1.* Royal Zoological Society of New South Wales.

McCulloch AR (1923a) Sea Dragons (*Phyllopteryx*). *Australian Museum Magazine* **1 (8)**, 231–232.

McCulloch AR (1923b) Notes on Fishes from Australia and Lord Howe Island. *Records of the Australian Museum* **14 (1)**, 1–17.

McCulloch AR (1923c) Fishes from Australia and Lord Howe Island, No. 2. *Records of the Australian Museum* **14 (2)**, 113–125.

McCulloch AR (1924a) Ichthyological Items, No. 2. *Memoirs of the Queensland Museum* **8 (1)**, 61-76.

McCulloch AR (1924b) Fishes and the Movies. *Australian Museum Magazine* **2 (3)**, 74.

McCulloch AR (1925a) Fishes of Australia. In *The Illustrated Australian Encyclopaedia.* Angus and Robertson, Sydney.

McCulloch AR (1925b) Stone Fishes and the Art of Camouflage. *Australian Museum Magazine* **2 (5)**, 159–162.

McCulloch AR (1925c) Questions to be considered at the Pan-Pacific Fisheries Conference. *Bulletin of the Pan-Pacific Union (new series)* **70**, 7–15.

McCulloch AR (1925d) Allan R. McCulloch's Last Article. *Bulletin of the Pan-Pacific Union (new series)* **71**, 10.

McCulloch AR (1926a) Studies in Australian Fishes. No. 8. *Records of the Australian Museum* **15 (1)**, 28–39.

McCulloch AR (1926b) Report on some Fishes obtained by the F.I.S. "Endeavour" on the Coasts of Queensland, New South Wales, Victoria, Tasmania, South and South-western Australia. Part v. *Zoological Results of the Fishing Experiments carried out by F.I.S. "Endeavour" 1909–1910* **5 (4)**, 157–216.

McCulloch AR (1929–1930) A Check-list of the Fishes Recorded from Australia. *Australian Museum Memoir* **5 (1-4)**, i–x, 1–534.

McCulloch AR (1934) *The Fishes and Fish-like Animals of New South Wales.* (third edition with supplement by Whitley GP) Royal Zoological Society of New South Wales, Sydney.

McCulloch AR and McNeill FA (1918) Some Australian Blennioid Fishes. *Records of the Australian Museum* **12 (2)**, 9–25.

McCulloch AR and McNeill FA (1923) Notes on Australian Decapoda. *Records of the Australian Museum* **14 (1)**, 49–59.

McCulloch AR and Ogilby JD (1919) Some Australian Fishes of the Family Gobiidae. *Records of the Australian Museum* **12 (10)**, 193–291.

McCulloch AR and Phillipps WJ (1923) Notes on New Zealand Fishes. *Records of the Australian Museum* **14 (1)**, 18–22.

McCulloch AR and Waite ER (1915) A Revision of the genus *Aracana* and its allies. *Transactions of the Royal Society of South Australia* **39**, 477–493.

McCulloch AR and Waite ER (1916) Additions to the Fish Fauna of Lord Howe Island, No. 5. *Transactions of the Royal Society of South Australia* **41**, 472–475.

McCulloch AR and Waite ER (1917) Results of the South Australian Museum Expedition to Strzelecki and Cooper Creeks, Pisces. *Transactions of the Royal Society of South Australia* **40**, 437–451.

McCulloch AR and Waite ER (1918a) Some New and Little-known Fishes from South Australia. *Records of the South Australian Museum* **1 (1)**, 39–78.

McCulloch AR and Waite ER (1918b) Descriptions of two new Australian Gobies. *Records of the South Australian Museum* **1 (1)**, 79–82.

McCulloch AR and Whitley GP (1925a) Some Little Known Australian Flat-fishes. *Records of the Australian Museum* **14 (4)**, 342–354.

McCulloch AR and Whitley GP (1925b) A List of the Fishes recorded from Queensland waters. *Memoirs of the Queensland Museum* **8 (2)**, 125–182.

McNeill FA (1929) Obituary. Thomas Whitelegge, 1850–1927. *Records of the Australian Museum* **17 (6)**, 265–277.

Mees GF (1961) Description of a New Fish of the family Galaxiidae from Western Australia. *Journal of the Royal Society of Western Australia* **44**, 33–38.

Mees GF (1962) The Subterranean Freshwater Fauna of Yardie Creek Station, North West Cape, Western Australia. *Journal of the Royal Society of Western Australia* **45**, 24–32.

Meyer FAA (1793) *Systematisch-summarische Uebersicht der neuesten zoologischen Entdeckungen in Neuholland und Afrika.* Leipzig in Verlage der Dykischen Buchhandlung.

Miklouho-Maclay N and Macleay WJ (1879) Plagiostomata of the Pacific, Part I. *Proceedings of the Linnean Society of New South Wales* **3**, 306–334.

Miller RG (1993) *A History and Atlas of the Fishes of the Antarctic Ocean.* Foresta Institute, Nevada.

Mitchell T (1838) *Three Expeditions into the Interior of Eastern Australia; with Descriptions of the Recently Explored Region of Australia Felix, and of the Present Colony of New South Wales.* 2 volumes. London.

Molineux A (1887) *Notes upon Additions to the Museum of the South Australian Public Library, Museum, and Art Gallery.* W.K. Thomas and Company, Adelaide.

Monod T (1963) Achille Valenciennes et l'Histoire Naturelle des Poissons. In *Mélanges Ichthyologiques dédiés a la Mémoire d'Achille Valenciennes (1794–1865) coauteur de l'Histoire Naturelle des Poissons (1828–1849). Mémoires de l'Institut Français d'Afrique Noire* **68**, 9–45.

Morris ER and Raff J (1909) Notes on the Structure of *Asymmetron bassanum*, Günther. *Proceedings of the Royal Society of Victoria (new series)* **22**, 85–90.

Morton A (1897) *Antennarius mitchellii,* sp. Nov. *Papers and Proceedings of the Royal Society of Tasmania 1896*, 98.

Moyal A (1986) *A Bright and Savage Land.* William Collins.

Müller JP and Henle FGJ (1838) Ueber die Gattungen der Plagiostomen. *Archiv für Naturgeschichte* **4**, 83–85.

Müller JP and Henle FGJ (1838–1841) *Systematische Beschreibung der Plagiostomen.* Berlin.

Mulvaney DJ, Morphy H and Petch A (Eds) (1997) 'My Dear Spencer' the Letters of F.J. Gillen to Baldwin Spencer. Hyland House, Melbourne.

Munro ISR (1956–1961) Handbook of Australian fishes. Nos. 1-42. *Fisheries Newsletter* pp. 1–172.

Munro ISR (1967) *The Fishes of New Guinea.* Department of Agriculture, Stock and Fisheries, Port Moresby.

Myers GS (1963) Foreword. In *Genera of Fishes* (DS Jordan). pp. vi–xvi. Stanford University, California.

Myers GS (1971) Regional Monographs on Extra-European Fishes, 1820–1870. In *The Fishes of Zanzibar* (RL Playfair and A Günther) pp. 5–21. Newton K. Gregg, U.S.A.

Nelson JS (2006) *Fishes of the World.* John Wiley & Sons.

Newland ED (1991) George Bennett and Sir Richard Owen: A case study of the colonization of early Australian science. In *International science and national scientific identity.* (Eds RW Home and SG Kohlstedt). Kluwer Academic, Holland.

Nichols JT and Raven HC (1928) A New Melanotaeniin Fish from Queensland. *American Museum Novitates* **296**, 1–2.

Nichols JT and Raven HC (1932) Australian Fresh-Water Fishes. *American Naturalist* **56**, 190–192.

Norman JR (1926) A Report on the Flatfishes (Heterosomata) collected by the F.I.S. "Endeavour", with a Synopsis of the Flatfishes of Australia and a Revision of the Subfamily Rhombosoleinae. *Biol. Res. Endeavour* **5 (5)**, 219–308.

Norman JR (1931) *A History of Fishes*. Benn, London.

Norman JR (1934) *A Systematic monograph of the Flatfishes (Heterosomata). Volume 1. Psettodidae, Bothidae, Pleuronectidae.* British Museum (Natural History), London.

Norman JR (1935) Coast Fishes, Part i. The South Atlantic. *Discovery Report* **12**, 1–58.

Norman JR (1937a) Coast Fishes, Part ii. The Patagonia Region. *Discovery Report* **16**, 1–150.

Norman JR (1937b) *Fishes. B.A.N.Z. Antarctic Research Expedition. 1929–1931. Reports* **(B) 1 (2)**, 51–88.

Norman JR (1938) Coast Fishes, Part iii. The Antarctic Zone. *Discovery Report* **18**, 1–105.

Norman JR (1966) *Draft Synopsis of the Orders, Families and Genera of Recent Fishes and Fish-like Vertebrates.* British Museum (Natural History), London.

Ogilby JD (1883) Catalogue of Birds obtained in Navarro County, Texas. *Scientific Proceedings of the Royal Dublin Society, (new series)* **3**, 169–249.

Ogilby JD (1885) Notes and Descriptions of some rare Port Jackson fishes. *Proceedings of the Linnean Society of New South Wales* **10 (1)**, 119–123.

Ogilby JD (1886) *Catalogue of the Fishes of New South Wales with their principal Synonyms.* Thomas Richards, Sydney.

Ogilby JD (1887a) On an undescribed *Pimelepterus* from Port Jackson. *Proceedings of the Zoological Society of London 1886,* 539–540.

Ogilby JD (1887b) *Catalogue of Fishes and other Exhibits at The Royal Aquarium, Bondi.* Maclardy, Sydney.

Ogilby JD (1888a) Catalogue of the Fishes in the collection of the Australian Museum. part i. Recent Palaeichthyan Fishes. *Australian Museum Catalogue* **14**, i–v, 1–26.

Ogilby JD (1888b) Note on the Cause of Death in Fishes in the National Park, N.S. Wales. *Proceedings of the Linnean Society of New South Wales* **(2) 3 (2)**, 890–891.

Ogilby JD (1889) The Reptiles and Fishes of Lord Howe Island. *Australian Museum Memoir* **2 (3)**, 51–74.

Ogilby JD (1890a) Descriptions of two new species of Australian Lophobranchiate fishes. *Records of the Australian Museum* **1(3)**, 55–56.

Ogilby JD (1890b) Re-description of *Anomalops palpebratus,* (Bodd.). *Records of the Australian Museum* **1(2)**, 69–71.

Ogilby JD (1892) Catalogue of Australian Mammals with introductory notes on general mammalogy. *Australian Museum Catalogue* **16**, 1–142.

Ogilby JD (1893a) *Edible Fishes and Crustaceans of New South Wales.* Government Printer, Sydney.

Ogilby JD (1893b) Description of a new Shark from the Tasmanian coast. *Records of the Australian Museum* **2 (5)**, 62–63.

Ogilby JD (1895) On two new Genera and Species of Fishes from Australia. *Proceedings of the Linnean Society of New South Wales* **(2) 10 (2)**, 320–324.

Ogilby JD (1896a) On a new Genus and Species of Fishes from Maroubra Bay. *Proceedings of the Linnean Society of New South Wales* **21 (1)**, 23–25.

Ogilby JD (1896b) A Monograph of the Australian Marsipobranchii. *Proceedings of the Linnean Society of New South Wales* **21 (3)**, 388–426.

Ogilby JD (1896c) Fish Industry In *New South Wales: "The Mother Colony of the Australias."* (Ed F Hutchinson) pp. 217–225. Government Printer, Sydney.

Ogilby JD (1897) New genera and species of Australian fishes. *Proceedings of the Linnean Society of New South Wales* **22 (1)**, 62–95.

Ogilby JD (1898a) New genera and species of fishes. *Proceedings of the Linnean Society of New South Wales* **23 (3)**, 280–299.

Ogilby JD (1898b) Additions to the Fauna of Lord Howe Island. *Proceedings of the Linnean Society of New South Wales* **23 (4)**, 730–745.

Ogilby JD (1899) Contributions to Australian Ichthyology. *Proceedings of the Linnean Society of New South Wales* **24 (1)**, 154–186.

Ogilby JD (1903) Studies in the Ichthyology of Queensland. *Proceedings of the Royal Society of Queensland* **18**, 7–27.

Ogilby JD (1906) *Honorary Curator's Report.* Amateur Fishermen's Association of Queensland Annual Report, 1905–1906.

Ogilby JD (1907a) Symbranchiate and Apodal Fishes new to Australia. *Proceedings of the Royal Society of Queensland* **20**, 1–15.

Ogilby JD (1907b) Some new Pediculate Fishes. *Proceedings of the Royal Society of Queensland* **20**, 17–25.

Ogilby JD (1907c) *Honorary Curator's Report.* Amateur Fishermen's Association of Queensland Annual Report,1906–1907.

Ogilby JD (1908a) On New Genera and Species of Fishes. *Proceedings of the Royal Society of Queensland* **21**, 1–26.

Ogilby JD (1908b) Descriptions of new Queensland Fishes. *Proceedings of the Royal Society of Queensland* **21**, 87–98.

Ogilby JD (1908c) New or Little Known Fishes in the Queensland Museum. *Annals of the Queensland Museum* **9**, 3–41.

Ogilby JD (1908d) Revision of the Batrachoididae of Queensland. *Annals of the Queensland Museum* **9**, 43–57.

Ogilby JD (1909a) *Honorary Curator's Report.* Amateur Fishermen's Association of Queensland Annual Report, 1908–1909.

Ogilby JD (1909b) *Report by J. Douglas Ogilby on a large fish destructive to oysters.* Report of the Marine Department of Queensland (1908–1909), Appendix **5**, 19–21.

Ogilby JD (1910a) On some new fishes from the Queensland Coast. *Endeavour Series* **1**, 85–139. Royal Society of Queensland, Brisbane.

Ogilby JD (1910b) *Honorary Curator's Report.* Amateur Fishermen's Association of Queensland Annual Report, 1909–1910.

Ogilby JD (1911a) On new or insufficiently described Fishes. *Proceedings of the Royal Society of Queensland* **23 (1)**, 1–55.

Ogilby JD (1911b) *List of the Edible Fishes of Moreton Bay and its Affluents.* Report of the Marine Department of Queensland, 1910-1911, Appendix **A**, 15–16.

Ogilby JD (1912a) On Some Queensland Fishes. *Memoirs of the Queensland Museum* **1**, 26–65.

Ogilby JD (1912b) *Honorary Curator's Report.* Amateur Fishermen's Association of Queensland Annual Report, 1911–1912.

Ogilby JD (1913a) Edible Fishes of Queensland, Part i – Family Pempheridae. Part ii – The Gadopseiform Percoids. *Memoirs of the Queensland Museum* **2**, 60–80.

Ogilby JD (1913b) On Six New or Rare Queensland Fishes. *Memoirs of the Queensland Museum* **2**, 81–89.

Ogilby JD (1913c) Ichthyological Notes. *Memoirs of the Queensland Museum* **2**, 90–93.

Ogilby JD (1915a) Edible Fishes of Queensland, Part iii – Carangidae (No. 1). *Memoirs of the Queensland Museum* **3**, 57–98.

Ogilby JD (1915b) Review of the Queensland Pomacanthinae. *Memoirs of the Queensland Museum* **3**, 99–116.

Ogilby JD (1915c) On some new or little-known Australian Fishes. *Memoirs of the Queensland Museum* **3**, 117–129.

Ogilby JD (1916a) *The Commercial Fishes and Fisheries of Queensland. An Essay.* Government Printer, Brisbane.

Ogilby JD (1916b) Check-list of the Cephalochordates, Selachians, and Fishes of Queensland. *Memoirs of the Queensland Museum* **5**, 70–98.

Ogilby JD (1916c) Edible Fishes of Queensland, Part iv - Synentognathi (No. 1). Part v - Heterosomata (No. 1). Part vi [misprinted iii] - Carangidae (No. 2). Part vii - Lethrinidae (No. 1). Part viii - Sparidae (No. 1). Part ix - Teuthoidea (No. 1). *Memoirs of the Queensland Museum* **5**, 127–177.

Ogilby JD (1916d) Ichthyological Notes, No. 3. *Memoirs of the Queensland Museum* **5**, 181–185.

Ogilby JD (1916e) Review of the Queensland Pomacanthinae. Supplement No. 1. *Memoirs of the Queensland Museum* **5**, 178–180.

Ogilby JD (1918) Edible Fishes of Queensland, Part x - Plesiopidae (No. 1). Part xi - Lutianidae (No. 1). Part xii - Nemipteridae (No. 1). Part xiii - Sciaenidae (No. 1). Part xiv - Balistidae (No. 1). *Memoirs of the Queensland Museum* **6**, 45–90.

Ogilby JD (1920) Edible Fishes of Queensland, Part xv - Serranidae (No. 1). Part xi - Lutianidae (No. 2). Part xvi - Opistognathidae (No. 1). *Memoirs of the Queensland Museum* **7 (1)**, 1–30.

Ogilby JD and McCulloch AR (1908) A Revision of the Australian Orectolobidae. *Proceedings of the Royal Society of New South Wales* **42**, 264–299.

Ogilby JD and McCulloch AR (1916) A Revision of the Australian Therapons with notes on some Papuan Species. *Memoirs of the Queensland Museum* **5**, 99–126.

Owen R (1846) *Lectures on the Comparative Anatomy and Physiology of the Vertebrate Animals. Part I. Fishes.* London.

Paepke H (1999) *Bloch's Fish Collection in the Museum für Naturkunde der Humboldt Universitat zu Berlin.* A.R.G. Gantner Verlag KG.

Pappenheim P (1912) *Die Fische der deutschen südpolar expedition 1901–1903. I. Die Fische der Antarktis und Subantarktis.* Deutsche Südpolar Expedition 1901–1903. *Zoologie V* **13**, 161–182.

Paradice WE (1924) 'A Report on the Sir Edward Pellew Group, with Special Reference to Biology and Physical Features.' Parliamentary Commission of Australia Report. Government Printer, Melbourne.

Paradice WE (1926) Fish and other Marine Animals of Australia, of Special Interest. *Quarterly Review of the Health Inspectors Association of Australia* **4 (3)**, 43–48.

Paradice WE and Whitley GP (1927) Northern Territory Fishes. An annotated list of fishes collected from the waters of the Northern Territory of Australia during the cruises of H.M.A.S. "Geranium", 1923–25. *Memoirs of the Queensland Museum* **9 (1)**, 76–106.

Parker TJ (1893) Preliminary Note on the Vesicula Seminalis and the Spermatophores of *Callorhynchus antarcticus. Transactions of the Australian Association for the Advancement of Science* **4**, 401–402.

Parrot J (2005) Franklin, Sir John (1786–1847) and Jane, Lady Franklin (1791–1875). In *The Companion to Tasmanian History.* (Ed A Alexander) pp. 145–146. Centre for Tasmanian Historical Studies, Hobart.

Paulin CD and Roberts CD (1992) *The Rockpool Fishes of New Zealand.* Museum of New Zealand.

Paxton JR and McGrouther MA (1997) A History of the Fish Collection at the Australian Museum (1860–1968), With a Summary of Current Australian Fish Collections. In *Collection Building in Ichthyology and Herpetology.* (Eds TW Pietsch and WD Anderson) pp. 183–205. American Society of Ichthyologists and Herpetologists.

Péron F (1807) *Voyage de découvertes aux terres australes, exécuté par ordre de Sa Majesté l'Empereur et Roi, sur les corvettes* le Géographe ... volume 1. Paris.

Péron F and de Freycinet L (1816) *Voyage de découvertes aux terres australes, exécuté par ordre de Sa Majesté l'Empereur et Roi, sur les corvettes* le Géographe ... volume 2. Paris.

Perry G (1810–1811) *Arcana, or Museum of Nature; containing Delineations of the most recent Discoveries, on the subject of Natural History: classical and descriptive explanations.* 21 parts. Smeeton, London.

Pescott RTM (1954) *Collections of a Century, the History of the First Hundred Years of the National Museum of Victoria.* National Museum, Melbourne.

Peters W (1864a) Über eine neue Percoidengattung, *Plectroperca*, aus Japan und eine neue Art von Haifischen, *Crossorhinus tentaculatus*, aus Neuholland. *Monatsbericht der Königlich Preussischen Akademie der Wissenschaften zu Berlin*, 121–126.

Peters W (1864b) Berichtete über einige neue Säugethiere ... Amphibien ... und Fische ... *Monatsbericht der Königlich Preussischen Akademie der Wissenschaften zu Berlin*, 381–399.

Peters W (1866) Mittheilung über Fische (*Protopterus, Auliscops, Labrax, Labracoglossa, Nematocentris, Serranus, Scorpis, Opisthognathus, Scomberesox, Acharnes, Anguilla, Gymnomuraena, Chilorhinus, Ophichthys, Helmichthys*). *Monatsbericht der Königlich Preussischen Akademie der Wissenschaften zu Berlin*, 509–526.

Peters W (1869) Über neue oder weniger bekannte Fische des Berliner Zoologischen Museums. *Monatsbericht der Königlich Preussischen Akademie der Wissenschaften zu Berlin*, 703–711.

Peters W (1877) Übersicht der während der von 1874 bis 1876 unter der Commando des Hrn. Capitän z. S. Freiherrn von Schleinitz ausgeführten Reise S. M. S. Gazelle gesammelten und von der Kaiserlichen Admiralität der Königlichen Akademie der Wissenschaften übersandten Fische. *Monatsbericht der Königlich Preussischen Akademie der Wissenschaften zu Berlin*, 831–854.

Phillip A (1789) *The Voyage of Governor Phillip to Botany Bay with an account of the establishment of the colonies of Port Jackson and Norfolk Island ...* John Stockdale, London.

Phillipps WJ (1927) Bibliography of New Zealand Fishes. *New Zealand Marine Department Fisheries Bulletin* **1**, 1–68.

Pietsch TW (Ed) (1995) *Fishes, Crayfishes, and Crabs: Louis Renard's "Natural History of the Rarest Curiosities of the Seas of the Indies."* 2 volumes. Johns Hopkins University Press, Baltimore and London.

Pietsch TW and Anderson WD (Eds) (1997) *Collection Building in Ichthyology and Herpetology.* American Society of Ichthyologists and Herpetologists.

Playfair RL and Gunther ACLG (1867) *The Fishes of Zanzibar, with a list of the fishes of the whole east coast of Africa.* John Van Voorst, London.

Press MM (1979) *Julian Tenison Woods.* Catholic Theological Faculty, Sydney.

Quoy JRC and Gaimard JP (1824–1825) Poissons. In *Voyage autour de Monde, entrepris par ordre du Roi, exécuté sur les corvettes de S.M. l'Uranie et la Physicienne, pendant les années 1817, 1818, 1819 et 1820.* (L. de Freycinet) Paris.

Quoy JRC and Gaimard JP (1834) Poissons. In *Voyage de découvertes de "l'Astrolabe," exécuté par ordre du Roi, pendant les années 1826–29, sous le commandement de M. J. Dumont d'Urville.* Paris.

Raff J (1912) A New Species of Asymmetron from the Great Australian Bight, South Australia. *Biol. Res. Endeavour* **1 (3)**, 301–308.

Rainbow WA (1922) Brief History of the Australian Museum. *Australian Museum Magazine* **1**, 167–170.

Ramsay EP (1871) Remarks on *Ceratodus forsteri. Proceedings of the Zoological Society of London* 1871, 7–8.

Ramsay EP (1876) Notes on the habits of some living Ceratodi in the Australian Museum, Sydney. *Proceedings of the Zoological Society of London* 1876, 698–699.

Ramsay EP (1881a) Description of a new species of *Hemerocoetes* from Port Jackson. *Proceedings of the Linnean Society of New South Wales* **6**, 575.

Ramsay EP (1881b) Description of a parasitic *Syngnatus. Proceedings of the Linnean Society of New South Wales* **5 (4)**, 494–495.

Ramsay EP (1882) On a new species of *Gobiesox* from Tasmania. *Proceedings of the Linnean Society of New South Wales* **2**, 148.

Ramsay EP and Ogilby JD (1886a) A contribution to the knowledge of the fish-fauna of New Guinea. *Proceedings of the Linnean Society of New South Wales* **(2) 1 (1)**, 8-20.

Ramsay EP and Ogilby JD (1886b) On specimens of the genus *Xiphasia*, Swainson, from Port Jackson. *Proceedings of the Linnean Society of New South Wales* **(2) 1 (3)**, 582–584.

Ramsay EP and Ogilby JD (1888a) On a new genus and species of labroid fish from Port Jackson. *Proceedings of the Linnean Society of New South Wales* **(2) 2 (4)**, 631–634.

Ramsay EP and Ogilby JD (1888b) On the Genus *Tetragonurus* of Risso. *Proceedings of the Linnean Society of New South Wales* **(2) 3 (1)**, 9–16.

Ramsay EP and Ogilby JD (1888c) Description of two new Australian fishes. *Proceedings of the Linnean Society of New South Wales* **(2) 3 (3)**, 1310–1312.

Randall JE (1983) *Red Sea Reef Fishes.* Immel Publishing, London.

Rasmussen C (2001) *A Museum for the People. A history of Museum Victoria and its predecessors 1854–2000.* Scribe Publications, Melbourne.

Regan CT (1903) On the classification of the fishes of the suborder Plectognathi; with notes and descriptions of new species from specimens in the British Museum Collection. *Proceedings of the Zoological Society of London* **2 (2)**, 284–303.

Regan CT (1906a) A collection of Fishes from the King River, Western Australia. *Annals and Magazine of Natural History* **(7) 18**, 450–453.

Regan CT (1906b) A revision of the fishes of the family Galaxiidae. *Proceedings of the Zoological Society of London* **2 (2)**, 363–384.

Regan CT (1908) A Revision of the Sharks of the Family Orectolobidae. *Proceedings of the Zoological Society of London* 1908 **3**, 247–364.

Regan CT (1909) A new Specific Name for an Orectolobid Shark. *Annals and Magazine of Natural History* **(8) 3 (18)**, 529.

Regan CT (1913a) A collection of fishes made by Professor Francisco Fuentes at Easter Island. *Proceedings of the Zoological Society of London* 1913 **3**, 368–374.

Regan CT (1913b) The Antarctic fishes of the Scottish National Antarctic Expedition. *Transactions of the Royal Society of Edinburgh* **(49) 2 (2)**, 229–292.

Regan CT (1914) Fishes. *British Antarctic ('Terra Nova') expedition, 1910. Natural History Report. Zoology* **1 (1)**, 1–54.

Regan CT (1925) Dwarfed males parasitic on the females in oceanic angler-fishes (Pediculati Ceratioidea). *Proceedings of the Zoological Society of London* Series B - Biological Science **97** (no. B684), 386–400.

Regan CT (1929) (Articles on Fishes and Selachians) In *Encyclopaedia Britannica* (14th edition). London & New York.

Rendahl CH (1921) Results of Dr. E. Mjöberg's Swedish Scientific Expeditions to Australia, 1910–1913, xxvii, Fische. *Kungl. Svenska Vet. Akad. Handl.* **61 (9)**, 1–24.

Rendahl CH (1922) A Contribution to the Ichthyology of North-West Australia. *Medd. Zool. Mus. Kristiania* **5**, 163–197.

Revell AL (1984) *Haslar the Royal Hospital.* Gosport Society Publication.

Reynolds J (1934) Clive E. Lord, F.L.S. 1889–1933. *Papers and Proceedings of the Royal Society of Tasmania 1933,* 99–102.

Richardson J (1823) Notice of the fishes. No. 6. In *John Franklin. Narrative of a journey to the shores of the Polar Sea in 1819–22 ... with an appendix on various subjects relating to science and natural history.* London.

Richardson J (1836) The Fish. In *Fauna Boreali-Americana; or the zoology of the northern parts of British America: containing descriptions of the objects of natural history collected on the late northern land expeditions, under the command of Sir John Franklin, R.N. Fauna Boreali-Americana* **3**, i–xv; 1–327.

Richardson J (1839) Account of a Collection of Fishes from Port Arthur, Van Diemen's Land. *Proceedings of the Zoological Society of London* **7**, 95–100.

Richardson J (1840) On some new species of fishes from Australia. *Proceedings of the Zoological Society of London* **8**, 25–30.

Richardson J (1841a) On some new or little known fishes from the Australian seas. *Proceedings of the Zoological Society of London* **9**, 21–22.

Richardson J (1841b) Descriptions of Australian Fish. *Transactions of the Zoological Society of London* **3**, 133–184.

Richardson J (1842) Contributions to the Ichthyology of Australia. *Annals and Magazine of Natural History (new series)* **9 (55)**, 15–31.

Richardson J (1843a) Contributions to the Ichthyology of Australia. *Annals and Magazine of Natural History* **(1) 11 (71)**, 352–359.

Richardson J (1843b) Report on the present state of ichthyology of New Zealand. *Report of the British Association for the Advancement of Science 1842,* 12–30.

Richardson J (1843c) *Icones Piscium or Plates of Rare Fishes.* London.

Richardson J (1844) Description of Australian Fish. (Part 2). *Transactions of the Zoological Society of London* **3**, 133-185.

Richardson J (1844–45) *Ichthyology*. In *The zoology of the voyage of H.M.S. Sulphur, under the command of Captain Sir Edward Belcher, R. N., C. B., F. R. G. S., etc., during the years 1836–42.* 3 parts. (Ed RB Hinds) Smith, Elder & Co., London.

Richardson J (1844–48) *Ichthyology.* In *The Zoology of the Voyage of H.M.S. "Erebus" & "Terror" under the command of Captain Sir James Clark Ross, R.N., F.R.S., during the years 1839–1843.* (Eds J Richardson and JE Gray) E.W. Janson, London.

Richardson J (1845) Catalogue of Reptiles and Fish found at King George's Sound, by Deputy Assistant Commissary-General Neill. In E.J. Eyre, *Journals of Expeditions of Discovery into Central Australia and Overland from Adelaide to King George's Sound, in the years 1840–1.* London.

Richardson J (1846a) Descriptions of six fish. In J. L. Stokes, volume 1. *Discoveries in Australia; with an account of the coasts and rivers explored and surveyed during the voyage of H.M.S. Beagle, between the years 1837 and 1843.* 2 volumes. London.

Richardson J (1846b) Report on the Ichthyology of the Seas of China and Japan. *Report of the British Association for the Advancement of Science* 15th meeting, 187–320.

Richardson J (1848) *Fishes.* In: A. Adams. *The zoology of the voyage of H.M.S. Samarang; under the command of Captain Sir Edward Belcher, during the years 1843–1846.* London.

Richardson J (1850) Notices of Australian Fish. *Proceedings of the Zoological Society of London* **18**, 58–77.

Richardson J (1856) *Ichthyology* in *Encyclopaedia Britannica.* 8th edition. London.

Richardson J (1858) On *Siphonognathus*, a new Genus of Fistulariidae. *Proceedings of the Zoological Society of London* 1857, 237–240.

Richardson J (1860) *Fishes.* The Museum of Natural History, London.

Richardson J and Gray JE (1843) *List of fish hitherto detected on the coasts of New Zealand, ...; with the description, by J. E. Gray, Esq., and Dr. Richardson, of the new species brought home by Dr. Dieffenbach.* Pp. 206–228. In E. Dieffenbach, *Travels in New Zealand,* London.

Roughley TC (1913) Fishes of Australia. Their Utilisation in Applied Art. *Technichal Gazette of New South Wales* **3 (1)**, 9–12; **3 (2)**, 10–20; **3 (3)**, 12–20.

Roughley TC (1916) *Fishes of Australia and their Technology.* Government Printer, Sydney.

Roughley TC (1933) *The Life History of the Australian Oyster (Ostrea commercialis). Proceedings of the Linnean Society of New South Wales* **58 (3, 4)**, 279–333.

Roughley TC (1936) *Wonders of the Great Barrier Reef.* Angus & Robertson, Sydney.

Roughley TC (1951) *Fish and Fisheries of Australia.* Angus & Robertson, Sydney.

Roule L (1913) *Deuxième expédition antarctique française. Poissons Antarctique.* 1–24. Paris.

Rüppell WPES (1828) *Atlas zu der Reise im nördlichen Africa. Fische des Rothen Meeres.* Frankfurt.

Russell B, Fraser TH and Larson HK (2010) Castelnau's Collection of Singapore Fishes described by Pieter Bleeker. *The Raffles Bulletin of Zoology 2010* **58 (1)**, 93–102.

Russell P (1803) *Descriptions and figures of two hundred fishes; collected at Vizagapatam on the coast of Coromandel.* 2 volumes. London.

Saville-Kent W (1880–82) *A Manual of the Infusoria.* D. Bogue, London.

Saville-Kent W (1884) *Fisheries of Tasmania: Report.* Parl. Tas. Rept. 78, 1–6.

Saville-Kent W (1885) '*Fisheries Department: Report for the Year ending 31st July, 1885.*' Parl. Tas. Rept. 90, 1–14.

Saville-Kent W (1886) Observations on a suspected hybrid species of Trumpeter, and upon other rare fish taken in Tasmanian Waters. *Proceedings of the Royal Society of Tasmania* 1886, xxxiii–xxxiv; 117–124.

Saville-Kent W (1889a) Preliminary Observations on a Natural History Collection made in connection with the Surveying Cruise of H.M.S. "Myrmidon," at Port Darwin and Cambridge Gulf. *Proceedings of the Royal Society of Queensland* **6 (5)**, 219–242.

Saville-Kent W (1889b) '*Preliminary Report on the Food-Fishes of Queensland*'. *Government Printer, Brisbane.*

Saville-Kent W (1892) Description of a new species of true Barramundi, *Osteoglossum jardinii*, from North Queensland. *Proceedings of the Royal Society of Queensland* **8**, 105–108.

Saville-Kent W (1893) *The Great Barrier Reef of Australia; its Products and Potentialities.* W.H. Allen, London.

Saville-Kent W (1897) *The Naturalist in Australia.* Chapman & Hall, London.

Schmidt J (1928) The fresh-water eels of Australia. With some remarks on the short-finned species of Anguilla. *Records of the Australian Museum* **16 (4)**, 179–210.

Seba A (2005) *Cabinet of Natural Curiosities. The Complete Plates in Colour 1734–1765.* Taschen.

Seba A (1734–1765) *Loccupletissimi Rerum Thesauri Accurata Descriptio*, 4 volumes.

Semon RW (1893) *Ceratodus forsteri. Zoolog. Forschungsreisen, i, Jena,* 1–50.

Semon RW (1899) *In the Australian bush and on the coast of the Coral Sea.* Macmillan, London & New York.

Shaw G and Nodder FP (1790–1813) *The Naturalist's Miscellany.* 24 parts in 12 volumes. London.

Shaw G (1794) *Zoology of New Holland.* Sowerby, London.

Shaw G (1801-13) *General Zoology.* 14 volumes. London.

Sheets-Pyenson S (1988) *Cathedrals of Science. The Development of Colonial Natural History Museums during the Late Nineteenth Century.* Kingston & Montreal: McGill-Queen's University Press.

Sherrard JE (1900) *Illustrated Handbook. Aquarium, Picture Salon, Cyclorama, Museum, and Technological Collections.* Government Printer, Melbourne.

Sherrin RAA (1886) *Handbook of the Fishes of New Zealand.* Wilsons & Horton, Auckland.

Smith A (1838–47) *Illustrations from the zoology of South Africa, consisting chiefly of figures and descriptions of the objects of natural history collected during an expedition into the interior of South Africa in 1834–36, Volume 4, Pisces.* London.

Smith JLB (1958) Fishes of the Families Tetrarogidae, Caracanthidae and Synanciidae, from the Western Indian Ocean with further notes on Scorpaenid fishes. *Ichthyological Bulletin of the J. L. B. Smith Institute of Ichthyology* **12**, 167–181.

Snow CP (1947) *The Light and the Dark.* Faber & Faber, London.

Spencer WB (1890) The Pineal Eye of *Mordacia mordax*. *Proceedings of the Royal Society of Victoria (new series)* **2**, 102–105.

Spencer WB (1892) Notes on the habits of *Ceratodus forsteri*. *Proceedings of the Royal Society of Victoria (new series)* **4 (1)**, 81–84.

Spencer WB (1893) *Contributions to our knowledge of Ceratodus*. Part I The Blood-Vessels, Macleay Memorial Volume Linnean Society of New South Wales, 1–34.

Spencer WB (1899) The Anatomy of *Amphiptyches urna* (Grube & Wagener). *Transactions of the Royal Society of Victoria* 1.

Spoehr FM (1963) *White Falcon. The House of Godeffroy and Its Commercial and Scientific Role in the Pacific*. Pacific Books, California.

Springer VG (1983) *Tyson belos*, New Genus and Species of Western Pacific Fish (Gobiidae, Xenisthminae), with Discussions of Gobioid Osteology and Classification. *Smithsonian Contributions to Zoology* **390**, 1–40.

Springer VG and Fraser TH (1976) Synonymy of the Fish Families Cheilobranchidae (= Alabetidae) and Gobiesocidae, with descriptions of two new species of *Alabes*. *Smithsonian Contributions to Zoology* **234**, 1–23.

Stanbury PJ and Holland J (Eds) (1988) *Mr. Macleay's Celebrated Cabinet: the History of the Macleays and Their Museum*. Macleay Museum, University of Sydney.

Stanbury PJ (1987) The Discovery of the Australian Fauna and the Establishment of Collections. In: *Fauna of Australia. Volume 1A*. (Eds GR Dyne and DW Walton) pp. 202–226. Australian Government Publishing Service.

Starks EC (1901) Synonymy of the Fish Skeleton. *Proceedings of the Washington Academy of Science* **3**, 507–539.

Stead DJ (1906) *Fishes of Australia*. William Brooks, Sydney.

Stead DJ (1907a) *Additions to the Fish-Fauna of New South Wales (No. 1)*, 1–28. Department of Fisheries New South Wales.

Stead DJ (1907b) *Notes on Eggs and Breeding-Habits of fishes*, William Brooks, Sydney.

Stead DJ (1908a) *The Edible Fishes of New South Wales*. Government Printer, Sydney.

Stead DJ (1908b) *New Fishes from New South Wales (No. 1)*, 1–12. Department of Fisheries New South Wales.

Stead DJ (1910) *A Brief Review of the fisheries of New South Wales: present and potential*. Government Printer, Sydney.

Stead DJ (1911) *On the Need for More Uniformity in the Vernacular Names of Australian Edible Fishes*. Government Printer, Sydney.

Stead DJ (1929) Development of Northern Australia with Special Reference to Australia's Tropical Fisheries. *Australian Geographer* **1 (2)**, 3–25.

Stead DJ (1933) *Giants and Pigmies of the Deep*. Shakespeare Head Press, Sydney.

Stead DJ (1963) *Sharks and Rays in Australian Seas*. Angus & Robertson, Sydney.

Steindachner F (1866) Zur Fischfauna von Port Jackson in Australien. *Sitzungsberichte der kaiserlichen Akademie der Wissenschaften Wien* **53**, 424–480.

Steindachner F (1867) Uber einige Fische aus dem Fitzroy-Flusse bei Rockhampton in Ost-Australien. *Sitzungsberichte der kaiserlichen Akademie der Wissenschaften Wien* **55**, 9–16.

Steindachner F (1879) Über einige neue und seltene Fisch-arten aus den zoologischen Museen zu Wien, Stuttgart und Warschau. *Anzeiger Akademie der Wissenschaften Wien* **16 (4)**, 29–34.

Steindachner F (1883) Ichthyologische Bëitrage (XIII) *Anzeiger Akademie der Wissenschaften Wien* **20 (22),** 194–197.

Steindachner F (1884) Ichthyologische Bëitrage (XIII). I. Bëitrage zur Kenntniss der Fische Australiens. *Sitzungsberichte der kaiserlichen Akademie der Wissenschaften Wien* **88 (4)**, 1065–1108.

Steindachner F (1898) Die Fische der Sammlung Plate. *Zoologischen Jahrbüchern* **4**, 281-338.

Steindachner F (1901) Fische aus dem Stillen Ocean. Ergebenisse einer Reise nach dem Pacific (Schauinsland, 1896–1897). *Denkschriften Akademie der Wissenschaften, Wien* **70**, 483-522.

Stewart DA (1931) Sir John Richardson, Surgeon, Physician, Sailor, Explorer, Naturalist, Scholar. *The British Medical Journal*, January 1931, 110-112.

Strahan R (1979) *Rare and Curious Specimens, an Illustrated History of the Australian Museum 1827–1979*. Australian Museum, Sydney.

Sumner R (1993) *A Woman in the Wilderness. The Story of Amalie Dietrich in Australia*. University of New South Wales Press, Sydney.

Swainston R (2011) *Swainston's Fishes of Australia. The complete illustrated guide.* Penguin Viking.

Temminck JC and Schlegel H (1842–1850) Pisces. In *Fauna Japonica* (P von Siebold). Lugdini Batavorum, Leiden.

Tench W (1789) *A Narrative of the Expedition to Botany Bay.* Debrett, London.

Tenison-Woods JE (1862) *Geological Observations in South Australia, principally in the district south-east of Adelaide.* Longmans, London.

Tenison-Woods JE (1865) *A History of the Discovery and Exploration of Australia; or an account of the progress of geographical discovery in that continent from the earliest period to the present day.* 2 volumes, London.

Tenison-Woods JE (1882) *Fish and Fisheries of New South Wales.* Government Printer, Sydney.

Tenison-Woods JE (1888) Fisheries of the Oriental Region. *Proceedings of the Linnean Society of New South Wales* **(2) 3,** 155–265.

Tosh JR (1902) On the Common Whiting of Moreton Bay (*Sillago bassensis*). *Proceedings of the Royal Society of Queensland* **17**, 175–184.

Tosh JR (1903) 'Marine Biologist's Report. Notes on the Habits, Development, etc., of the Common Food Fishes of Moreton Bay'. Appendix to the Parliamentary Report of the Marine Department of Queensland for 1902–03.

Vallance TG (1977) Gilbert Percy Whitley, 1903–1975. *Proceedings of the Linnean Society of New South Wales* **101 (4)**, 256–260.

Verco J (1935) *Combing the Southern Seas.* (Ed Cotton BC) Mail Newspapers, Adelaide.

Waite ER (1891) Vertebrates on the Western Ainsty. *The Naturalist, 1891*, 81–114.

Waite ER (1893) Notes on Australian Typhlopidae. *Records of the Australian Museum* **2**, 57–62.

Waite ER (1894) New or Rare Fishes from Maroubra, New South Wales. *Proceedings of the Linnean Society of New South Wales* **(2) 9**, 215–227.

Waite ER (1895) On the egg-cases of some Port Jackson Sharks. *Journal of the Linnean Society of New South Wales* **25**, 325–329.

Waite ER (1896) Muridae of the Horn Expedition. In *Report on the Work of the Horn Scientific Expedition to Central Australia. Part 2. Zoology* (Ed WB Spencer) pp. 176–180. Melville, Mullin and Slade, Melbourne.

Waite ER (1897) The Mammals, Reptiles and Fishes of Funafuti. *Memoirs of the Australian Museum* **3**, 165–201.

Waite ER (1898a) Report on the Fishes. In *Report upon Trawling Operations off the Coast of New South Wales between the Manning River and Jervis Bay, carried on by H.M.C.S. Thetis.* pp. 23–62. Government Printer, Sydney.

Waite ER (1898b) *A Popular Account of Australian Snakes, with a complete list of the species and an introduction to their habits and organisation.* Thomas Shine, Sydney.

Waite ER (1899a) *Stegostoma tigrinum* Gmel. An Addition to the Fauna of New South Wales. *Records of the Australian Museum* **3 (5)**, 133–134.

Waite ER (1899b) *Regalecus glesne*. An Addition to the Fauna of New South Wales. *Records of the Australian Museum* **3 (6)**, 163-165.

Waite ER (1899c) *Lampris luna.* Its recurrence in New Zealand Waters. *Records of the Australian Museum* **3 (6)**, 166–167.

Waite ER (1899d) The Fishes of Funafuti (Supplement). *Memoirs of the Australian Museum* **3 (9)**, 540–546.

Waite ER (1899e) Introduction. In *Scientific Results of the Trawling Expedition of H.M.C.S. "Thetis". Australian Museum Memoir* **4 (1) 1**, 1–23.

Waite ER (1899f) Fishes. In *Scientific Results of the Trawling Expedition of H.M.C.S. "Thetis". Australian Museum Memoir* **4 (1) 1**, 25–132.

Waite ER (1900a) Notes on Fishes from Western Australia and Description of a new Species. *Records of the Australian Museum* **3**, 210–216.

Waite ER (1900b) The Card-Catalogue System Adapted to Museum Requirements. *Records of the Australian Museum* **3**, 217–218.

Waite ER (1901) Studies in Australian Sharks with Diagnosis of a New Family. *Records of the Australian Museum* **4**, 28–35.

Waite ER (1902a) Studies in Australian Sharks, No. 2. *Records of the Australian Museum* **4**, 175–178.

Waite ER (1902b) Notes on Fishes from Western Australia, No. 2. *Records of the Australian Museum* **4**, 179–194.

Waite ER (1902c) New Records or Recurrences of Rare Fishes from Eastern Australia. *Records of the Australian Museum* **4**, 263–273.

Waite ER (1902d) Skeleton of *Luvarus imperialis*, Rafinesque. *Records of the Australian Museum* **4**, 292–297.

Waite ER (1903) Studies in Australian Sharks No. 3. *Records of the Australian Museum* **6**, 226–229.

Waite ER (1904a) Additions to the Fish Fauna of Lord Howe Island, No. 4. Catalogue of the fishes of Lord Howe Island. *Records of the Australian Museum* **5**, 135–230.

Waite ER (1904b) New Records and Recurrences of Rare Fishes from Eastern Australia. No. 3. *Records of the Australian Museum* **5**, 231–244.

Waite ER (1904c) A Review of the Eleotrids of New South Wales. *Records of the Australian Museum* **5**, 277–286.

Waite ER (1904d) The Breeding Habits of the Fighting Fish. *Records of the Australian Museum* **5**, 293–295.

Waite ER (1904e) Synopsis of the Fishes of New South Wales. *Memoirs of the New South Wales Naturalists' Club* **2**, i–iv; 1–59.

Waite ER (1905a) The Breeding Habits of the Paradise Fish. *Records of the Australian Museum* **6**, 1–4.

Waite ER (1905b) Notes on Fishes from Western Australia, No.3. *Records of the Australian Museum* **6**, 55–82.

Waite ER (1906) Descriptions of and Notes on some Australian and Tasmanian Fishes. *Records of the Australian Museum* **6**, 194–210.

Waite ER (1907) Basic List of the Fishes of New Zealand. *Records of the Canterbury Museum* **1**, 1–39.

Waite ER (1909a) Vertebrata of the Subantarctic Islands of New Zealand. Pisces. *Subantarctic Islands of New Zealand* **25**, 543–598.

Waite ER (1909b) Scientific Results of the New Zealand Government Trawling Expedition, 1907. Introduction. *Records of the Canterbury Museum* **1**, 45–64.

Waite ER (1909c) Scientific Results of the New Zealand Government Trawling Expedition, 1907. Pisces. Part 1. *Records of the Canterbury Museum* **1 (2)**, 131–156.

Waite ER (1910a) A List of the Known Fishes of Kermadec and Norfolk Islands. *Transactions of the New Zealand Institute* **42**, 370–383.

Waite ER (1910b) Notes on New Zealand Fishes. *Transactions of the New Zealand Institute* **42**, 384–391.

Waite ER (1911a) Scientific Results of the New Zealand Government Trawling Expedition, 1907. Pisces. Part 2. *Records of the Canterbury Museum* **1 (3)**, 157–272.

Waite ER (1911b) Antarctic Fishes. In *British Antarctic Expediton 1907–9 under the command of Sir E. H. Shackleton, C.V.O. Report of Scientific Investigations* **2 (2)**, 11–16.

Waite ER (1912a) Notes on New Zealand Fishes: No. 2. *Transactions of the New Zealand Institute* **44**, 194–202.

Waite ER (1912b) Additions and Corrections to the Basic List of the Fishes of New Zealand. *Records of the Canterbury Museum* **1 (4)**, 313–322.

Waite ER (1913a) Notes on New Zealand Fishes: No. 3. *Transactions of the New Zealand Institute* **45**, 215–224.

Waite ER (1913b) Fishes of the Genus *Tripterygion* and Related genera in New Zealand. *Records of the Canterbury Museum* **2 (1)**, 1–16.

Waite ER (1914) Notes on New Zealand Fishes: No. 4. *Transactions of the New Zealand Institute* **46**, 127–131.

Waite ER (1916a) Fishes of the Australasian Antarctic Expedition. *Australian Antarctic Expedition Report series c* **3**, 1–92.

Waite ER (1916b) A List of the Fishes of Norfolk Island and Indication of their Range to Lord Howe Island, Kermadec Island, Australia, and New Zealand. *Transactions of the Royal Society of South Australia* **40**, 452–458.

Waite ER (1921) Illustrated Catalogue of the Fishes of South Australia. *Records of the South Australian Museum* **2 (1)**, 1–208.

Waite ER (1922a) Description of a New Australian Fish of the Genus *Congiopus. Records of the South Australian Museum* **2 (2)**, 215–217.

Waite ER (1922b) Studies in Australian Sharks, No. 4. *Records of the South Australian Museum* **2 (2)**, 219–220.

Waite ER (1923a) The Flora and Fauna of the Nuyts Archipelago and the Investigator Group. No. 7. - The Fishes. *Transactions of the Royal Society of South Australia* **47**, 95–96.

Waite ER (1923b) *The Fishes of South Australia.* South Australian Branch of the British Science Guild, Adelaide.

Waite ER (1924) Illustrations of and Notes on some Australian Fishes. *Records of the South Australian Museum* **2 (4)**, 479–487.

Waite ER (1927) Supplement to the Catalogue of the Fishes of South Australia. *Records of the South Australian Museum* **3 (3)**, 223–234.

Waite ER (1928) A Catalogue of the Marine Fishes of South Australia. *Journal of the Pan-Pacific Research Institute* **3 (1)**, 3–10.

Waite ER (1929) *The Reptiles and Amphibians of South Australia.* Government Printer, Adelaide.

Waite ER and Hale HM (1921) Review of the Lophobranchiate Fishes of South Australia. *Records of the South Australian Museum* **1 (4)**, 293–324.

Waite ER and McCulloch AR (1915) The Fishes of the South Australian Government Trawling Cruise, 1914. *Transactions of the Royal Society of South Australia* **39**, 455–476.

Watts P, Pomfrett JA and Mabberley D (1998) *An Exquisite Eye. The Australian Flora and Fauna Drawings 1801–1820 of Ferdinand Bauer.* Historic Houses Trust of New South Wales.

Weber M (1895) *Fische von Ambon, Java, Thursday Island, dem Burnett-Fluss und von der Süd-Küste von Neu-Guinea.* In: Zoologische Forschungsreisen in Australien und dem malayischen Archipel; mit Unterstützung des Herrn Dr. Paul von Ritter ausgeführt ... Jahren 1891–1893 von Dr. Richard Semon. *Fische Malayan Archipelago* **5**, 259–276.

Weber M (1907) *Süsswasserfische von Neu-Guinea ein Beitrag zur Frage nach dem früheren Zusammenhang von Neu-Guinea und Australien.* In *Nova Guinea. Résultats de l'expédition scientifique Néerlandaise à la Nouvelle-Guinée* **5 (2)**, 210–267.

Weber M (1909) Diagnosen neuer Fische der Siboga-Expedition. *Notes Leyden Museum* **31 (4)**, 143–169.

Weber M (1910) Neue Fische aus Niederländisch Süd-Neu-Guinea. *Notes Leyden Museum* **32**, 225–240.

Weber M (1913) *Die Fische der Siboga-Expedition.* E. J. Brill, Leiden.

Weber M and de Beaufort LF (1911–1962) *Fishes of the Indo-Australian Archipelago*, 11 volumes. E.J. Brill, Leiden.

Weber M and de Beaufort LF (1964) *Fishes of the Indo-Australian Archipelago*, volume 1 (second impression). E.J. Brill, Leiden.

Weber WA (Ed) (2000) *The American Cockerell: A Naturalist's Life, 1866–1948.* University Press of Colorado.

Welsby T (1967) Schnappering. In *The Collected Works of Thomas Welsby* (Ed AK Thomson) pp. 31–242. Jacaranda Press, Brisbane.

Wheeler AC (1958) The Gronovius Fish Collection: A Catalogue and Historical Account. *Bulletin of the British Museum (Natural History), Historical Series* **1 (5)**, 187–249.

Wheeler AC (1962) Life and Work of Peter Artedi In *Ichthyologia with an Introduction by A.C. Wheeler.* pp. vii–xx. J. Cramer, Weinheim.

Wheeler AC (1986) Catalogue of the Natural History Drawings Commissioned by Joseph Banks on the Endeavour Voyage 1768–1771 held in the British Museum (Natural History). *Bulletin of the British Museum (Natural History), Historical Series* **13**, 1–172.

White J (1790) *Journal of a voyage to New South Wales with sixty-five plates of non descript animals, birds, lizards, serpents, curious cones of trees and other natural productions.* Debrett, London.

Whitehead PJP and Talwar PK (1976) Francis Day (1829–1889) and his collections of Indian fishes. *Bulletin of the British Museum (Natural History), Historical Series* **5 (1)**, 1–190.

Whitehead PJP (1968) *Forty Drawings of Fishes made by the Artists who accompanied Captain James Cook on his Three Voyages to the Pacific 1768–71 1772–75* 1776–80 some being used by authors in the description of new species. British Museum (Natural History), London.

Whitehead PJP (1969) *Reeves Collection of Chinese Fish Drawings. Bulletin of the British Museum (Natural History), Historical Series* **3 (7)**, 193–233.

Whitley GP (1926a) Allan Riverstone McCulloch, 1885-1925. Bibliography. *Records of the Australian Museum* **15 (2)**, 143–148.

Whitley GP (1926b) Bibliography of James Douglas Ogilby. *Records of the Australian Museum* **15 (2)**, 149–156.

Whitley GP (1929a) Some Fishes of the order Amphiprioniformes. *Memoirs of the Queensland Museum* **9 (3)**, 207–246.

Whitley GP (1929b) Studies in Ichthyology No. 3. *Records of the Australian Museum* **17 (3)**, 102–143.

Whitley GP (1932) Marine Zoogeographical Regions of Australasia. *The Australian Naturalist* **8 (8)**, 166–167.

Whitley GP (1935) Fishes from Princess Charlotte Bay, North Queensland. *Records of the South Australian Museum* **5 (3)**, 345–365.

Whitley GP (1940) *Sharks, Rays, Devil-Fish, and other Primitive Fishes of Australia and New Zealand.* Royal Zoological Society of New South Wales, Sydney.

Whitley GP (1944) New sharks and fishes from Western Australia. *Australian Zoologist* **10 (3)**, 252–273.

Whitley GP (1945) New Sharks and Fishes from Western Australia. Part 2. *Australian Zoologist* **11 (1)**, 1–42.

Whitley GP (1948) A List of the Fishes of Western Australia. *Western Australian Fisheries Department Bulletin* **2**, 1–35.

Whitley GP (1956) Archibald Menzies and the Fishes of King George's Sound. *Western Australian Naturalist* **5 (3)**, 57-59.

Whitley GP (1960) *Native Freshwater Fishes of Australia.* Jacaranda Press, Brisbane.

Whitley GP (1962) *Marine Fishes of Australia.* 2 volumes. Jacaranda Press, Brisbane.

Whitley GP (1963) *History of the Australian Museum.* Unpublished manuscript dated 1963.

Whitley GP (1964) A Survey of Australian Ichthyology. *Proceedings of the Linnean Society of New South Wales* **89 (1)**, 11–127.

Whitley GP (1966) T.J. Lempriere, an Early Tasmanian Naturalist. *Australian Zoologist* **13**, 350–355.

Whitley GP (1968) Check-List of the Fishes Recorded from the New Zealand Region. *Australian Zoologist* **15 (1)**, 1–102.

Whitley GP (1970) *Early History of Australian Zoology.* Royal Zoological Society of New South Wales, Sydney.

Whitley GP (1975) *More Early History of Australian Zoology.* Royal Zoological Society of New South Wales, Sydney.

Whitley GP and Allan J (1958) *The Sea-horse and its Relatives.* Georgian House, Melbourne.

Wilkinson I (1996) The Battle for the Museum: Frederick McCoy and the Establishment of the National Museum of Victoria at the University of Melbourne. *Historical Records of Australian Science* **11 (1)**, 1–11.

Willughby F and Ray J (1686) *De Historia Piscium* ... Oxford.

Wood-Jones F (1923) *The Mammals of South Australia, Part I, containing the Monotremes and the Carnivorous Marsupials (The Ornithodelphia and the didactylous Didelphia.).* Government Printer, Adelaide.

Woodward BH (1901) Class V. - Pisces. In *Western Australian Yearbook for 1900-1901*, 1 (2) Descriptive. 7. Fauna, 269–273.

Yarrell W (1835–36) *A history of British fishes, illustrated by nearly 400 wood-cuts.* 2 volumes. London.

Zietz AHC (1896) Pisces. In *Report on the Work of the Horn Scientific Expedition to Central Australia. Part 2. Zoology* (Ed WB Spencer) pp. 176–180. Melville, Mullin and Slade, Melbourne.

Zietz AHC (1908a) A Synopsis of the fishes of South Australia: Part I. *Transactions of the Royal Society of South Australia* **32**, 288–293.

Zietz AHC (1908b) A Synopsis of the fishes of South Australia: Part II. *Transactions of the Royal Society of South Australia* **32**, 294–299.

Zietz AHC (1908c) Description of a Hitherto Undescribed Species of Shark from Investigator Strait. *Transactions of the Royal Society of South Australia* **32**, 287.

Zietz AHC (1909) A Synopsis of the fishes of South Australia: Part III. *Transactions of the Royal Society of South Australia* **33**, 263–269.

Index

Page references in bold text denote a plate; colour plates are distinguished by the prefix '**col. pl.**'.
Page references in italics indicate a principal reference.
Col. pl. (caption) refers to information in the caption of a colour plate.

www.ingramcontent.com/pod-product-compliance
Ingram Content Group UK Ltd.
Pitfield, Milton Keynes, MK11 3LW, UK
UKHW060219061026
15625UKWH00007B/404

9 780643 106703